Topics in Fluorescence Spectroscopy

Volume 1
Techniques

Topics in Fluorescence Spectroscopy

Edited by JOSEPH R. LAKOWICZ

Volume 1: Techniques
Volume 2: Principles
Volume 3: Biochemical Applications

Topics in Fluorescence Spectroscopy

Volume 1
Techniques

Edited by

JOSEPH R. LAKOWICZ

Center for Fluorescence Spectroscopy
Department of Biological Chemistry
University of Maryland School of Medicine
Baltimore, Maryland

PLENUM PRESS • NEW YORK AND LONDON

Library of Congress Cataloging-in-Publication Data

Topics in fluorescence spectroscopy / edited by Joseph R. Lakowicz.
p. cm.
Includes bibliographical references and index.
Contents: v. 1. Techniques
ISBN 0-306-43874-7 (v. 1)
1. Fluorescence spectroscopy. I. Lakowicz, Joseph R.
QD96.F56T66 1991
543'.0858--dc20 91-32671
CIP

A Division of Plenum Publishing Corporation
233 Spring Street, New York, N.Y. 10013

Printed in the United States of America

Contributors

David J. S. Birch • Department of Physics and Applied Physics, Strathclyde University, Glascow G4 ONG, Scotland

Michael Edidin • Department of Biology, The Johns Hopkins University, Baltimore, Maryland 21218

Ignacy Gryczynski • Center for Fluorescence Spectroscopy, Department of Biological Chemistry, University of Maryland School of Medicine, Baltimore, Maryland 21201

Robert E. Imhof • Department of Physics and Applied Physics, Strathclyde University, Glascow G4 ONG, Scotland

Karl J. Kutz • Georgia Instruments, Inc., Atlanta, Georgia 30366

Joseph R. Lakowicz • Center for Fluorescence Spectroscopy, Department of Biological Chemistry, University of Maryland School of Medicine, Baltimore, Maryland 21201

Margaret M. Martin • SERC Daresbury Laboratory, Warrington WA4 4AD, England

Lászlo Mátyus • Department of Biology, The Johns Hopkins University, Baltimore, Maryland 21218; *permanent address*: Department of Biophysics, University Medical School of Debrecen, H-4012 Debrecen, Hungary

Ian H. Munro • SERC Daresbury Laboratory, Warrington WA4 4AD, England

Thomas M. Nordlund • Department of Biophysics and Department of Physics and Astronomy, University of Rochester, Rochester, New York 14642; *present address*: Department of Physics, University of Alabama at Birmingham, Birmingham, Alabama 35294

Enoch W. Small • Department of Biochemistry and Biophysics, Oregon State University, Corvallis, Oregon 97331-6503

Nancy L. Thompson • Department of Chemistry, University of North Carolina at Chapel Hill, Chapel Hill, North Carolina 27599-3290

John E. Wampler • Department of Biochemistry, University of Georgia, Athens, Georgia 30602

Robert A. White • Department of Biochemistry, University of Georgia, Athens, Georgia 30602

Preface

Fluorescence spectroscopy and its applications to the physical and life sciences have evolved rapidly during the past decade. The increased interest in fluorescence appears to be due to advances in time resolution, methods of data analysis, and improved instrumentation. With these advances, it is now practical to perform time-resolved measurements with enough resolution to compare the results with the structural and dynamic features of macromolecules, to probe the structure of proteins, membranes, and nucleic acids, and to acquire two-dimensional microscopic images of chemical or protein distributions in cell cultures. Advances in laser and detector technology have also resulted in renewed interest in fluorescence for clinical and analytical chemistry.

Because of these numerous developments and the rapid appearance of new methods, it has become difficult to remain current on the science of fluorescence and its many applications. Consequently, I have asked the experts in particular areas of fluorescence to summarize their knowledge and the current state of the art. This has resulted in the initial two volumes of *Topics in Fluorescence Spectroscopy*, which is intended to be an ongoing series which summarizes, in one location, the vast literature on fluorescence spectroscopy. The third volume will appear shortly.

The first three volumes are designed to serve as an advanced text. These volumes describe the more recent techniques and technologies (Volume 1), the principles governing fluorescence and the experimental observables (Volume 2), and applications in biochemistry and biophysics (Volume 3).

Additional volumes will be published as warranted by further advances in this field. I welcome your suggestions for future topics or volumes, offers to contribute chapters on specific topics, or comments on the present volumes.

Finally, I thank all the authors for their patience with the delays incurred in release of the first three volumes.

Joseph R. Lakowicz

Baltimore, Maryland

Contents

1. Time-Domain Fluorescence Spectroscopy Using Time-Correlated Single-Photon Counting

David J. S. Birch and Robert E. Imhof

1.1. Introduction 1
1.1.1. What Is a Fluorescence Lifetime? 2
1.1.2. Methods of Measuring Fluorescence Lifetimes 4
1.1.3. Analytical Applications 6
1.2. The Single-Photon Technique 8
1.2.1. Basic Principles 9
1.2.2. Statistics and Pileup 11
1.2.3. Instrumental Response and Convolution 14
1.3. Data Analysis 16
1.3.1. Formulation of the Problem 17
1.3.2. Least-Squares Fitting 19
1.3.3. Fitting Algorithms 22
1.3.4. Anisotropy Analysis 26
1.4. Instrumentation 28
1.4.1. Pulsed Light Sources 28
1.4.2. Optical Components 36
1.4.3. Detectors 41
1.4.4. Electronics and Computer Hardware 47
1.5. Applications and Performance 52
1.5.1. Exponential Decay 52
1.5.2. Nonexponential Decay 59
1.5.3. Anisotropy 64
1.5.4. Time-Resolved Spectra 71
1.6. Multiplexing Techniques 73
1.6.1. Differential Pulse Fluorometry 75
1.6.2. T-Format Studies 81
1.6.3. Array Fluorometry 83
References 88

2. Laser Sources and Microchannel Plate Detectors for Pulse Fluorometry

Enoch W. Small

2.1. Introduction ... 97
2.2. Laser Sources ... 99
2.2.1. Laser Fundamentals ... 100
2.2.2. The Argon Ion, Nd:YAG, and Dye Lasers—Energy Levels and Physical Layout ... 123
2.2.3. Laser Pulse Generation ... 132
2.3. The Microchannel Plate (MCP) Photomultiplier ... 146
2.3.1. How an MCP Photomultiplier Works ... 147
2.3.2. Photomultiplier Performance ... 155
2.4. Tuning a Monophoton Decay Fluorometer for High Performance 168
2.4.1. Judging the Quality of Data ... 168
2.4.2. Tuning ... 170
2.4.3. Performance of a Tuned Instrument ... 174
2.5. Conclusion ... 178
References ... 180

3. Streak Cameras for Time-Domain Fluorescence

Thomas M. Nordlund

3.1. Introduction ... 183
3.1.1. What Is a Streak Camera? ... 183
3.1.2. Scope of the Chapter ... 185
3.1.3. Basic Features and Abilities of a Photoelectronic Streak Camera ... 186
3.1.4. History ... 187
3.2. System Building Blocks ... 190
3.2.1. Modes of Operation ... 190
3.2.2. Image Converter Tube ... 194
3.2.3. Image Intensifiers ... 205
3.2.4. Signal Readout and Storage ... 208
3.2.5. Computer and Data Manipulation ... 212
3.3. Streak Camera Capabilities ... 213
3.3.1. Detector Characteristics ... 213
3.3.2. Signal Averaging ... 222
3.3.3. Streak Cameras and Time-Resolved Single-Photon Counting ... 223
3.4. Time Dependence of Fluorescence ... 225
3.4.1. Decay Times ... 225
3.4.2. Rise Times ... 231
3.4.3. Physical Processes and Characteristic Times ... 233

3.5. Wavelength Dependence of Fluorescence 233
3.5.1. Measurement of Time and Wavelength Dependence of Signals: Additional Time Dispersion 234
3.5.2. Time Dependence of Emission Spectra 237
3.6. Two-Dimensional Detection: Space and Time Coordinates 239
3.6.1. Position and Time Detection 239
3.6.2. Two-Dimensional Spatial Detection 240
3.7. Fluorescence Anisotropy 241
3.7.1. Simultaneous versus Separate Measurements 242
3.7.2. Signal-to-Noise Ratio 244
3.8. Applications 245
3.8.1. Laser Diagnostics 245
3.8.2. Solid-State Systems 246
3.8.3. Solutions: Molecular Relaxation in Solvents 247
3.8.4. Biological Systems 249
3.8.5. X-Ray Streak Cameras: Fusion Reactions 252
3.9. The Future 254
3.9.1. The 0.1–20-ps Time Regime 254
3.9.2. Integrated Detection Systems 255
3.9.3. Time-Resolved Photon-Counting Streak Cameras (TRPCSC) 255
References 256

4. Time-Resolved Fluorescence Spectroscopy Using Synchrotron Radiation

Ian H. Munro and Margaret M. Martin

4.1. Introduction 261
4.2. Proteins and Peptides 262
4.2.1. Lumazine 263
4.2.2. Angiotensin II 266
4.2.3. Lactate Dehydrogenase 270
4.3. Photosynthesis 272
4.4. Membranes 276
4.4.1. Diphenylhexatriene 276
4.4.2. Triazine Dyes 279
4.5. Excited State Processes 281
4.5.1. Host/Guest Complexes 282
4.5.2. Stilbene 283
4.5.3. The Vacancy in Diamond 286
References 289

5. Frequency-Domain Fluorescence Spectroscopy

Joseph R. Lakowicz and Ignacy Gryczynski

5.1. Introduction ... 293
5.2. Comparison of Time- and Frequency-Domain Fluorometry ... 294
5.2.1. Intensity Decays ... 294
5.2.2. Anisotropy Decays ... 298
5.3. Theory of Frequency-Domain Fluorometry ... 302
5.3.1. Decays of Fluorescence Intensity ... 302
5.3.2. Decays of Fluorescence Anisotropy ... 304
5.4. Instrumentation and Applications ... 305
5.4.1. FD Instruments with Intensity-Modulated Light Sources ... 305
5.4.2. A 2-GHz Harmonic Content FD Instrument ... 309
5.4.3. Resolution of Anisotropy Decays ... 316
5.5. Future Developments ... 326
5.6. Summary ... 331
References ... 331

6. Fluorescence Correlation Spectroscopy

Nancy L. Thompson

6.1. Introduction ... 337
6.2. Conceptual Basis and Theoretical Background ... 337
6.3. Experimental Apparatus and Methods ... 343
6.4. Analysis of Autocorrelation Function Magnitudes ... 351
6.4.1. Number Densities ... 351
6.4.2. Molecular Weights ... 352
6.4.3. Molecular Aggregation and Polydispersity ... 353
6.5. Analysis of Autocorrelation Function Temporal Decays ... 354
6.5.1. Translational Diffusion ... 354
6.5.2. Flow and Sample Translation ... 357
6.5.3. Kinetic Rate Constants ... 359
6.6. Special Versions of FCS ... 361
6.6.1. Nonideal Solutions ... 361
6.6.2. Rotational Diffusion ... 362
6.6.3. Total Internal Reflection Illumination ... 364
6.6.4. High-Order Autocorrelation ... 369
6.6.5. Cross-Correlation ... 373
6.7. Summary and Future Directions ... 374
References ... 374

7. Fundamentals of Fluorescence Microscopy

Robert A. White, Karl J. Kutz, and John E. Wampler

7.1. Introduction 379
7.1.1. Sensitivity and Its Limitations 382
7.1.2. Visualization versus Quantitation 382
7.2. The Illumination Light Path of the Fluorescence Microscope 384
7.2.1. Lamps 386
7.2.2. Lamp Housing 387
7.2.3. Auxiliary and Alternative Optics 389
7.2.4. The Objective Lens as Condenser 397
7.2.5. Specimen and Mount 397
7.3. The Imaging Light Path in Fluorescence Microscopy 398
7.3.1. The Specimen as an Optical Component 398
7.3.2. The Objective 399
7.3.3. Optics between Objective and Eyepiece 402
7.3.4. Role and Position of Ocular 403
7.3.5. Detector Placement for Electronic Imaging and Photometry 403
References 407

8. Flow Cytometry and Cell Sorting

László Mátyus and Michael Edidin

8.1. Introduction/History 411
8.2. Operation of Flow Cytometers 413
8.2.1. Sample Handling and Delivery Systems 416
8.2.2. The Nozzle and the Sheath Fluid 416
8.2.3. Light Sources 417
8.2.4. The "Intersection Point" 418
8.2.5. Detectors 420
8.2.6. Electronics 423
8.2.7. Data Analysis and Storage 424
8.2.8. Slit-Scan Flow Cytometry 425
8.2.9. Cell Sorting 427
8.3. Parameters of Flow Cytometry 428
8.3.1. Light Scatter 428
8.3.2. Fluorescence 428
8.4. Conclusion 439
References 440

Index 451

1

Time-Domain Fluorescence Spectroscopy Using Time-Correlated Single-Photon Counting

David J. S. Birch and Robert E. Imhof

1.1. Introduction

Although the phenomenon of molecular fluorescence emission contains both spectral (typically UV/visible) and time (typically 10^{-9} s) information, it is the former which has traditionally been more usually associated with routine analytical applications of fluorescence spectroscopy. In contrast, since the early seventies, the most important strides in fluorescence spectroscopy have been by and large concerned with the development and application of new time-resolved techniques. Applications of these now span not only the traditional areas of photochemistry but also research in polymers, membranes, liquid crystals, low-dimensional structures, solar energy, and semiconductors, to name but a few. Such diverse areas are linked by their common instrumentation needs, which have greatly benefited over the past decade from quite radical progress in optoelectronics and on-line data analysis. There is now widespread recognition of the advantages of studying and using fluorescence in the time domain. Quite simply, fluorescence measurements in the time domain (fluorometry) possess a much greater information content about the rates and hence kinetics of intra- and intermolecular processes than is afforded by the wavelength spectroscopy of the energy domain (fluorimetry). A simple analogy to the greater detail shown by time domain as compared to steady-state fluorescence measurements is that of motion pictures in comparison to still photographs.

The rising importance of fluorescence measurements in the time domain has been reflected during the past decade in several texts devoted exclusively

---

David J. S. Birch and Robert E. Imhof • Department of Physics and Applied Physics, Strathclyde University, Glasgow G4 ONG, Scotland.
Topics in Fluorescence Spectroscopy, Volume 1: Techniques, edited by Joseph R. Lakowicz. Plenum Press, New York, 1991.

to this topic.[1–4] Nevertheless, the field has not stood still, and, if anything, the rate of innovation has probably increased over the past few years. The purpose of this chapter then is to illustrate some of the capabilities of time-resolved fluorescence techniques in general by means of a description of the fundamental principles, applications, and new developments concerning the most widely used method of studying fluorescence lifetimes, namely, time-correlated single-photon counting. As such, it is hoped that, in addition, this coverage might also serve as a useful introduction to some of the topics covered in more detail in later chapters.

1.1.1. What Is a Fluorescence Lifetime?

In terms of reaction kinetics the fluorescence lifetime or decay time of a molecule can be defined in the time domain in terms of the rate of depopulation of the first excited singlet state following δ-function (i.e., impulse) optical excitation from the ground state. Using the notation popularized by Birks,[5] as will be done throughout this chapter, this is given in the case of a simple monoexponential decay by

$$\frac{d[^1\mathrm{M}^*]}{dt} = -\frac{[^1\mathrm{M}^*]}{\tau_\mathrm{M}} \tag{1.1}$$

which upon integration gives a fluorescence response function of the form

$$[^1\mathrm{M}^*] = [^1\mathrm{M}^*]_0 \, e^{-t/\tau_\mathrm{M}} \tag{1.2}$$

where $[^1\mathrm{M}^*]$ and $[^1\mathrm{M}^*]_0$ represent the excited state molar concentration at time t and $t = 0$, respectively, and τ_M is the molecular fluorescence lifetime.

Equation (1.2) describes the simplest fluorescence response to δ-function excitation. Other more complex forms will be considered later. Because the excited state population is proportional to the fluorescence quantum intensity, the fluorescence lifetime can be determined experimentally by measuring the time taken for the fluorescence intensity to fall to $1/e$ of its initial value following δ-function excitation. This observation forms the basis of the time-correlated single-photon counting technique whereby the quantum nature of light enables the time distribution of individual photons within the decay profile to be recorded.

The use of the fluorescence lifetime of a molecule as a probe of microenvironment can be illustrated by considering the dependence of τ_M on the rates of competing decay pathways, for example,

$$\tau_\mathrm{M} = 1/(k_\mathrm{FM} + k_\mathrm{IM}) = 1/k_\mathrm{M} \tag{1.3}$$

where k_{FM} is the radiative rate parameter (in s^{-1}), and k_{IM} is the nonradiative rate parameter and k_M the total decay rate. Both k_{FM} and k_{IM} can depend on the fluorophore's environment in a manner determined by such properties as refractive index, polarity, and temperature.

The additional measurement of the fluorescence quantum yield ϕ_{FM} ($=k_{FM}/k_M$) enables k_{FM} and k_{IM} to be determined. The radiative rate parameter k_{FM} and the purely radiative lifetime τ_{FM} are described in quantum-mechanical terms by the Einstein A coefficient and the transition dipole moment operator M via

$$k_{FM} = 1/\tau_{FM} = \sum_m A_{u0 \to lm} \propto |\langle \psi_l^* | \mathrm{M} | \psi_u \rangle|^2 \tag{1.4}$$

where the Einstein A coefficient is summed over the complete fluorescence spectrum attributed to transitions from the zeroth vibrational level of the upper singlet state (u) to different vibrational levels (m) of the lowest (ground) singlet state (l).

Equation (1.4) leads to an approximate but useful expression[5] which relates τ_{FM} for a molecule to the maximum extinction coefficient $\varepsilon_{max}(\lambda)$ determined from the absorption spectrum as a function of wavelength (λ):

$$\tau_{FM} = 10^{-4}/\varepsilon_{max}(\lambda) \tag{1.5}$$

This predicts a value of $\sim 10^{-9}$ s or 1 ns for τ_{FM} for strong absorbers, for which $\varepsilon_{max} \approx 10^5$ liter/mol $\cdot$ cm^2. A measured fluorescence lifetime τ is usually less than the value predicted by Eq. (1.5) due to the presence of nonradiative decay rates which are not necessarily of intramolecular origin. For example, the fluorescence lifetime is also sensitive to intermolecular interactions (and hence can be used to study them) including collisional quenching and energy transfer.

Oxygen, which occurs at concentrations of $\sim 10^{-3}$ M in commonly used spectroscopic solvents, is an efficient quencher of fluorescence, as are many other impurities. Unless removed, it causes a further reduction in the measured fluorescence lifetime via

$${}^1\mathrm{M}^* + \mathrm{Q} \to {}^1\mathrm{M} + \mathrm{Q} \tag{1.6}$$

with

$$\frac{d[{}^1\mathrm{M}^*]}{dt} = -(k_{FM} + k_{IM} + k_{QM}[\mathrm{Q}])[{}^1\mathrm{M}^*] \tag{1.7}$$

giving a measured fluorescence lifetime $\tau < \tau_M$ of

$$\tau = 1/(k_{FM} + k_{IM} + k_{QM}[\mathrm{Q}]) \tag{1.8}$$

where k_{QM} is the quenching rate parameter (in $mol^{-1}\,s^{-1}$), and [Q] is the quencher concentration.

1.1.2. Methods of Measuring Fluorescence Lifetimes

There are essentially two types of methods for measuring fluorescence lifetimes that are in widespread use today. These are pulse fluorometry, which relates to measurements performed in the time domain, and phase and modulation fluorometry, relating to the frequency domain. Figure 1.1 summarizes the two different approaches.

The basic principle of pulse fluorometry, whereby a sample is excited by a fast pulse of light from a spark source or laser and the time dependence of the fluorescence decay is then recorded, has already been outlined. Phase and modulation fluorometry traditionally incorporates a modulated excitation source such that the finite fluorescence lifetime of the sample causes the fluorescence emission waveform to be phase shifted and of different amplitude

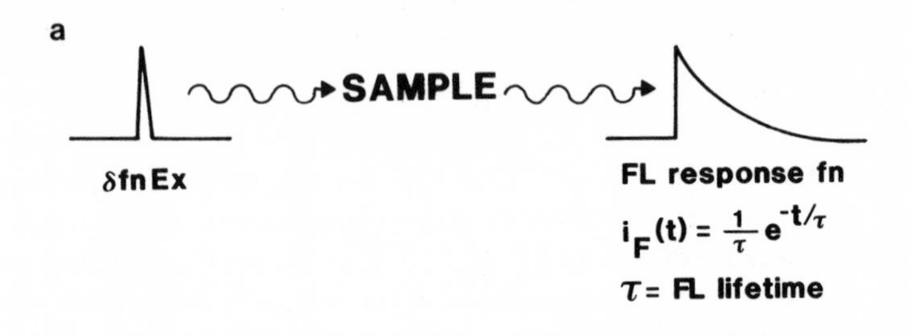

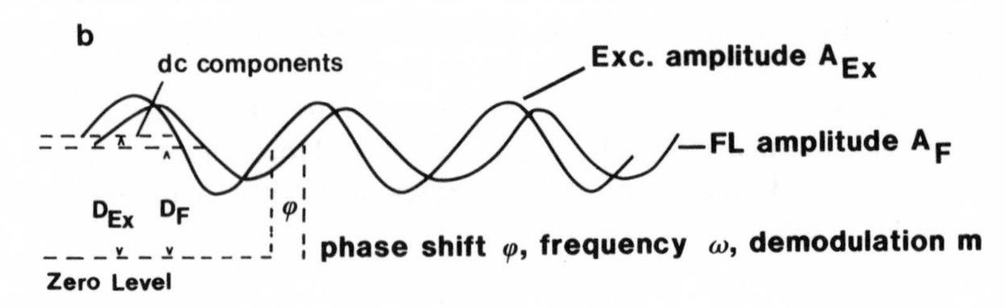

$$\tan \varphi = -\omega\tau$$

$$m = \frac{A_F/D_F}{A_{Ex}/D_{Ex}} = \frac{1}{(1+\omega^2\tau^2)^{1/2}}$$

Figure 1.1. Principles of (a) pulse and (b) phase–modulation techniques for measuring fluorescence lifetimes.

with respect to the excitation waveform. That this is so can be easily seen by considering the effect on Eq. (1.1) of a sinusoidally modulated excitation signal $D_{\mathrm{Ex}} + A_{\mathrm{Ex}} \exp(i\omega t)$ producing a modulated fluorescence signal $D_{\mathrm{F}} + A_{\mathrm{F}} \exp i(\omega t + \phi)$. By substitution in Eq. (1.1), which describes a single-exponential decay, we obtain the phase relationship

$$\tan \phi = -\omega\tau \tag{1.9}$$

and the expression for the demodulation factor

$$m = \frac{A_{\mathrm{F}}/D_{\mathrm{F}}}{A_{\mathrm{Ex}}/D_{\mathrm{Ex}}} = \frac{1}{(1 + \omega^2\tau^2)^{1/2}} \tag{1.10}$$

Traditionally, phase and modulation fluorometry has been performed using electro-optic (Kerr or Pockels cell) and acousto-optic modulation of continuous-wave lamp or laser excitation. Direct modulation of discharge lamps is also sometimes used. The upper limit of modulation frequency is an important factor in determining the shortest decay time which can be measured (Eq. 1.9), and current modulation methods operate at a maximum frequency of $\sim$200 MHz. Phase shifts and demodulation factors can be determined using cross-correlation with a phase-sensitive detector. The methodology of phase fluorometry is developed in more detail in subsequent chapters, and a comprehensive introduction to how frequency-domain measurements fit into the general area of fluorescence spectroscopy is given in the text by Lakowicz.[6]

Much has been written about the relative merits of using time-domain and frequency-domain methods to determine fluorescence lifetimes, and the reader can form a personal view from this volume and other texts cited here. In being evenhanded, it should be recognized that both approaches are of course complementary as they contain a common information content, namely, the sample's fluorescence decay characteristics. Moreover, the fluorescence response in the frequency domain can be obtained from the Fourier transform of the impulse response in the time domain and vice versa for the inverse transform. This very fact has recently been used to good effect in the development of multifrequency phase fluorometry, whereby the high harmonic content rather than the short pulse duration of picosecond mode-locked laser pulses is utilized.[7] Although the optical configuration is essentially identical to that developed some years earlier for laser pulse fluorometry, the fluorescence decay data in such instruments are analyzed in the frequency rather than the time domain. The approach was originally developed using synchrotron radiation by Gratton and co-workers,[8] but it is the implementation using mode-locked, synchronously pumped, cavity-dumped dye lasers which has enabled the effective frequency range of phase fluorometry to be extended into the gigahertz region.[9]

Traditional phase fluorometers using sinusoidally modulated excitation only operated at a few frequencies, which restricted applications to simple fluorescence decay kinetics, that is, one or two exponentials. The use of pulse excitation with mode-locked lasers has overcome this limitation by making a wider range of frequencies, and hence a better description of the frequency response of sample and apparatus, more readily available.

With regard to the comparison between time- and frequency-domain techniques, it is interesting to highlight the principal conclusion from some recent work using a hybrid time-correlated single-photon/multifrequency phase fluorometer.[10] This instrument was configured in a T-format with a mode-locked laser as the excitation source and detection arms operating as separate time-correlated and phase fluorometers. The results obtained with the two methods were found to be consistent for a number of standard compounds and single-tryptophan proteins. This result is perhaps not surprising given the theoretical relationship between the two approaches. Any differences only result from a lack of equivalence in the technology and methods of implementation.

1.1.3. Analytical Applications

The word *analytical* in a spectroscopy context tends to imply identification and quantification of an unknown sample. The word is often also associated with routine techniques. The whole area of time-resolved fluorescence spectroscopy does not fit easily into either of these categories. Fluorescence lifetime measurements alone tend not to be specific enough for identification purposes with unknown samples. More usually, the decay characteristics of a known fluorophore are studied for one of three reasons:

1. For intrinsic interest in the intramolecular photophysics of the fluorophore, for example, to determine the rates of competitive deexcitation pathways.
2. To determine reaction kinetics, for example, the study of molecular diffusion dynamics.
3. To probe local environment, for example, to determine conformation in proteins or fluidity in membranes using the fluorescence depolarization.

There are however a number of applications of time-resolved fluorescence techniques which touch on analytical use and which are worthy of mention here. These applications span both frequency- and time-domain measurements and, to a large extent, the choice of which would seem to often be of lesser significance than the analytical objective. For example, fluorescence lifetime measurements have recently been applied to liquid chromatography.[11] This can add to the specificity of mixture analysis by complementing the information afforded by the usual recording of retention time. The measurement of the fluorescence lifetimes of transient samples presents problems experimentally as most methods involve some kind of signal averaging. In this regard, the speed of traditional phase and modulation fluorometry has some advantage. With enhanced data collection rates, further beneficial applications to chromatography might become possible, for example, time-resolved fluorescence spectral analysis to identify elements of similar retention time. An interesting aside is that laser-induced fluorescence has only recently started to be used for detection in chromatography. The restriction that all current UV laser sources are pulsed rather than continuous gives rise to special problems with regard to detector saturation, and these will have to be overcome before the benefits of the high average laser powers can be realized in chromatography. When these problems are eventually overcome, the pulsed nature of a UV laser source may prove to have additional benefits along the lines indicated.

The combination of picosecond laser excitation and time-correlated single-photon counting also finds applications in Raman spectroscopy as a means of rejecting rather than recording fluorescence.[12] The Raman signal is instantaneous whereas the fluorescence decay occurs typically on the nanosecond time scale. Hence, simple time-gating of the detected signal enables the fluorescence to be discriminated against.

The potential for developing optical fiber sensors based on time-resolved fluorescence methods has not been lost on the analytical community. Optrodes based on rhodamine B, rhodamine 6G, and eosin immobilized in resin have been developed[13] and shown to be capable of detecting iodide ion concentrations down to below 1 m*M* via the Stern–Volmer quenching of the dye fluorescence.[14] Similar techniques have been used to detect down to 0.01 % volume concentrations of sulfur dioxide in air.[15] These techniques are still in the early stages of development, with much of the work concerned with steady-state fluorescence studies. Nevertheless, time-resolved optrodes using pulsed laser excitation with sampling oscilloscope detection[16] and multi-frequency phase fluorometry[17] have recently been reported.

The combination of optical fiber sensing, time-gated detection, Stern–Volmer theory, and fluorescence decay time measurements has found a very useful application in detecting low levels of uranium in the environment.[18] In the absence of quenching, the uranyl ion has an intense green fluorescence with a decay time on the order of a few hundred microseconds. This fluorescence is strongly quenched in natural water by organics dissolved from vegetation. However, the organics have a much faster fluorescence decay time (on the order of nanoseconds) than that of the uranyl ion, and hence organic emission can be discriminated against using pulsed excitation and time-gating to reject the initial part of the decay (the converse of time-resolved Raman spectroscopy). Moreover, if we integrate Eq. (1.7), the effect of quenching can be corrected to enable accurate uranium concentrations to be determined, viz.,[19]

$$\ln[\mathrm{U}^*]_t = \ln[\mathrm{U}^*]_0 - t/\tau \tag{1.11}$$

and similarly for a reference sample we obtain from Eq. (1.2)

$$\ln[\mathrm{U}^*_\mathrm{R}]_t = \ln[\mathrm{U}^*_\mathrm{R}]_0 - t/\tau_\mathrm{M} \tag{1.12}$$

Hence, if the fluorescence decay of the quenched sample (τ) of unknown uranium concentration and a reference sample (τ_M) of known concentration are measured, the uranyl concentration ($[\mathrm{U}^*]_0$) can be determined from the intercept of the two decay profiles. Detection limits for uranium as low as a few parts per trillion have been reported with this technique when combined with pulsed laser excitation and photon counting.[19]

The analytical potential for time-resolved fluorescence methods outside the traditional disciplines is also starting to be appreciated. For example, the use of fluorescence decay time measurements to identify irradiated food from the presence of *ortho*-tyrosine has recently been proposed.[20] *para*-Tyrosine is a common amino acid, and *ortho*-tyrosine is produced on irradiating food. The fluorescence decay times of the two isomers are close ($\sim$3.4 and 2.9 ns, respectively), but a biexponential analysis of the decay can in principle lead to a determination of the relative concentration.

There is no doubt that the list of analytical applications of time-resolved fluorescence so far mentioned is not exhaustive. Nevertheless, the examples chosen do convey some of the opportunities for further development. Indeed, all such analytical applications are still very much in their infancy. In contrast, the usual research applications of fluorescence decay spectroscopy as outlined in points 1–3 at the beginning of this subsection are very well established. Clearly, the whole area of time-resolved fluorescence spectroscopy is expanding rapidly, but for the rest of this chapter we will concentrate on the technique of time-correlated single-photon counting itself, although, where appropriate, bearing in mind any wider implications.

1.2. The Single-Photon Technique

The most comprehensive text yet produced on the time-correlated single-photon counting technique is that written by O'Connor and Phillips.[3] There are earlier reviews[21–23] dedicated to this method of determining fluorescence lifetimes and some which consider it along with other methods.[24, 25] All these texts are still relevant today and lack only the coverage of subsequent work.

For this reason we have chosen to emphasize the recent work of significance through drawing on fundamental aspects from the past.

Historically, the first report in the scientific literature of time-correlated single-photon counting as we know it today was by Bollinger and Thomas in 1961.[26] These workers reported the measurement of the scintillation response of various crystals and glasses using gamma, neutron, and alpha excitation. The basic method that Bollinger and Thomas developed was quite similar to the delayed-coincidence technique pioneered in nuclear physics. To quote from their original paper,

> In general terms, the method for determining the time dependence of the scintillation consists of measuring the distribution of the difference in time between the excitation of the scintillator and the formation of individual photoelectrons at the cathode of a photomultiplier that views the scintillation. That is the rate of formation of photoelectrons within a narrow interval of time difference is recorded as a function of time difference. Because individual photoelectrons are counted, the yields of many different pulses may be added until the required statistical accuracy is achieved. Thus an accurate measurement can be made even when the average number of photoelectrons per pulse is very small. Indeed, for purely practical reasons the photon detection efficiency on the photomultiplier used in the experiments reported here was deliberately reduced to such an extent that the probability of detecting even one photoelectron per pulse is small.

These days the single-photon timing technique (as it is also sometimes called) is more usually associated with the decay of fluorescence stimulated by optical excitation rather than scintillation, but the basic principle remains the same as Bollinger and Thomas first described. Indeed, most advances associated with the technique have been the result of technological improvements in the instrumentation rather than in the measurement principle itself. A major exception to this is the recent introduction of the multiplexing methods discussed in Section 1.6. These overcome some of the previous limitations of single-photon timing and are now opening up new application areas. But first let us look at the basic measurement technique.

1.2.1. Basic Principles

Figure 1.2 shows the layout of a typical time-correlated single-photon counting fluorometer. A pulsed light source, usually a flashlamp or mode-locked laser, generates multiphoton excitation pulses which stimulate absorption in an assembly of sample molecules. At low levels of excitation power, each sample molecule absorbs one photon at the most, on a time scale which is effectively instantaneous. The subsequent deexcitation of these molecules via the emission of fluorescence photons occurs with a distribution of time delays normally described by Eq. (1.2), that is, exponential. The single-photon timing technique records this distribution by measuring the time delays of individual

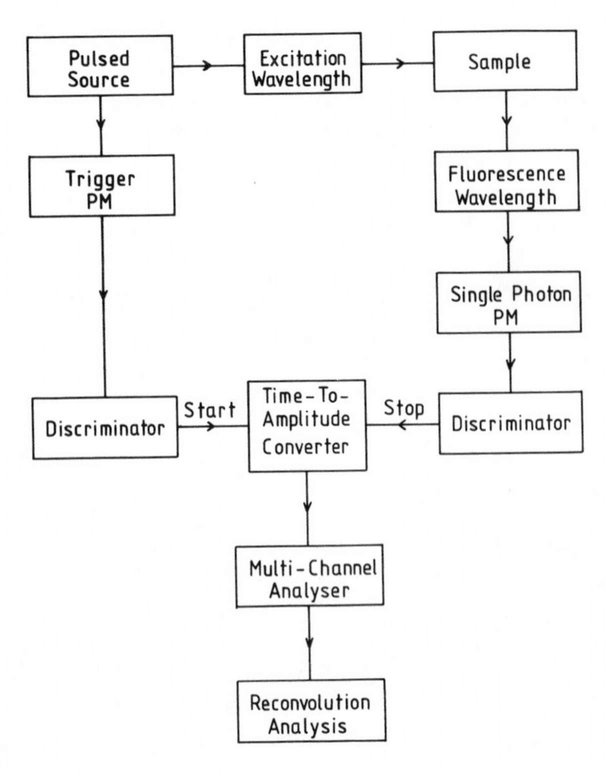

Figure 1.2. A typical time-correlated single-photon counting arrangement.

fluorescence photons with respect to the excitation pulse. The heart of the technique lies in what is known as the time-to-amplitude converter (TAC). When the excitation pulse occurs, a synchronization pulse or "start" timing pulse triggers the charging of a capacitor in the TAC. The voltage on the capacitor increases linearly until either a preselected time range is reached or a "stop" timing pulse is detected. The latter is initiated by the detection of a fluorescence photon, and the "start"–"stop" time interval generates a proportional voltage across the capacitor. This voltage pulse is stored according to amplitude using an analog-to-digital converter within a multichannel analyzer (MCA) and thereby allocated a proportional channel number. On repeating the "start"–"stop" cycle many times, a histogram representative of the fluorescence decay is acquired in the MCA memory such as that shown later on in Figure 1.4. The decay parameters can then be extracted using on-line data analysis of the histogram using numerical and statistical procedures (see Section 1.3). Photomultipliers are the most widely used single-photon timing devices, and in order to minimize the registering of noise pulses and ensure that the timing definition of the "start" and "stop" pulses is largely independent of the signal pulse height, discriminators are used. These shape and provide the time definition for the pulses in the "start" and "stop" channels.

The nature of the TAC operation is such as to only register the first "stop" pulse detected after a "start" pulse. Consequently, the "stop" rate has to be low enough for the probability of detecting more than one photon per excitation pulse to be negligible. This means that the probability of detecting one photon also has to be very small, and in effect the "start" rate has to be much greater than the "stop" rate. If this restriction is not adhered to, then the TAC will preferentially detect photons which occur at shorter times, and the fluorescence decay time distribution will not be sampled correctly, such that decay time will appear to be faster than it really is. This photon pileup effect makes the single-photon technique inefficient in cases where the fluorescence signal is intense because some of the available signal has to be wasted in order to ensure the absence of pileup. In practical terms, though, it is a shortage rather than the abundance of fluorescence signal that can limit the measurement capability of single-photon timing, and the pileup restriction is rarely onerous.

1.2.2. Statistics and Pileup

From the previous discussion, it will be obvious that the time-correlated single-photon technique is a digital rather than an analog technique. This gives it certain advantages with regard to a high dynamic range ($>10^6$) and the measured decay profile being independent of fluctuations in the excitation pulse intensity. In addition, the theory of the single-photon detection is based on well-known statistics for which the precision, data weights, goodness-of-fit criteria, etc., can be easily calculated and have a sound mathematical basis. This solid theoretical base enables the qualitative description already given on photon pileup to be supported by a more quantitative approach.[3, 21–26]

If we consider any channel i contained within a fluorescence decay which is being accumulated in the MCA, then this channel corresponds to a mean time delay t and interval Δt, that is, the time calibration corresponding to $t_{i-(1/2)}$ to $t_{i+(1/2)}$.

If, following one excitation pulse, the registering of a coincident event in channel i corresponds to an average number of $\bar{n}_i$ photons reaching the photocathode of the "stop" photomultiplier and liberating an average number of $\bar{x}_i$ photoelectrons, then

$$\bar{x}_i = \bar{n}_i q \tag{1.13}$$

where q is the photocathode quantum efficiency. If there are a large number of excitation pulses for every count registered in channel i, then the probability $P_x(i)$ of liberating other than $\bar{x}_i$ photoelectrons is given by the Poisson distribution, that is,

$$P_x(i) = \frac{(\bar{x}_i)^x e^{-\bar{x}_i}}{x!} \tag{1.14}$$

with

$$\sum_{x=0}^{\infty} P_x(i) = 1 \tag{1.15}$$

Therefore, the probability per excitation pulse of at least one photoelectron pulse being detected is given by

$$P_{x \geqslant 1}(i) = 1 - P_0(i) = 1 - e^{-\bar{x}_i} = 1 - \left(1 - \bar{x}_i + \frac{\bar{x}^2}{2} + \cdots\right) \tag{1.16}$$

which, if $\bar{x}_i >> \bar{x}^2/2$, that is, at very low light levels, gives

$$P_{x \geqslant 1}(i) \simeq \bar{x}_i \simeq \bar{n}_i q \tag{1.17}$$

Hence, under these conditions, the probability of detecting a coincident event in channel i is proportional to the fluorescence intensity at a delay time t. This is the general condition which is essential for accurate single-photon counting measurements. Consequently, if the repetition rate of the source is S_t ("start" rate) and the total rate of photomultiplier anode pulses due to fluorescence photons over all delay times is S_p ("stop" rate), then provided

$$S_p/S_t = \alpha \gtrsim 0.01 \tag{1.18}$$

the probability of detecting two fluorescence pulses per excitation pulse is negligible. For δ-function excitation, the number of counts Y_i acquired in channel i after excitation, in a measurement time T is given, for a single exponential decay time τ, by

$$Y_i = \alpha S_t T \frac{\Delta t}{\tau} e^{-i\Delta t/\tau} \tag{1.19}$$

Equation (1.19) demonstrates another advantage of the single-photon technique whereby the measurement precision can be enhanced simply by increasing the measurement run time.

The level of fluorescence count rate under which Eq. (1.19) breaks down and methods by which higher count rates can be achieved were the subject of a number of investigations some years ago.[27-29] Methods to correct for distortion at higher count rates include theoretical corrections, discrimination of counted events in favor of a single photoelectron pulse height distribution, and inhibiting the TAC conversion if more than one "stop" pulse is detected during a TAC cycle. For practical purposes most workers ensure that $\alpha \gtrsim 0.01$, which puts the absence of pileup distortion beyond doubt. However, as was already mentioned, the pileup restriction is not really a problem in most cases. Even with flashlamps, which typically have a repetition rate of ~50 kHz, a 500-counts/s ($\alpha = 0.01$) fluorescence rate enables nanosecond decay times to be measured to ~1% precision in a matter of a few minutes at the most.

Indeed, measurements obtained in our laboratory indicate that the widely practiced 1% "stop" to "start" rate ratios may in some cases be unnecessarily severe. Table 1.1 shows that for a dilute solution of 1,6-diphenyl-1,3,5-hexatriene (DPH) in cyclohexane the systematic error due to pileup only starts to exceed the statistical error at ~10% "stop:start" ratio. In some cases, working at such high "stop:start" ratios is helpful. An example is in the use of the low repetition rate of flashlamps to determine rotational parameters from anisotropy decay curves. Because the rotational information is contained in the difference between two decay curves, the curves have to be measured to a far greater number of counts than for straighforward lifetime measurements (50,000 counts in the peak of the decay being quite common). A second example where working at high count rates is necessary is in the study of transient samples,[11] which might only be available for measurement for a few seconds.

In Section 1.6, we describe the techniques and benefits associated with multiplexing time-correlated single-photon signals from different detection channels. One of the attractions of one of these methods is the enhanced duty cycle of the TAC when time-shared between the different channels. By this

Table 1.1. Effect of Pileup on a Measured Fluorescence Lifetime[a]

"Stop"/"Start" (%)	τ (ns)
1.25	8.07
2.35	8.09
4.1	8.04
9.4	7.97
18.4	7.85
35.7	7.36
52.0	7.26

[a] Sample: Aereated DPH in cyclohexane. Statistical error of 3 std. dev. = ±0.06 ns. Flashlamp repetition rate, 30 kHz.

means, decays free from the effect of pileup distortion have been accumulated with total "stop:start" ratios in excess of 30% (see Table 1.7). Indeed, the factor controlling the maximum count rate in time-correlated single-photon counting is now no longer a theoretical limit but rather an experimental one governed by the dead time of the electronic timing components.

1.2.3. Instrumental Response and Convolution

The assumption made in Section 1.1.1 that a δ-function optical source is available experimentally does effectively hold in many cases where the duration of the exciting pulse $\Delta t_e << \tau$. Picosecond optical pulses are now readily available from mode-locked lasers, but even if such sources are used to study nanosecond fluorescence lifetimes, a true δ function is not necessarily measured. The reason for this is that timing jitter in the other components of the single-photon timing arrangement causes the measured excitation pulse to be broader than the pure optical component. The chief sources of broadening are:

1. the single-photon timing detector
2. the timing electronics, particularly the single-photon timing discriminator
3. optical components, for example, the monochromator

If the impulse response of the ith component in the system has a full width at half-maximum (fwhm) Δt_i, and the excitation fwhm is Δt_e, then the measured instrumental fwhm Δt_m will be given approximately by

$$\Delta t_m \simeq \left[\Delta t_e^2 + \sum_i (\Delta t_i^2) \right]^{1/2} \tag{1.20}$$

Typical values are, for a linear focused photomultiplier, 500 ps, for a constant fraction discriminator, 50 ps, and for an optical spectrometer, 50 ps. These mean that a 700-ps optical pulse would give a measured instrumental profile of ca. 860 ps fwhm.

The effect of having an instrumental pulse of finite duration (i.e., $\Delta t_m \geqslant \tau$) is that the measured fluorescence decay form departs from the true fluorescence response function, that is, Eq. (1.2). Nevertheless, the measured fluorescence decay $F(t)$ can still be analyzed, since it can be expressed as the convolution of the measured instrumental prompt response $P(t)$ and the theoretical fluorescence response function $i(t)$ which would be obtained for δ-function excitation; that is,

$$F(t) = \int_0^t P(t')\, i(t-t')\, dt' \tag{1.21}$$

where t' defines the variable time delays (in practice, channel numbers) of the infinitesimally small time widths dt' (i.e., channel widths) of which $P(t)$ is composed.

By measuring $P(t)$ experimentally over i channels of data, the convoluted form of $F(t)$ can be obtained from Eq. (1.21) by assuming a functional form for $i(t)$, for example one, two, or three exponentials. Figure 1.3 demonstrates the effect of convoluting $P(i)$ with a single exponential decay to give the fluorescence function $F(i)$ with which to compare the measured decay. By iterating the values of the decay components until good agreement is obtained, the best fit values are determined. The standard statistical procedure used to assess the goodness of fit is the χ^2 test, which is described in Section 1.3.2.

It will be noted that the process of analyzing a decay measurement requires a knowledge of what the form of the fluorescence response function is likely to be. That is to say, the analysis is one of *re*convolution and compare rather than determination of an unknown decay law by deconvolution. A very comprehensive review of convolution procedures with an emphasis on fluorescence lifetime spectroscopy is given in the proceedings of the conference held on this topic in Nancy in 1982.[30]

If $F(t)$ determined experimentally and via convolution do not agree within accepted statistical limits, then two explanations are possible:

1. The theoretical model for $F(t)$ is inappropriate.
2. A systematic experimental error is present.

In this case, the immediate recourse is to think again about the kinetics and try an alternative model. Typical systematic errors which might be encountered include stray light, changes in the excitation pulse profile, and filter fluorescence. In our experience, it is unusual to be able to correct for the effect on the goodness of fit of a systematic error by convoluting with a

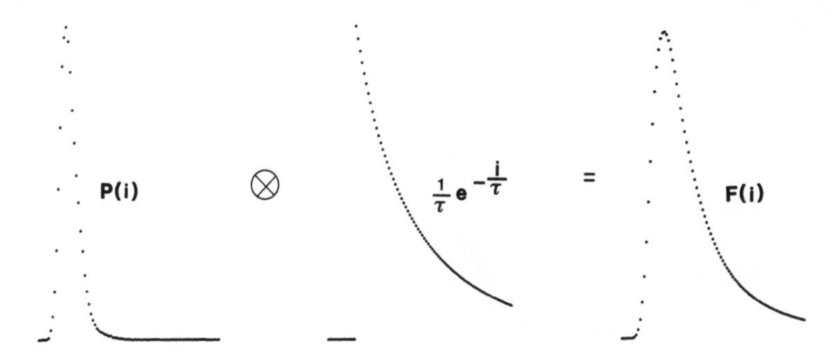

Figure 1.3. Convolution of a measured instrumental prompt response $P(i)$ with a monoexponential fluorescence impulse response to give a decay function $F(i)$ with which to fit to the measured fluorescence decay.

more complex decay form. The two possibilities are thus usually clearly distinguishable. As a rule of thumb, we find that data giving $\chi^2 < 1.2$ does not support the investigation of either more complex models or systematic errors.

For the convolution procedure to be valid, as much care is needed when measuring the instrumental pulse $P(t)$ as is required for the decay $F(t)$. $P(t)$ is usually measured by tuning the analysis wavelength to the same value as that for excitation and recording the excitation pulse. For this purpose, the sample is replaced by a scatterer capable of simulating the sample's optical density and the isotropic nature of fluorescence. The most commonly used scatterer is Ludox (colloidal silica, available from Dupont), though other particle suspensions, for example, milk are also used. Both are capable of emitting fluorescence when excited in the UV, which can be particularly troublesome when no analyzing monochromator is present. Nevertheless, in our experience, careful use of such scattering media consistently gives better fits than are obtained with the use of highly reflective surfaces.

The primary reason for attempting to simulate identical conditions when measuring the excitation and fluorescence is to ensure that the optical images of both are identical at the photocathode of the single-photon timing detector. This is because the timing response of all single-photon detection depends on the point of illumination, some more than others. For example, with the linear focused photomultiplier, the Philips XP2020Q, the photoelectron transit time varies by ~ 14 ps/mm.[31] The corresponding figure for the side-window Hamamatsu R928 is ~ 450 ps/mm.[32]

Another difference between the measurement of $F(t)$ and $P(t)$ which may be significant is that of the photon wavelength. However, careful choice of the photomultiplier and the associated voltage divider network can make such effects insignificant for most practical purposes.

An interesting question is what is the smallest fluorescence lifetime which can be measured for a given instrumental width. In theory, the convolution integral gives no clues as there is no theoretical limit. In practice, numerous workers have reported 200-ps decay times using a spark source with an instrumental width of ~ 1.5 ns.[33, 34] Most workers agree that a factor somewhere between 5 and 10 is not at all unrealistic.

The whole question of how to assess the confidence which can be ascribed to accepting that a particular kinetic model describes a decay in the light of experimental errors is an important one which hinges to a large extent on the experience of the user. In an earlier review[35] we have sought to give some guidelines toward this important decision-making process.

1.3. Data Analysis

The aim of decay time experiments is to study decay kinetics. This is an indirect process: The measurement system yields fluorescence decay curves, consisting of intensity values at some hundreds or thousands of consecutive short time periods (channels) after the time of pulsed excitation. The parameters describing the kinetics are then obtained by reducing these data using statistical methods of data analysis. Decay analysis software therefore plays a key role in decay time determination.

Many approaches to the general problem of decay analysis have been explored, including reconvolution least-squares,[36] Fourier transform,[37, 38] Laplace transform,[39] modulation function,[40] and moment analysis.[41] Much argument exists among software authors as to the merits of the different approaches. We restrict our discussion to reconvolution least-squares analysis, partly because it is the most widely used approach, partly because it is the area in which we have most experience, and partly because many of the concepts and limitations are shared by other techniques.

1.3.1. Formulation of the Problem

The data analysis problem is defined by the nature of the measurement method and the associated errors of measurement, both systematic (e.g., non-linearity, finite response speed) and random (e.g., noise). The instrumentation itself clearly needs to be designed to reduce these errors to a minimum. Thereafter, the data analysis software aims to extract the kinetic information as precisely as possible, by correcting, as far as practical, the remaining systematic errors of measurement. This can be simply illustrated for the measurement of a unimolecular decay, analyzed using least-squares reconvolution. The complete kinetic model in terms of the impulse response is:

$$i(t) = \begin{cases} \dfrac{1}{\tau} e^{-t/\tau} & t \geqslant 0 \\ 0 & t < 0 \end{cases} \tag{1.22}$$

Relating this to a measured decay curve such as that shown in Figure 1.4 is not straightforward, because several instrumental factors need to be taken into account:

(a) The measurement quantizes the smooth fluorescence intensity versus time signal into discrete data channels.
(b) The signal is noisy.
(c) The signal is superimposed on background (dark) noise.
(d) The instrument response distorts the decay curve.
(e) The location of the $t = 0$ point of Eq. (1.22) is unknown.

Point (a) can be accommodated by evaluating the function as stepped averages over individual data channels. The noise background (c) can be included in the fitting function as an unknown parameter. The main instrumental distortion (d) in such measurements is smearing, caused by the finite response speed of the instrument. This is represented mathematically by the convolution integral given by Eq. (1.21).

Of course, the prompt response needs to be known, if we are to use Eq. (1.21) in the analysis of fluorescence decays. In practice, we normally replace the sample with a scatterer, retune the monochromator, and measure an *excitation profile*. This, of course, is an imperfect measure of $P(t)$—it is measured at the *wrong* wavelength, it is quantized and noisy, and so on. Nevertheless, the use of *excitation profiles* in decay curve analysis does bring very substantial benefits in terms of improved time resolution and the ability to fit meaningfully to complex decay models. In practice, the excitation profile needs to be measured alongside each fluorescence decay measurement, to minimize another measurement error: instrumental response drift.

In reconvolution analysis, the convolution is applied to the kinetic model, prior to comparison with the decay data, as illustrated in Figure 1.3. Of course, sums have to be used (numerical convolution) rather than integrals, because of the quantization of the data. This procedure is used, rather than deconvoluting the fluorescence decay data because of its greater stability of convolution. Convolution is smearing. Deconvolution is an ill-posed problem, because you cannot unsmear by reversing the action.

The calculation of the fitting function, as in Figure 1.3, also determines the position among the data channels of the time-zero point, since this is directly related to the position of the measured excitation profile, $P(i)$. However, a small mismatch in positioning is inevitable, because of quantization, noise, wavelength effects, etc., inherent in the less than perfect representation of the instrumental response by the excitation profile. Since the kinetic function (e.g., Eq. 1.22) has its most rapid change at $t = 0$, a small positioning error can have a significant effect on the fit. For this reason, a time shift parameter is sometimes included in such data analysis programs. This can be used either as a fixed offset or as an unknown parameter, to be determined as part of the analysis. In practice, shift iteration brings very substantial benefits to the analysis of decay curves, especially when kinetic information in the fast-rising edge of a decay curve is sought.

Finally, what started out as a simple exponential decay now has the form

$$F(i) = P(i) \otimes \frac{1}{\tau} \exp(-i/\tau) \tag{1.23}$$

from the numerical convolution (represented by $\otimes$), and

$$F_Y(i) = B + A \cdot F(i + \Delta) \tag{1.24}$$

for comparing with the data. The integer i is used to denote data channels. The parameter B is the noise background, A is a scaling factor, Δ is the shift parameter and τ in this case is measured in channels.

1.3.2. Least-Squares Fitting

Equation (1.24) now needs to be fitted to the fluorescence decay data $Y(i)$ of Figure 1.4. The most common method used is the least-squares method (see Ref. 42 for a readable introduction to statistical data analysis ideas). This uses a quantity χ^2 as a measure of mismatch between data and fitted function. In the above example,

$$\chi^2 = \chi^2(A, B, \tau, \Delta) \qquad (1.25)$$

The method of least squares strives to determine the *best-fit parameters* A', B', τ', and Δ' that will yield the lowest possible value of χ^2. By definition,

$$\chi^2 = \sum_{\text{DATA}} \left[\frac{Y(i) - F_Y(i)}{\sigma(i)} \right]^2 = \sum_{\text{DATA}} [W(i)]^2 \qquad (1.26)$$

where $Y(i)$ is the fluorescence decay datum value, $F_Y(i)$ is the fitting function value, and $\sigma(i)$ is the statistical uncertainty of the datum value $Y(i)$ (standard deviation). Note that $\sigma^{-2}(i)$ is referred to as the data weight.

The reason for this choice of fitting criterion is not quite arbitrary: it is equivalent to the more rigorous, though less easy to apply, method of maximum likelihood, for Gaussian and, asymptotically, for Poisson statistics.[42]

The quantity inside the brackets of Eq. (1.26), $W(i)$, is called a *weighted residual*. For any value of i, the numerator is the actual deviation between the datum value $Y(i)$ and the corresponding fitting function value $F_Y(i)$. The denominator represents the deviation expected from statistical (noise) considerations. Equation (1.26) can therefore be written as

$$\chi^2 = \sum_{\text{DATA}} \left[\frac{\text{actual deviation}}{\text{expected deviation}} \right]^2 \qquad (1.27)$$

Now, if the fitting function is appropriate, then the minimum should be characterized by equality between actual and expected deviations. The weighted residual is then unity for each term of the sum, and

$$\chi^2 = \text{no. of data points } (N) \qquad (1.28)$$

This is not quite right for two reasons:

(a) The *expected deviation* is a statistical expectation, and therefore the mean value of a statistical distribution. This necessarily makes the weighted residuals and χ^2 statistical quantities, with associated distribution functions.

(b) The number of fitted parameters, ν, needs to be taken into account. The quantity $(N-\nu)$ is called the number of *degrees of freedom* and should be used in place of N.

A more accurate, though less precise, statement of Eq. (1.28) therefore is

$$\chi^2 \approx (N-\nu) \qquad \text{for a good fit} \tag{1.29}$$

It is customary and convenient to normalize χ^2, so that its value is independent of $(N-\nu)$:

$$\chi_N^2 = \frac{\chi^2}{(N-\nu)} \approx 1 \qquad \text{for a good fit} \tag{1.30}$$

In photon counting experiments, the *expected deviation*, $\sigma(i)$, which characterizes the random noise, can be estimated from the data, using the Poisson distribution [Eq. 1.14]. Thus,

$$\sigma(i) = [Y(i)]^{1/2} \tag{1.31}$$

In Eq. (1.26), χ^2 was defined as a sum over the individual data points of a fluorescence decay. It is straightforward to extend this definition to include several such local sums in a global χ^2:

$$\chi_G^2 = \chi_{L1}^2 + \chi_{L2}^2 + \cdots \tag{1.32}$$

Global, rather than individual, χ^2 minimization makes sense for families of decay curves with family characteristics. Monomer/excimer decays, for example, should show amplitude changes but not decay time changes with emission wavelength. One can therefore fit a whole family of excimer decay curves at different wavelengths using the same decay times for all of them, but allowing for individual amplitudes. Constraining the decay times in this way can often make the difference between "curve representation" and kinetic sense!

A typical set of data and the result of reconvolution analysis are shown in Figure 1.4. The points are fluorescence data $Y(i)$ and prompt $P(i)$, and the line through the decay is the best-fit function $F_Y(i)$. This type of presentation has become a standard in the field, and a logarithmic plot is usually chosen for convenient visual inspection of the degree of exponentiality. Such a

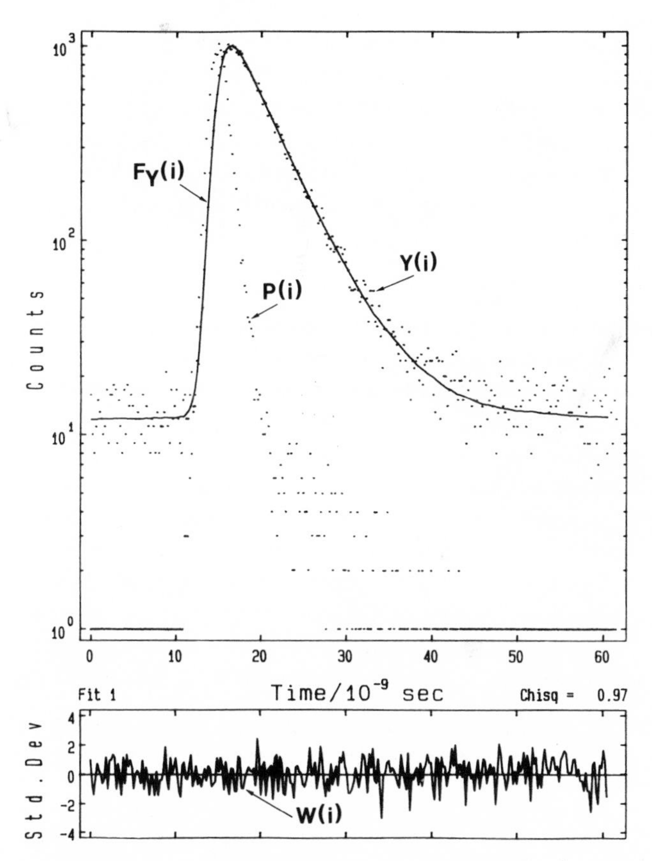

Figure 1.4. A typical measured fluorescence decay $Y(i)$, instrumental response $P(i)$, fitted function from reconvolution $F_Y(i)$, and weighted residuals $W(i)$. The sample is fluorescein in water, and the best-fit decay time is 4.22 ± 0.11 ns.

plot tends to exaggerate features in the lower part of the decay at the expense of those around the peak. The lower display box shows the weighted residuals, which are an important means of assessing the goodness of fit for the following reasons:

(a) They show *where* misfits occur.
(b) Their normalization compensates for the varying data precision within the data set and from data set to data set.
(c) The deviations are expressed in a statistically meaningful way—in terms of the standard deviations of the associated data noise.
(d) The relationship between residuals and χ^2, the fitting criterion, is straightforward.

It should be noted that there are further methods of analyzing and comparing information contained within the weighted residuals, for example, the autocorrelation function and Durbin–Watson parameter.[42] However, these are just alternative representations of the same information content, and the weighted residual analysis is usually adequate for most practical purposes. In fact, the simple pictorial display of kinetics afforded by plots such as Figure 1.4 is one of the key advantages of the single-photon technique.

1.3.3. Fitting Algorithms

The fitting procedure consists of finding those values of the parameters (e.g., Eq. 1.25) that will yield the lowest possible value of χ^2. The difficulty is that even the simplest kinetic models yield least-squares equations that are nonlinear in the parameters, for which no analytical solutions exist. Such problems have to be solved by trial-and-error methods, driven by a minimization algorithm. Many such algorithms have been devised. They all require user-supplied estimates of the unknown parameters (starting values), which are then refined in the systematic minimization search of the algorithm. One immediate problem with all such algorithms is that they are prone to trapping by local minima, with lower minima in other regions of parameter space remaining unexplored. Other problems that are encountered, especially with the more ambitious kinetic models, include effects of numerical precision (we

Table 1.2. Comparison of Marquardt and Hybrid Algorithms[a]

Initial estimates			Marquardt		Hybrid	
τ_1	τ_2	τ_3	Run speed (s)	Result	Run speed (s)	Result
Section 1						
62.5	26.4	11.3	21	OK	19	OK
60	20	10	24	OK	30	OK
100	10	1	51	OK	82	OK
32	8	2	380	OK	112	OK
Section 2						
15	10	5	110	$\chi^2 = 27$!	99	OK
80	8	0.8	$\Rightarrow \infty$		62	OK
Section 3						
4	2	1	59	$\chi^2 = 431$!	120	$\tau_3 \Rightarrow \infty$[b]

[a] The experimental decay data were analyzed without iterative shifting, with reconvolution analysis from the peak, over 205 channels. Decay times (uncalibrated) are 62.5, 26.4, and 11.3 channels, with relative amplitudes of 14, 45, and 41 %, giving $\chi^2 = 0.98$. The table shows program run speeds and results of analyses for different initial parameter estimates. OK means that a correct result has been obtained.

[b] Program warns of false minimum.

use 64-bit number representations in key sections of code), numerical range, and search path distortions caused by constraints imposed on the parameters.

In its simplest form, an algorithm can step through all combinations of parameter values, within preset bounds, then offer the parameters yielding the lowest χ^2 value as the solution. This is certainly a safe and stable procedure, whose precision can be high, if the parameter increments of the mesh are chosen small enough. However, the computation time increases as N^{ν}, where ν is the number of parameters and N is the number of parameter increments used. More refined algorithms use techniques such as steepest descent, ravine search, function linearization, spline representation, gradient expansion, parabolic extrapolation, and less rigorous convergence criteria to reduce computation time. However, since these shortcuts rely on assumptions about the χ^2 hypersurface that may not apply in practice, a loss of stability is the price paid for increased speed.

Table 1.2 compares two fitting algorithms, the popular Marquardt algorithm (PRA Inc. implementation) and our own hybrid, using a refined grid search strategy. The same data were fitted to three exponential terms, without shift iteration, using various starting values, to measure speed of convergence and stability. The timings refer to a standard PC-AT computer, with 6-MHz clock, equipped with a mathematics coprocessor. The table is divided into three sections, depending on the outcome:

- Section 1 produced convergence to the correct decay times with both algorithms. The first line used the correct values as starting values, to give a measure of the maximum speed of convergence.
- Section 2 used starting values for which the Marquardt algorithm failed to converge in two different ways. In the first line, the program reports a normal convergence, which is clearly wrong. This is caused by a failure of the convergence criterion, which looks for a below-threshold change of χ^2 (first-derivative criterion), rather than a true χ^2 minimum (first- and second-derivative criterion) as required in the hybrid algorithm. In the second line, the starting values resulted in the Marquardt algorithm being trapped in an infinite loop. This is a failure of the parameter increment vectoring calculations.
- Section 3 used starting values for which both algorithms failed. The Marquardt algorithm again converged to a false minimum. The hybrid algorithm failed because of the properties of the fitting function, rather than the algorithm itself. Given the values of τ_1 and τ_2, the χ^2 value becomes independent of τ_3 as τ_3 tends to infinity. The second derivative convergence criterion can therefore not be satisfied. The program continues, but warns of this convergence failure.

In conclusion, while the Marquardt algorithm generally produced faster convergence, it was more prone to convergence failure which, by the nature of the algorithm, appears to the user as normal convergence. It must be emphasized that these false minima were not local minima, but were failures of convergence recognition.

The previous discussion leads us to the interesting and frequent question as to what is the smallest ratio of decay components which can be separated with least-squares analysis of single-photon data. The answer is not simple as it depends on a number of factors in addition to the decay times, namely, relative amplitudes, instrumental width, etc. Lakowicz[43] has presented a good example of what can be achieved using flashlamp excitation and the hybrid grid search algorithm. This is shown in Figure 1.5 for a mixture of anthracene ($\tau = 4.2$ ns) and 9-vinylanthracene ($\tau = 7.6$ ns). Although the raw decay looks exponential, a single-exponential fit gives a χ^2 value of 8.88. A two-component fit gives a χ^2 value of 0.89 with best-fit decay parameters of 4.08 and 7.74 ns, which are in good agreement with those values measured separately. Ultimately, the effect of correlations between the best-fit parameters will dominate, the χ^2 parabola becomes extremely shallow, and the two components are not resolved. Gardini *et al.* [44] have investigated this effect for simulated two-component decays. Some of their results are shown in Table 1.3. Examples where good resolution of two components is achieved include $\tau_1 = 6$ ns, $\tau_2 = 9$ ns, with relative amplitude $A_1/A_2 = 1$; it is not achieved when $A_1/A_2 = 0.25$ or when $\tau_1 = 8$ ns, $\tau_2 = 9.6$ ns, and $A_1/A_2 = 1$. In cases where it is not possible to resolve two decay components in a single measurement at up to 10^4 counts in the peak, there is usually little to be gained from just acquiring more counts. A knowledge of one of the decay parameters, which can then be fixed in the iterations, clearly enhances the

Table 1.3. Analysis of Simulated Biexponential Decay[a]

Simulated parameters			Fitted parameters			
τ_1 (ns)	τ_2 (ns)	A_1/A_2	τ_1 (ns)	τ_2 (ns)	A_1/A_2	χ^2
8.0	9.6	1		Not resolved		
24.0	36.0	1	21.56	33.54	0.42	0.99
12.5	20.0	1	12.56	19.56	0.83	0.97
6.0	9.0	0.25		Not resolved		
6.0	9.0	1	6.15	8.52	0.44	1.18
6.0	9.0	4	5.33	7.89	0.81	0.95
6.0	12.0	0.25		Not resolved		
6.0	12.0	1	6.33	11.97	0.92	1.38
6.0	12.0	4	5.75	11.01	2.36	0.93

[a] Data obtained by reconvolution fit to synthetic decays using a measured excitation pulse of ~3 ns fwhm and a fluorescence response function of $A_1 \exp(-t/\tau_1) + A_2 \exp(-t/\tau_2)$. (From Ref. 44.)

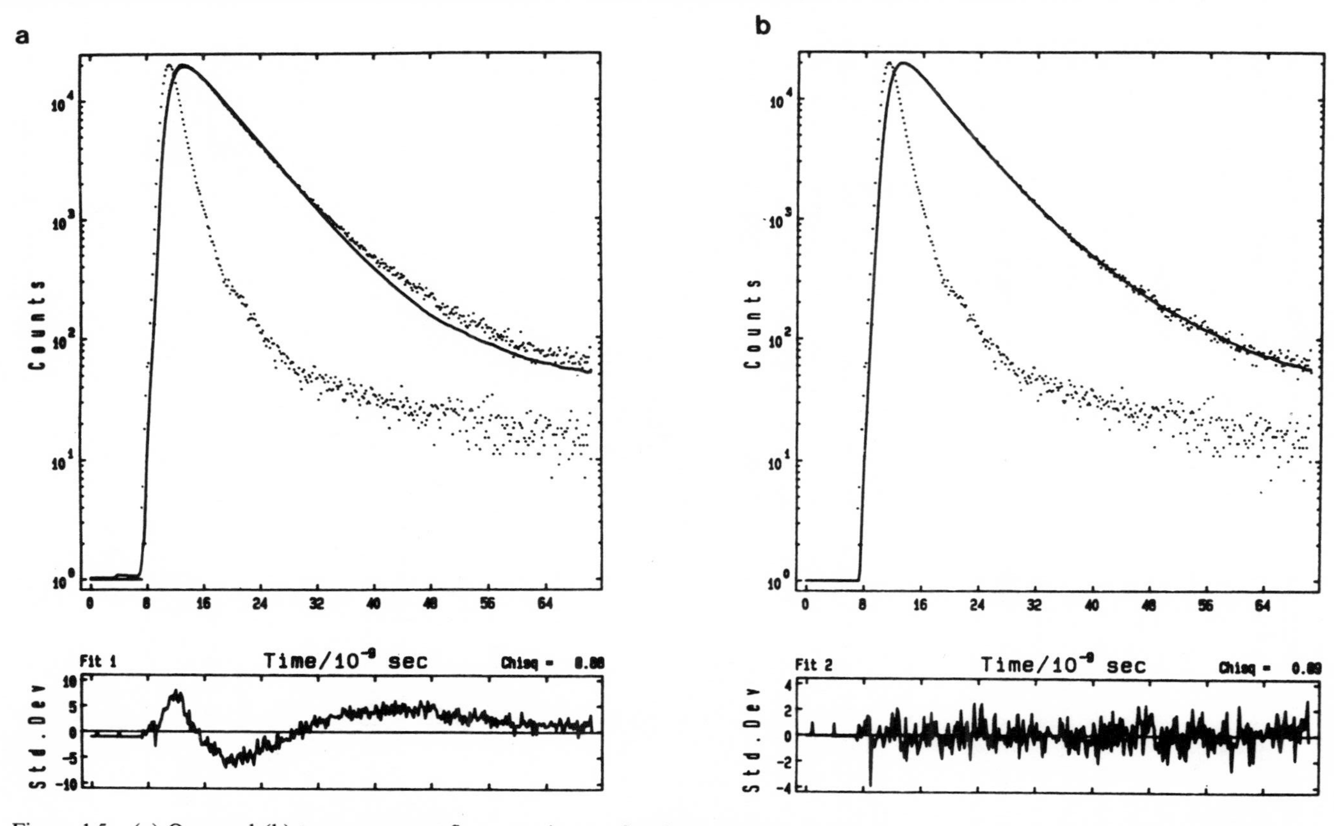

Figure 1.5. (a) One- and (b) two-component fits to a mixture of anthracene and 9-vinylanthracene. The decay times of 4.08 and 7.74 ns are close but well resolved and in good agreement with the measurements taken separately. (From Ref. 43.)

precision. Similarly, where a biexponential decay relates to two species of, say, overlapping but dissimilar emission spectra, there is an advantage in bringing more information to bear on the problem by performing decay measurements at several emission wavelengths and performing a global analysis of the whole data set with a single χ^2 criterion. By this means and using decay curves measured at six emission wavelengths, decay components of 4.6 and 3.2 ns, relating to quenched 9-cyanoanthracene and anthracene, are typical of what has been shown to be separable.[45] This whole approach has now blossomed to enable the study of more complex kinetics (decay data linked to other results,[46] anisotropy,[47] etc.), but the basic methodology of using as much information as possible to constrain the analysis remains the same.

1.3.4. Anisotropy Analysis

The theory and application of using time-resolved fluorescence anisotropy to study molecular rotation is discussed more fully in Section 1.5.3. In this section we will just consider some of the general considerations concerning the analysis of anisotropy data. The anisotropy impulse response, Ir(t), can be defined by

$$\mathrm{Ir}(t) = \frac{I_p(t) - I_x(t)}{I_p(t) + 2I_x(t)} = \frac{\mathrm{ID}(t)}{\mathrm{IS}(t)} \tag{1.33}$$

where $I(t)$ represents the fluorescence impulse response, and the subscripts p and x refer to parallel and crossed polarizers, respectively. The numerator ID(t) and denominator IS(t) are often referred to as the *difference* and the *sum* functions, respectively. $I_x(t)$ and $I_p(t)$ need to be measured under identical conditions for Eq. (1.33) to make sense. In practice, we observe $Y(i)$, not $I(t)$. If we combine $Y_x(i)$ and $Y_p(i)$ as in Eq. (1.33), the resulting anisotropy decay curve $r(i)$ is distorted by the instrumental response. Since reconvolution is inapplicable to the ratio function, a more sophisticated approach is required to correct for this. In addition, anisotropy curves calculated from Poisson decay data are not themselves characterized by Poisson statistics—hence, correct data weights need to be calculated. This is illustrated later in Figure 1.22, where the anisotropy decay data show a rapidly increasing noise in the tail. These data points contribute little to the fit, because their data weights, calculated by a propagation of errors technique from the parallel and perpendicular decay curves, are very low.

There are basically four alternative methods of analyzing anisotropy decay data,[35] which can be summarized as follows:

1. Tail Fit. By fitting to the tail of $r(i)$ (Eq. 1.33) using an exponential function without reconvolution, an estimate of the rotational correlation time

τ_R and residual anisotropy r_∞ can be obtained. This is useful for a quick estimate but is only likely to be accurate if $\tau_R >> \Delta t_m$, the measured instrumental response.

2. *Vector Reconvolution.* The theoretical forms of $I_p(t)$ and $I_x(t)$ can be fitted to directly in a global analysis using a single χ^2 criterion. The appropriate expressions for a spherical rotor are[1, 3, 6]

$$I_p(i) = B_p + A_p(1 - r_0 e^{-i/\tau_R})e^{-i/\tau} \tag{1.34}$$

$$I_x(i) = B_x + A_x(1 + 2r_0 e^{-i/\tau_R})e^{-i/\tau} \tag{1.35}$$

with

$$\chi^2 = \sum_i \left[\frac{Y_p(i) - I_p(i) \otimes P_p(i)}{[Y_p(i)]^{1/2}}\right]^2 + \sum_i \left[\frac{Y_x(i) - I_x(i) \otimes P_x(i)}{[Y_x(i)]^{1/2}}\right]^2 \tag{1.36}$$

where B_x and B_p refer to the background count, A_x and A_p are amplitude scaling factors, r_0 is the initial anisotropy, τ_R is the rotational correlation time, and τ is the fluorescence decay time.

This method of analysis is generally applicable and is suited to analyzing anisotropy data from *T*-format fluorometers without matched detection channels. Otherwise, in general, the excitation profile $P_p(i) = P_x(i)$.

3. *Impulse Reconvolution.* This method involves correcting for the instrumental response by fitting to the difference function $D(i)$ using impulse reconvolution. It was first used by Wahl,[48] and it probably represents the best overall method of handling anisotropy data. The procedure is as follows.

Rewriting Eq. (1.33) in terms of the impulse response functions appropriate to an actual measurement, we obtain

$$\mathrm{Ir}(i) \times \mathrm{IS}(i) = \mathrm{ID}(i) \tag{1.37}$$

The objective is to determine $\mathrm{Ir}(i)$ in terms of τ_{R1}, τ_{R2}, etc. Now, $S(i)$ represents the fluorescence decay only, and hence $\mathrm{IS}(i)$ can be easily obtained separately by reconvolution analysis with $P(i)$ using fluorescence decay parameters τ_1, τ_2, etc., as are required to obtain a good fit. $\mathrm{ID}(i)$ does not need to be known since $D(i)$ is obtained from the measurement of $Y_p(i) - Y_x(i)$. Hence, $\mathrm{Ir}(i)$ can be obtained by convoluting $P(i)$ with $\mathrm{Ir}(i) \times \mathrm{IS}(i)$ using a trial $\mathrm{Ir}(i)$ function and fitting to $D(i)$ in a manner analogous to that in Eq. (1.26):

$$\chi^2 = \sum_i \left\{\frac{D(i) - [\mathrm{Ir}(i) \times \mathrm{IS}(i)] \otimes P(i)}{[D(i)]^{1/2}}\right\}^2 \tag{1.38}$$

The predetermination of $\mathrm{IS}(i)$ thus acts as a constraint in the evaluation of Eq. (1.38), thereby enabling the determination of $\mathrm{Ir}(i)$ from the difference data. In theory, any number of fluorescence and anisotropy decay parameters can be analyzed by this method. In practice, a sum of five decay components in total is probably at the limit of what can be supported by experimental data.

4. Impulse Response Manipulation. This involves determination of Ir(i) using IS(i) and ID(i) in Eq. (1.33). The latter impulse response functions are determined from $I_p(i)$ and $I_x(i)$ using as many exponential decay components as are needed to obtain a good fit. Ir(i) can then be analyzed in order to obtain the rotational information. Like method 2, this approach is also suitable for T-format instruments without a matched impulse response. The approach is the nearest to deconvolution that can be sensibly obtained. However, because all the noise is removed from the data, it is important that the numerical analysis includes a full variance propagation of errors.

1.4. Instrumentation

In this section we will look in more detail at the individual performance of the equipment components which together constitute a working time-correlated single-photon counting spectrometer. The two most important criteria with which to assess the system performance are probably time resolution and sensitivity, which, at the end of the day, depend on the individual components. To a large extent, the performance of such systems depends on that of components which are not specifically designed for this application, though this is changing as manufacturers recognize the growing interest in fluorescence lifetime studies. Not surprisingly, there is some diversity of opinion as to the optimum choice of detector, monochromator, pulsed source, etc. Unfortunately, the ideal concept of maximizing the performance-to-price ratio does not always lend itself to objective comparisons. As a result, we will seek to outline most of the equipment options, though at the same time commenting from our own experience.

1.4.1. Pulsed Light Sources

There are basically three choices of pulsed source for time-correlated single-photon counting. These are:

1. flashlamps
2. mode-locked lasers
3. synchrotrons or storage rings

The first two are by far the most commonly used, with flashlamps probably being the most abundant of all. Mode-locked lasers and synchrotrons are discussed in detail in subsequent chapters in this series so we will choose to mainly discuss flashlamps, although a brief survey of the alternatives is relevant here.

The introduction of commercially available mode-locked synchronously pumped dye lasers in the 1970s[49] ensured that for the first time the ideal of an optical δ-function excitation would become widely available. These days mode-locked laser systems producing femtosecond pulses are not uncommon.[50] However, as far as single-photon timing systems are concerned, it is still the detector rather than picosecond laser pulses which limits the instrumental pulse width. The advantages of mode-locked laser-based systems are a high repetition rate (MHz) and pulse intensity (mW–W average power depending on wavelength) which can give several orders of magnitude higher sensitivity than that obtained with flashlamps. The pulse profile is also extremely stable, routinely ~ 5 ps, and spectrally pure from < 300 nm to 1 μm. The drawbacks of such large-frame laser systems are the high expense (> 10 times that of a flashlamp and monochromator), complexity of operation, and a spectral output that is anything but routine to tune over a wide range. The early mode-locked laser systems were based primarily on argon ion lasers, but Nd:YAG lasers have now become more popular in this application, chiefly due to the lower running costs.

The recent introduction of pulsed semiconductor diode lasers offers a laser source for single-photon timing which is comparable in cost to flashlamps. Imasaka *et al.* were the first to demonstrate the use of diode lasers for single-photon timing fluorescence lifetime applications in a study on polymethine dyes.[51] Although originally limited to near-infrared emission for optical fiber communications, these devices are now available in the red, and Hamamatsu Photonics has recently introduced a frequency-doubled version to obtain 410 nm. Typical pulse widths are 50 ps, repetition rate is up to 10 MHz, and peak power is ~ 100 mW. The peak power available after frequency doubling is at present ~ 1 mW. This corresponds to $\sim 10^5$ photons per pulse, which should be very workable although there are as yet no published fluorescence lifetimes measured with frequency-doubled diode lasers. Although currently of lower power than synchronously pumped laser systems, diode lasers seem to promise much for the future with single-photon timing fluorescence lifetime studies.

Synchrotron radiation is composed of a broad continuum which peaks in the X-ray region but extends through the UV and into the IR. The radiation emitted is due to the acceleration of relativistic electrons when contained within a circular orbit. The pulse profiles are Gaussian and typically ~ 100 ps fwhm at best, determined by the length of the bunch of electrons in orbit. Storage rings tend to be preferred to synchrotrons because of their better intensity stability although both are used. A significant number of papers reporting the use of synchrotron radiation for fluorescence lifetime work have been published,[52–54] although the number is restricted by the small number of such centers and the fact that they are more frequently operated for optimum performance for X-ray and vacuum UV spectroscopists. This usually means operation in multibunch mode at ~100-MHz pulse repetition rate. With many longer fluorescence lifetimes, this produces a sawtooth decay form due to sample reexcitation occurring before the decay is complete. The same effect is observed with mode-locked laser systems if cavity dumping to reduce the repetition rate is not included. Data analysis methods have been proposed to handle such distorted decay rate,[55] but with sawtooth waveforms an extra level of complexity is introduced which can mask important kinetic features, for example, long-lived decay components.

Nevertheless, the high repetition rate of both synchrotron and mode-locked lasers means that decay curves can often be accumulated in a matter of seconds, whereas hours might be needed with flashlamps. Generally speaking, mode-locked lasers offer significantly higher intensity than synchrotrons apart from wavelengths below ~250 nm, where suitable lasers are not yet available. A good example of a case in which synchrotrons have the edge over lasers for UV work is reported in a study of the first ion-pair state in IBr by Donovan *et al.* using 188-nm excitation.[56] Interestingly enough, hydrogen flashlamps are also quite suitable for operation in the vacuum UV. Ware and Lyke have reported flashlamp fluorescence lifetime studies of saturated hydrocarbons at excitation wavelengths down to 120 nm.[57, 58] However, this spectral region is of considerably less interest than the region above 250 nm, appropriate to aromatic fluorophores.

Flashlamps operate by achieving a spark discharge in a gaseous environment, usually hydrogen or nitrogen. The first such device was developed by Malmberg in 1957[59] and used for fluorescence lifetime measurements in the same year by Brody.[60] There are basically two types of flashlamp: a free-running discharge operating in the mode of a relaxation oscillator, and a gated discharge operated by externally controlled switching of a hydrogen thyratron.

Although simple in principle, flashlamps have a number of potential drawbacks which arguably for many years held back the widespread use of the single-photon technique. The discharge between two electrodes is an erosive process capable of causing an unstable excitation pulse profile, the low repetition rate (<10 kHz) and weak pulse intensity of early designs led to tedious measurements, and, perhaps most serious of all, flashlamps can generate significant radio-frequency interference. The latter is also generated by synchrotrons and in the mode locker and cavity dumper of laser systems. Figure 1.6 shows the drastic effect which rf pickup can have on a decay curve in terms of an oscillation superposed on the decay curve. In the case of the measurement of relatively long decay times (>5 ns), the effect of rf pickup is quite obvious, but for fast decay times (<1 ns) it can be quite subtle and only

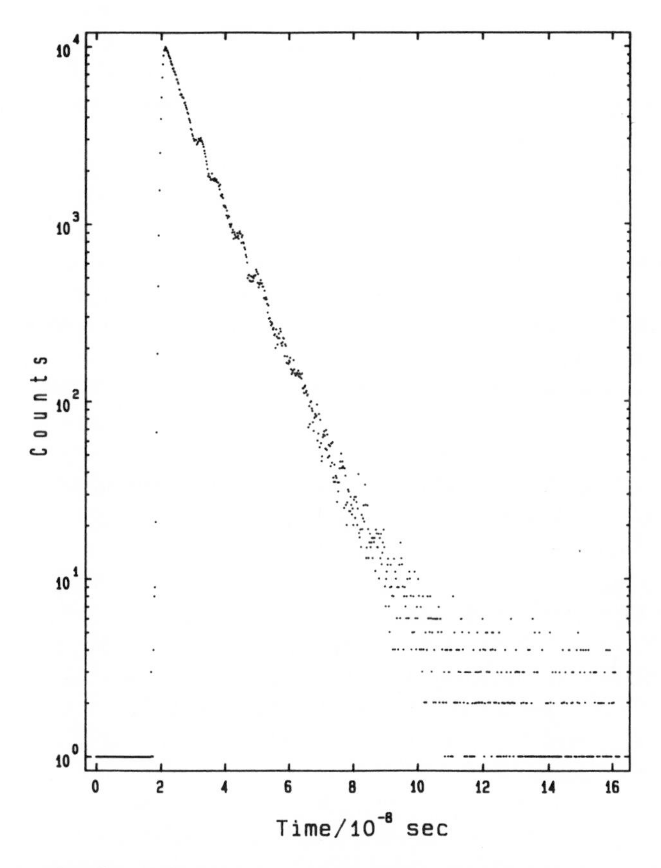

Figure 1.6. The effect of rf pickup on a time-correlated single-photon decay measurement.

detected from the persistence of a high χ^2 value. A correction curve can be used to rectify the effect of rf pickup,[61] but this has been found to be somewhat unsatisfactory.[62] We would also recommend that this effect be eliminated at source rather than through subsequent analysis. Standard approaches to this include having metallic shielding with a good single earth and using filters on the electrical connections. Nevertheless, once present, rf problems often seem to defy science, are frequently intermittent, and are difficult to eliminate without design changes.

With eliminating rf pickup in mind, the first flashlamp we developed was based on an all-metal brass construction with a thyratron-gated coaxial transmission line geometry to minimize electrical reflection and maximize optical output.[63] In this case, the rf shield was the gaseous enclosure itself, and our subsequent version[64] has proved to be very successful, having been distributed worldwide. Other successful designs include those of Ware and co-workers,[58, 65] which were also commercialized, but as a ceramic construction. The greater degree of control achievable with these second-generation flashlamps based on gated designs has led to their having much wider usage than free-running constructions.

Figure 1.7 shows our flashlamp design, and we will briefly consider some of its features and performance as they are typical of what is now widely achieved. The flashlamp operates by charging the anode electrode to a high voltage which can be switched across the spark gap on gating the hydrogen thyratron, whereupon a spark occurs. The spark is extinguished passively by limiting the current which flows by means of a charging resistor of a few megaohms. A key element in obtaining a fast pulse is to minimize the discharge capacitance. Hess[66] reported a theoretical approach to optimizing the design of nanosecond flashlamps, but in practice we have found that the optimum conditions for our lamp differ markedly from these predictions. The experimental variables are electrode separation d, repetition rate R, switching voltage V, and gas pressure P. Ideally, one would like to obtain the maximum excitation pulse intensity in the minimum duration. However, some compromise is always necessary. We have found that in hydrogen the overall optimum operating conditions, at which we usually work, are $d \approx 1$ mm, $R \approx 40$ kHz, $V \approx 7$ kV, and $P \approx 0.5$ B (bar). This typically gives an overall instrumental pulse duration of ~ 1.2–1.5 ns fwhm when measured with a Philips XP2020Q photomultiplier. Under these conditions, $\sim 10^8$ photons per pulse are generated.[64] A further reduction in pulse width can be most simply achieved by reducing d, R, and V, with perhaps a slight increase in P.

The pulse widths obtainable with nitrogen are generally broader than with hydrogen, and it is often necessary to work with nitrogen at higher pressures (~ 1.5 B) and small gaps (~ 0.25 mm) in order to obtain a fwhm of < 2 ns.[64] Both gases have their uses. Hydrogen always gives the fastest pulses possible. Figure 1.8 shows the pulse from the hydrogen flashlamp measured in our laboratory using a single-photon avalanche photodiode (SPAD) detector.[67] This pulse has a fwhm of 730 ps and a full width at 1% of maximum of 2.6 ns and, to the best of our knowledge, is the fastest instrumental pulse yet reported with a spark source. Because the SPAD has an optical impulse response of < 100 ps fwhm, this pulse effectively represents the flashlamp's optical pulse profile (Eq. 1.20). Hydrogen has advantages over nitrogen by giving better reconvolution fits, higher pulse stability, negligible afterglow, and a pulse profile which is much less dependent on wavelength. As a consequence of the latter, any stray light exciting the sample at a different wavelength from that selected will still give the same profile in the fluorescence decay. Nitrogen does however have spectral lines at 316, 337, and 358 nm which, although of varying pulse profile, are more than an order of magnitude more intense than with hydrogen at the same pulse duration. By and large, though, we make every effort to perform measurements with

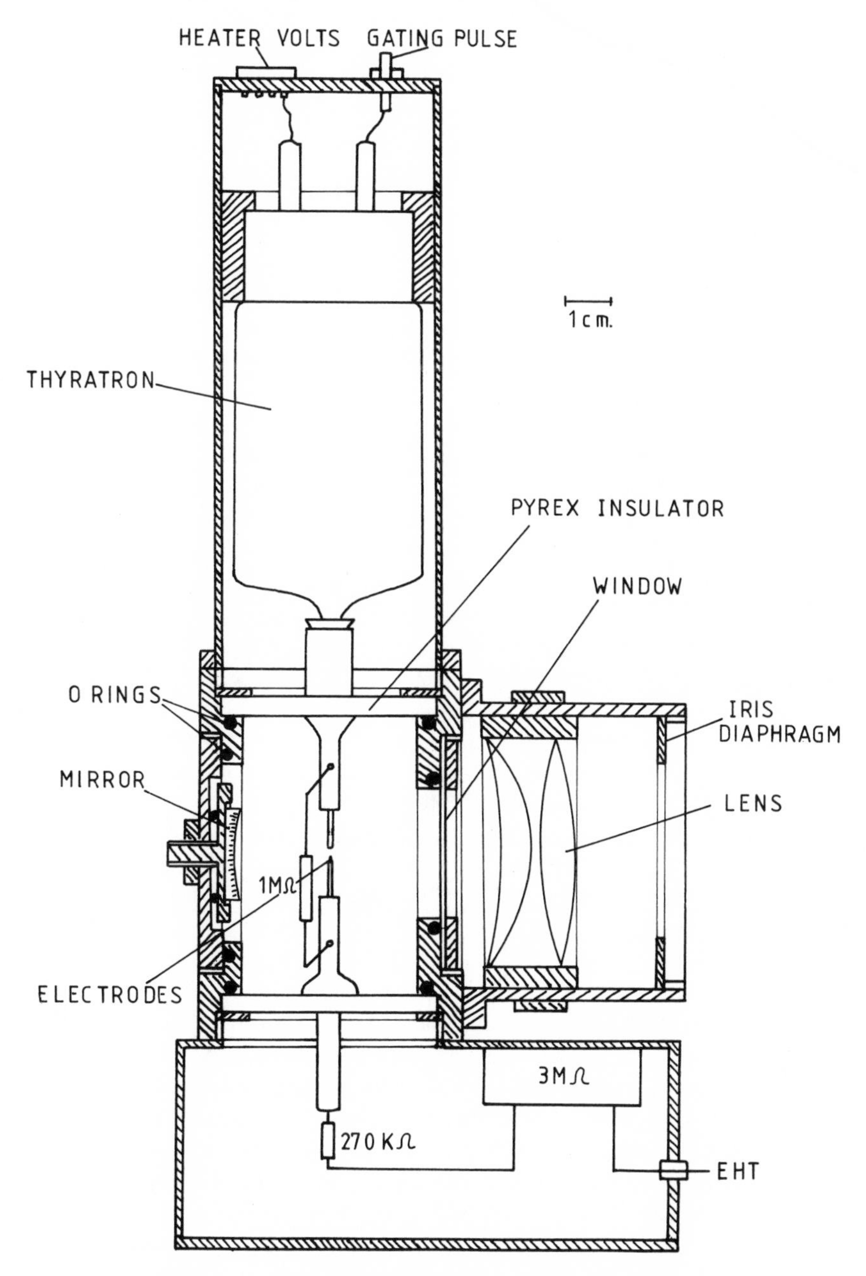

Figure 1.7. The coaxial nanosecond flashlamp. (From Ref. 64.)

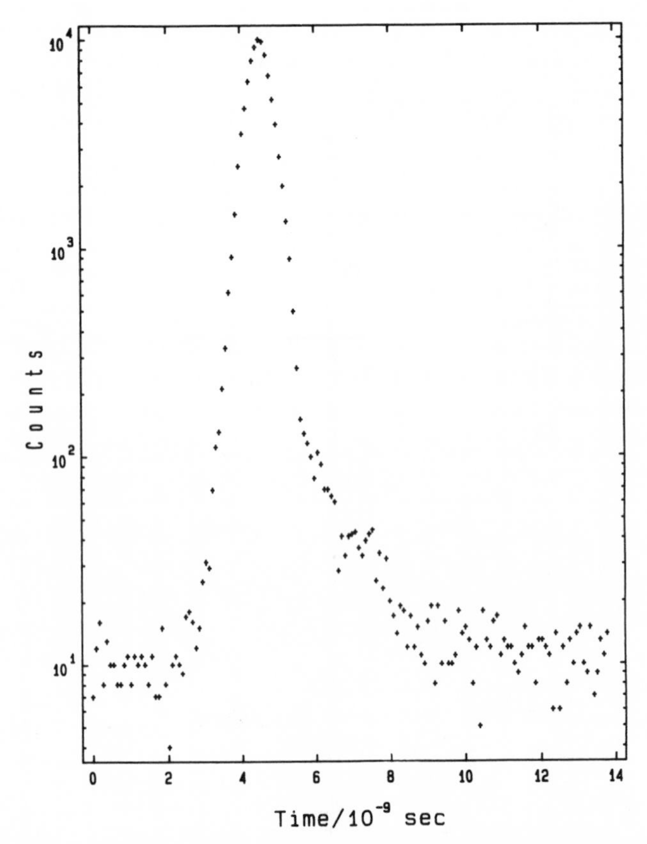

Figure 1.8. Coaxial flashlamp optical pulse measured using SPAD detector with flashlamp operating conditions of 50-kHz repetition rate, 0.7 mm electrode separation, 0.75-Bar pressure of hydrogen and 7-kV switching voltage. The fwhm is 730 ps, and the full width at 1% of maximum is 2.6 ns.

hydrogen, as even when it means a longer run time we have found it to facilitate a better test of the kinetic model. Deuterium gives up to twice the intensity of hydrogen but at the expense of an increase in pulse width.

Much has been written about the stability of spark sources. Certainly this was a problem with the early flashlamp designs. As far as the single-photon technique is concerned, what matters is the reproducibility of the time-averaged instrumental pulse profile although both the pulse-to-pulse intensity and time profile can fluctuate. With the flashlamp shown in Figure 1.7, we have found it to be a routine matter to obtain good fits for decays measured for up to ~4 h with the lamp pulse measured sequentially. We have even measured subnanosecond decay times with up to 18 h run time,[68] but for the study of such weak emissions we would generally advocate the use of differential methods to actively correct for any possible spark instability rather than relying on passive stability (see Section 1.6.1). The limiting factor with regard to the time-averaged stability of the spark seems to be the gradual buildup of a "whiskerlike" deposit on the tungsten electrodes, which has most effect on the optical pulse at around the 1% level as shown in Figure 1.9. The effect becomes more dominant as the electrode separation is reduced. We generally clean the electrodes on average once per week before the effect becomes significant. For this reason, it is important to have easy access for the removal of the electrodes. A useful monitor of the stability and electrode degradation is to compare the optical pulse rate with that of the gating signal. We have developed a continuous method of monitoring this by using LEDs to warn of any free-run or misfire events.[64]

One criticism which can be levied fairly at spark sources is their low intensity beyond 500 nm. The longest excitation wavelength which we have

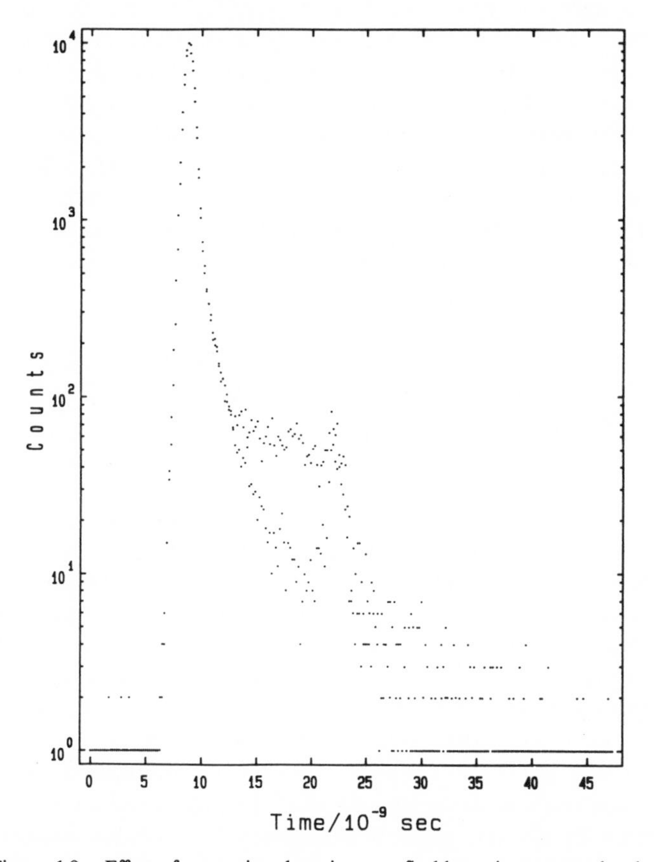

Figure 1.9. Effect of sputtering deposits on a flashlamp instrumental pulse.

used is 620 nm in a study of the laser dye IR140.[68] Certainly, this is a region where laser systems offer an enormous gain in sensitivity, though no doubt flashlamps could be optimized for use with gases emitting at longer wavelengths, for example, neon. Nevertheless, for all-around versatility, cost effectiveness, and convenience, there is as yet no source comparable to the humble spark for single-photon fluorescence studies, even 30 years after its inception.

1.4.2. Optical Components

Apart from the source and detector, probably the most crucial component in a single-photon counting spectrometer is the monochromator(s). This is because, assuming a stable excitation pulse, the dominant factor in limiting time resolution (i.e., the smallest lifetime which can be measured) is often the level of stray light. Stray light which is scattered by the sample and is detected has the time profile associated with the excitation pulse and causes a misfit which is pronounced around the rising edge and peak of a fluorescence decay. This can be corrected to some extent in the data analysis by fitting an additional term to the measured fluorescence decay which mixes in a variable fraction of the excitation light.[35] An alternative approach is to try and fit an additional decay component with a fixed but very short lifetime (e.g., 0.1 channel). In our experience, it is very rare for the presence of stray light to take on the cloak of an additional decay component, and some simple experimental tests can usually be performed to explore its presence, for example, repeating the experiment under the same conditions but with the pure solvent replacing the sample. Stray light is usually present to some extent in most lifetime measurements. It only has a significant effect on the χ^2 value if

$$Y_i^s \geqslant Y_i^{1/2} \tag{1.39}$$

that is, where the number of stray light counts in channel i, Y_i^s, is comparable to or exceeds the expected statistical error in the decay datum value, $Y_i^{1/2}$. This is because as the number of counts increases, the fractional statistical error decreases but the fractional systematic error stays the same. The form of Eq. (1.39) is also applicable to other systematic errors, for example, TAC nonlinearities. Hence, although increasing the number of counts will give a more precise lifetime measurement, the χ^2 value worsens.

Cutoff filters are often useful in reducing stray light levels, but the problem of filter fluorescence has already been mentioned. Because of their monochromatic nature, laser sources offer distinct advantages as compared to flashlamps in respect of minimizing stray light. A particularly difficult measurement condition arises when using nitrogen in a flashlamp to study fluorescence which coincides with one of the intense nitrogen lines, for example, excitation at 297 nm and fluorescence at 358 nm. This problem is eased somewhat with hydrogen, which has an overall decrease in intensity with increasing wavelength.

The stray light performance of holographic gratings is better than that of ruled gratings; however, the latter have a higher efficiency. There is a good case for using double monochromators for analyzing fluorescence, particularly when using laser excitation, in which case the loss in throughput can be more easily tolerated. With double monochromators, the stray light reduction might be typically a factor of 10^8 as compared to 10^4 for the equivalent single monochromator. If a double monochromator is preferred, then it makes sense to use one which is subtractively dispersive in order to also correct for pulse broadening across the grating.[69] This can be expressed for a single grating as[70]

$$\Delta t_g = (D/c)(M\lambda/d) \tag{1.40}$$

where D is the width of grating illuminated, c the velocity of light, M the diffraction order, λ wavelength, and d the groove spacing.

For a 1200-lines/mm grating, with $D = 5$ cm and $\lambda = 400$ nm, Δt_g is 80 ps. This is clearly not insignificant, particularly when used with mode-locked laser pulses of a few picosecond duration and microchannel plate photomultipliers capable of having an instrumental response of ~ 50 ps. The difference in pulse broadening across the grating between the excitation and fluorescence wavelengths can thus be an important source of error in reconvolution when measuring decay times of < 100 ps. Unfortunately, it is at present inconvenient to tune the wavelength over a wide range in mode-locked laser systems in order to overcome the problem by measuring the instrumental profile at the fluorescence wavelength. The case for using a subtractively dispersive monochromator with picosecond single-photon timing systems is thus very strong although, surprisingly enough, most such systems still use single monochromators.

All optical monochromators possess the potential for generating artifacts at certain wavelengths, and these are often much easier to observe in the time domain than in the steady-state or the frequency domain. One important effect is retrodiffracted light,[71] which is present in all monochromators but is more easily observed and therefore a greater potential source of error in T-format spectrometers as shown in Figure 1.10. The simple laws of diffraction allow for not only forward diffraction toward the exit slit but also retrodiffraction toward the entrance slit. If δ is the Ebert angle of the monochromator and θ the angle of rotation of the grating, then the grating equation leads to the following expressions for a Czerny–Turner monochromator:

$$\sin(\delta + \theta) - \sin(\delta - \theta) = M_T N \lambda_T \tag{1.41}$$

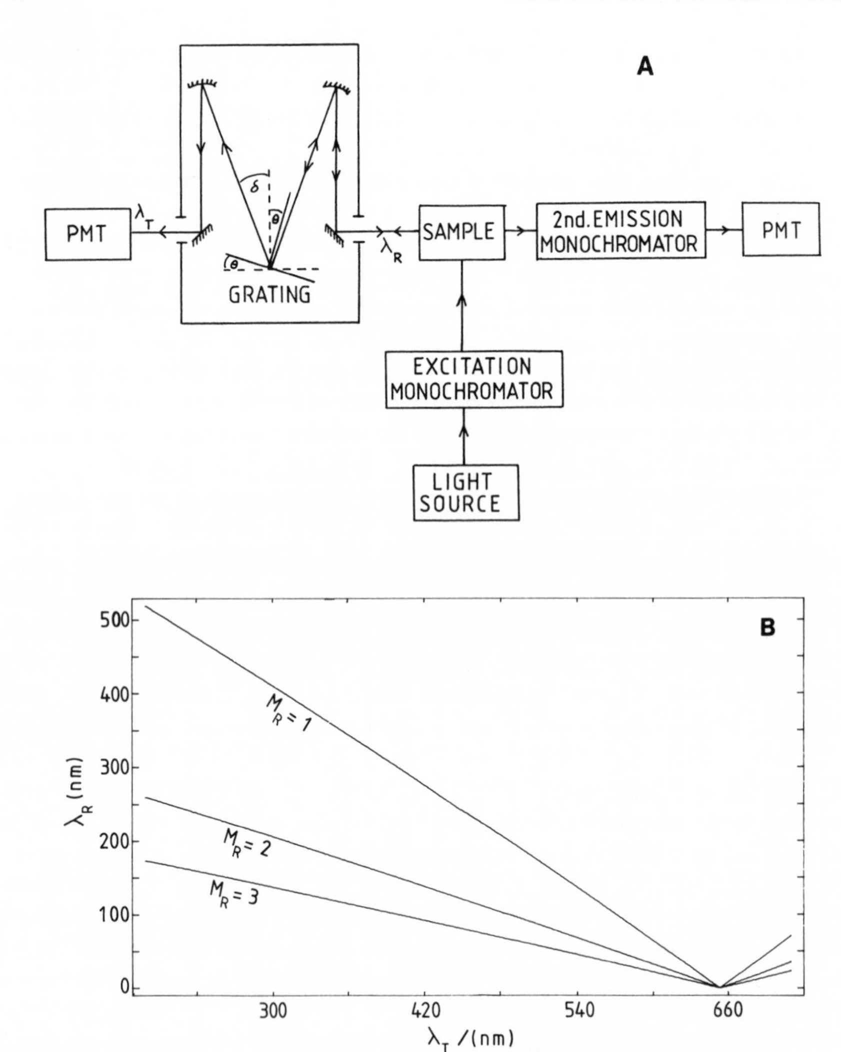

Figure 1.10. (A) Generation and detection of retro light in a T-format spectrometer. (B) Typical relationship between retro wavelength λ_R and transmitted wavelength λ_T in a Czerny–Turner monochromator. (From Ref. 71.)

for the transmitted wavelength λ_T at the exit slit and

$$2\sin(\delta - \theta) = M_R N \lambda_R \tag{1.42}$$

for the emergent retro wavelength λ_R at the entrance slit, where N is the number of grooves per meter, and M_T and M_R are the orders of diffraction.

Solving Eqs. (1.41) and (1.42) leads to the following relationship between λ_R and λ_T:

$$\lambda_R = M_R^{-1}\{[4(\cos^2\delta)/N^2 - M_T^2\lambda_T^2]^{1/2}\tan\delta \pm M_T\lambda_T\} \tag{1.43}$$

where the + and − refer to the grating facing toward the exit or entrance slit, respectively.

The main point about retrodiffraction is that at certain wavelength combinations in a T-format it can cause distortion of a fluorescence decay because retro light is time shifted. The conditions at which $\lambda_T = \lambda_R$ and where all incoming wavelengths are retrodiffracted at $\theta = \delta$ are of particular concern. Figure 1.10 shows a typical dependence of λ_R on λ_T, in this case for a Spex Minimate model 1650 monochromator, with $N = 1.2 \times 10^6$ and $\delta = 25.6°$.

We would recommend that the retrodiffraction characteristics always be taken into consideration in the designing of T-format fluorometers.

A recent development in time-correlated single-photon counting systems has been the interfacing to microscope optics. Steady-state fluorescence microscopy is a widely used analytical technique in bacteriology in particular. It has also gained eminence in recent years in semiconductor diagnostics. Most of the early research into time-resolved fluorescence microscopy was performed using nitrogen laser excitation and a sampling oscilloscope to record the decay.[72, 73] The combination of mode-locked laser excitation and single-photon timing promises much for the future in the areas of single-cell photophysics and the topography of surfaces, particularly semiconductors. In some early work, Rodgers and Firey reported single-photon timing microscopy measurements on algal chloroplasts,[74] and Minami *et al.*[75] have studied clover leaves stained with Acridine Orange (see Figure 1.11). In all this type of work, a spatial resolution of $\sim 1\ \mu m$ is typical, with confocal microscopes offering the highest spatial resolution and increased depth of field.

Other optical components which are widely used with single-photon timing systems include polarization accessories for time-resolved anisotropy studies. Although anisotropy studies are discussed in Sections 1.3.4 and 1.5.3, some aspects are appropriate to mention here. The time-resolved fluorescence anisotropy function given by Eq. (1.33) yields the rotational correlation time pertaining to the fluorophore and its environment. Most usually, anisotropy decays are measured using a fluorometer configured in an L-format, although the T-format has advantages which are discussed in Section 1.6.2. In L-format instruments, the parallel and crossed polarization components of fluorescence are usually acquired alternately with a period of a few minutes in separate segments of the MCA in order to correct for fluctuations in the source intensity. Both dichroic sheet and prism polarizers are used although the latter offer a higher extinction ratio and wider spectral range in the UV (typically > 280 nm).

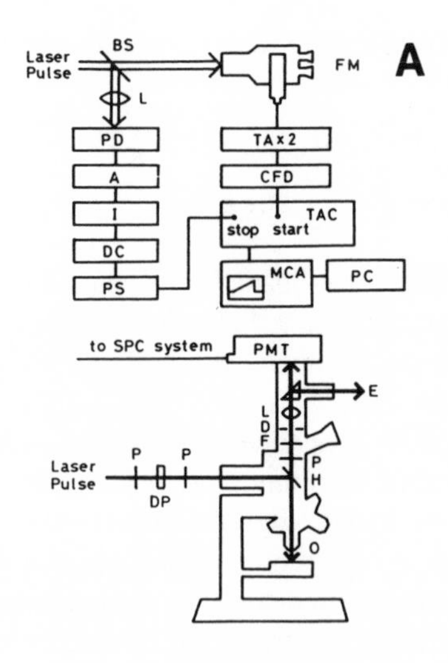

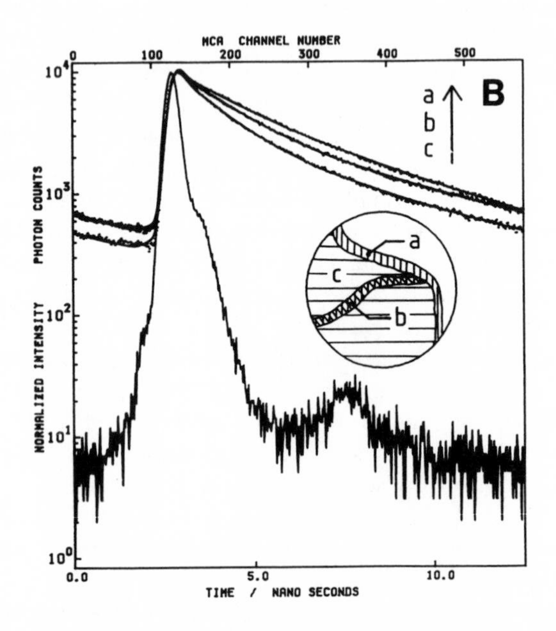

Figure 1.11. (A) Single-photon timing fluorescence microscope system. BS, beam splitter; PD, pin photodiode; A, amplifier; I, inverter; DC, discriminator; PS, pre-scaler; TA, timing amplifier; CFD, constant fraction discriminator; TAC, time-to-amplitude converter; MCA, multi-channel analyzer; PC, personal computer; FM, fluorescence microscope; P, polarizer; DP, depolarizer; PMT, photomultiplier; E, visual observation; L, lens; D, diaphragm; F, glass filter; H, half mirror; O, objective. (B) Heterogenous fluorescence decays observed for a clover leaf stained with Acridine Orange. Labels a, b, and c on the decay curves refer to the different sections of the leaf studied with $\sim$2.5-μm spatial resolution (see insert). (From Ref. 75.)

The effect of rotational depolarization itself can be a source of error in measurement of the fluorescence lifetimes if the lifetime is comparable to the rotational time of the molecule.[76] The effect can be present even with unpolarized light. However, it should be remembered that diffraction gratings impose a degree of polarization on both the excitation and fluorescence signals. Rotational effects are manifested as a misfit around the rising edge and peak of a decay, often described by a rise time. In low-viscosity solvents the effect of the rotational depolarization of small molecules is insignificant on the nanosecond time scale. It can, in any case, be eliminated by a variety of methods. O'Connor and Phillips[3] provided a summary of these. The most widely used approach is to excite with vertically polarized light and detect with the emission polarizer at the so-called magic angle of 54.75° to the vertical.

The lens system in single-photon spectrometers is a crucial component which is often given insufficient attention. The difference between using high and low solid angles throughout can often be an order of magnitude in sensitivity. The lenses should also be externally adjustable in position for optimization of the fluorescence signal. This is particularly important with samples of high optical density.

1.4.3. Detectors

Although the photon-counting aspects of detectors have been well covered in reviews (see, for example, those of Meade[77] and Candy[78]), the overview of photon detectors for time-correlated experiments has received much less attention.

Three general classes of detectors are available for single-photon timing fluorescence measurements. In order of abundance in the field, these are:

1. photomultiplier tubes
2. microchannel plate photomultipliers
3. avalanche photodiodes

Each has its own virtues, but the conventional photomultiplier is undoubtedly the workhorse of the technique. The principal features of a good single-photon timing detector are low timing jitter, low timing dependence on wavelength and point of illumination, low intensity after pulsing, high amplification, low noise, and wide spectral range. The alternatives listed above will now be examined with regard to performance in these and other areas.

Two types of photomultiplier are in widespread use: the end-on linear focused construction and the side-window cage construction.

The Philips XP2020Q photomultiplier and its variants are probably the most widely used of the linear focused type so we will concentrate on these. These devices have a gain of up to 10^7, a pulse rise time of 1.5 ns, and a photoelectron transit time distribution of 250-ps standard deviation.[31] Probably the fastest instrumental response obtained with a linear focused photomultiplier is the 235 ps fwhm reported by Bebelaar for the XP2020.[79] At this width, it should be possible to determine lifetimes down to ~30 ps using reconvolution. One of the main advantages of the linear focused construction over side-window photomultipliers is the lower transit time dependence on the point of photocathode illumination. Calligaris *et al.*[80] have reported this to be ~13 ps/mm, and the manufacturers quote 14 ps/mm.[31] This means effectively that there is no need to sacrifice sensitivity by reducing the illuminated area in order to obtain the maximum time resolution.

The wavelength dependence of the photoelectron transit time due to the photoelectric effect itself in linear focused photomultipliers has been investigated by many groups. Lewis *et al.*[65] were the first to demonstrate this potential source of error in reconvolution. Around the same time, Wahl *et al.*[81] reported large shifts in photomultiplier response of ~6 ps/nm. Subsequently, considerable efforts have been devoted to eliminating this effect. These include using an iterative shift of the fitted function in reconvolution[82] and additional reconvolution using measurements on reference compounds.[83–86] However, before any of these steps are considered, the problem can be minimized experimentally by working with as high a voltage as possible between the photocathode and first dynode. This minimizes the fraction of the total photoelectron energy at the first dynode due to the photoelectric effect. Given such an experimental precaution, maximum time shifts of 1 ps/nm[87] and 0.3 ps/nm[35] have been reported for the XP2020Q. Such small shifts are usually easily corrected for using an iterative time shift of the fitted fluorescence decay function. However, a change in the shape of the photomultiplier impulse response can also occur with wavelength. In practice, an iterative shift of the fitted function can provide a good first-order correction for both the shape change and time shift dependence on wavelength. In our experience, the combination of a well-designed voltage distribution applied to the dynode chain and an iterative time shift means that, in practice, wavelength effects are negligible for decay times down to 200 ps.

We have found that a modified voltage distribution network applied to the XP2020Q as compared to that suggested by Philips offers some decided advantages with regard to reduced noise and late pulsing.[68] Figure 1.12 shows the modified design, and Figure 1.13 demonstrates the improved noise and late pulse performance. The improved noise performance is achieved simply by increasing the relative photocathode to grid 1 voltage with respect to that of grid 1 to the first dynode. This has the effect of reducing the area of collection of photoelectrons by approximately an order of magnitude. Most of the detected noise in photomultipliers is due to thermionic emission of

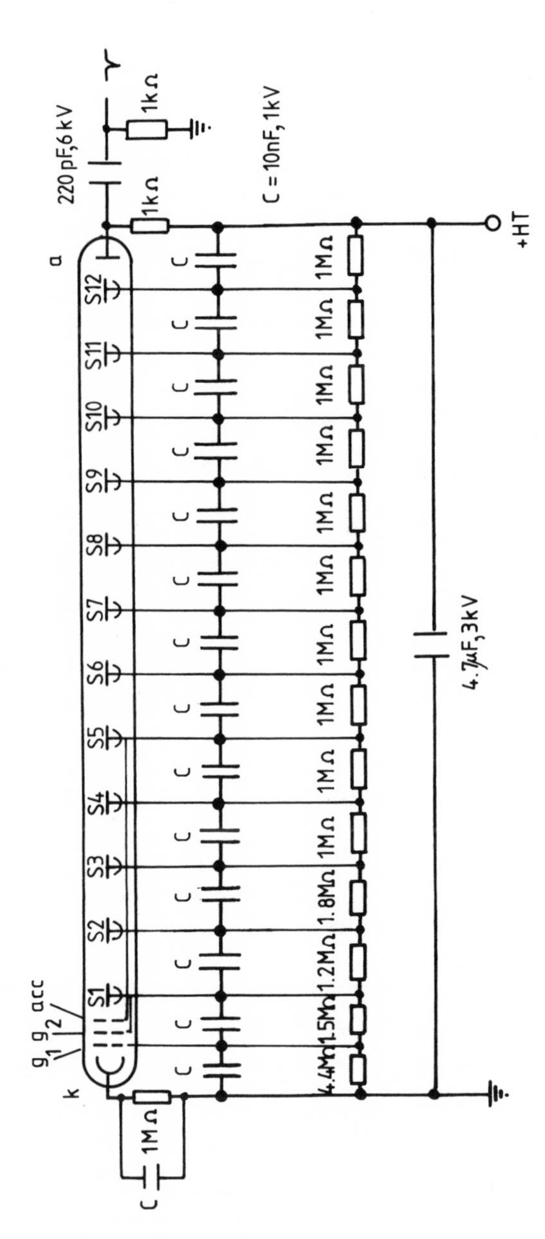

Figure 1.12. Modified voltage divider network for the Philips XP2020Q photomultiplier, giving reduced noise and late pulsing. (From Ref. 68.)

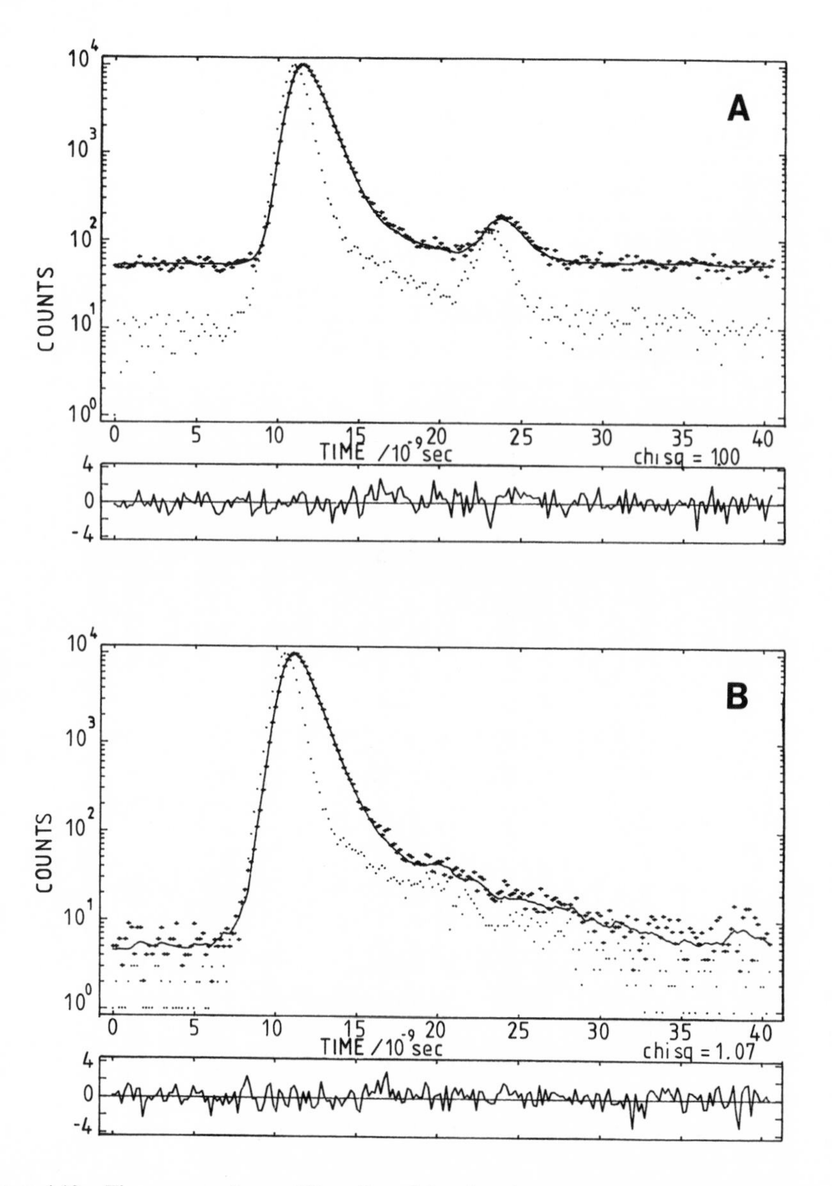

Figure 1.13. Fluorescence decay of Rose Bengal in ethanol at room temperature using excitation and fluorescence wavelengths of 555 and 580 nm, respectively. The photomultiplier is the XP2254B. Data accumulation time, ~15 min. (A) With Philips recommended divider network type A, fitting gives $\tau_M = 772 \pm 14$ ps. $\chi^2 = 1.00$. (B) With low-noise modification, fitting gives $\tau_M = 775 \pm 14$ ps. $\chi^2 = 1.07$. (From Ref. 68.)

single photoelectrons from the photocathode, which is thus reduced by the same factor as the area. The active photocathode area remaining of ~1 cm diameter is still ample to ensure only slight loss in sensitivity provided focusing optics are used. Noise levels below 10 counts/s can be obtained for the XP2020Q with this voltage distribution, without cooling.

The origin of afterpulsing in linear focused photomultipliers has received quite a lot of attention.[65, 88–90] Figure 1.14 is typical of what can be observed in a single-photon timing experiment. Late peaks are observed at ~10 ns

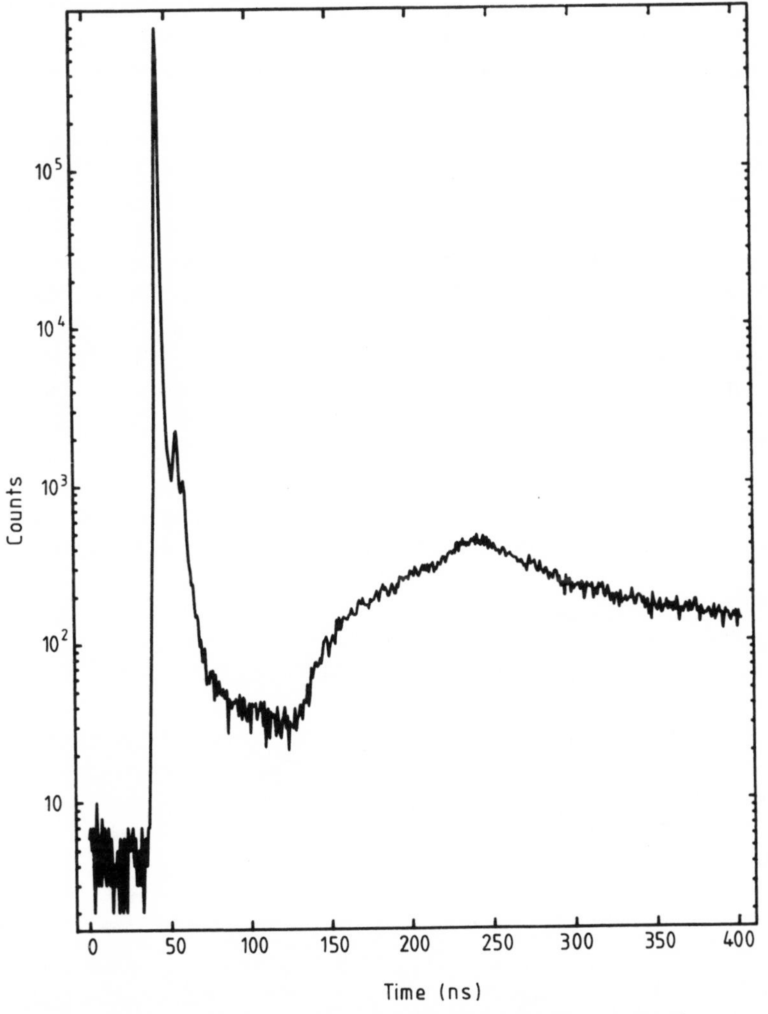

Figure 1.14. Impulse response of a linear focused photomultiplier measured over a high dynamic range in order to observe late pulsing.

and, within a broader structure, ~200 ns after the main pulse. The latter is undoubtedly due to residual gas causing ion feedback to the photocathode,[89, 90] and because of its weak intensity it does not usually present any problem in fitting data, even when just the tail of a decay is analyzed without reconvolution. The origin of the peak at 10 ns is still not clear though electron[79] and optical feedback[91] to the photocathode have been suggested as the cause. Figure 1.13 clearly shows the reduction in the late peak at 10 ns resulting from using the divider network sown in Figure 1.12. In practice, we find that reconvolution is not affected by this peak provided its intensity is less than 2% of that of the main peak.

As regards the useful spectral range, we have performed single-photon timing experiments with the XP2020Q and its variants at fluorescence wavelengths from 230 up to 930 nm.[68]

In comparison to linear focused photomultiplier tubes, side-window constructions have only been applied to fluorescence lifetime studies relatively recently. Kinoshita and Kushida[32] first reported this development, obtaining an instrumental response to 10-ps mode-locked laser excitation of 160 ps fwhm for the Hamamatsu R928 side-window photomultiplier. These workers also found that this detector has a high spatial time dependence (~450 ps/mm shift) and low wavelength time dependence (0.19 ps/nm). Subsequent work generally agreed with these findings.[92–94] There can be no doubt that such side-window photomultipliers are faster by a factor of ~2 than presently available linear focused photomultipliers. Canonica *et al.*[93] have reported an overall instrumental response of 112 ps for the R928. However, in order to obtain the fastest pulses, it is necessary to irradiate only a very small area of the photocathode (Kinoshita and Kushida used an area of $\sim 8 \times 10^{-3}$ mm^2). This considerably restricts the sensitivity of the measurements to a level which can only really be tolerated when using mode-locked laser excitation. With flashlamp excitation, the area has to be increased. This seems to cause problems in fitting the rising edge due to differences in the spatial intensity distribution between the measurement of the excitation and that of the fluorescence.

The low cost of side-window photomultipliers is certainly attractive for fluorescence decay measurements. However, cooling is usually necessary whereas, when the modified divider network shown in Figure 1.12 is used, no cooling is required for linear focused photomultipliers, which, if anything, makes them more attractive financially. The spectral range of side-window photomultipliers is comparable to that of linear focused tubes.

Microchannel plate photomultiplier (MCP-PM) tubes have been around for some time but only quite recently made an impact in fluorescence lifetime studies. A microchannel plate operates as a continuous dynode structure, and as such the early designs had problems associated with the impulse response being dependent on count rate, saturation at high count rates, positive ion feedback damaging the photocathode, and afterpulsing.[95] Most of these problems seem to have been overcome, and the microchannel plate photomultiplier seems now to be the detector of choice with most mode-locked laser systems. This is simply because of its extremely fast impulse response. Typical of the performance now widely achieved is that reported by Yamazaki *et al.*[96] for a conventional MCP-PM, giving a fwhm of 140 ps, and a proximity focused MCP-PM, giving a fwhm of 60 ps. Proximity focused MCP-PMs offer superior timing by virtue of close coupling between the current amplification channels and the photocathode. Some late pulsing is experienced with these devices, but nevertheless there seem to be few problems with reconvolution analysis. This is facilitated by the low spatial dependence (<1 ps/mm for the proximity device) and negligible wavelength dependence even on picosecond time scales.[96]

Like microchannel plate photomultipliers, avalanche photodiodes with single-photon timing capability were demonstrated some time ago.[97] The application of single-photon avalanche photodiodes (SPADs) to fluorescence decay studies has been pursued mainly by Cova and co-workers.[67, 98] These devices operate on the principle of photoinduced breakdown in a *p–n* junction. Very fast instrumental response widths down to 70 ps fwhm have been demonstrated with single-photon timing. Apart from this advantage, SPADs seem to offer good opportunities for measurements in the infrared where S-1 photomultiplier photocathodes are notoriously noisy. Silicon SPADs can be used up to 1 μm, but with other fabrications, for example, indium gallium arsenide or germanium, it should be possible to perform single-photon timing experiments at even longer wavelengths. It has even been possible to detect single photons at 20 μm, albeit with microsecond time resolution, by using avalanche breakdown in silicon epitaxially doped with arsenic.[99] Perhaps the current drive in research into multi-quantum well superlattices will lead to a new generation of photomultipliers based on solid-state technology. However, for the moment the SPAD, although not widely used, probably represents the state of the art as far as solid-state single-photon timing detectors are concerned. Unfortunately, the small area of such devices, which are around 10 μm in diameter, considerably limits their practical application. In addition, the SPAD impulse response contains a relatively long tail due to carriers generated in the neutral *p*-region beneath the depletion layer. This tail depends on wavelength, becoming more pronounced at longer wavelengths, and consequently can cause problems in reconvolution analysis.

1.4.4. Electronics and Computer Hardware

The major items of hardware in single-photon fluorometers concern the timing electronics and computer, both almost exclusively bought-in items.

Together, these enable the acquisition, storage, and display of fluorescence decays.

The timing electronics are at present still largely based on Nuclear Instrumentation modules (NIM) developed originally for nuclear physics. Major suppliers include EG&G Ortec, Tennelec, and Canberra. There has been little customization of NIM modules to suit fluorescence decay applications, and many of the NIM features are inappropriate. The key NIM modules are the TAC, "start" and "stop" discriminators, ratemeters, and calibrated delay lines.

The principle of timing with a TAC has already been explained in Section 1.2.1. It is important that the TAC linearity over the range of interest be checked before making decay measurements. This can be simply done by using a fixed frequency "start" signal and random "stop" signals, which, because they are uncorrelated, should provide a flat histogram. At least 10^4 counts per channel should be acquired in order to give a standard deviation of $\pm 1\%$ ($\sqrt{10^4}$). Low levels (care!) of room light are a good source of random photomultiplier pulses. The TAC can be easily calibrated in terms of nanoseconds per channel in the MCA by using either a commercially available module designed for the task or, more inexpensively, by using calibrated coaxial cables. In the case of the latter, the "start" discriminator output should be split, and, after passing through a variable delay, used as "stop" pulses in order to provide coincidence markers in the MCA. It has been found to improve accuracy if the TAC is calibrated under "start"–"stop" conditions similar to those for a typical measurement.[100] In addition to the "start"–"stop" inputs and output to the MCA, a TAC usually will provide and receive other useful signals. For example, the output can be inhibited externally, and a single-channel analyzer (SCA) output is usually available for measuring conversion count rate within a preselected time (i.e., voltage) window.

With high-repetition-rate sources (mode-locked lasers or synchrotrons), it is usual to reverse the functions of the "start" and "stop" signals in order to reduce the dead time of the TAC. Biased amplifiers are sometimes used with TACs in order to enable a higher time dispersion to be selected.

The discriminators provide a timing definition which is independent of pulse height and also discriminate against low-amplitude noise. The signal for the "start" discriminator can be derived by a number of methods. The minimum time jitter is usually obtained by viewing a fraction of the excitation pulse with a side-window photomultiplier or fast photodiode. In the case of synchronously pumped lasers, the electrical signal from the cavity dumper can also be used. The timing jitter on the "start" synchronization signal is much less than that of the "stop" photomultiplier due to the greater spread in pulse height from the latter. Constant-fraction discriminators (CFDs) offer a good method for minimizing the timing jitter from single photoelectron pulses of different amplitude. The timing walk of CFDs is typically ~ 50 ps. Leading edge discriminators give a greater timing walk of up to 1 ns and are not generally used for single-photon timing. Figure 1.15 shows the principle of constant-fraction and leading edge discrimination. The inherently high timing jitter of leading edge discriminators is clearly evident. The constant-fraction principle is more complicated and is based on generating a zero-crossing signal which is independent of pulse height. With both types of discriminator, a threshold can usually be adjusted down to ~ 5 mV in order to discriminate against low-amplitude noise. Such low thresholds mean that preamplification

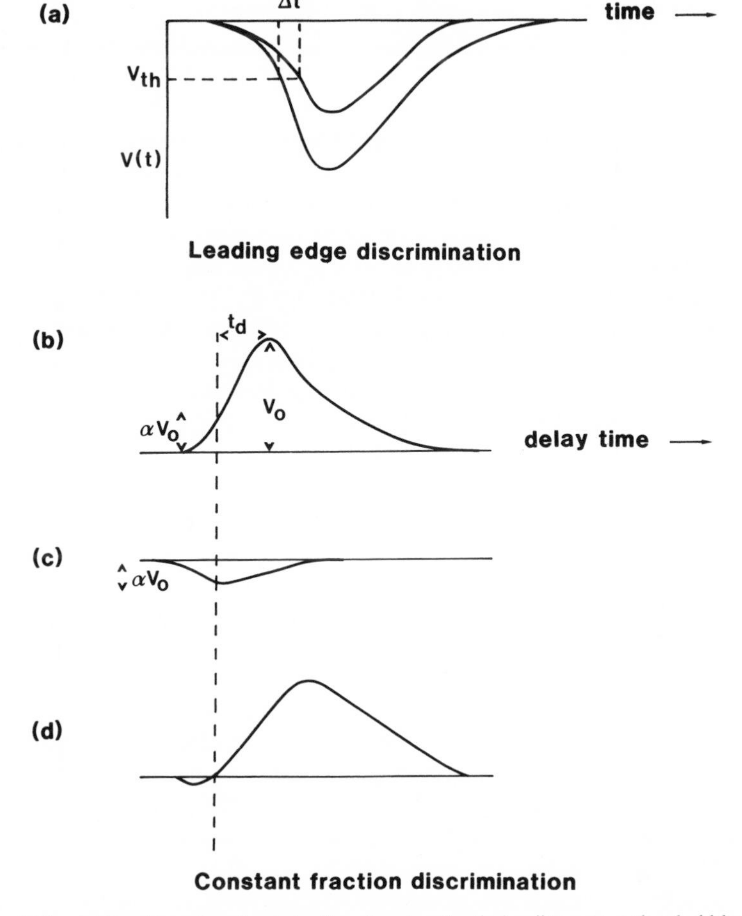

Figure 1.15. (a) Leading edge discrimination showing the timing jitter Δt at threshold level V_{th}. (b) With constant-fraction discrimination the negative input pulse from the photomultiplier is inverted and delayed by t_d. (c) Part of the original input pulse of amplitude V_0 is undelayed and attenuated to a fraction α. (d) The waveforms shown in (b) and (c) are added to produce the zero-crossing time signal.

should not be necessary as it may only introduce an additional time spread. The setting of this threshold is important. When the threshold is too high, multiphoton signals are preferentially accepted, giving an anomalously narrow instrumental width. When it is too low, the detected noise level is higher than is necessary. If anything, it is better to have a threshold too low than too high. However, there is a simple method of setting the level, which does not seem to be widely practiced. If the shortest TAC range is selected (typically 50 ns) and the SCA output monitored with the time window set to comfortably include the instrumental pulse, then the detected noise level will be negligible. A discriminator threshold of, say, -10 to -30 mV is selected, and the photomultiplier voltage is slowly increased until no further increase in the SCA output count due to the excitation pulse is observed. This voltage is then set on the threshold of detecting the majority of single-photon events. Any further voltage increase will then only serve to unnecessarily increase the number of low-amplitude photomultiplier noise pulses that are detected.

Other NIM modules that are sometimes useful include gate and delay generators, which are necessary when operating a TAC on time ranges of $\geqslant 1\ \mu$s. When the TAC range is increased, an offset in the timing origin also increases. On the lower ranges ($< 1\ \mu$s), this can be easily handled by increasing the delay due to coaxial cables in the "stop" channel. However, eventually the "stop" signal is attenuated below the TAC threshold and an electronic delay unit is necessary.

Important aspects of MCA performance for single-photon fluorometry include a low dead time (typically 20 μs) for work at the highest possible count rates and high analog-to-digital converter (ADC) linearity, as for the TAC.

Two types of MCA are at present in widespread use:

(i) free standing
(ii) PC-controlled systems

The latter are a fairly recent innovation and have done much to reduce the cost of single-photon systems. Several companies now supplying such systems include EG&G Ortec, Nucleus, and Canberra.

In the 1960s the slide rule was the key tool of kinetic analysis. The first attempts at numerical analysis relied on central computers, often using paper tape produced by a teletype (20 minutes for 1024 channels of data!) as data transfer medium. At that time, we were happy to have a fit provided the next day. Now it is available the next minute. It was the advent of affordable laboratory microcomputers that eventually forged the link between measurement and analysis. This close coupling of question and answer has done much to advance both the instrument development and application of fluorescence lifetime spectroscopy.

Both memory size and number-crunching speed are important for data analysis. Dr. A. Hallam of IBH Consultants Ltd. has kindly provided us with run times for a typical single exponential fit to 255 channels of data, with shift iteration, on various processors using the hybrid grid search algorithm. These are shown in Figure 1.16. The PDP-11/2 (Digital Equipment Corporation) was the first of the affordable laboratory microcomputers, using large-scale integrated circuits. Memory size was restricted to 64 kilobytes, which limited the size of the data sets we could analyze to 512 channels. An optional floating-point instruction set chip increased the calculating speed of the main processor. The micro-PDP then evolved to the 11/23 and 11/73 processors, with extended memory, faster processors, and more powerful mathematics chips. A 256-kilobyte memory allowed our data sets to expand to 4095 channels for all but the largest (global) analysis programs. In the 1990s, the run time of the PDP-11/2 looks very pedestrian indeed, but in the late 1970s just the convenience of having an on-line computer system was sufficient in itself, without thought to speed. Certainly, anything more complicated than a two-exponential fit is a tedious task with such a processor.

The advent of the IBM-PC marked an important change, from purpose-designed laboratory microcomputers to office computers, which, because of

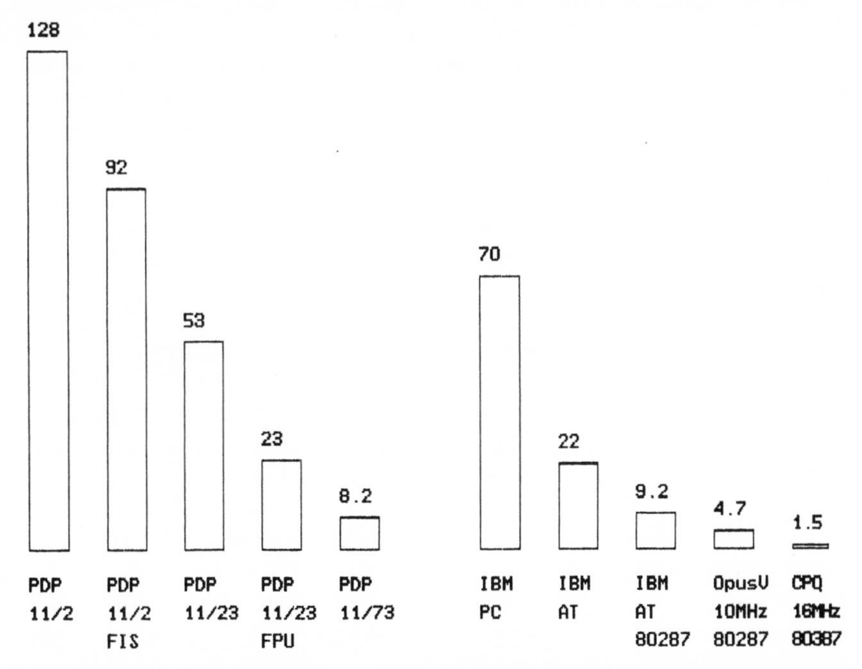

Figure 1.16. Run times obtained for a range of computer systems when analyzing a typical one-exponential fit to 255 channels of data with shift iteration and reconvolution. Numbers represent time (in seconds) for FIT 1 reconvolution analysis. (Courtesy of Dr. A. Hallam of IBH Consultants Ltd.)

their much larger markets, produced strong downward pressure on price. While the performance of the main processor in the original PC-XT and PC-AT computers was modest, a considerable boost was available in a low-cost mathematics coprocessor. The evolution of fast clones and 386-based machines is now offering quite remarkable number-crunching performance. The expansion of memory to a standard 640 kilobytes, together with powerful software tools for further memory expansion, allows very large data areas to be processed.

The data-harvesting power of modern instrumentation is making increasing demands on fast and powerful data display capabilities. Full-color displays in two and three dimensions are important tools in data and fit assessment. This is greatly facilitated in the IBM-PC environment by standardization of the screen protocols and the availability of low-cost hardware. Digital plotters in particular have had a great impact on data presentation and thereby aided kinetic interpretation.

Finally, a laboratory computer needs to interact with the outside world in complex and flexible ways. The availability within the IBM-PC of a number of standard interfacing slots has resulted in the development of a wide variety of interfacing cards for analog and digital data, for driving stepping motors, etc. Of particular interest in time-domain spectroscopy are the low-cost multichannel analyzer cards already mentioned. Not only do those replace expensive stand-alone analyzers, but they also offer the added advantages of rapid transfer of data to disk and full analysis functions integrated into the same hardware.

Throughout the 1980s, more and more users of the single-photon technique were able to study increasingly complex kinetics because of the decreasing cost of more powerful computers. Long may this continue!

1.5. Applications and Performance

In this section, we will demonstrate some of the applications and the performance of time-correlated single-photon fluorometry in the study of kinetics, for which this technique is frequently used.

1.5.1. Exponential Decay

Exponential fluorescence response functions are the ones which are most commonly encountered and are of the form

$$i(t) = \sum_j A_j e^{-t/\tau_j} \tag{1.44}$$

An important feature of Eq. (1.44) is that the existence of j decay components always confirms the existence of j excited states. However, it should always be remembered that the number of decay components chosen for least-squares analysis should always be the minimum necessary to give a satisfactory fit (i.e., $\chi^2 < 1.2$).

The shortest single exponential decay time which can be measured with the single-photon technique is constantly being reinvestigated. As early as 1976, Hartig *et al.*[101] reported a decay time of 90 ± 30 ps for erythrosin in water. However, this is probably at the limit of time resolution for spark sources. By using mode-locked laser excitation and photomultiplier detection

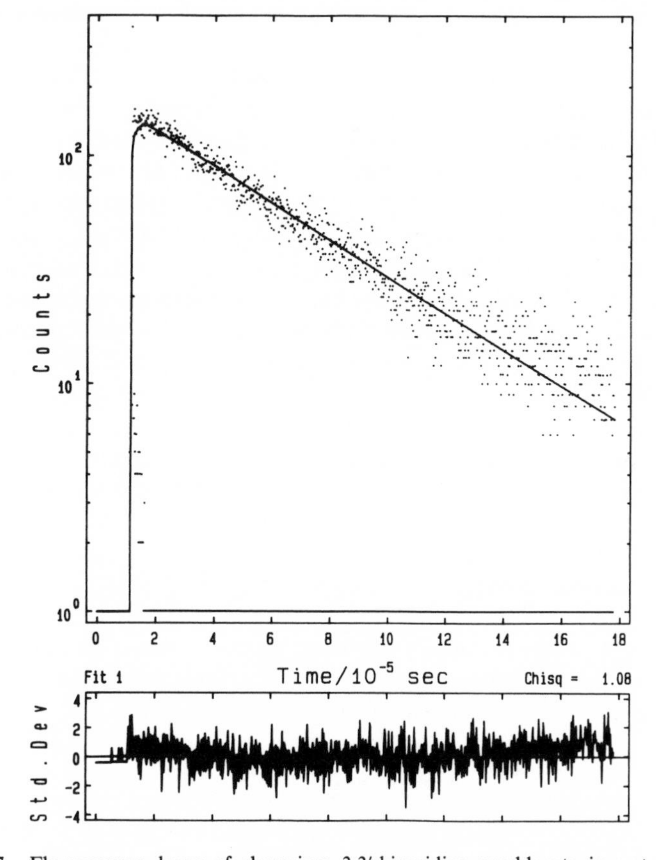

Figure 1.17. Fluorescence decay of chromium 3,3′-bipyridine perchlorate in water measured using single-photon timing. The decay time is 53 μs. The repetition rate of the hydrogen spark source was reduced to 4.7-kHz repetition rate in order to prevent reexcitation during the fluorescence decay. Excitation wavelength, 340 nm. Fluorescence detected at wavelengths > 560 nm. Channel width, 197.8 ns. (Sample by courtesy of Professor M. Maestri of the University of Bologna.)

with an instrumental width of ~0.5 ns, Small *et al.* tentatively reported a decay time of ~64 ps for erythrosin in ethanol quenched by potassium iodide.[102] More recently, using a mode-locked laser and microchannel plate detector with an instrumental response width of 140 ps, Visser *et al.*[103] have demonstrated that several pseudoazulenes have decay times in the range 1–10 ps, although the measurement precision did not permit more definitive values. Using a similar configuration but with an instrumental width of ~28 ps, Kume *et al.* have resolved a fluorescence decay time of ~6 ps for stiff-stilbene in isooctane.[104] It would appear that, with present technology, decay times of ~10 ps are at the limit of resolution of the single-photon technique, with 50 ps being fairly routine using picosecond laser excitation, the limit then being determined primarily by wavelength and spatial effects and in some cases stray light.

Although most attention is usually focused on the shortest decay time that can be measured with the single-photon technique, one useful advantage that single-photon fluorometers have over phase fluorometers is the additional capability for measuring long lifetimes on the same instrument. Figure 1.17 demonstrates this for the 53-μs decay of chromium 3,3′-bipyridine perchlorate. Hence, a range in decay times of over five orders of magnitude time is accessible with single-photon fluorometry.

Numerous standards have been suggested for testing the performance of fluorescence lifetime instruments.[105, 106] We have found 2,5-diphenyloxazole (PPO) to be useful in this regard. It has a high yield and negligible temperature dependence of its decay time and can be obtained off the shelf with high purity. Typical values that we have measured are a decay time of 1.61 ± 0.02 ns in degassed ethanol[91] and 1.27 ± 0.02 ns in aerated cyclohexane.[64] Care must taken to keep the concentration dilute ($\leqslant 10^{-5}$ M) or excimer formation may occur, although in itself this is a useful biexponential standard. In fact, PPO provides a good demonstration of a number of photophysical phenomena, so we will use it in several places throughout the chapter.

There are many examples of biexponential decays in photophysics. Probably the most well known is that of excimer formation. The simple kinetic scheme as derived by Birks[5] is

$$\begin{array}{ccc} {}^1\mathrm{M}^* + {}^1\mathrm{M} & \rightleftharpoons & {}^1\mathrm{D}^* \\ \downarrow \lambda_F & & \downarrow \lambda_D \\ {}^1\mathrm{M} & & 2\,{}^1\mathrm{M} \end{array} \tag{1.45}$$

where the ${}^1D^*$ excimer emission λ_D is shifted to longer wavelengths with respect to the ${}^1M^*$ monomer emission at λ_F. In these two spectral regions, the fluorescence response functions of the monomer and excimer emissions are given by

$$i_M(t) = A_1 e^{-t/\tau_1} + A_2 e^{-t/\tau_2} \quad (1.46)$$

and

$$i_D(t) = A_3[e^{-t/\tau_1} - e^{-t/\tau_2}] \quad (1.47)$$

Because excimer formation is diffusion-controlled in fluid media, it is frequently used to study diffusion coefficients in membranes. Pyrene in particular is often used as a probe. Not only does it readily form an excimer with a high fluorescence yield, but because the fluorescence lifetime is several hundred nanoseconds, diffusion can be studied over ~ 100 Å displacement. However, other molecules also show excimer effects which are of different interest. In the case of PPO, the unquenched monomer has a lifetime of only ~ 1 ns, which means that short-time diffusional effects can be investigated (see Section 1.5.2). Figure 1.18 shows the fluorescence decay of an excimer-forming

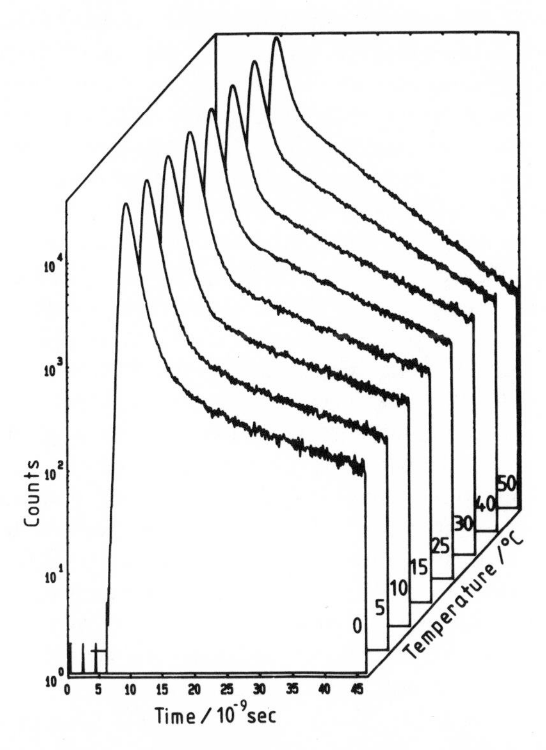

Figure 1.18. Fluorescence decay at 360 nm for 0.1 *M* PPO in ethanol. The biexponential behavior is caused by excimer formation, and the two decay components are clearly resolved. (From Ref. 107.)

solution of 0.1 *M* PPO in ethanol as a function of temperature.[107] In this case, the two decay components of ~1 and 8–18 ns are clearly distinguishable and easy to analyze.

By and large, the determination of most biexponential decay parameters is routine with both phase and pulse fluorometers. However, the addition of a third decay component has the potential to complicate matters considerably. In our experience, many other kinetic models can be described quite well by using a three-component model, and often additional knowledge is required before an unambiguous conclusion can be made as to the true kinetics. Three exponential decays often possess a good deal of "curvature," with the separate decay components not obvious on a simple visual inspection. However, there are many interesting examples of the presence of three exponentials being interpreted kinetically. In many cases though, the interpretation is still being questioned. Ross *et al.*[46] used linked-function analysis to describe the decay of tyrosine emission in simple analogues and polypeptides in terms of ground state rotamers, with decay components, in the case of *N*-acetyltyrosinamide, of 2.33, 1.73, and 0.54 ns. Three-component decays have been widely interpreted as representing the number of reaction centers in photosynthetic pigments of chloroplasts and green algae.[108] Another example concerns the case of excimer formation where kinetically isolated (and hence unquenched) monomer molecules can occur. This was first proposed by Phillips and co-workers for intramolecular excimers in polymers, though subsequent interpretations in terms of nonexponential functions are acknowledged to describe the decay data equally well.[109, 110] An example of a case in which the presence of isolated monomer sites in an excimer-forming system looks to be established concerns PPO in liposomes.[111] Table 1.4 shows this for the two-component analysis of this system over a range of temperatures above and below the phase transition at 41°C at different fluorescence wavelengths, studied using single-photon timing array fluorometry (see Section 1.6.3). The monomer decay component of 1.6 ns for PPO is clearly evident for nearly all the measurements. However, although the largest decay component clearly relates to the excimer emission, the data set shows a number of anomalous trends, for example, decay parameters increasing with λ_F, which suggests that the kinetics are more complicated. In this case, a distribution of decay components may be present due to either a single excimer conformation in a distribution of local environments or a distribution of conformations in a single environment or both. Clearly, caution is needed when interpreting such complex systems. James and Ware[112] were the first to point out the potential ambiguity in assigning complex kinetics to discrete lifetimes rather than a distribution of lifetimes with a mean value τ_m and a variance σ_m; that is,

$$i(t) = \text{fn}(\tau_m, \sigma_m) \tag{1.48}$$

Table 1.4. Isolated Monomer Sites in an Excimer System: Biexponential Best-Fit Parameters for the Fluorescence Decay of 500:1 DPPC: PPO Measured Using Single-Photon Array Detection[a]

λ_F (nm)	T (°C)	τ_1 (ns)	$A_1{}^b$ (%)	τ_2 (ns)	$A_2{}^b$ (%)	$\chi^2_{2\text{exp}}$	$\chi^2_{1\text{exp}}$
370	25	5.76 ± 0.9	6.2	1.45 ± 0.06	93.8	2.21	5.11
	30	7.86 ± 1.2	4.2	1.51 ± 0.03	95.8	2.02	4.83
	35	9.30 ± 1.8	2.8	1.60 ± 0.03	97.2	1.52	3.60
	40	9.02 ± 2.4	2.2	1.62 ± 0.03	97.8	1.49	2.81
	45	8.52 ± 3.2	1.2	1.66 ± 0.03	98.8	1.18	1.62
	50	7.78 ± 2.7	1.5	1.62 ± 0.03	98.5	1.24	1.73
	55	6.07 ± 1.2	2.1	1.58 ± 0.03	97.9	1.19	1.75
398	25	10.4 ± 0.9	6.9	1.51 ± 0.03	93.1	1.71	7.43
	30	10.7 ± 1.2	6.2	1.57 ± 0.03	93.8	1.65	6.77
	35	9.6 ± 0.9	5.3	1.61 ± 0.03	94.7	1.25	5.09
	40	8.7 ± 0.9	4.2	1.63 ± 0.03	95.8	1.16	3.86
	45	4.1 ± 0.6	5.2	1.60 ± 0.06	94.8	1.20	1.77
	50	5.4 ± 1.2	2.8	1.61 ± 0.03	97.2	1.11	1.60
	55	5.9 ± 1.2	2.7	1.60 ± 0.03	97.3	0.99	1.68
430	25	17.3 ± 0.9	22.1	1.62 ± 0.06	77.9	1.41	16.18
	30	16.6 ± 0.9	19.1	1.64 ± 0.06	80.9	1.41	14.06
	35	15.8 ± 0.9	16.0	1.70 ± 0.03	84.0	1.16	11.40
	40	13.6 ± 0.9	12.3	1.70 ± 0.03	87.7	1.08	7.16
	45	13.9 ± 1.5	5.8	1.74 ± 0.03	94.2	1.01	4.02
	50	10.7 ± 1.2	6.5	1.67 ± 0.03	93.5	1.09	3.92
	55	10.7 ± 1.2	6.3	1.67 ± 0.03	93.7	1.11	3.83
461	25	19.1 ± 0.6	49.4	1.67 ± 0.09	50.6	1.47	26.70
	30	18.5 ± 0.6	45.3	1.72 ± 0.09	54.7	1.19	23.00
	35	18.0 ± 0.6	39.6	1.83 ± 0.06	60.4	1.12	20.05
	40	16.7 ± 0.6	30.8	1.80 ± 0.06	69.2	1.10	11.45
	45	13.5 ± 0.9	18.5	1.72 ± 0.06	81.5	1.12	7.41
	50	12.9 ± 1.2	18.1	1.68 ± 0.06	81.9	1.28	7.38
	55	12.2 ± 0.9	16.9	1.69 ± 0.06	83.1	1.20	5.79

[a] From Ref. 111.
[b] A_1 and A_2 are normalized to give the percentage contribution of each decay component to the total emission, i.e., $A_1\tau_1/(A_1\tau_1 + A_2\tau_2)$ and $A_2\tau_2/(A_1\tau_1 + A_2\tau_2)$.

Other potential cases in which Eq. (1.48) might be appropriate include polymers, molecules absorbed on surfaces, and proteins. However, there is a danger in generalizing from what is undoubtedly often a possible, though only a possible, interpretation: general in the sense that all lifetimes arise from distributions, and dangerous in the sense that does a distribution get us anywhere in our kinetic understanding? Certainly, what we are sometimes interested in are trends, and in these cases the kinetic model may well be irrelevant!

Notwithstanding the potential for misinterpretation or overinterpretation with three and a distribution of exponential functions, some researchers have been bold enough to venture forth into the realms of four and even five exponentials! Holzwarth *et al.* reported a four-exponential fit of 26 ps, 112 ps, 1.38 ns, and 4.23 ns to the decay of a tetrapyrrole pigment, biliverdin dimethyl ester.[113] Figure 1.19 shows this fit and the weighted residuals for three and four components. In this case, the decay components are well separated, but nevertheless with this high level of parametrization an improved fit may

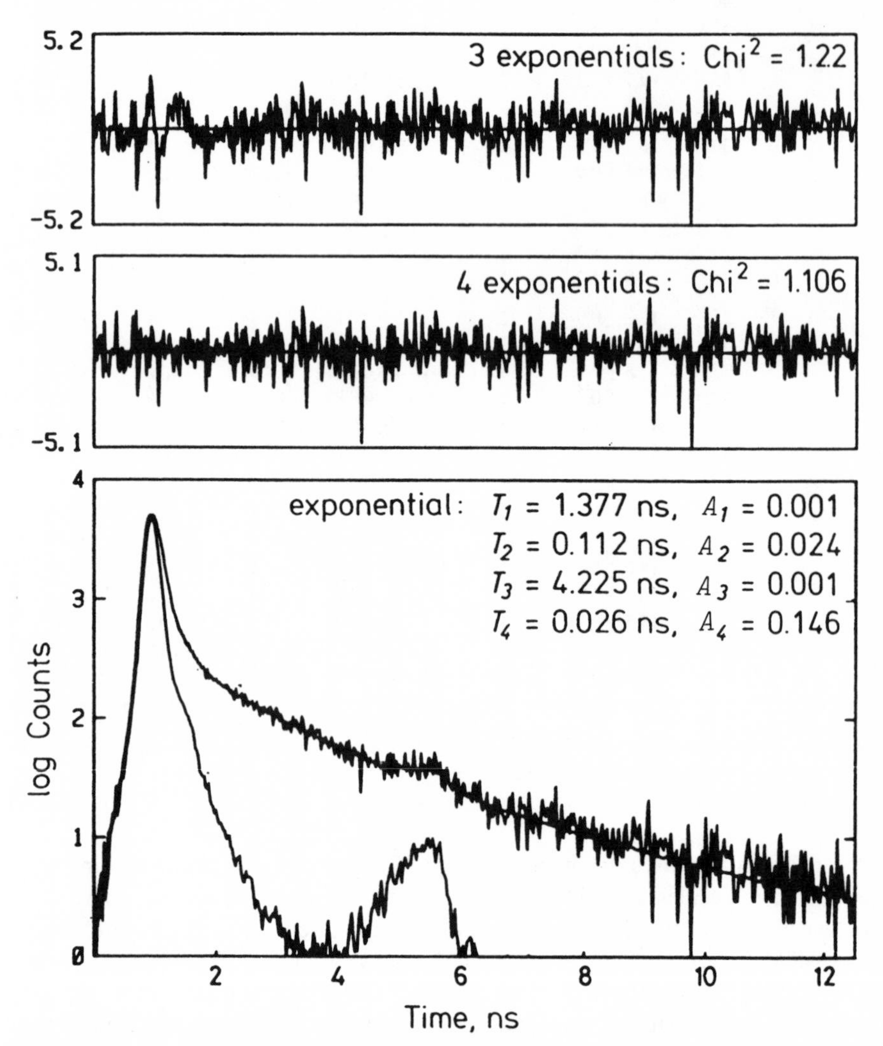

Figure 1.19. Three- and four-exponential fit to the decay of a tetrapyrrole pigment, biliverdin dimethyl ester. (From Ref. 113.)

sometimes not mean a better kinetic interpretation. In a recent paper, Itaya *et al.*[114] have reported studies on excimer emission in films of poly(*N*-vinyl-carbazole) which give four and five exponential decays using mode-locked laser excitation and MCP detection. The instrumental pulse width was ~60 ps, and the four-component fit to the decay at 475 nm included three rise times of 33 ps, 275 ps, and 1.55 ns as shown in Figure 1.20. The decay time was 35 ns. The five-exponential fit at 380 nm gave decay components of 13 ps, 255 ps, 1.43 ns, 7,14 ns, and 35 ns. There is a reassuring consistency between the rise times and decay times of their results [see Eqs. (1.46) and (1.47)], and the data have been interpreted in terms of two excimer conformations with some degree of interconversion.

1.5.2. Nonexponential Decay

There are a number of interesting cases in which nonexponential decay forms are encountered and lead to the determination of useful physical parameters. Three of the most commonly encountered are Förster energy transfer, transient diffusion effects, and quenching in micelles, all of which we will outline here.

In the case of spectral overlap between a donor molecule ($^1M^*$) emission and an acceptor molecule (^1Y) absorption, nonradiative dipole–dipole energy

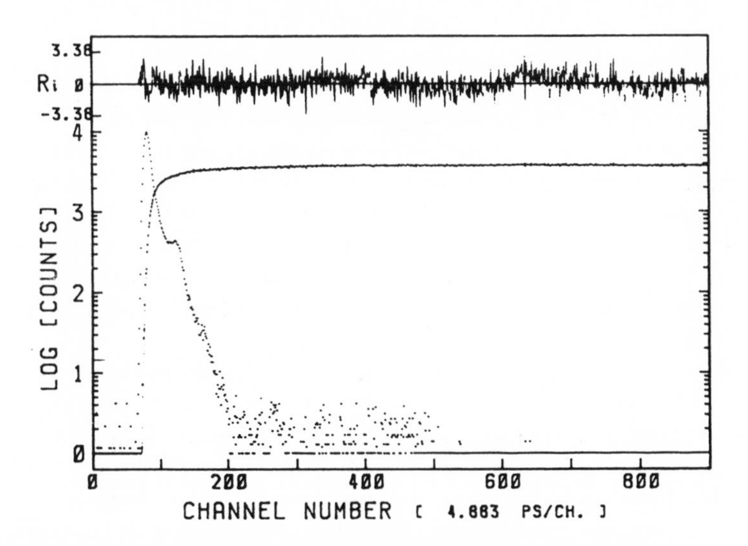

Figure 1.20. Four-exponential fit containing three rise times of 33 ps, 275 ps, and 1.55 ns and a decay time of 35 ns. The sample is a film of poly(*N*-vinylcarbazole). The χ^2 value is 1.08. (From Ref. 114.)

transfer can occur. In this case, the transfer rate is time-dependent, and hence a nonexponential decay of donor and acceptor fluorescence is observed.

For a random distribution of molecules, considered to be stationary, Förster derived the expression which defines the generic nature of all subsequent models applicable under different conditions, for example:

$$^1M^* + {}^1Y \rightarrow {}^1M + {}^1Y^* \tag{1.49}$$

described by

$$i(t) = A_1 e^{[-t/\tau_M - 2\gamma(t/\tau_M)^{1/2}]} \tag{1.50}$$

where

$$\gamma = [Y]/[Y]_0 \tag{1.51}$$

and

$$[Y]_0 = 3000/2\pi^{3/2} N R_0^3 \tag{1.52}$$

is the critical acceptor concentration, N is Avogadro's number, R_0 is a critical transfer distance, and τ_M is the donor's fluorescence decay time in the absence of the acceptor.

The rate of energy transfer k_{YM} depends on the separation r of the donor and acceptor according to

$$k_{YM} = \frac{1}{\tau_M}\left[\frac{R_0}{r}\right]^6 \tag{1.53}$$

The power of using Förster and related energy transfer kinetics frequently lies in applications as a "spectroscopic ruler." There are many applications in the study of bio- and synthetic polymers in particular where energy transfer between donor and acceptor pairs is used to determine distance of separation, site location, mobility, etc. (See, for example, Lakowicz.[6])

Nonradiative energy transfer and indeed radiative energy transfer between the same species, which can occur at high concentrations, are of course a potential source of error in both fluorescence[5] and anisotropy[115] decay measurements.

An interesting development of Eq. (1.50) is encountered in ordered media. Tamai *et al.*[116] have used single-photon timing to study Förster type energy transfer in Langmuir–Blodgett monolayers containing carbazole as a donor and anthracene as acceptor. In this case, a two-dimensional version of Eq. (1.50) was found to describe the kinetics well at high acceptor concentrations (>3 mol%), provided an additional component due to unquenched donor fluorescence was added; that is,

$$i(t) = A_1 e^{[-t/\tau_M - 2\gamma(t/\tau_M)^{1/3}]} + A_2 e^{-t/\tau_M} \tag{1.54}$$

The same workers have also studied two-dimensional energy transfer between rhodamine 6G (donor) and Malachite Green (acceptor) absorbed on the surface of dihexadecyl phosphate vesicles.[117] In this case, the dimensionality of the acceptor molecules surrounding a donor was found to be best described by a superposition of 2D and 3D terms due to penetration of the dye molecules into the vesicles; that is,

$$i(t) = A_1 e^{[-t/\tau_M - 2\gamma(t/\tau_M)^{1/3}]} + A_2 e^{[-t/\tau_M - 2\gamma(t/\tau_M)^{1/2}]} \tag{1.55}$$

Subsequent analysis of Eq. (1.55) gave a critical transfer distance of 62.3 Å for this system. Clearly, Eqs. (1.54) and (1.55) offer a high degree of curve parametrization, and some additional knowledge of the likely kinetics is needed in order to exclude other models (e.g., three or four exponentials) which might give an equally good fit. In the case of nonexponential functions, where the transients are more noticeable at short times, this whole process is helped if the experimental arrangement has as narrow an instrumental pulse as possible. Tamai *et al.* used a synchronously pumped dye laser and MCP detector to obtain an instrumental pulse of ~40 ps fwhm. However, many time-dependent rates and transient phenomena can be studied satisfactorily using flashlamp excitation. The time dependence of the rate of quenching of fluorescence by diffusion-induced collisions (Eq. 1.6) is a good example of this.

The solution of Fick's second law of diffusion leads to the following expression for the rate of quenching k_{QM}:

$$k_{QM} = \frac{4\pi N\, DpR}{1000}\left[1 + \frac{pR}{(\pi\, Dt)^{1/2}}\right] \tag{1.56}$$

where D is the sum of the diffusion coefficients of the quencher and the excited molecule, R is the sum of their interaction radii, N is Avogadro's number, and p is the reaction probability per collision. At $t > p^2R^2/\pi D$, the time dependence of k_{QM} can be neglected, and a simple exponential decay form is observed. However, if the whole decay form is analyzed, then the correct rate equation to be considered is

$$\frac{d[^1\mathrm{M}^*]}{dt} = -[^1\mathrm{M}^*]\left\{\frac{1}{\tau_M} + k_{QM}[\mathrm{Q}]\right\} \tag{1.57}$$

This leads to a solution analogous to Eq. (1.50); that is,

$$i(t) = A_1 e^{(-t/\tau - 2bt^{1/2})} \tag{1.58}$$

where

$$\frac{1}{\tau} = \frac{1}{\tau_M} + \frac{4\pi R\, DN[\mathrm{Q}]}{1000} \tag{1.59}$$

and

$$b = \frac{4(\pi D)^{1/2} R^2 N[\mathrm{Q}]}{1000} \tag{1.60}$$

By fitting to Eq. (1.58), $1/\tau$ and b, and hence D and R, can be determined. The application of Eq. (1.58) to a range of aromatic fluorophores and heavy-atom quenchers (e.g., CBr_4) was extensively studied by Ware and coworkers[118, 119] in high-viscosity solvents (e.g., 1,2-propanediol), where the transient effects are more readily observed with flashlamp excitation. However, transient diffusion effects have also been studied in low-viscosity solvents. Using single-photon counting with mode-locked laser excitation and MCP detection with an instrumental response of $\sim$150 ps fwhm, Wijnaendts van Resandt demonstrated the transient quenching of N-acetyltryptophanamide by iodide ions in water.[120] Such effects can in fact be observed with flashlamp excitation in low-viscosity solvents provided enough counts are accumulated and the channel width is narrow enough.

Self-quenching due to the excimer formation of pyrene (Eq. 1.45) is widely used to determine diffusion constants in lipid media. In the absence of excimer emission and reverse dissociation of the excimer to form excited monomer, the response of the monomer is given by Eq. (1.58). In this case, D and R are simplified by only representing one species. If we analyze the decay curves from the data shown in Figure 1.18 for a 0.1 M solution of PPO in ethanol in terms of the biexponential model for monomer fluorescence given by Eq. (1.46), we should expect to obtain good fits. Visually, the data certainly look to show two decay components. However, such is the time resolution of present-day flashlamp fluorometers that conclusive evidence of a departure from the expected kinetics is revealed quite easily.[107] The weighted residuals for the biexponential model as a function of temperature are shown in Figure 1.21. For these data, the channel width was 90 ps, the peak count 40,000, and the instrumental width using flashlamp excitation $\sim$1.3 ns fwhm. Under these conditions, the temperature dependence of transient quenching effects is easily observed. However, in this case analyzing them is complicated by the time dependence of the excimer reverse dissociation, precluding an analytical solution. The addition of a second decay component to Eq. (1.58) to correct for this can only be an approximation. At such high quencher concentrations as 0.1 M and on the very short time scale of 1 ns fluorescence decay, as is involved with PPO, the simple predictions of continuum diffusion theory break down. Nevertheless, the example serves to illustrate the usefulness of "residual spectroscopy" such as that shown in Figure 1.21, whereby trends with respect to some variable (in this case temperature) bring additional kinetic information.

Notwithstanding the complications in isotropic solvents, excimer kinetics

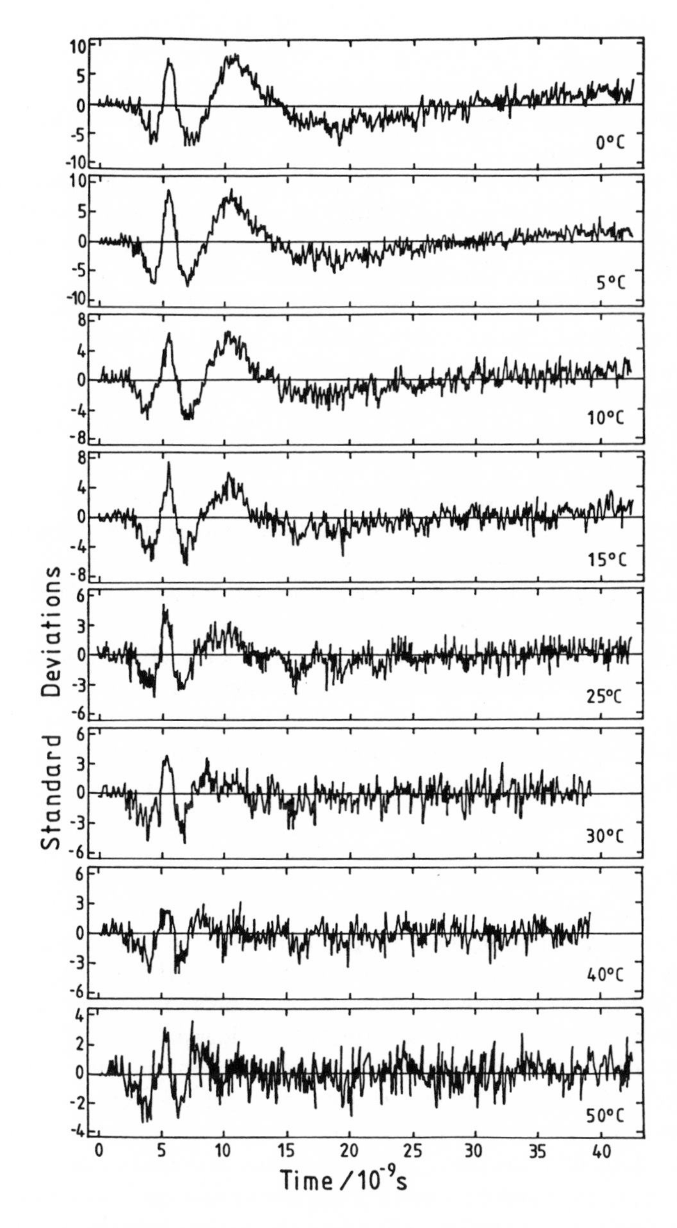

Figure 1.21. Weighted residuals for a biexponential fit to the decays shown in Figure 1.18 for 0.1 *M* PPO in ethanol. The deviations from the expected kinetics increase with decreasing temperature due to transient quenching effects. (From Ref. 107.)

are still widely used to determine diffusion coefficients in the anisotropic environment of a bilayer membrane.[121] Recently, fractal modeling using a non-integer space dimension has been used to analyze single-photon fluorescence decay data for pyrene excimer formation in liposomes.[122] Perhaps this approach can be developed further. However, the evidence for a serious departure from the assumed excimer diffusion-controlled kinetics, due to probe aggregation in membranes, is now mounting.[111, 123]

For the sake of completeness we should briefly mention in passing one other type of kinetics where nonexponential behavior has been widely studied by single-photon fluorometry. This concerns quenching of fluorophores bound in micelles[124] and microemulsions.[125]

In these cases, a fluorescence response of the following form is often appropriate when there is at most one fluorophore per micelle and the quenchers are free to move between micelles:

$$i(t) = A_1 e^{\{-t/\tau_M + \langle N_Q \rangle [\exp(-k_{QM} t) - 1]\}} \tag{1.61}$$

where τ_M is the unquenched molecular fluorescence lifetime, $\langle N_Q \rangle$ is the mean number of quencher molecules per micelle, and k_{QM} is the intramicellar quenching rate constant assuming one quencher.

From analysis using Eq. (1.61), micelle properties such as aggregation number and fluidity can be determined.[124]

The main conclusion as far as single-photon fluorometry is concerned is that in the study of nonexponential phenomena, where the information of key interest occurs at short times, there can be no substitute for working with as narrow instrumental pulse as is possible and fitting with reconvolution over the rising edge as well as the decay.

1.5.3. Anisotropy

In terms of single-photon timing measurements, the study of the rotational properties of fluorescent probe molecules is probably second only in importance to that of fluorescence lifetimes. Certainly, in recent years the whole area of molecular rotation has attracted considerable interest, from the point of view of both fundamental studies as well as applications.

The crux of measuring molecular rotation with fluorescent molecules hinges on the change in polarization of fluorescence which occurs with molecular rotation. With the use of polarized excitation, the intensity of fluorescence at parallel (F_p) and crossed (F_x) emission polarizer orientations will depend on the rotation of the fluorophore in the excited state. In the simple case of no rotation occurring in the excited state, and if the absorption and emission dipoles are parallel, $F_x = 0$ and $F_p \neq 0$. These measurements can of course be performed using continuous excitation. However, provided the molecular rotational correlation time τ_R is comparable to τ, the fluorescence lifetime, then the depolarization of emission as a function of time can be taken to be representative of the time-resolved molecular rotation.

The time-dependent depolarization of fluorescence can be expressed conveniently with reference to Eq. (1.33) by the anisotropy function $r(t)$:

$$r(t) = \frac{F_p(t) - F_x(t)}{F_p(t) + 2F_x(t)} = \frac{D(t)}{S(t)} \tag{1.62}$$

where $D(t)$ and $S(t)$ represent the difference and sum functions, respectively. The factor of 2 arises because there is an additional plane of fluorescence polarization orthogonal to $F_p(t)$ and $F_x(t)$ which is not detected. Hence, $S(t)$ represents the decay of total fluorescence and is given in the simplest case by

$$S(t) = S(0)e^{-t/\tau} \tag{1.63}$$

where τ is the fluorescence lifetime.

Consequently, it is $D(t)$ that contains the rotational information, and in the case of an unbound spherically symmetric rigid molecular rotor in an isotropic medium:

$$r(t) = r(0)e^{-t/\tau_R} \tag{1.64}$$

with $r(0) = 0.4$ (theoretical value).

In terms of simple Stokes–Einstein theory, τ_R is related to the medium by

$$\tau_R = \frac{\eta V}{kT} = \frac{1}{6D} \tag{1.65}$$

where η is the viscosity of the medium, V is the molecular volume, k is Boltzmann's constant, T is absolute temperature, and D is the diffusion coefficient.

In real cases, this simplified approach is not strictly accurate because of asymmetric rotations and molecular flexibility, which bring in additional rotational times. In this case,

$$r(t) = \sum_j r_j(0)e^{-t/\tau_{R_j}} \tag{1.66}$$

Nevertheless, the assumption of a single rotational correlation time is still very useful in a wide range of applications and is always a good first step when trying to model a free rotation. The procedures for analyzing anisotropy decay data have already been outlined in Section 1.3.4. Here we will just consider some results.

Figure 1.22 shows typical measurements of $F_p(t)$ and $F_x(t)$ along with the derived anisotropy function $r(t)$ for the commonly used probe DPH in a white oil. A characteristic of free rotation is that at long times $F_x(t) = F_p(t)$ and hence $r(t) = 0$. This property is in itself useful in checking if the polarizers are aligned. In addition, Eq. (1.62) may need to be corrected for any differences in the detection efficiencies of the two planes of polarization. This

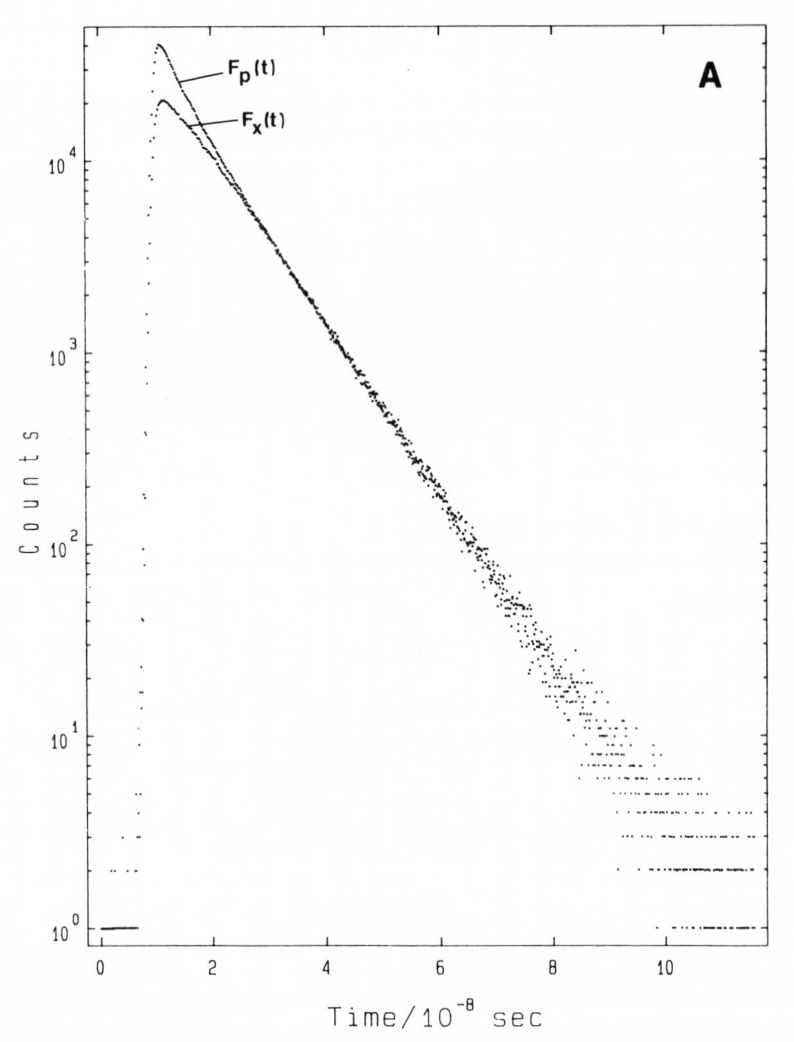

Figure 1.22. (A) Typical polarized fluorescence decays for a freely rotating fluorophore. $F_p(t)$ and $F_x(t)$ relate to the decay measurements at parallel and crossed emission polarizers, respectively. (B) Associated anisotropy function. The sample is DPH in a white oil at 20°C, and the rotational correlation time τ_R is 6.7 ± 0.7 ns. The excitation source is a hydrogen flashlamp.

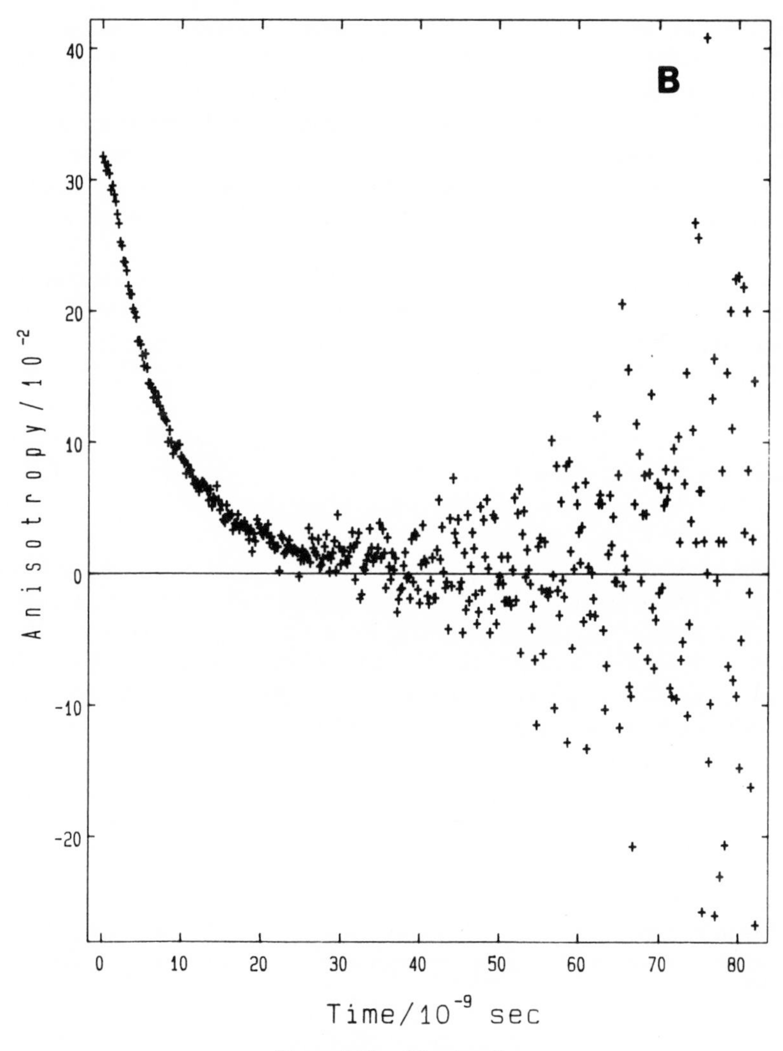

Figure 1.22. (*Continued*)

so-called *g*-factor can be easily obtained by putting the excitation polarizer horizontal rather than vertical whereupon, with the sample in place, the ratio of $F_x(t)$ and $F_p(t)$ can be used as a scaling factor in Eq. (1.62).[126]

Because most of the information content in the anisotropy decay for a free rotation occurs around the peak, one of the main limitations in determining time resolution is the width of the instrumental response. In addition, high statistical precision is required if the difference between $F_p(t)$ and $F_x(t)$ is to be significant. For both these reasons, there are few examples in the literature of subnanosecond rotational correlation times measured with flashlamp excitation. However, there have been examples of two rotational times being resolved with flashlamps. Barkley *et al.*[127] demonstrated the anisotropic rotation of perylene in glycerol in terms of two rotational times down to 9 and 1.1 ns at 40°C using an instrumental fwhm of ~2 ns. These measurements were subsequently corroborated using mode-locked laser excitation and linear focused photomultiplier detection[128] (instrumental width ~150 ps fwhm). In the latter study perylene rotational correlation times of 0.56 and 2.5 ns were determined for an 80% (v/v) glycerol:water mixture at 25°C. Again using picosecond laser excitation, but this time with a side-window photomultiplier to give an instrumental width of ~250 ps, Ware *et al.*[92] reported a rotational correlation time for Rose Bengal in water of ~50 ps. This seems to be about the limit reported for rotational correlation times measured using conventional photomultiplier detection and single-photon counting. Using mode-locked laser excitation and MCP detection, Chen *et al.*[129] recently reported rotational correlation times of 12 ps for 3-methylindole and ~30 ps for tryptophan, though with substantial uncertainty in the precision.

The basic anisotropy theory for a free rotor has been developed further to describe the restricted rotation of a fluorophore. This behavior is encountered in two important areas, namely, for a fluorophore covalently bound to a macromolecule (e.g., protein or vinyl polymer) and for a fluorophore rotating in an anisotropic environment (e.g., a lipid bilayer or liquid crystal). This restricted motion is characterized by a finite anisotropy value r_∞ at long times and can be modeled in the simplest form from Eq. (1.64); that is,

$$r(t) = (r_0 - r_\infty)e^{-t/\tau_R} + r_\infty \tag{1.67}$$

where r_0 is the anisotropy at $t = 0$.

Further analysis is based on a model whereby the fluorophore undergoes a wobbling motion with a diffusion coefficient D_w within a cone angle θ such that[130]

$$\frac{r_\infty}{r_0} = \left[\frac{1}{2}(\cos\theta)(1+\cos\theta)\right]^2 \tag{1.68}$$

and

$$\tau_R = \frac{\sigma}{D_w} \tag{1.69}$$

where values for σ, which depend on θ, have been given by Kinosita *et al.*[130, 131]

In the case of anisotropic media, Heyn[132] was the first to point out that r_∞/r_0 was related to the order parameter of the medium by

$$S^2 = \frac{r_\infty}{r_0} \tag{1.70}$$

Expressions (1.62) to (1.70) are widely used to interpret single-photon timing anisotropy decay data. Typical characteristic behavior for a restricted rotation is shown in Figure 1.23 for DPH in small unilamellar liposomes of DMPC. The advantage that time-domain data have over frequency-domain data in giving a visual display of the kinetics in terms of the raw data is probably nowhere better demonstrated than in anisotropy studies. The differences between the rotation of a free and a restricted rotor are clearly shown in Figures 1.22 and 1.23.

The additional complication imposed by r_∞ for restricted rotors certainly degrades the time resolution which can be obtained routinely. This is caused by a given peak count having less counts in the anisotropy decay component and correlation between the iterated parameters. Certainly, this effect is much more noticeable in anisotropy analysis than it is for fluorescence decay data with a high noise background.

Using hydrogen flashlamp excitation with an instrumental pulse of ~1.4 ns fwhm, we typically find that the lower limit of rotational correlation time which can be measured routinely using impulse reconvolution analysis is ~0.7 ns.[133] This was for 20,000 counts in the peak of $F_p(t)$ at a channel width of ~0.5 ns. If the peak count is increased to 40,000 and the channel width reduced to 0.2 ns, then rotational correlation times down to 0.4 ns can be measured with reasonable precision. Our experience is that this figure seems to be about the lowest achievable with flashlamp excitation and photomultiplier detection, that is, approximately twice that for fluorescence decay times. However, in the case of restricted rotation with $r_\infty \geqslant 0.1$, we would put the lower limit of rotational time which can be routinely measured with this arrangement as nearer 1 ns,[134] which also seems to be consistent with other work.[135] Again in such cases our experience is that two rotational times can still be resolved with flashlamp excitation, but conditions have to be favorable and the time resolution is undoubtedly degraded further. The use of mode-locked laser excitation and MCP detection again extends these limits. Typical of the performance of the single-photon technique in combination with such a configuration are measurements on polypeptides labeled with a single tryptophan recently reported by Chen *et al.*[136] Using an instrumental pulse of 90 ps fwhm, these workers reported fits to two rotational correlation times of 0.15 and 0.83 ns. Measurements of complex rotations with such high time resolution is not possible with flashlamp sources, and the greater sensitivity and time resolution of mode-locked lasers, has, in our opinion, still not yet been made the most of in single-photon timing anisotropy studies. Nevertheless, the development of the multiplexing techniques discussed in Section 1.6 does extend the time resolution of flashlamps for anisotropy studies by permitting long run times without fear of the effect of changes in the pulse profile.

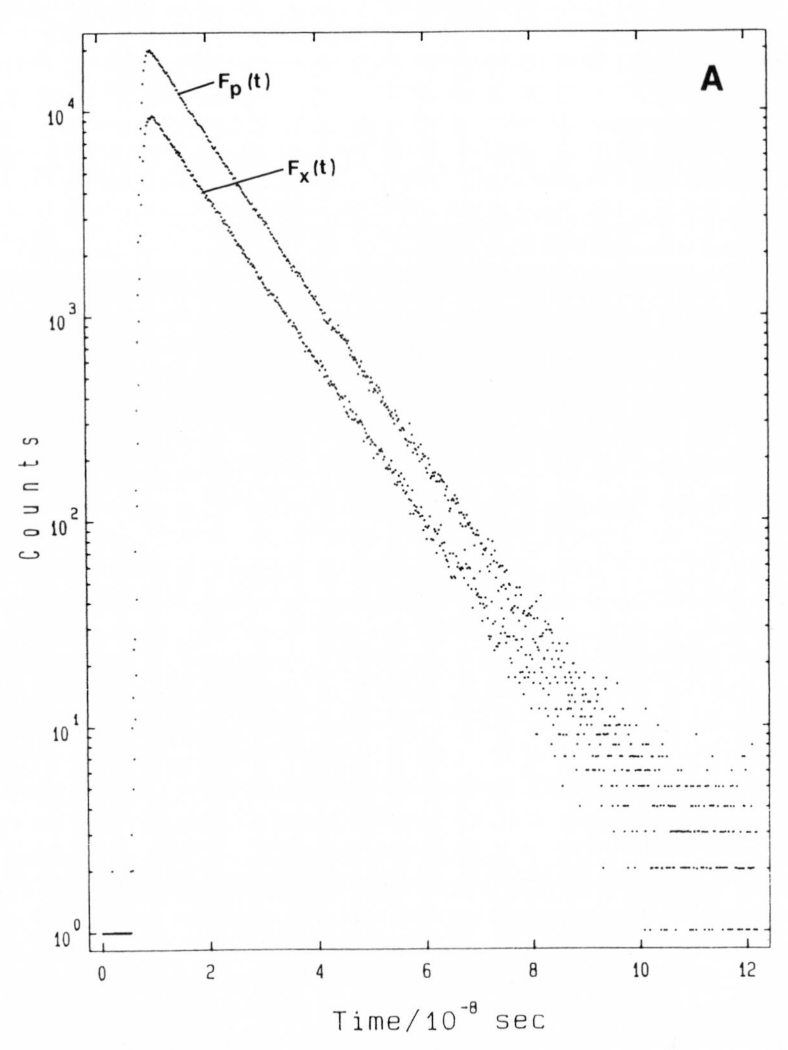

Figure 1.23. (A) Typical polarized fluorescence decays $F_p(t)$ and $F_x(t)$ for a restricted fluorophore rotation. (B) Associated anisotropy function. The sample is DPH in small unilamellar liposomes of DMPC at a probe-to-lipid ratio of 1:500 and below the lipid phase transition at 15°C. Impulse reconvolution analysis gave $\tau_R = 0.55 \pm 0.36$ ns, $r_0 = 0.369$, and $r_\infty = 0.246$.

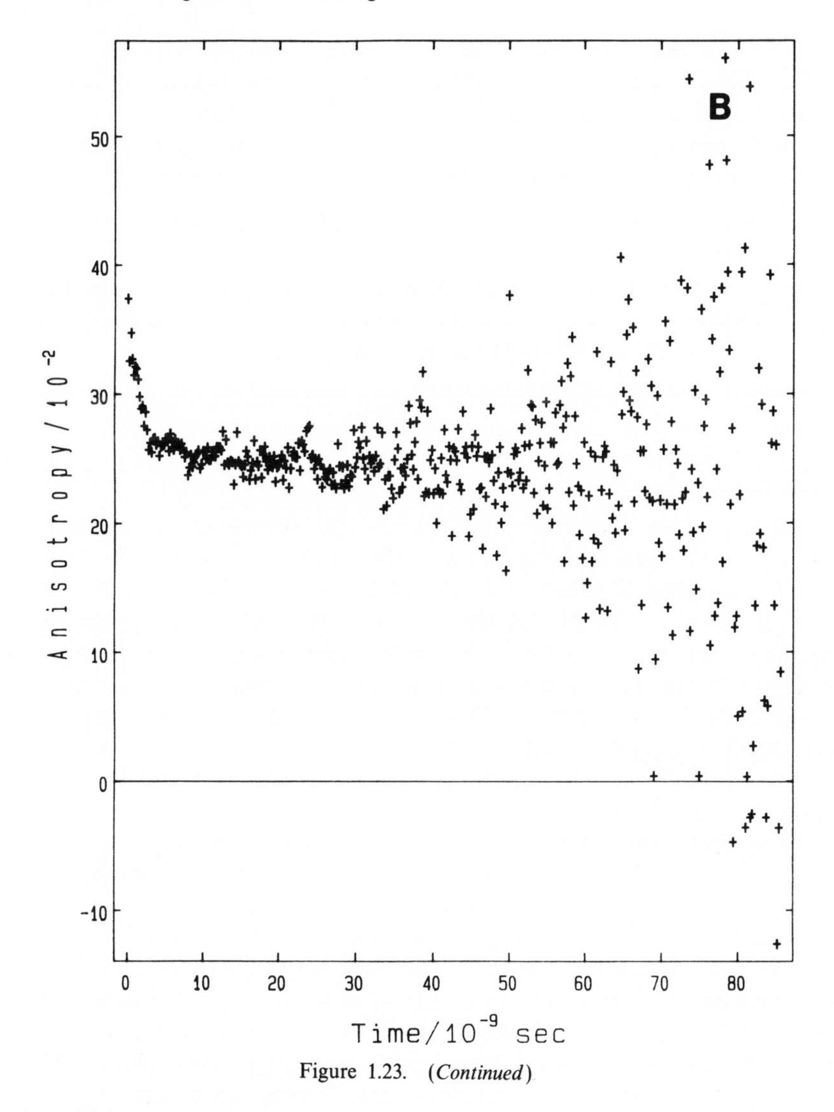

Figure 1.23. (*Continued*)

1.5.4. Time-Resolved Spectra

The general objective of time-resolved spectroscopy is to record the fluorescence (or indeed excitation) spectrum of the sample of interest as a function of wavelength at varying delay times following excitation with a narrow pulse of light. By this means, the evolution of kinetics in terms of a spectral representation can be obtained. The information content of such measurements is, like a fluorescence decay curve, just one small part of the whole fluorescence contour of intensity, time, and wavelength.

Time-resolved measurements were originally performed with fluorescence by gating a photomultiplier to obtain the necessary time delay following excitation with a nanosecond flashlamp.[137] However, single-photon timing lends itself very well to these types of measurements. There are basically two methods currently used for obtaining the information of interest[138]:

1. Fluorescence decay data are accumulated within a preselected time window and time delay as the wavelength is scanned.
2. Fluorescence decay curves are accumulated consecutively at different wavelengths, and slices at different delay times are assembled (i.e., computed) to form the time-resolved spectrum.

Method 1 has the advantages of being a direct measurement and simple to perform, and it is relatively quick for spectra at a few delay times to be obtained. However, one potential drawback is that the data are uncorrected for the finite duration of the instrumental pulse.

Method 2 has the advantage of eventually providing the whole time-domain contour, and it is possible to assemble the spectra from the impulse response functions obtained from the reconvolution analysis of the decay curves at each wavelength. By this means, correction for the effect of the duration of instrumental pulse can be made. Nevertheless, obtaining the decay curve at every wavelength of interest is a tedious process. A third method has recently been demonstrated.[111] This is based on single-photon timing array detection, and it promises the advantages of method 2 but with much more rapid data acquisition. (See Section 1.6.3.) However, for the time being, method 1 is undoubtedly the most popular method.

In single-photon fluorometers, the time window of interest can be easily selected using the SCA output of the TAC.[139] The time window is simply obtained by setting the voltage window of two potentiometers found on the front of most TACs. By switching the SCA function on and off, the time delay and time window can be located on the whole decay curve. However, rather than operating the MCA in a pulse height analysis mode, the multichannel scaling (MCS) function is selected. This increments the channel in which data are accumulated as the monochromator is scanned in synchronism. Many MCAs have a maximum channel dwell time of only a few seconds so it is usual to drive the MCS externally. In most cases, it is advantageous to combine this with a mechanism for correcting for any intensity fluctuations of the source during the measurement. A widely adopted method for this is to have a reference detector monitoring the source intensity. This analog signal is passed to a voltage-to-frequency converter, and the digital count then drives a scaler/timer such that the wavelength and channel are only incremented when a preset number of counts (i.e., light flux) have been accumulated. Although data accumulation from one time window at a time is usually adequate, methods using more (e.g., four) have been developed.[140]

Time-resolved fluorescence spectra are useful for such applications as investigations of mixtures of samples, solvent relaxation, and excimer and exciplex kinetics. With reference to the PPO excimer example mentioned already, Figure 1.24 shows the time-resolved spectra for PPO in small unilamellar liposomes of α-dipalmitoylphosphatidylcholine (DPPC) at a lipid-to-probe ratio of 50:1.[111] Here several features are revealed which illustrate well the diagnostic power of time-resolved spectral information:

(a) The rising edge spectrum is identical to that for the unimolecular species in dilute solution. A ground state dimer would have an emission spectrum shifted to longer wavelength as in the case of the excimer. Hence, negligible ground state dimer formation is occurring.

(b) The spectra at longer delay times are shifted to longer wavelengths with respect to that for the rising edge. These are thus representative of excimer emission.

(c) The spectra at 30 and 40 ns are separated. This suggests that more than one excimer conformation or environment is present. In solution, the excimer spectrum is independent of delay time.

As well as specific information, what the data of Figure 1.24 show are the important trends. This is often where time-resolved spectra are most useful, and, in this sense, correction for the finite duration of the excitation pulse or correction for the spectral response of the instrument is not usually worthwhile. Perhaps for this reason also, flashlamps are a perfectly adequate source for many applications. Indeed, like anisotropy measurements, time-resolved spectral studies are not as routine as the measurement of fluorescence lifetimes, irrespective of whether a laser or flashlamp is used.

1.6. Multiplexing Techniques

Nearly all the single-photon timing instruments which have been mentioned so far were based on what could be called a simplex approach. That is to say, only a single time-correlated decay is acquired during a measurement. The exception to this already mentioned is the alternate acquisition of parallel and crossed planes of fluorescence polarization when measuring anisotropy decay. This approach can be appropriately termed a multiplexed measurement though only recently have some of the much wider benefits of such an approach been demonstrated.

Time-correlated single-photon counting lends itself well to multiplexed measurements, and much of its potential in this direction has undoubtedly still to be realized. By means of multiplexing, much higher data acquisition rates and more kinetic information can be brought to bear than is possible with the simplex approach.

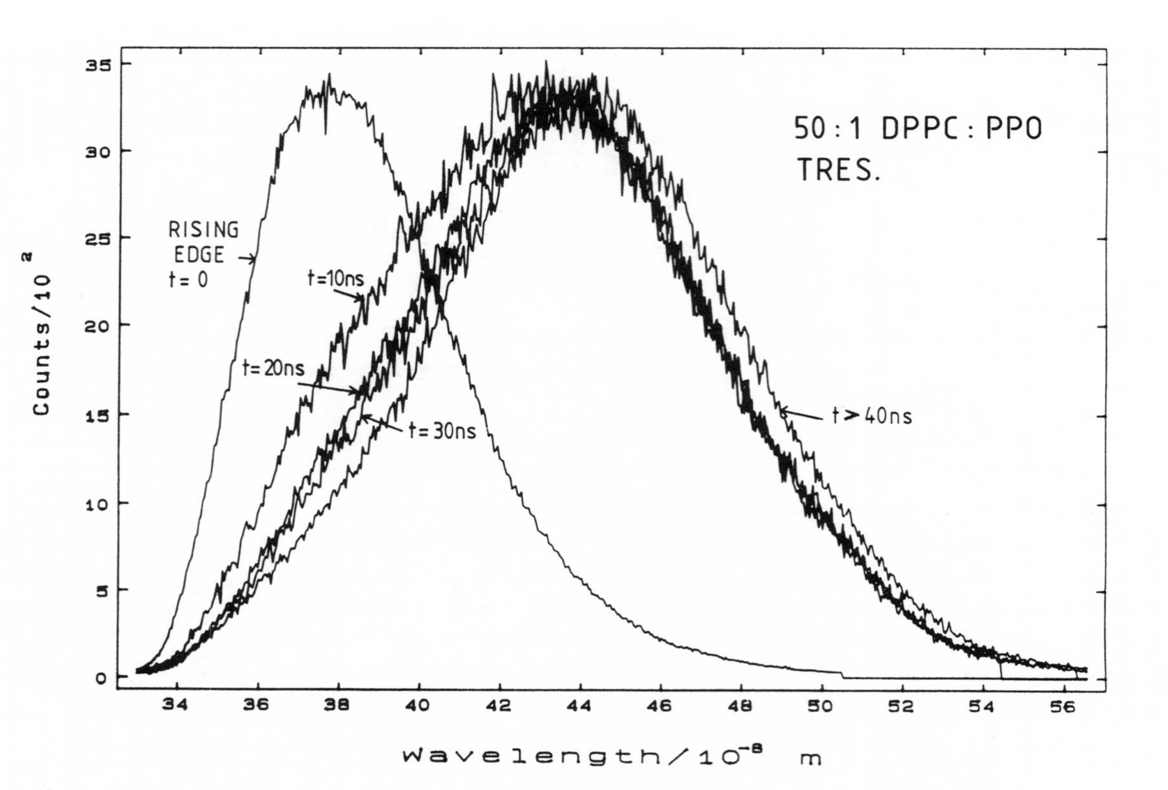

Figure 1.24. Time-resolved spectra for 50:1 DPPC: PPO at 25°C and different delay times t. The spectral bandwidth is 10 nm. The time windows are 15 ns for $t = 10$, 20, and 30 ns and 5 ns for $t = 0$. $t > 40$ ns refers to the tail of the decay. The spectra are normalized to that for $t > 40$ ns. The excitation source is the coaxial flashlamp filled with hydrogen. (From Ref. 111.)

The main restriction on data acquisition rate considered so far is that of the pileup of "stop" pulses discussed in Section 1.2.2. However, the TAC and MCA-ADC have intrinsic dead times when processing data which total $\sim 10\ \mu s$ at the very minimum with current technology. Hence, only count rates to a maximum of 100,000 counts/s can be registered irrespective of the pileup restriction of the single-photon technique itself. Of course, such a high data rate is a luxury and could only be tolerated at source repetition rates approaching 10 MHz. However, this leads to two complementary criticisms often laid at the door of the single-photon technique, namely, that the pileup restriction means that measurements are slow and that data are wasted in satisfying it. Interestingly enough, if a fluorescence decay has 10^6 counts accumulated in 1 h, then the actual data processing time during which the instrument is working productively, at a 20-μs TAC/MCA dead time, is only 20 s. This means that more than 99% of the available measurement time is unused. A very lazy technique indeed! However, in one of the multiplexing methods to be discussed, statistical time-sharing, this spare time is used to process events from many detection channels and thereby circumvent the pileup restriction. This is the most efficient multiplexing technique currently in use and hence we will discuss it in some detail. Summarizing all the multiplexing methods so far reported, we have:

1. Alternate acquisition of different signals using a single set of data timing and acquisition electronics.
2. Simultaneous acquisition of different signals by using a time delay to separate the signals in a single set of electronics.
3. Simultaneous data acquisition using multiple sets of electronics.
4. Simultaneous statistical multiplexing of several detection channel signals using a single set of electronics.

We will now describe and compare these methods in more detail with regard to the different implementations.

1.6.1. Differential Pulse Fluorometry

In differential pulse fluorometry, the excitation pulse and the fluorescence decay are measured together. This automatically corrects the convolution integral (Eq. 1.21) for fluctuations in the temporal profile of the excitation pulse. It is thus of most use in extending the capability of flashlamps. The correction can be seen by applying the principle of linear superposition to Eq. (1.21):

$$\sum_j F_j(t) = \sum_j \int_0^t P_j(t)\, i(t-t')\, dt' \tag{1.71}$$

where j excitation pulse profiles produce a superposition of j fluorescence decays.

Rearranging Eq. (1.71),

$$\sum_j F_j(t) = \int_0^t \sum_j P_j(t)\, i(t-t')\, dt' \tag{1.72}$$

Equation (1.72) basically shows that provided the measured instrumental pulse $P(t)$ and the fluorescence decay $F(t)$ are measured simultaneously, they are still related by the convolution integral even if they originate from differently shaped excitation profiles. In the usual simplex approach, $F(t)$ and $P(t)$ are measured consecutively and may indeed originate (in fact, with spark sources they most certainly do) from excitation pulses of different shapes. However, what is then required for the convolution integral to be valid is that the *time-averaged* excitation pulse profile remains constant over both measurements; that is, the instrument is passively stabilized such that

$$F(t) = \sum_j F_j(t) \qquad \text{and} \qquad P(t) = \sum_j P_j(t) \tag{1.73}$$

Equation (1.72) offers the ideal differential arrangement whereby $F(t)$ and $P(t)$ are measured simultaneously.

The first step toward achieving this was originally developed by Hazan *et al.*[(141)] with a fluorometer incorporating a sampling scope, and the same principle was subsequently used in a number of single-photon fluorometers.[(83)] Basically, the approach is to alternate the sample and scatterer in the detection plane using a motor-driven twin sample holder such that the fluorescence and excitation signals are detected for a selected time period. The wavelength of the emission monochromator also needs to be alternated accordingly. Such tasks are fairly straightforward with microprocessor technology. The two time-correlated signals can be routed to separate memory segments of the MCA using simple logic levels; for example, 11 might denote the first segment, and 10 the second segment.

The approach of alternate collection of lamp and decay has become less important with the development of stable flashlamps which permit consecutive measurements for up to ~4 h routinely and even longer in some cases. Nevertheless, with alternate data acquisition, the performance of any flashlamp can be usefully extended, for example, for overnight runs. The main disadvantages of this approach are that a finite data acquisition time has to be chosen for each period (typically 30 s) and hence not all changes in the excitation pulse are necessarily corrected. In addition, the overall measurement time is longer than with the conventional simplex method due to the additional dead time between the measurement of the excitation and of the decay.

The use of differential methods has not only been applied to flashlamps however. Wijnaendts van Resandt *et al.*[142] have used an optical delay to separate the excitation from fluorescence pulses with synchronously pumped cavity-dumped dye laser excitation and microchannel plate detection. The experimental arrangement and typical results are shown in Figure 1.25. In this case there is no need to fear about fluctuations in the time profile of the excitation pulses. However, the timing response of a microchannel plate detector can depend on the count rate, and this type of arrangement ensures that the detector response is identical for the measurement of the excitation and of the fluorescence. Here the excitation pulse was not measured directly, but rather the fluorescence decay of a reference compound was detected instead in order to match the wavelengths. The optical delay technique is elegant but only practicable with lasers, and for long lifetimes the optical delay would be unworkable. Because different parts of the time range are used for the two measurements, any differences in nonlinearity might also impair successful reconvolution. Also, because the two time-correlated signals are measured simultaneously in the same MCA segment, their summed rather than separate count rates are limited by the usual pileup restriction. Hence, there are no dead-time savings to be gained by this method of multiplexing, but the convenience of the simultaneous measurement itself is undoubtedly worthwhile.

The use of multiple sets of timing electronics, such that one set of TACs is used to measure the excitation pulse and a different set of TACs is used to measure the decay, seems to have been originally proposed by Hara.[143] Separate photomultipliers are used for simultaneous acquisition of fluorescence and excitation (SAFE). The TACs are arranged in each set in parallel but stopped sequentially by consecutive "stop" pulses. The TAC conversions are multiplexed to a single ADC interfaced to a microcomputer for data acquisition. By this means the usual pileup limit is rather cleverly overcome and hence the run time usefully reduced. The statistics of this arrangement are complicated, and the technique does not seem to be widely used. We have no knowledge as to the effect on time calibration, pileup limit, level of detector matching, etc.

The main problems with simply multiplying the sets of timing electronics and MCAs in proportion to the number of channels are the increased cost and possible differences between the components. However, provided the pileup limitation is satisfied, replication does facilitate operation of each channel at a count rate limited only by the TAC/MCA dead time. Nevertheless, replication does seem rather inefficient at channel count rates at which the TAC and MCA are not used to the maximum duty cycle.

By using a T-format with one arm detecting fluorescence and the other excitation scattered from a triangle cuvette, Birch and co-workers were the first to report successful reconvolution in a SAFE configuration.[91, 144] The discriminator output pulses from the two channels were simply mixed

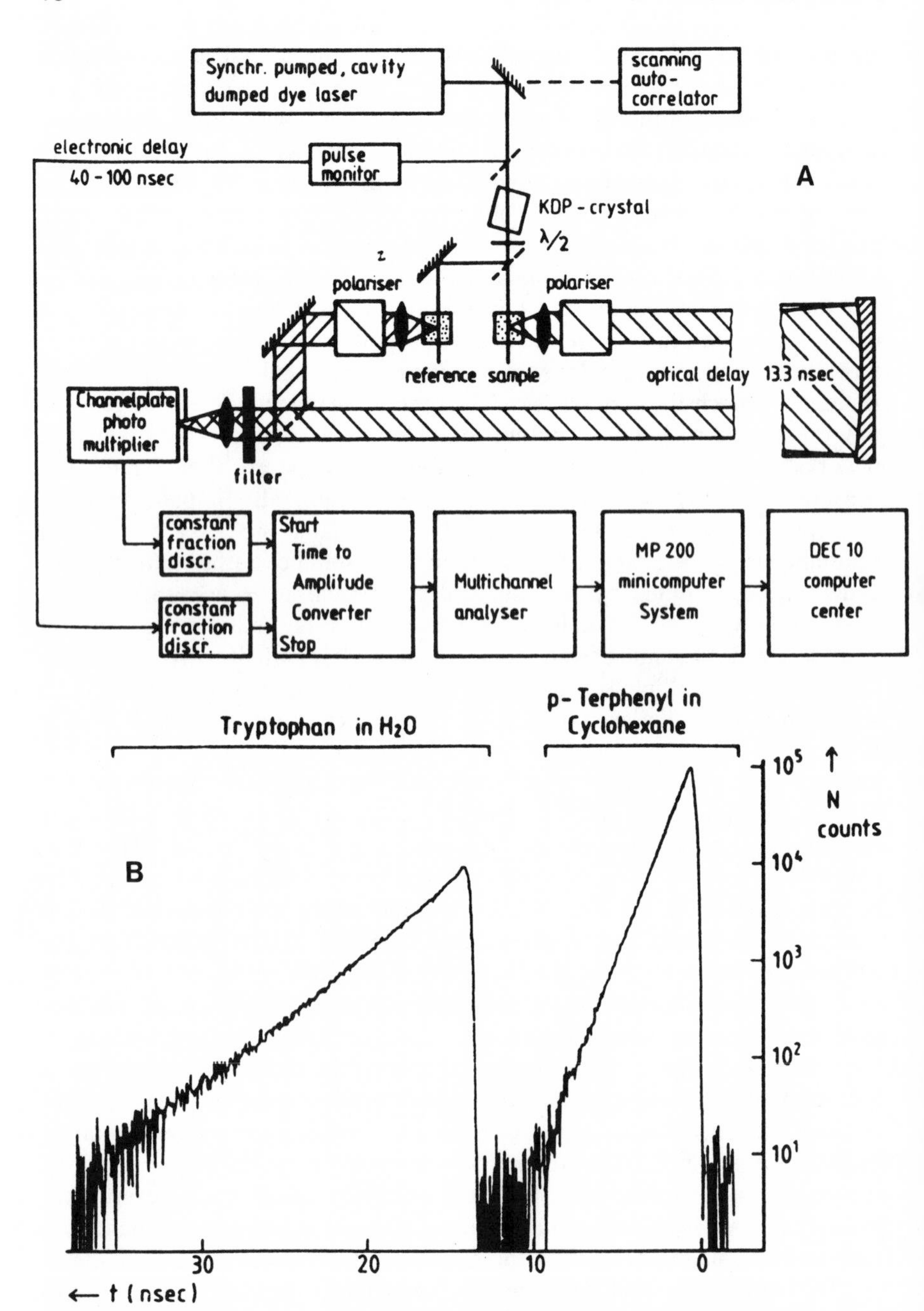

Figure 1.25. (A) Differential pulse fluorometry using an optical delay to separate fluorescence and excitation. (B) Measured fluorescence decay at 340 nm of tryptophan and reference compound excited at 295 nm. (From Ref. 142.)

together using a BNC T-connector to provide a summed rate of "stop" pulses to the TAC. The TAC conversions from these pulses were separated into separate segments of the MCA memory according to the channel of origin by using a coincident discriminator pulse from each channel to "label" the "stop" pulses. These label pulses were used to generate the required logic levels for selection of the different MCA memory segments. This approach forms the basis of statistical multiplexing. Various sources of misrouted signals are possible in such systems, and a purpose-designed NIM routing module had to be constructed. The basic functions of the routing are[91]:

(a) Enable time window. This provides a time window defined by the TAC "true start" and SCA outputs. The time window is typically ~500 ns on a 200-ns TAC range.
(b) Route. The −ve output label pulses from the discriminators are used to generate logic levels which select the appropriate MCA segment, for example, 11, 10, etc. These logic signals are the same as are necessary when alternately collecting fluorescence and excitation. However, in this case the data are acquired in a statistical sampling manner. The routing signal is reset with each "true start" signal.
(c) Inhibit. The MCA "dead-time" signal is used to inhibit the reset of a routing signal until an event has been routed and stored in the MCA. If this is not done, misrouting occurs at "start" rates greater than the reciprocal of the MCA dead time, due to the reset of the routing signal required in (b).
(d) Reject. If pulses are detected in both detection channels during the "enable" time window, the TAC is reset. This prevents any possible misrouting due to such dual coincidences.

With this type of arrangement, it is possible to completely eliminate misrouted signals.

At first glance, it might be reasoned that in such a system the sum of the "stop" rates from the two channels still has to be kept less than the pileup limit, that is, <2% say. However, the reject function (d) ensures that the two-time correlated signals are sampled separately such that each channel can in fact be operated at the pileup limit. Thus, for two channels a saving in time of up to a factor of 2 can be obtained over the conventional consecutive approach because some of the usual 99% dead time of the single-photon technique is now being used more efficiently. The time savings, though not large for two channels, are still nevertheless worth having. Perhaps more important than this implementation in itself is the demonstration of the principle of multiplexing based on statistical sharing of the timing electronics between detection channels.

Successful data analysis in SAFE measurements with different detection channels requires that the impulse responses of the channels are matched. Even with unmatched channels, single exponential lifetimes down to 1.6 ns have been measured to within 3% precision, though the χ^2 values were too high to test the kinetic model.[(91)] By adjusting potentiometers in the voltage divider network, the goodness of fit can be made comparable to that obtained with consecutive measurements in the same detection channel. Table 1.5 illustrates the level of agreement between simultaneous and consecutive measurements which can be achieved with flashlamp excitation. The goodness of fit for the simultaneous measurements is comparable to that for the consecutive measurements, despite slight differences in the matching of the two channels and the decay time values agree within the statistical error. In addition, the high wavelength separation between the matching and fluorescence wavelengths (~330-nm separation) shows that the matching is reproducible over a wide spectral range.

This kind of differential arrangement extends the performance of flashlamps to the measurement of extremely weak emitters simply by allowing long run times. The active correction for source instabilities using Eq. (1.72) has been demonstrated by deliberately changing the excitation pulse profile during a differential measurement.[(91)] Because the sample remains undisturbed in a differential measurement, the approach is also useful when studying samples under carefully controlled conditions, for example, with a flow cell or at precise temperatures.

The reason why differences in matching can be tolerated with a flashlamp is because any differences can be made less than other sources of error. Nevertheless, for lifetimes of ~200 ps such differences can become significant.

Table 1.5. A Comparison between Simultaneous and Consecutive Excitation and Decay Measurements[a,b]

Molecular symmetry	Simultaneous		Consecutive	
	τ_M (ns)	χ^2	τ_M (ns)	χ^2
C_{2v}	8.48	1.12	8.42	1.20
C_{2v}	8.71	1.28	8.65	1.30
D_{2h}	8.86	1.31	8.80	1.26
C_{2v}	8.77	1.32	8.71	1.30
D_{4h}	9.65	1.26	9.57	1.38

[a] From Ref. 145.

[b] The samples are aminotetraphenylporphyrins. The different substitution symmetries are reflected in small changes in the fluorescence lifetime. For the simultaneous measurements the excitation and fluorescence were measured using XP2020Q and XP2254B photomultipliers, matched at 350 nm. The excitation wavelength was 420 nm, and the fluorescence wavelength 660–680 nm. The statistical error in each measurement of 3 std. dev. is ~50 ps.

In the early work, the main obstacle to better matching was the late peak in the photomultiplier response at ~ 10 ns, but the improved voltage divider network shown in Figure 1.12 has overcome this problem by reducing the intensity of the late peak as shown in Figure 1.13.

Probably none of the differential methods described is ideal, but most workers would agree that they do point to the way ahead. Future developments as well as existing ones wait to be tried, and they will undoubtedly bring better performance. For example, two detection channels will be matched identically if the detector response time is much less than the duration of the excitation. This means that MCPs or SPADs could already be used with a flashlamp to obtain perfectly matched detection channels.

1.6.2. T-Format Studies

The earliest report of a time-correlated single-photon counting fluorometer in a T-format is that by Schuyler and Isenberg in 1971.[146] Subsequently, Harvey and Cheung[147] reported the use of a T-format to record decays at different fluorescence polarizations for anisotropy studies on myosin rods with fluorescent labels attached. However, it is fair to say that until recently the virtues of using a T-format had been much more widely appreciated for frequency-domain measurements. Nevertheless, with single-photon counting a T-format is an elegant way of performing both dual-wavelength and anisotropy studies, and in the latter it continuously corrects for the effect of fluctuations in the excitation intensity on the measured anisotropy function. Dual-channel instruments do not necessarily need to be arranged in a T-format for anisotropy studies, but this is the usual configuration. However, there is some virtue with optically dense or heterogeneous samples in being able to view both planes of polarization from the same side of the sample. This can be accomplished using a polarizing beam splitter such as a Wollaston prism. The instrument shown in Figure 1.25 has been used in such a configuration to measure the rotational times of rhodamine 6G in various solvents, yielding a value of 95 ± 3 ps in methanol.[148]

Figure 1.26 shows a three-channel fluorometer which combines a T-format with a differential arrangement for SAFE.[133] The two polarized fluorescence decays and the excitation pulse are measured simultaneously using the statistical multiplexing technique. Each channel can be operated at the pileup limit such that the run time is reduced to one-third of that with consecutive measurements. Table 1.6 compares test results obtained with the usual simplex approach and this multiplex method. The agreement is good down to rotational correlation times of ~ 0.7 ns.

Analysis of anisotropy data from T-format instruments can be performed using all the alternative methods described in Section 1.3.4. If the detection

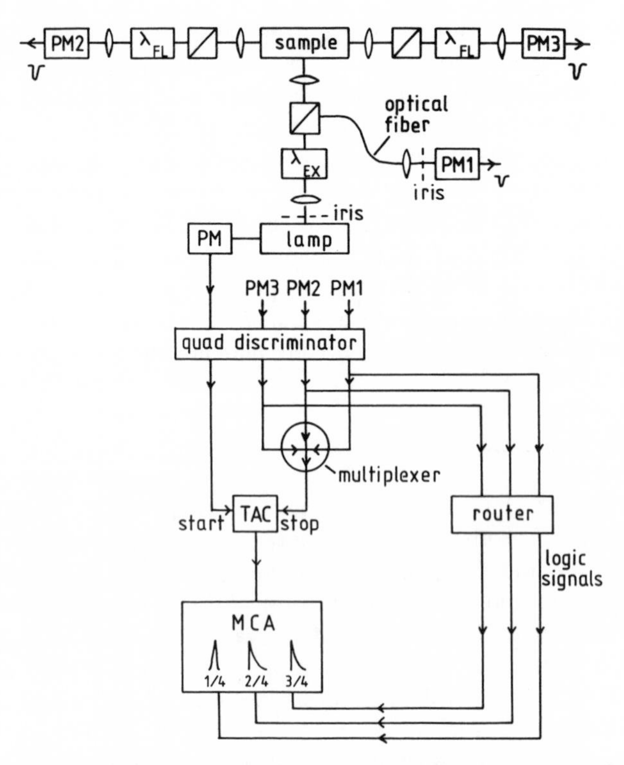

Figure 1.26. T-format differential fluorometer using statistical multiplexing for anisotropy studies. This arrangement provides a complete correction for both temporal and intensity fluctuations of the source. (From Ref. 133.)

Table 1.6. Comparison of Anisotropy Techniques[a,b]

	Multiplex		Simplex	
T (°C)	τ_R (ns)	χ^2	τ_R (ns)	χ^2
5	15.40 ± 2.70	1.31	15.20 ± 2.10	1.04
20	6.58 ± 0.66	1.37	6.69 ± 0.69	1.12
23	5.38 ± 0.54	1.19	5.71 ± 0.57	1.07
35	2.93 ± 0.32	1.18	2.85 ± 0.30	1.03
50	1.48 ± 0.24	1.14	1.31 ± 0.24	1.09
65	0.67 ± 0.30	1.24	0.74 ± 0.27	0.94
80	0.15 ± 0.30	1.34	0.48 ± 0.48	1.07

[a] From Ref. 134.
[b] The sample is DPH in a white oil. Channel width, 0.49 ns. Multiplex run times, <10 min. Data analyzed using impulse reconvolution.

channels are not matched, then the impulse response analysis or global method is necessary. However, if the channels are matched, then there is a wider choice of data analysis methods available.

In general, T-formats offer a better all-around research capability and greater flexibility than a conventional L-format. For example, the *g*-factor can be simply tuned in a T-format just by altering the collection efficiencies of the two channels, for example, by moving lenses, iris adjustment, etc.

1.6.3. Array Fluorometry

All the measurements we have described so far have involved in different ways the study of just one small part of the whole multidimensional topographical contour of fluorescence intensity, time, wavelength, polarization, and space. This contour is the fluorescence signature of a species, and the more of it we know about, the higher is the measurement specificity. If the contour is to be constructed with conventional techniques, then the process is extremely laborious, involving a wide range of sequential measurements. The only exception to this mentioned so far is the T-format, but clearly two detection channels with separate monochromators is not far away from the maximum number of such channels that is practicable.

Array detection methods and the future development of integrated optics

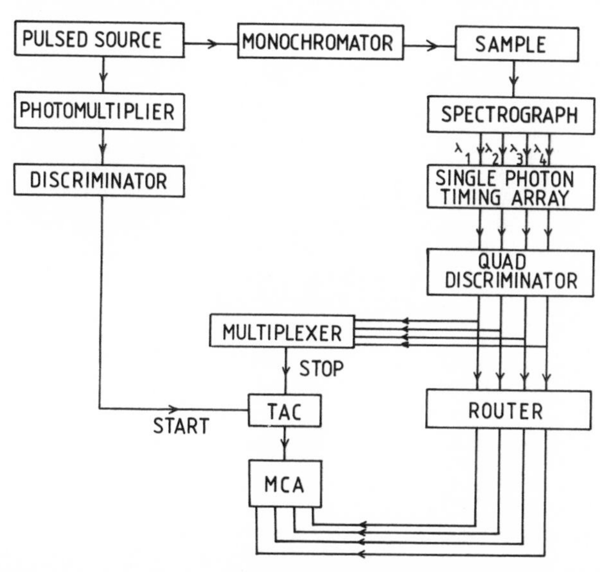

Figure 1.27. Fluorometer with multiplexed single-photon timing array detection. Fluorescence decays are measured simultaneously at four emission wavelengths. (From Ref. 149.)

offer the potential for measuring, if not all of the fluorescence contour, then a large part of it in a single measurement. In addition, on-line computer power capable of handling such quantities of data is now becoming available at reasonable prices. This approach has recently been demonstrated with single-photon timing at multiple fluorescence wavelengths for the first time by Birch

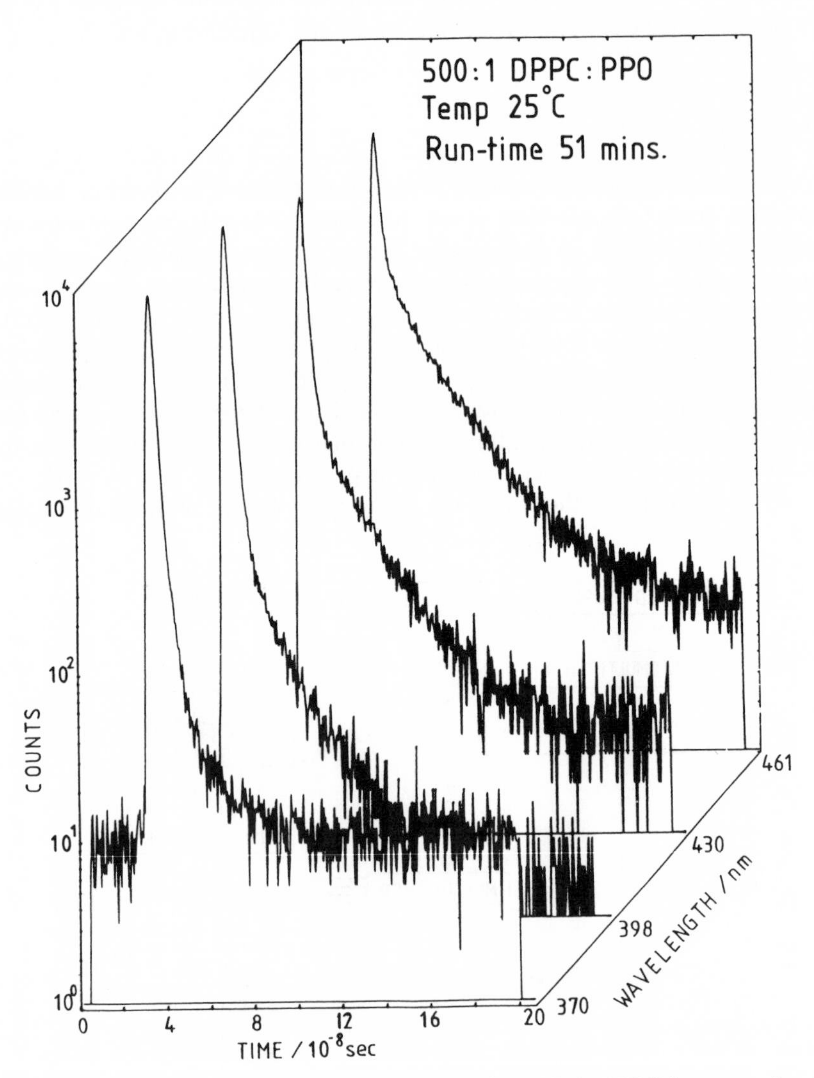

Figure 1.28. Single-photon timing array measurement for PPO in DPPC liposomes, showing the presence of a longer decay component which increases in intensity with increasing fluorescence wavelength. This is caused by excimer emission from ground state aggregates. (From Ref. 111.)

and co-workers.[149, 111, 134] Table 1.4 demonstrates the higher level of information available with this approach, and Figure 1.27 shows the layout of the array instrument used. The decay data corresponding to the four fluorescence wavelengths were acquired simultaneously in separate memory segments of the MCA using the statistical multiplexing technique. The fluorescence was dispersed using a spectrograph and detected using four photomultipliers, coupled to the spectrograph using an array of optical fibers. Figure 1.28 shows a typical data set obtained with this arrangement.

Having four multiplexed detection channels has enabled us to check more fully how the pileup limitation is overcome with statistical multiplexing. Table 1.7 shows that an increase in the "stop" rate from 4.1 to 35.7% of the "start" rate in a single channel causes the apparent decay time to be shortened from ~8.04 to 7.36 ns. However, in the multiplexed case at a total "stop" to "start" rate ratio of 4 × 8.85% with a 3.3% reject (i.e., ~30% net) the correct decay time of ~8.03 ns is still obtained. These results are also in good agreement with our theory describing statistical multiplexing with several detection channels.[150]

The theory of statistical multiplexing for m channels can be described by the binomial distribution. In this case the probability of k channels containing single-photon pulses per excitation pulse is given by

$$P_k(m) = \frac{m!}{k!(m-1)!} (\alpha)^k (1-\alpha)^{m-k} \tag{1.74}$$

Table 1.7. Demonstration of the Pileup Advantage with Statistical Multiplexing[a]

Multiplexed channel			Simplexed channel	
"Stop" rate[b] (% "start" rate)	τ^c (ns)	Total reject rate[d] (% "start" rate)	"Stop" rate[e] (% "start" rate)	τ (ns)
1.25	8.07	0.02	4.1	8.04
2.35	8.09	0.2	9.4	7.97
4.50	8.10	0.8	18.4	7.85
5.50	8.13	1.3	22.5	7.79
7.15	8.10	2.4	29.1	7.69
8.85	8.03	3.3	35.7	7.36
12.70	7.95	4.5	44.4	7.48
14.30	7.90	7.4	52.0	7.26

[a] Sample is DPH in aerated cyclohexane. Flashlamp repetition rate, 30 kHz.
[b] Relates to the count rate in one channel. Total stop rate is ~4 times this value.
[c] Best fit τ for channel used.
[d] Reject count rate due to coincident "stop" pulses in all four channels. Note that one reject pulse involves the rejection of at least two stop pulses.
[e] Stop rate in the channel used for the simplex measurement is arranged to be approximately equal to the sum count for the four multiplexed channels.

where in order to simplify the mathematics α is assumed to be a constant "stop" to "start" rate ratio per channel. In practice, α is likely to vary (e.g., over different emission wavelengths), but a more complex binomial treatment can still be applied.

If coincident events in more than one channel are rejected electronically in order to prevent misrouting (Section 1.6.1), then the probability of obtaining a TAC output pulse is given by the probability of detecting a coincident event in just one channel; that is,

$$P_1(m) = m\alpha(1-\alpha)^{m-1} \tag{1.75}$$

with the total rate of TAC conversions of "stop" pulses over all channels, S_p^c, then being given by

$$S_p^c = S_t m\alpha(1-\alpha)^{m-1} \tag{1.76}$$

where S_t is the "start" rate. The rate of reject events due to coincident "stop" pulses, S_p^r, is then given by

$$S_p^r = S_t[1 - P_0(m) - P_1(m)] \tag{1.77}$$

$$= S_t[1 - (1-\alpha)^m - m\alpha(1-\alpha)^{m-1}] \tag{1.78}$$

Note that the rate of reject events is less than the rate of "stop" pulses which are rejected, because the reject events refer to multiple "stop" pulses.

Equation (1.75) describes the acquisition of data from a number of single-photon counting channels in cases where coincident events in more than one channel interfere with one another and are rejected. If $m = 1$, we of course have the well-known expression for a single-channel instrument, and for $m = 2$ the theory can be used to describe a T-format. In the latter case, the number of reject events is small under the usual single-photon condition, that is, $\alpha \sim 0.01$.

For large arrays the multiplexing theory predicts that the reject rate can become significant. By using Eq. (1.75), single-photon timing arrays can be optimized for maximum total count rate accepted, with regard to the number of channels, channel count rate, and reject rate. Figure 1.29 shows some of the contours which relate these parameters. Even allowing for the fact that some of the single-channel count rates used in Figure 1.29 would lead to single-channel pileup, it is obvious that TAC conversion rates much greater than that which is permissible with a single channel are possible. For example, for $m = 64$ and $\alpha = 0.02$, a $P_1(64)$ value of 0.358 is predicted, that is, a TAC conversion rate of effectively 35.8% "start" rate. Such a high count rate would lead to significant pileup distortion in single-channel instruments, but, as Table 1.7 shows, is quite acceptable with a multiplexed array.

The implications of the enhanced data collection rate obtainable with

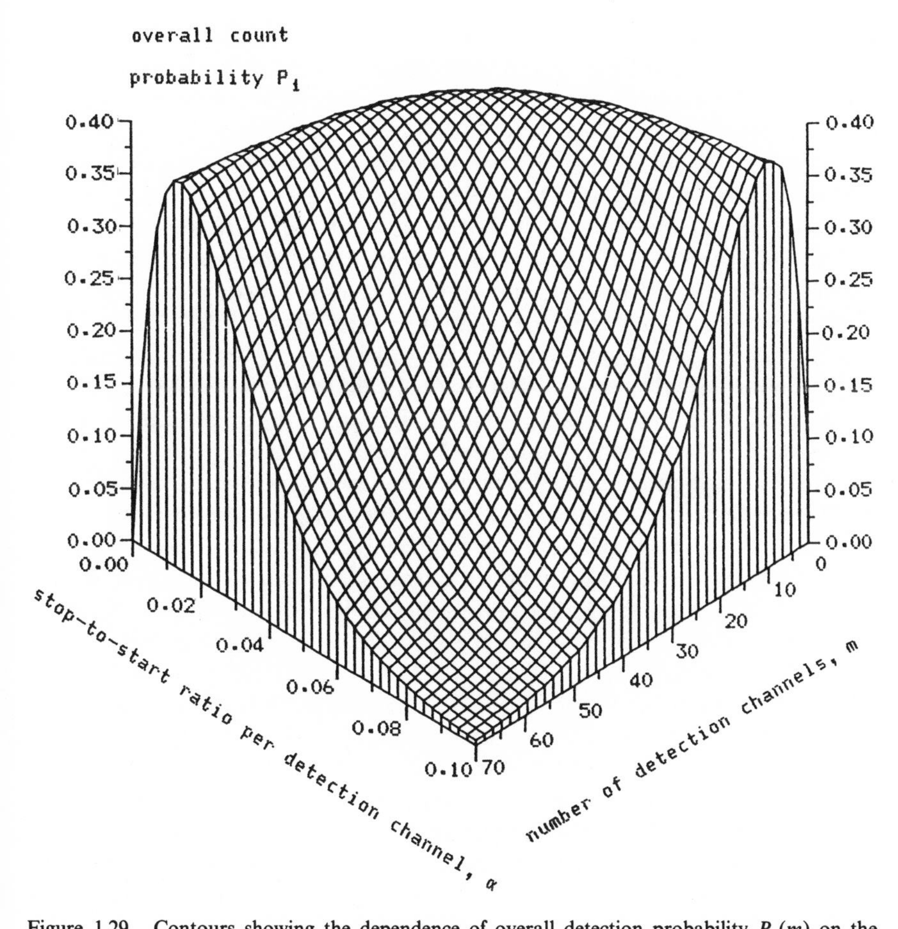

Figure 1.29. Contours showing the dependence of overall detection probability $P_1(m)$ on the number of channels m and "stop" to "start" rate ratio α for a single-photon timing array. (From Ref. 150.)

multiplexing have interesting consequences in the use of both low- and high-repetition-rate sources. With nitrogen or excimer lasers operating at, say, ~100 Hz, it should now be more practicable to measure fluorescence decays using single-photon timing. At the high-repetition-rate end, using mode-locked lasers, the dead time of the TAC and MCA limits the total data collection rate to 100 kHz. The introduction of gallium arsenide technology could enhance this tenfold such that more might be made of the multiplexing advantage. However, even at present, a 64-element single-photon timing array could span a 200-nm range with sufficient spectral resolution, and at a mean count rate of ~1 kHz per channel the whole intensity–time–wavelength contour of a sample could probably be accurately measured in a few minutes. Flashlamps offer an intermediate consideration. At say 60-kHz repetition rate and 30% total "stop" to "start" rate ratio, a 64-element array with a mean channel count rate of ~300 Hz would be workable.

The area of array fluorometry with single-photon counting is clearly only just beginning. There are already opportunities for array detection using solid-state arrays or multianode MCPs.[104] In the case of solid-state arrays fabricated on a single chip, it should be straightforward to ensure that the timing and spectral properties of each element are identical. Time-domain imaging by applying array detection techniques to microscopy and anisotropy or combinations thereof, with or without wavelength dispersion, would now seem possible. Phase fluorometry has yet to be used for array measurements but no doubt this will come. In fact, the only alternative which has been used in fluorescence lifetime measurements is of course the streak camera. These devices have higher time resolution (<1 ps) but lack the dynamic range and sensitivity of photon counting. They are also more expensive.

To conclude our chapter, it is interesting to speculate on what single-photon counting instruments of the future might look like. The user's ideal would probably be based on solid-state laser excitation, tunable over a wide spectral range, with array detection and simultaneous excitation measurement, PC control, and rapid on-line data analysis. All these features are already here in one form or another; it is just the combined performance that is missing. Whatever the future holds though, there can be no real substitute in this field, as in others, for the skilled experimenter. Twenty years ago, in the preface to his book *The Photophysics of Aromatic Molecules*, John Birks quoted Leonardo: "Experiments never deceive. It is our judgement that sometimes deceives itself because it expects results which experiment refuses. We must consult experiment, varying the circumstances, until we have deduced reliable rules." What continues to make fluorescence lifetime spectroscopy so interesting is that many of the rules are still being discovered.

References

1. R. B. Cundall and R. E. Dale (eds.), *Time Resolved Fluorescence Spectroscopy in Biochemistry and Biology*, Plenum, New York (1983).
2. J. N. Demas, *Excited State Lifetime Measurements*, Academic Press, New York (1983).
3. D. V. O'Connor and D. Phillips, *Time-Correlated Single Photon Counting*, Academic Press, London (1984).
4. A. J. W. G. Visser (ed.), *Time-resolved fluorescence spectroscopy*, *Anal. Instrum. 14*, 193–565 (1985).
5. J. B. Birks, *Photophysics of Aromatic Molecules*, Wiley, London (1970).
6. J. R. Lakowicz, *Principles of Fluorescence Spectroscopy*, Plenum, New York (1983).
7. J. R. Alcala, E. Gratton, and D. M. Jameson, A multifrequency phase fluorometer using the harmonic content of a mode-locked laser, *Anal. Instrum. 14*, 225–250 (1985).

8. E. Gratton, D. M. Jameson, N. Rosato, and G. Weber, Multifrequency cross-correlation phase fluorometer using synchrotron radiation, *Rev. Sci. Instrum. 55*, 486–494 (1984).
9. J. R. Lakowicz, G. Laczko, and I. Gryczynski, 2-GHz frequency-domain fluorometer, *Rev. Sci. Instrum. 57*, 2499–2506 (1986).
10. J. Hedstrom, S. Sedarous, and F. G. Prendergast, Measurements of fluorescence lifetimes by use of a hybrid time-correlated and multifrequency phase fluorometer, *Biochemistry 27*, 6203–6208 (1988).
11. D. J. Desilets, P. T. Kissinger, and F. E. Lytle, Measurement of fluorescence lifetimes during liquid chromatography, *Anal. Chem. 59*, 1830–1834 (1987).
12. J. Howard, N. J. Everall, R. W. Jackson, and K. Hutchinson, Fluorescence rejection in Raman spectroscopy using a synchronously pumped, cavity dumped dye laser and gated photon counting, *J. Phys. E: Sci Instrum. 19*, 934–943 (1986).
13. W. A. Wyatt, G. E. Poirier, F. V. Bright, and G. M. Hieftje, Fluorescence spectra and lifetimes of several fluorophores immobilized on nonionic resins for use in fibre optic sensors, *Anal. Chem. 59*, 572–576 (1987).
14. W. A. Wyatt, F. V. Bright, and G. M. Hieftje, Characterization and comparison of three fibre-optic sensors for iodide determination based on dynamic fluorescence quenching of rhodamine 6G, *Anal. Chem. 59*, 2272–2276 (1987).
15. O. S. Wolfbeiss and S. Sharma, Fibre-optic fluorosensor for sulphur dioxide, *Anal. Chim. Acta. 208*, 53–58 (1988).
16. G. H. Vickers, R. M. Miller, and G. M. Hieftje, Time-resolved fluorescence with an optical fibre probe, *Anal. Chim. Acta 192*, 145–153 (1987).
17. F. V. Bright, Remote sensing with a multifrequency phase-modulation fluorometer, in: *Time-Resolved Laser Spectroscopy in Biochemistry* (J. R. Lakowicz, ed.), *Proc. SPIE 909*, 23–28 (1988).
18. R. A. Malstrom and T. Hirschfeld, in: *Analytical Spectroscopy* (W. S. Lyon, ed.), pp. 25–30, Elsevier, Amsterdam (1983).
19. B. A. Bushaw, in: *Analytical Spectrosocpy* (W. S. Lyon, ed.), pp. 57–62, Elsevier, Amsterdam (1983).
20. H. Pal, D. Palit, T. Mukherjee, and J. P. Mittal, The fluorescence lifetimes of isomeric tyrosines, *Chem. Phys. Lett. 151*, 75–78 (1988).
21. A. E. W. Knight and B. K. Selinger, Single photon decay spectroscopy, *Austr. J. Chem. 26*, 1–27 (1972).
22. J. Yguerabide, *Methods Enzymol. 26C*, 498–578 (1972).
23. T. H. Binkert, H. P. Tschanz, and P. E. Zinsli, The measurement of fluorescence decay curves with the single-photon counting method and the evaluation of rate parameters. *J. Lumin. 5*, 187–217 (1972).
24. J. B. Birks and I. H. Munro, in: *Progress in Reaction Kinetics*, Vol. 4, pp. 239–303, Pergamon Press, Oxford (1967).
25. W. R. Ware, in: *Creaton and Detection of the Excited State* (A. A. Lamola, ed.), Vol. 1A, pp. 213–302, Marcel-Dekker, New York (1971).
26. L. M. Bollinger and G. E. Thomas, Measurement of the time-dependence of scintillation intensity by a delayed coincidence method, *Rev. Sci. Instrum. 32*, 1044–1052 (1961).
27. P. B. Coates, The correction for photon "pile-up" in the measurement of radiative lifetimes, *J. Phys. E: Sci. Instrum. 1*, 878–879 (1968).
28. C. C. Davis and T. A. King, Correction methods for photon pile-up in lifetime determination by single-photon counting, *J. Phys. A. 3*, 101–109 (1970).
29. D. E. Donohue and R. C. Stern, Correction of single photon or particle timing measurements for multiparticle events, *Rev. Sci. Instrum. 43*, 791–796 (1972).
30. Proceedings of the Conference on *Deconvolution and Reconvolution of Analytical Signals* (M. Bouchy, ed.), pp. 411–423, printed by ENSIC-INPL, Nancy (1982).

31. Philips Data Handbook, Philips Publications, pp. 75–78 (1987).
32. S. Kinoshita and T. Kushida, High performance, time-correlated single photon counting apparatus using a side-on type photomultiplier, *Rev. Sci. Instrum. 53*, 469–472 (1982).
33. D. J. S. Birch and R. E. Imhof, The origin of fluorescence from trans-trans diphenylbutadiene, *Chem. Phys. Lett. 88*, 243–247 (1982).
34. H. Leismann, H.-D. Scharf, W. Strassburger, and A. Wollmer, Determination of subnanosecond fluorescence decays of chlorobenzene, tryptophan and the benzene–triethylamine exciplex using a nanosecond flashlamp, *J. Photochem. 21*, 275–280 (1983).
35. D. J. S. Birch and R. E. Imhof, Kinetic interpretation of fluorescence decays, *Anal. Instrum. 14*, 293–329 (1985).
36. A. Grinvald and I. Z. Steinberg, On the analysis of fluorescence decay kinetics by the method of least squares, *Anal. Biochem. 59*, 583–598 (1974).
37. S. W. Provencher, A Fourier method for the analysis of exponential decay curves, *Biophys. J. 16*, 27–41 (1976).
38. J. C. Andre, C. M. Vincent, D. V. O'Connor, and W. R. Ware, Applications of fast Fourier transform to deconvolution in single photon counting, *J. Phys. Chem. 83*, 2285–2294 (1979).
39. A. Gafni, R. L. Modlin, and L. Brand, Analysis of fluorescence decay curves by means of the Laplace transformation, *Biophys. J. 15*, 263–279 (1975).
40. B. Valeur, Analysis of time-dependent fluorescence experiments by the method of modulating functions with special attention to pulse fluorometry, *Chem. Phys. 30*, 85–93 (1978).
41. I. Isenberg and R. D. Dyson, The analysis of fluorescence decay by a method of moments, *Biophys. J. 9*, 1339–1350 (1969).
42. P. R. Bevington, *Data Reduction and Error Analysis for the Physical Sciences*, McGraw-Hill, New York (1969).
43. J. R. Lakowicz, in: *Applications of Fluorescence in the Biomedical Sciences* (D. L. Taylor, A. S. Waggoner, R. F. Murphy, F. Lanni, and R. R. Birge, eds.), pp. 29–67, Alan R. Liss, New York (1986).
44. E. Gardini, S. Dellonte, L. Flamigni, and F. Barigelletti, The reliability of iterative re-convolution in fitting two exponential fluorescence decay curves, *Gazz. Chim. Ital. 110*, 533–537 (1980).
45. J. R. Knutson, J. M. Beecham, and L. Brand, Simultaneous analysis of multiple fluorescence decay curves: A global approach, *Chem. Phys. Lett. 102*, 501–507 (1983).
46. J. B. Ross, W. R. Laws, J. C. Sutherland, A. Buku, P. G. Katsoyannis, I. L. Schwartz, and H. R. Wyssbrod, Linked-function analysis of fluorescence decay kinetics: Resolution of side-chain rotamer populations of a single aromatic amino acid in small polypeptides, *Photochem. Photobiol. 44*, 365–370 (1986).
47. S. R. Flom and J. H. Fendler, Global analysis of fluorescence depolarization experiments, *J. Phys. Chem. 92*, 5908–5913 (1988).
48. P. Wahl, Analysis of fluorescence anisotropy decays by a least-squares method, *Biophys. Chem. 10*, 91–104 (1979).
49. C. K. Chan, Synchronously Pumped Dye Lasers, Spectra-Physics Laser Technical Bulletin No. 8 (February 1978).
50. G. R. Fleming, Subpicosecond spectroscopy, *Annu. Rev. Phys. Chem. 37*, 81–104 (1986).
51. T. Imasaka, A. Yoshitake, K. Hirata, Y. Kawabata, and N. Ishibashi, Pulsed semiconductor laser fluorometry for lifetime measurements, *Anal. Chem. 57*, 947–949 (1985).
52. R. Lopez-Delgado, A. Tramer, and I. H. Munro, A new pulsed light source for lifetime studies and time-resolved spectroscopy: The synchrotron radiation from an electron storage ring, *Chem. Phys. 5*, 72–83 (1974).
53. W. R. Laws and J. C. Sutherland, The time-resolved photon-counting fluorometer at the national synchrotron light source, *Photochem. Photobiol. 44*, 343–348 (1986).
54. R. Rigler, O. Kristensen, J. Roslund, P. Thyberg, K. Oba, and M. Eriksson, Molecular

structures and dynamics: Beamline for time-resolved spectroscopy at the MAX Synchrotron in Lund, *Phys. Scr. T17*, 204–208 (1987).
55. Y. Sakai and S. Hirayama, A fast deconvolution method to analyse fluorescence decays when the excitation pulse repetition period is less than the decay times, *J. Lumin. 39*, 145–151 (1988).
56. R. J. Donovan, G. Gilbert, M. MacDonald, I. Munro, D. Shaw, and G. R. Mant, Determination of absolute quenching rates and fluorescence lifetime for IBr(D) using synchrotron radiation, *Chem. Phys. Lett. 109*, 379–382 (1984).
57. W. R. Ware and R. L. Lyke, Fluorescence lifetimes of saturated hydrocarbons, *Chem. Phys. Lett. 24*, 195–198 (1974).
58. R. L. Lyke and W. R. Ware, Instrument for vacuum ultraviolet lifetime measurements, *Rev. Sci. Instrum. 48*, 320–326 (1976).
59. J. H. Malmberg, Millicrosecond duration light source, *Rev. Sci. Instrum. 28*, 1027–1029 (1957).
60. S. S. Brody, Instrument to measure fluorescence lifetimes in the millimicrosecond region, *Rev. Sci. Instrum. 28*, 1021–1026 (1957).
61. G. F. W. Searle, A. van Hoek, and T. J. Schaafsma, in: *Picosecond Chemistry and Biology* (T. A. M. Doust and M. A. West, eds.), pp. 35–67, Science Reviews Ltd., London (1983).
62. A. van Hoek and A. J. W. G. Visser, Artifact and distortion sources in time-correlated single-photon counting, *Anal. Instrum. 14*, 359–378 (1985).
63. D. J. S. Birch and R. E. Imhof, A single photon counting fluorescence decay-time spectrometer, *J. Phys. E: Sci. Instrum. 10*, 1044–1049 (1977).
64. D. J. S. Birch and R. E. Imhof, Coaxial nanosecond flashlamp, *Rev. Sci. Instrum. 52*, 1026–1212 (1981).
65. C. Lewis, W. R. Ware, L. J. Doemeny, and T. L. Nemzek, The measurement of short lived fluorescence decay using the single photon counting method, *Rev. Sci. Instrum. 44*, 107–114 (1973).
66. H. Hess, On the theory of the spark plasma in nanosecond light sources and fast-gap switches, *J. Phys. D: Appl. Phys. 8*, 685–689 (1975).
67. S. Cova, R. Ripamonti, and A. Lacaita, Avalanche semiconductor detector for single optical photons with a time resolution of 60ps, *Nucl. Instrum. Methods Phys. Res. A253*, 482–487 (1987).
68. D. J. S. Birch, G. Hungerford, B. Nadolski, R. E. Imhof, and A. Dutch, Time-correlated single-photon counting fluorescence decay studies at 930nm using spark source excitation, *J. Phys. E: Sci. Instrum. 21*, 857–862 (1988).
69. R. E. Imhof and D. J. S. Birch, Distortion of Gaussian pulses by a diffraction grating, *Opt. Commun. 42*, 83–86 (1982).
70. W. H. Schiller and R. R. Alfano, Picosecond characteristics of a spectograph measured by a streak camera/video readout system, *Opt. Commun. 35*, 451–454 (1980).
71. A. Dutch, D. J. S. Birch, and R. E. Imhof, Retro-diffracted light in fluorescence spectrometers, *Chem. Phys. Lett. 125*, 57–63 (1986).
72. A. Andreoni, C. A. Sacchi, S. Cova, G. Bottiroli, and G. Prenna, in: *Lasers in Physical Chemistry and Biophysics* (J. Joussot-Dubien, ed.), pp. 413–424, Elsevier, Amsterdam (1975).
73. G. Bottiroli, G. Prenna, A. Andreoni, C. A. Sacchi, and O. Svelto, Fluorescence of complexes and quinacrine mustard with DNA. Influence of the DNA base composition on the decay time in bacteria, *Photochem. Photobiol. 29*, 23–28 (1979).
74. M. A. J. Rodgers and P. A. Firey, Instrumentation for fluorescence microscopy with picosecond time-resolution, *Photochem. Photobiol. 42*, 613–616 (1985).
75. T. Minami, M. Kawahigashi, Y. Sakai, K. Shimamoto, and S. Hirayama, Fluorescence lifetime measurements under a microscope by the time-correlated single-photon counting technique, *J. Lumin. 35*, 247–253 (1986).

76. A. H. Kalanter, Isotropic rotational relaxation of photoselected emitters and systematic errors in emission decay times, *J. Phys. Chem. 72*, 2801–2805 (1968).
77. M. L. Meade, Instrumentation aspects of photon counting applied to photometry, *J. Phys. E: Sci. Instrum. 14*, 909–918 (1981).
78. B. Candy, Photomultiplier characteristics and practice relevant to photon counting, *Rev. Sci. Instrum. 56*, 183–193 (1985).
79. D. Bebelaar, Time response of various types of photomultipliers and its wavelength dependence in time-correlated single-photon counting with an ultimate resolution of 47ps FWHM, *Rev. Sci. Instrum. 57*, 1116–1125 (1986).
80. F. Calligaris, P. Ciuti, I. Gobrielli, R. Giacomich, and R. Mosetti, Wavelength dependence of timing properties of the XP2020 photomultiplier, *Nucl. Instrum. Methods 157*, 611–613 (1978).
81. P. Wahl, J. C. Auchet, and B. Donzel, The wavelength dependence of the response of a pulse fluorometer using the single photoelectron counting method, *Rev. Sci. Instrum. 45*, 28–32 (1974).
82. D. J. S. Birch and R. E. Imhof, Fluorescence lifetimes and relative quantum yields of 9,10-diphenylanthracene in dilute solutions of cyclohexane and benzene, *Chem. Phys. Lett. 32*, 56–58 (1975).
83. D. M. Rayner, A. E. McKinnon, A. G. Szabo, and P. A. Hackett, Confidence in fluorescence lifetime determination: A ratio correction for the photomultiplier time response variation with wavelength, *Can. J. Chem. 54*, 3246–3259 (1976).
84. D. R. James, D. R. M. Demmer, R. E. Verrall, and R. P. Steer, Excitation pulse-shape mimic technique for improving picosecond-laser-excited time-correlated single-photon counting deconvolutions, *Rev. Sci. Instrum. 54*, 1121–1130 (1983).
85. L. J. Libertini and E. W. Small, F/F deconvolution of fluorescence decay data, *Anal. Biochem. 138*, 314–318 (1984).
86. M. Zuker, A. G. Szabo, L. Bramall, D. T. Krajcarski, and B. Selinger, Delta function convolution method (DFCM) for fluorescence decay experiments, *Rev. Sci. Instrum. 56*, 14–22 (1985).
87. R. J. Robbins, G. R. Fleming, G. S. Beddard, G. W. Robinson, P. J. Thistlethwaite, and G. J. Woolfe, Photophysics of aqueous tryptophan: pH and temperature effects, *J. Am. Chem. Soc. 102*, 6271–6279 (1980).
88. S. S. Stevens and J. W. Longworth, Late output pulses from fast photomultipliers, *IEEE Trans. Nucl. Sci. NS-19*, 356–359 (1972).
89. P. B. Coates, The origins of afterpulses in photomultipliers, *J. Phys. D: Appl. Phys. 6* 1159–1166 (1973).
90. S. Torre, T. Antonioli, and P. Benetti, Study of afterpulse effects in photomultipliers, *Rev. Sci. Instrum. 54*, 1777–1780 (1983).
91. D. J. S. Birch, R. E. Imhof, and A. Dutch, Pulse fluorometry using simultaneous acquisition of fluorescence and excitation, *Rev. Sci. Instrum. 55*, 1255–1264 (1984).
92. W. R. Ware, M. Pratinidhi, and R. K. Bauer, Performance characteristics of a small side-window photomultiplier in laser single-photon fluorescence decay measurements, *Rev. Sci. Instrum. 54*, 1148–1156 (1983).
93. S. Canonica, J. Forrer, and U. P. Wild, Improved timing resolution using small side-on photomultipliers in single photon counting, *Rev. Sci. Instrum. 56*, 1754–1758 (1985).
94. I. E. Meister, U. P. Wild, P. Klein-Bolting, and A. Holzwarth, Time response of small side-on photomultiplier tubes in time-correlated single photon counting, *Rev. Sci. Instrum. 59*, 499–501 (1988).
95. G. Pietri, Contribution of the channel electron multiplier to the race of vacuum tubes towards picosecond resolution time, *IEEE Trans. Nucl. Sci. NS-24*, 228–232 (1977).
96. I. Yamazaki, N. Tamai, H. Kume, H. Tsuchiya, and K. Oba, Microchannel-plate

photomultiplier applicability to the time-correlated photon-counting method, *Rev. Sci. Instrum.* *56*, 1187–1194 (1985).

97. W. Fichtner and W. Hacker, Time resolution of Ge avalanche photodiodes operating as photon counters in delayed coincidence, *Rev. Sci. Instrum.* *47*, 374–377 (1976).
98. S. Cova, A. Longoni, A. Andreoni, and R. Cubeddu, A semiconductor detector for measuring ultraweak fluorescence decays with 70ps FWHM resolution, *IEEE J. Quantum Electron.* *QE-19*, 630–634 (1983).
99. M. D. Petroff, M. G. Stapelbroek, and W. A. Kleinhaus, Detection of individual 0.4–28 μm wavelength photons via impurity-impact ionization in a solid state photomultiplier, *Appl. Phys. Lett.* *51*, 406–408 (1987).
100. A. Hallam and R. E. Imhof, Performance tests of a time-to-amplitude converter at high conversion rates, *J. Phys. E: Sci. Instrum.* *13*, 520–521 (1980).
101. P. R. Hartig, K. Sauer, C. C. Lo, and B. Leskovar, Measurement of very short fluorescence lifetimes by single-photon counting, *Rev. Sci. Instrum.* *47*, 1122–1129 (1976).
102. E. W. Small, L. J. Libertini, and I. Isenberg, Construction of a monophoton decay fluorometer with high resolution capabilities, *Rev. Sci. Instrum.* *55*, 879–885 (1984).
103. A. J. W. G. Visser, T. Kulinski, and A. van Hoek, Fluorescence lifetime measurements of pseudoazuleues using picosecond resolved single photon counting, *J. Mol. Struct.* *175*, 111–116 (1988).
104. H. Kume, K. Koyama, K. Nakatsugawa, S. Suzuki, and D. Fatlowitz, Ultrafast microchannel plate photomultipliers, *Appl. Opt.* *27*, 1170–1178 (1988).
105. R. A. Lampert, L. A. Chewter, D. Phillips, D. V. O'Connor, A. J. Roberts, and S. R. Meech, Standards for nanosecond fluorescence decay time measurements, *Anal. Chem.* *55*, 68–73 (1983).
106. R. A. Velapoldi, in: *Advances in Standards and Methodology in Spectrophotometry* (C. Burgess and K. D. Mielenz, eds.), pp. 175–193, Elsevier, Amsterdam (1987).
107. D. J. S. Birch, A. D. Dutch, R. E. Imhof, and B. Nadolski, The effect of transient quenching on the excimer kinetics of 2,5-diphenyloxazole, *J. Photochem.* *38*, 239–254 (1987).
108. A. R. Holzwarth, Fluorescence lifetimes in photosynthetic systems, *Photochem. Photobiol.* *43*, 707–725 (1986).
109. D. Phillips, Time-resolved fluorescence of excimer-forming polymers in solution, *Br. Polym. J.* *19*, 135–149 (1987).
110. D. Phillips, in: *Photophysics of Polymers* (C. E. Hoyle and J. M. Torkelson, eds.), pp. 308–322, American Chemical Society, Washington, D.C. (1987).
111. D. J. S. Birch, A. S. Holmes, R. E. Imhof, and J. Cooper, PPO excimers in lipid bilayers studied using single-photon timing array detection, *Chem. Phys. Lett.* *148*, 435–444 (1988).
112. D. R. James and W. R. Ware, A fallacy in the interpretation of fluorescence decay parameters, *Chem. Phys. Lett.* *120*, 455–459 (1985).
113. A. R. Holzwarth, J. Wendler, K. Schaffner, V. Sundstrom, and T. Gillbro, in: *Picosecond Chemistry and Biology* (T. A. M. Doust and M. W. West, eds.), pp. 82–107, Science Reviews, Ltd., London (1983).
114. A. Itaya, H. Sakai, and H. Masuhara, Excimer dynamics of poly(*N*-vinylcarbazole) films revealed by time-correlated single photon counting measurements, *Chem. Phys. Lett.* *138*, 231–236 (1987).
115. P. A. Anfinud, D. E. Hart, J. F. Hedstrom, and W. S. Straine, Fluorescence depolarization of rhodamine 6G in glycerol: A photon-counting test of three-dimensional excitation transport theory, *J. Phys. Chem.* *90*, 2374–2379 (1986).
116. N. Tamai, T. Yamazaki, and I. Yamazaki, Two-dimensional excitation energy transfer between chromophoric carbazole and anthracene in Langmuir–Blodgett monolayer films, *J. Phys. Chem.* *91*, 841–845 (1987).

117. N. Tamai, T. Yamazaki, I. Yamazaki, A. Mzuma, and N. Matoga, Excitation energy transfer between dye molecules absorbed on a vesicle surface, *J. Phys. Chem. 91*, 3503–3508 (1987).
118. T. L. Nemzek and W. R. Ware, Kinetics of diffusion controlled reactions: Transient effects in fluorescence quenching, *J. Phys. Chem. 15*, 477–489 (1975).
119. M. H. Hui and W. R. Ware, Exciplex photophysics IV. Effect of diffusion-controlled quenching on exciplex photokinetics, *J. Am. Chem. Soc. 98*, 4712–4717 (1976).
120. R. W. Wijnaendts van Resandt, Picosecond transient effect in the fluorescence quenching of tryptophan, *Chem. Phys. Lett. 95*, 205–208 (1983).
121. D. Daems, M. Van den Zegel, N. Boens, and F. C. De Schryver, Fluorescence decay of pyrene in small and large unilamellar L,α-dipalmitoylphosphatidylcholine vesicles above and below the phase transition temperature, *Eur. Biophys. J. 12*, 97–105 (1985).
122. G. Duportail and P. Lianas, Fractal modelling of pyrene excimer quenching in phospholipid vesicles, *Chem. Phys. Lett. 149*, 73–78 (1988).
123. M. F. Blackwell, K. Gouranis, and J. Barber, Evidence that pyrene excimer formation in membranes is not diffusion-controlled, *Biochim. Biophys. Acta 858*, 221–234 (1986).
124. M. Almgren and J. E. Lofroth, Determination of micelle aggregation numbers and micelle fluidities from time-resolved fluorescence quenching studies, *J. Colloid Interface Sci. 84*, 486–499 (1981).
125. N. J. Bridge and P. D. I. Fletcher, Time-resolved studies of fluorescence quenching in water-in-oil microemulsion, *J. Chem. Soc., Faraday Trans. 1 79*, 2161–2169 (1983).
126. D. M. Jameson, G. Weber, R. D. Spencer, and G. Mitchell, Fluorescence polarization: Measurements with a photon-counting photometer, *Rev. Sci. Instrum. 49*, 510–514 (1978).
127. M. D. Barkley, A. Kowalczyk, and L. Brand, Fluorescence decay studies of anisotropic rotations of small molecules, *J. Chem. Phys. 75*, 3581–3593 (1981).
128. R. L. Christensen, R. C. Drake, and D. Phillips, Time-resolved fluorescence anisotropy of perylene, *J. Phys. Chem. 90*, 5960–5967 (1986).
129. L. X.-Q. Chen, R. A. Engh, and G. R. Fleming, Reorientation of tryptophan and simple peptides: Onset of internal flexibility, in: *Time-Resolved Laser Spectroscopy in Biochemistry* (J. R. Lakowicz, ed.), *Proc. SPIE 909*, 223–230 (1988).
130. K. Kinosita, S. Kawato, and A. Ikegami, A theory of fluorescence polarisation decay in membranes, *Biophys. J. 20*, 289–305 (1977).
131. K. Kinosita, A. Ikegami, and S. Kawato, On the wobbling-in-cone analysis of fluorescence anisotropy decay, *Biophys. J. 37*, 461–464 (1982).
132. M. P. Heyn, Determination of lipid order parameters and rotational correlation times from fluorescence depolarization experiments, *FEBS Lett. 108*, 359–364 (1979).
133. D. J. S. Birch, A. S. Holmes, J. R. Gilchrist, R. E. Imhof, S. M. Al-Alawi, and B. Nadolski, A multiplexed single-photon instrument for routine measurement of time-resolved fluorescence anisotropy, *J. Phys. E.: Sci. Instrum. 20*, 471–473 (1987).
134. D. J. S. Birch, A. S. Holmes, R. E. Imhof, B. Z. Nadolski, and J. C. Cooper, Multiplexed time-correlated single photon counting, in: *Time-Resolved Laser Spectroscopy in Biochemistry* (J. R. Lakowicz, ed.), *Proc. SPIE 909*, 8–14 (1988).
135. K. Hildenbrand and C. Nicolau, Nanosecond fluorescence anisotropy decays of 1,6-diphenyl-1,3,5-hexatriene in membranes, *Biochim. Biophys. Acta 553*, 365–377 (1979).
136. Lin X. Q. Chen, J. W. Petrich, G. R. Fleming, and A. Perico, Picosecond fluorescence studies of polypeptide dynamics: Fluorescence anisotropies and lifetimes, *Chem. Phys. Lett. 139*, 55–61 (1987).
137. W. R. Ware, P. Chow, and S. K. Lee, Time-resolved nanosecond emission spectroscopy: Spectral shifts due to solvent–solute relaxation, *Chem. Phys. Lett. 2*, 356–358 (1968).
138. S. R. Meech, D. V. O'Connor, A. J. Roberts, and D. Phillips, On the construction of nanosecond time-resolved emission spectra, *Photochem. Photobiol. 33*, 159–172 (1981).
139. W. R. Ware, S. K. Lee, G. J. Brant, and P. P. Chow, Nanosecond time-resolved emission

spectroscopy: Spectral shifts due to solvent-excited solute relaxation, *J. Chem. Phys. 54*, 4729–4737 (1971).
140. S. Canonica and U. P. Wild, Single photon counting with synchronously pumped dye laser excitation, *Anal. Instrum. 14*, 331–357 (1985).
141. G. Hazan, A. Grinvald, M. Maytal, and I. Z. Steinberg, An improvement of nanosecond fluorimeters to overcome drift problems, *Rev. Sci. Instrum. 45*, 1602–1604 (1974).
142. R. W. Wijnaendts van Resandt, R. H. Vogel, and S. W. Provencher, Double beam fluorescence lifetime spectrometer with subnanosecond resolution: Application to aqueous tryptophan, *Rev. Sci. Instrum. 53*, 1392–1397 (1982).
143. K. Hara, Measuring apparatus of light emission of life of sample, Japanese Patent Appl. No. 57-183244 (1984).
144. D. J. S. Birch, R. E. Imhof, and A. Dutch, Differential pulse fluorometry using matched photomultipliers—a new method of measuring fluorescence lifetimes, *J. Phys. E: Sci. Instrum. 17*, 417–418 (1984).
145. D. J. S. Birch, R. E. Imhof, and C. Guo, Fluorescence decay studies using multiplexed time-correlated single-photon counting: Application to aminotetraphenylporphyrins, *J. Photochem. Photobiol. A: Chem. 42*, 223–231 (1988).
146. R. Schuyler and I. Isenberg, A monophoton fluorometer with energy discrimination, *Rev. Sci. Instrum. 42*, 813–817 (1971).
147. S. C. Harvey and H. C. Cheung, Fluorescence depolarization studies on the flexibility of myosin rod, *Biochemistry 16*, 5181–5187 (1977).
148. R. W. Wijnaendts van Resandt and L. De Maeyer, Picosecond rotational diffusion by differential single-photon fluorescence spectroscopy, *Chem. Phys. Lett. 78*, 219–229 (1981).
149. D. J. S. Birch, A. S. Holmes, R. E. Imhof, B. Z. Nadolski, and K. Suhling, Multiplexed array fluorometry, *J. Phys. E: Sci. Instrum. 21*, 415–417 (1988).
150. D. J. S. Birch, K. Suhling, A. S. Holmes, A. D. Dutch, and R. E. Imhof, Array fluorometry: the theory of the statistical multiplexing of single photon timing, in: *Time-Resolved Laser Spectroscopy in Biochemistry II* (J. R. Lakowicz, ed.), *Proc. SPIE 1204*, 26–34 (1990).

2

Laser Sources and Microchannel Plate Detectors for Pulse Fluorometry

Enoch W. Small

2.1. Introduction

In the past several years, technical advances in light sources and detectors have revolutionized pulse fluorometric instrumentation. Commercially available synchronously pumped dye laser systems can now provide high-repetition-rate light flashes with temporal half-widths of a few picoseconds and thus are nearly an ideal light source for fluorescence decay measurements. These sources, when used in conjunction with a frequency doubler, are tunable over wide spectral ranges from the UV to the near infrared. Also, microchannel plate photomultipliers now provide extremely fast response to the fluorescence and can even be used for monophoton discrimination to eliminate the distorting effects of photon pileup. In this chapter, I attempt to shed some light on the functioning of these light sources and detectors. These descriptions are largely intuitive rather than mathematical. Any design details given apply specifically to Spectra-Physics laser systems and Hamamatsu photomultiplier tubes. I have restricted myself to descriptions of these particular components, because they are the ones with which my laboratory has experience. Other components are commercially available and will be mentioned.

The main section of this chapter concerns lasers. Beginning in Section 2.2.1, I introduce the basic physical principles behind the laser source and attempt to describe in simple terms what a laser is and how it works. This section begins with a brief description of Einstein's coefficients and the need for a population inversion. It describes optical resonance, the use of Brewster's

---

Enoch W. Small • Department of Chemistry and Biochemistry, Eastern Washington University, Cheney, Washington 99004.
Topics in Fluorescence Spectroscopy, Volume 1: Techniques, edited by Joseph R. Lakowicz. Plenum Press, New York, 1991.

angle to polarize the light of the laser, and how different physical phenomena in the lasing medium give rise to different properties of the emitted light. Two somewhat extraneous processes, frequency doubling by means of nonlinear optical effects and the interactions between light and sound waves, are also discussed.

In Section 2.2.2, particular laser features are illustrated through descriptions of three lasers: an argon ion gas laser, a solid-state Nd:YAG laser, and a liquid dye laser. These three lasers have been chosen because they are common components of synchronously pumped dye laser systems. Typically, the green output of an argon ion laser or a frequency-doubled Nd:YAG laser is used to pump a dye laser, in the configuration shown in the block diagram of Figure 2.1. While the discussion is directed toward an understanding of the synchronously pumped systems, each of these three laser types has a bright future in other applications of biophysics as well.

The last part of the section on lasers, Section 2.2.3, focuses on laser pulse generation and how it is accomplished. The lasers described in this section all produce light pulses through a process called mode locking. This is done in the pumping laser using an acousto-optic device, and in the dye laser by the process of synchronous pumping. Light pulses are extracted for use from the dye laser using an acousto-optic cavity dumper. This device will be described as well. In general, Section 2.2 is meant to be useful to a biophysicist who does not want to study lasers but wishes to effectively use them as research tools.

In Section 2.3, I describe microchannel plate (MCP) photomultipliers and how they work. The section emphasizes the fast events that occur in the photocathode and the electron multiplication that occurs within the channels of the plate. A method used for the construction of microchannel plates is described. There is also a discussion of several properties of these tubes which are important for decay fluorometers.

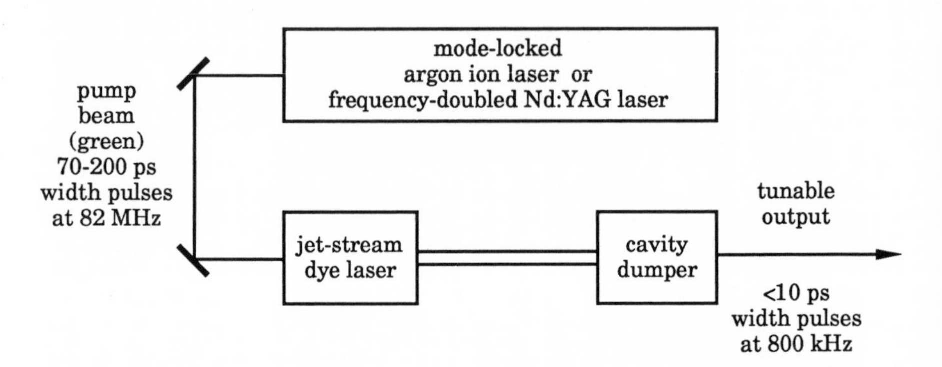

Figure 2.1. Block diagram of the synchronously pumped laser system.

Finally, in Section 2.4, the laser source and MCP photomultiplier are combined into a decay fluorometer. When this is done, several artifacts, which would be thoroughly hidden on an instrument with less resolving power, emerge. The user must then "tune" the decay instrument to eliminate these artifacts. I will describe some of the changes that we have made in our instrument to improve the quality of the data and will give some examples of the performance of the tuned instrument.

This chapter is ment to be a didactic introduction to the use of lasers and microchannel plate photomultipliers for pulse fluorometry. There is much more information available elsewhere. For those who wish to pursue laser theory further, there are a number of good introductory books. My laboratory has found the texts by O'Shea *et al.*[1] and Beesley[2] to be particularly illuminating. Those who wish a basic intuitive approach to understanding lasers may enjoy the book by Hitz[3] or, if more detailed information is needed, the much more rigorous treatment by Siegman.[4] There are also several recent chapters[5–8] and two books[9, 10] which discuss the use of pulsed lasers for fluorometric measurements. For general information on microchannel plate photomultipliers, I recommend two articles by Leskovar[11, 12] and the technical literature sent out by Hamamatsu Corporation (Middlesex, New Jersey) and Galileo Electro-Optics Corporation (Sturbridge, Massachusetts). Finally, I might add, one of the best texts my laboratory has found for general information on optics is not a text at all, but the sales catalog of Melles Griot (Irvine, California). It contains many well-polished explanations of optical phenomena.

2.2. Laser Sources

In the relatively short time since lasers were invented, an entire engineering field has built up around them. People who work on lasers seem to have their own language. We are biophysicists trying to solve biological problems, and we often find that lasers provide a nearly ideal light source for a particular experiment. When we use lasers, we would like to know how they work, but the laser jargon is obscure. We buy the lasers from laser companies, which are more than happy to provide information explaining their products, but such information is generally incomplete. The laser literature itself is vast and intimidating, and since much of laser development has been done commercially, it is often difficult to find the kinds of details that we need (such as dimensions and other specifications) to answer our questions. If we work with lasers for some time, over the years we talk to numerous technicians and others who are knowledgeable about them, and we read what we can. My qualification for writing this section is not that I work *on* lasers, but that I have been working *with* lasers for a number of years, as excitation sources first for Raman spectroscopy and then for fluorescence spectroscopy. Over this time, as a biophysicist, I have acquired my own understanding of how lasers work and hope that I might be in a good position to explain them to someone with a background similar to mine. The explanations given will not be rigorous, but intuitive. I will attempt to clarify the "buzz" words of the laser field.

2.2.1. Laser Fundamentals

> What is light? The bald fact is that no one knows what light really consists of. Textbooks on light give no answer—instead they hedge. The more advanced the book, the more carefully the author avoids saying what light is.
>
> William A. Shurcliff and Stanley S. Ballard in *Polarized Light*[13]

Many of us encounter this little paperback book early in our training. We then move on to quantum-mechanical and electromagnetic treatments of light, often to find ourselves back where we started. Understanding lasers requires understanding light, but light defies simple intuitive descriptions. We begin here by treating light as particles with energy $h\nu$ (Planck's constant times the frequency in Hz or s^{-1}). At our convenience, we will oscillate back and forth, using either particles or waves to describe the light of the laser.

As originally formulated in 1958 by Schawlow and Townes,[14] the laser depends on two processes involving light and its interactions with matter: stimulated emission and optical resonance. At the time the laser was announced, both phenomena were well known and had actually been combined before to make the maser, a device which operates with microwave radiation. The first of these processes, *stimulated emission*, was postulated as far back as 1917 by Einstein in his classic paper on the quantum theory of radiation.[15,16] A quantum of light interacts with a molecule in an excited state, inducing it to emit another quantum of the same energy and direction. When this phenomenon was postulated, it had not yet been observed, but Einstein argued that it is required by momentum conservation. To describe the second phenomenon, *optical resonance*, we switch to the wave properties of light. Optical resonance was also known at the time the laser was developed, because it is simply the principle behind the Fabry–Perot interferometer,[17] a device in which light oscillates between two reflective surfaces at frequencies which derive from having an integral number of half-wavelengths between them. It is the combination of stimulated emission and optical resonance that constitutes the laser.

To make a laser, one places an active medium (capable of stimulated emission) in an optical cavity between two mirrors and introduces energy by a process called pumping. Lasing is initiated by a small amount of spontaneous emission along the axis of the cavity joining the two mirrors. As it travels along, the light is amplified by the process of stimulated emission. Light which travels off the axis between the mirrors is lost, but that which strikes the mirror and reflects back on itself becomes reinforced. Regenerative feedback (Section 2.2.1.4) between the mirrors favors highly directional resonant light within the cavity. Optical resonance occurs in the laser cavity between the mirrors. By using a partially silvered mirror at one end, a fraction of the light is permitted to escape. The resulting laser beam has a high degree of spatial coherence (extended regions of the light show oscillations in step with one another), high monochromaticity, and low divergence (a property which defines the directionality of the light). In later sections we will see why these properties arise. It is possible to make either pulsed lasers, which produce repetitive bursts of light, or continuous-wave (cw) lasers.

Subsequent development has thoroughly tested the limits of lasing action. Active media can be gases, liquids, or solids. Some pulsed lasers, such as diode lasers and nitrogen lasers, have active media with such high initial gains that true optical resonance is not really needed. A diode laser achieves sufficient optical feedback simply off the surfaces of the lasing crystal, whereas nitrogen lasers achieve the directionality of their light with a single mirror combined with the small reflectivity of a simple exit window. Both of these lasers have seen limited use for pulse fluorometry, but the three lasers which will primarily concern us here, the gaseous argon ion laser, the liquid dye laser, and the solid-state Nd:YAG laser, all use standard resonant cavities bounded by mirrors.

The properties of a particular laser are determined largely by the lasing medium. A means must be found to create a population inversion, where most of the active component is in an excited state. For the visible and near-IR lasers discussed here, the excited state requires the promotion of an electron into a higher electronic level.

2.2.1.1. The Lasing Medium and the Need for a Population Inversion

Einstein's hypotheses[15] regarding the interactions of radiation with matter were elegantly straightforward. Consider an isolated molecule with two possible states possessing energies ε_1 and ε_2 and energy difference ($\varepsilon_2 - \varepsilon_1 = h\nu$) equal to that of a quantum of radiation (see Figure 2.2). Einstein postulated three forms of energy exchange: spontaneous emission, absorption, and stimulated emission.

In spontaneous emission, a molecule in the excited state can emit a quantum of radiation and undergo a transition down from energy level 2 to level 1. If the radiation is at or near visible wavelengths, spontaneous emission is commonly called luminescence. Further, if both states have the same multiplicity (in terms of electronic spin; usually both singlet states), then it is called fluorescence. Einstein argued, by analogy to radioactive decay, that

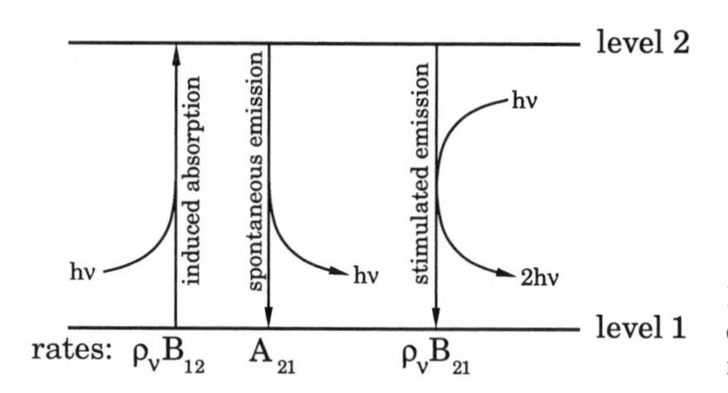

Figure 2.2. The three forms of energy exchange between matter and radiation postulated by Einstein.

spontaneous emission should be a first-order process. The rate constant for spontaneous emission is denoted A_{21}, which those in the field of fluorescence will recognize as the reciprocal of τ_0, the intrinsic lifetime. Although necessary for initiating lasing, spontaneous emission does not have the directional properties of stimulated emission and therefore represents a *loss* in the laser cavity.

In absorption, a photon transfers its energy to the molecule and induces a transition from level 1 to level 2. In stimulated emission, on the other hand, a photon interacts with a molecule already in level 2 and induces the emission of another photon with a transition down to level 1. Both absorption and stimulated emission require the initial presence of radiation and have rates proportional to the radiation density, $\rho_\nu(\mathrm{J \cdot s/m^3})$, at frequency ν. The rates of absorption and stimulated emission (also assumed by Einstein to be first-order) can be expressed as $\rho_\nu B_{12}$ and $\rho_\nu B_{21}$. A_{21}, B_{12}, and B_{21} are now called the Einstein coefficients for spontaneous emission, absorption, and stimulated emission, respectively.

In his paper, Einstein presented a new and remarkably simple derivation of Planck's radiation law and also showed the relationships between the three rate constants. At thermal equilibrium the populations, N_1 and N_2, of levels 1 and 2 can be related by Boltzmann statistics:

$$\frac{N_2}{N_1} = \frac{g_2 \exp(-\varepsilon_2/kT)}{g_1 \exp(-\varepsilon_1/kT)} \tag{2.1}$$

where g_1 and g_2 are the degeneracies of the states, k is Boltzmann's constant, and T is the absolute temperature. Also, at equilibrium the number of transitions from level 1 to 2 must equal those from 2 to 1:

$$g_1 \exp(-\varepsilon_1/kT)\rho_\nu B_{12} = g_2 \exp(-\varepsilon_2/kT)(\rho_\nu B_{21} + A_{21}) \tag{2.2}$$

Einstein further assumed (as one would expect for a blackbody radiator) that ρ_ν will increase to infinity with T. This leads to:

$$g_1 B_{12} = g_2 B_{21} \tag{2.3}$$

Of course, this is little more than a mathematical trick as far as we are concerned, since lasers do not operate at temperatures adequate to significantly populate level 2 by the Boltzmann relationship, but it does show the simple relationship between the Einstein B coefficients.

Equation (2.3) has important implications for the laser. Imagine the light beam traveling down through the active medium of the laser. As the beam travels through, we want it to increase in intensity by stimulated emission. Equation (2.3) says that even if we could obtain extreme conditions (e.g., temperatures approaching infinity) which result in equal populations of states 1 and 2, and choose a system which does not undergo spontaneous emission, the very best we can hope for is an even competition between absorption and stimulated emission. In order to have the light amplified in the laser, it is necessary to create an artificial condition in which level 2 has a higher population than level 1. This situation is called a *population inversion* and is defined [taking into consideration the degeneracies of the states—Eq. (2.1)] as the condition where

$$\frac{N_2}{g_2} > \frac{N_1}{g_1} \tag{2.4}$$

A population inversion is generated by a process called *pumping*, which will be described in Section 2.2.1.2. Obviously, a laser will require more than a simple population inversion, due to the competing processes of spontaneous emission, nonradiative transitions, and optical losses within the cavity (including, of course, the output beam).

Einstein also related the A_{21} and B_{21} coefficients:

$$A_{21} = (8\pi h\nu^3/c^3) B_{21} \tag{2.5}$$

The ratio of A_{21} to B_{21} is proportional to ν^3, and the shorter the wavelength, the more effectively spontaneous emission will compete with stimulated emission for the energy of level 2. This is the reason for the shortage of laser sources in the UV and the extreme difficulty of achieving laser action at X-ray wavelengths.

2.2.1.2. Three- and Four-Level Lasers

Molecules, of course, have more than two electronic energy levels, and it is by making use of levels besides the two responsible for lasing that the population inversion is generated. The first working laser, a ruby laser reported by Maiman in 1960,[18] makes use of a three-level scheme such as that illustrated on the left side of Figure 2.3. A flashlamp is used to optically excite Cr^{3+} ions in a ruby crystal from the ground state into a series of energy

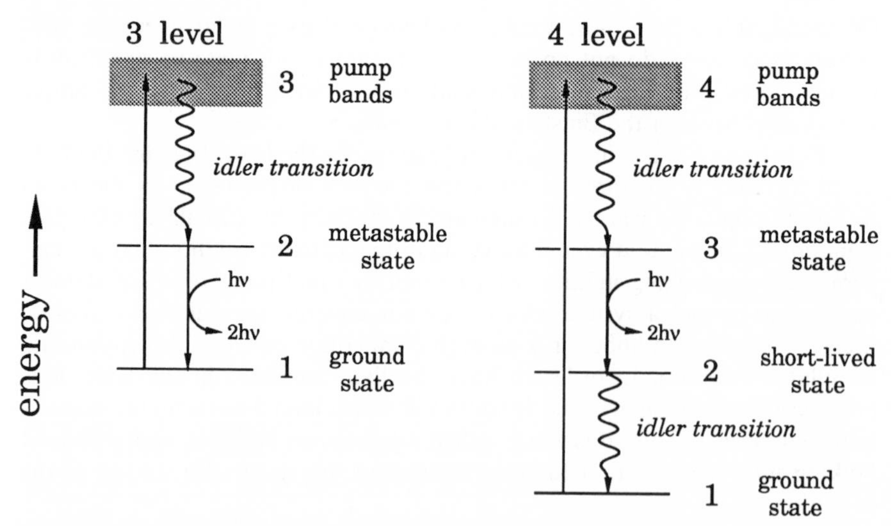

Figure 2.3. A schematic comparison between three- and four-level lasers. A wavy line represents a radiationless transition.

levels called the pump bands. Rapid radiationless transitions (lifetime of about 50 ns) result in the population of a metastable state with a much longer lifetime (about 5 ms). This initial nonradiative drop is called an idler transition. If the flashlamp is sufficiently bright, pumping can create a population inversion, and lasing can occur by stimulated emission back to the ground state.

Obviously, the degree of population inversion, and consequently the efficiency of a laser, can be enhanced either by increasing the population of the upper metastable state or by decreasing the population of the lower state reached after stimulated emission. The population of the upper level might be increased by the expenditure of additional pumping energy. The difficulty with a three-level laser is that the only means of depopulating the lower level is by the same relatively inefficient pumping process. Some molecules can be pumped up to the higher level, but a large number will remain. Not only does a high lower level population diminish the population inversion, but it also leads to losses due to simple absorption of the laser beam, because the lower level can absorb photons at the energy of the laser beam. (Other levels of the atom probably cannot.) A much greater population inversion could be obtained if another more effective means could be found to decrease the lower level population.

Most lasers now depend on transitions to short-lived energy levels which are still above the ground state, in a scheme such as that shown on the right side of Figure 2.3. Such lasers are called four-level lasers. Molecules in the lower level of the lasing transition quickly decay off to the ground state by another idler transition. This greatly increases the degree of population inversion without the further expenditure of pumping energy.

The use of the three-level ruby laser has declined with the development of more efficient solid-state lasers which depend on a four-level scheme. As we will see below, both the Nd:YAG and the dye laser are four-level lasers, although, strictly speaking, more than four distinct levels are involved in their function. The argon ion laser is not a four-level laser, but a special case, since, instead of simple pumping, it depends on a cycle of ionization and neutralization of argon atoms.

2.2.1.3. Light Amplification by Stimulated Emission

If one shines light into an absorbing medium, the intensity of the light is attenuated exponentially with the path length, *l*, according to Lambert's contribution to the familiar Beer–Lambert law:

$$I_l = I_0 \exp[-\alpha(\nu) l] \tag{2.6}$$

where I_0 is the initial intensity of the light, I_l is the intensity after trasversing a distance of l through the medium, and $\alpha(\nu)$ is the absorption coefficient, a function of the frequency of the light. As mentioned above, stimulated emission occurs in the same direction as the incident light. If a sufficient population inversion exists, then light of the proper frequency passing through the material will not be attenuated but will actually increase in intensity. The light is amplified, not absorbed. Equation (2.6) still applies, but the actual absorption coefficient becomes negative. For low intensities, well before the laser medium becomes saturated, a quantity called the *small-signal gain coefficient*, $\beta(\nu)$, is simply the resulting negative absorption coefficient:

$$\beta(\nu) = -\alpha(\nu) \tag{2.7}$$

$\beta(\nu)$ is proportional to the degree of population inversion, $(g_1/g_2)N_2 - N_1$.

The atomic and molecular transitions that are used for lasing are not infinitely narrow in wavelength but are broadened by various physical processes. As we will see later, some of the properties of a laser will depend on which particular physical processes dominate the broadening. The frequency breadth of the transition is called the *lineshape function*. The small-signal gain coefficient has the same dependence on frequency as the lineshape function. It is within the frequency range of the lineshape function that an adequate small-signal gain coefficient permits lasing to occur.

Stimulated emission has another important feature. It occurs, not only in the same direction as the incident light, but in phase with it as well. It is this property which is responsible for the coherent nature of the light of the laser beam.

2.2.1.4. Regenerative Feedback

If one places the lasing medium between two mirrors and optically pumps it to produce a population inversion, the mirrors can introduce regenerative feedback. This is much like the regenerative feedback that is often heard through public address systems. A microphone will pick up amplified sound, amplify it, and send it out through the speakers just to be picked up again. The level of the sound quickly builds up to produce a signal which saturates the amplifier. Due to the delay introduced between the pickup of the sound and the next amplified version, a periodic signal is generated which rapidly builds up to a disconcerting whine.

In the laser, a small amount of spontaneous emission stimulates further emission. If this light is traveling in the appropriate direction, it can reflect off a mirror and pass back through the laser medium. If the gain (the ability of the lasing medium to amplify the light) is sufficient to overcome energy losses from the cavity, electromagnetic oscillations can be set up between the mirrors and will build up in intensity until they saturate the gain.

To examine the conditions under which this will occur, we assume for convenience that the active medium of the laser fills the region between the mirrors and that pumping excitation is uniform. We will also assume that only one frequency of light is involved. In traveling the distance L from one mirror to the other, the beam intensity increases from its initial value, I_0, to a value I_L given by

$$I_L = I_0 \exp[(\beta_\nu - \alpha_{\text{loss}})L] \tag{2.8}$$

β_ν is the small-signal gain coefficient at frequency ν, and α_{loss} is the loss per unit distance resulting from scattering and absorption due to nonactive components of the medium. After striking the first mirror, the beam intensity is reduced by the reflectivity of the mirror, R_1. The returning beam passes back through the active medium to be reflected off the second mirror with a reflectivity of R_2. Typically, one of the mirrors is highly reflective (>99%) and the other has a lower reflectivity (perhaps 90%) in order to permit a fraction of the beam to exit the laser cavity. In laser jargon, the partially transmitting mirror is called the *output coupler*.

After the round trip between the mirrors, the ratio of the intensity of the beam at the end to that at the start is given by

$$G_\nu = R_1 R_2 \exp[2(\beta_\nu - \alpha_{\text{loss}})L] \tag{2.9}$$

G_ν is called the round trip gain. As long as $G_\nu > 1$ when the laser is turned on, light will begin to oscillate in the cavity and its intensity will quickly increase. As the population inversion is depleted, the small-signal gain begins to decrease, bringing down the value of G_ν. In cw lasers, the light intensity reaches a steady-state value in which the round trip gain of the laser medium exactly compensates for the sum of the losses in the laser. Under these conditions, the round trip gain becomes pinned to a value of 1. The value of the small-signal gain for which $G_v = 1$ is called the threshold gain, β_{th}.

In pulsed lasers, on the other hand, the round trip gain does not stabilize at 1. Instead, the intensity of the light first increases to a very high value, rapidly depleting the population inversion. Instead of reaching a steady-state value, lasing then stops until the population inversion can be regenerated. It is this process which generates the high-intensity fast pulses which are so useful as an excitation source for fluorescence decay measurements.

The combination of optical components and active medium within which the laser light oscillates is called the *optical resonator*. Resonance occurs between the mirrors, in the *laser cavity*.

2.2.1.5. The Optical Resonator

The reader should recall that a continuous traveling light wave, such as the output beam of a laser, has the form

$$E_t(t) = E_0 \sin(\omega t - ks) \tag{2.10}$$

where E_0 is the maximum amplitude, $\omega = 2\pi v$ is the angular frequency (radians/s), t is the time, s is a distance coordinate, and $k = 2\pi/\lambda$ is the propagation constant. The wavelength, λ, is given by $\lambda = c/v$, where c is the speed of light. For a traveling light wave, the sinusoidal oscillations of the electric vector move in the direction of increasing s at velocity c.

In the optical resonator of a laser, the light travels back and forth, reinforcing itself to the form of a standing wave with nodes at the surface of each mirror. Such a wave has the form

$$E_s(t) = E'_0 \cos \omega t \sin ks \tag{2.11}$$

Now, the points of maximum fluctuation of the electric vector of the light, as well as the nodes between them, do not move.

Another device, besides the laser, which uses standing light waves between two mirrors is the Fabry–Perot interferometer. If one directs a light beam toward a single partially reflective mirror, part of the light is reflected and part passes through. If one then takes a second partially reflective mirror and positions it in the beam behind the first and parallel to it, then light at very specific frequencies does not reflect off the first mirror at all, but passes through both mirrors unimpeded. Standing light waves are set up between the partially reflective surfaces. The frequencies which pass through the two surfaces are those for which there is an integral number of half-wavelengths between them.

The laser is a similar device, except that the second mirror at the back of the laser is almost totally reflective and the light originates from the population inversion inside. Imagine a laser operating with an output coupler of 10% transmittance, with an intracavity beam of 10 W. Since 10% of the light would escape, the laser would produce an output beam of 1 W. If one were to reverse this process by removing the active medium from the laser while maintaining the optimum spacing of the mirrors for resonance, and instead directed an equivalent 1-W beam of light toward the resonator from the outside, one would observe a peculiar phenomenon. Although the output coupler would have an apparent reflectance by itself of 90%, because of its relationship with the mirror at the back of the resonator, *no* light would be reflected back. Instead, the light would pass into the cavity where it would reflect back and forth between the mirrors until it reinforced itself up to an intracavity power of 10 W. It sounds almost like a free lunch. Perhaps it would help in understanding this phenomenon to realize that if light had been reflected from the output coupler of the laser (or from the Fabry–Perot interferometer), then it would be perfectly out of phase with transmitted light which would leak out of the cavity through the coupler.[†] The appearance of additional light would violate energy conservation.

Since the resonator supports waves which have an integral number of half-wavelengths between the mirrors, in a cavity of length L, resonance occurs only at the discrete wavelengths, λ_n, given by

$$\lambda_n = 2L/n \qquad n = \text{an integer} \tag{2.12}$$

This is called the *longitudinal resonance condition.* The frequencies which oscillate in the laser are called either the *longitudinal* modes or the *axial* modes. Of course, only those frequencies for which there is adequate gain in the lasing medium and for which Eq. (2.12) is satisfied will be active in the laser. They must lie within the lineshape function.

Since ν_n, the temporal frequency of the light (in s^{-1} or Hz), is equal to c/λ_n,

$$\nu_n = nc/2L \tag{2.13}$$

and the frequency difference between two axial modes is

$$\Delta\nu = \nu_{n+1} - \nu_n = c/2L \tag{2.14}$$

[†] This may not be immediately obvious to the reader. It might help to remember that the standing wave in the laser cavity can be considered to be a sum of two waves of equal amplitude traveling in opposite directions. One of these traveling waves is in phase with the light entering the cavity through the output coupler, and the other is in phase with the light that would be passing out of the cavity. These waves must be perfectly out of phase at the reflective surface, where the intracavity standing wave has a node.

The axial modes of a laser are very close in wavelength. A Spectra-Physics model 171 argon ion laser, for example, has an effective cavity length of 1.83 m, and from Eq. (2.14) the frequency separation between axial modes or *mode spacing* can be calculated to be 82 MHz. Near the green output of the laser at 514.5 nm, an 82-MHz separation is equivalent to only about 7.2×10^{-5} nm. The lineshape function of the argon ion transition, and hence the gain curve of the laser, is broad enough so that, unless something is done to limit their number, about 20 different axial modes will oscillate simultaneously. Twenty adjacent axial modes span only about 0.0014 nm, and thus the laser output, with all of the modes active, could still be regarded as highly monochromatic. Having different axial modes active in the laser at the same time is essential for our application. As we will see below in Section 2.2.3.1, these modes are necessary in order to obtain mode-locked pulses out of the laser. In fact, the more modes oscillating, the narrower is the pulse which can be obtained.

Lasers will oscillate in different *transverse modes* as well. These transverse modes are apparent in the cross section of the oscillating light within the cavity and can be seen as well in the cross section of the beam exiting the laser. Some possible transverse modes are shown schematically in Figure 2.4. They are specified by the term TEM_{ij} (transverse electromagnetic), where i and j specify the number of nodes in two directions. TEM_{00} is the simplest beam without nodes. For the higher order oscillations, there is no consistent system for specifying the directions i and j. One person's TEM_{21} mode is another's TEM_{12}, and some authors use even a different system with three subscripts. TEM_{10*}, or the "doughnut mode," is a special mode structure

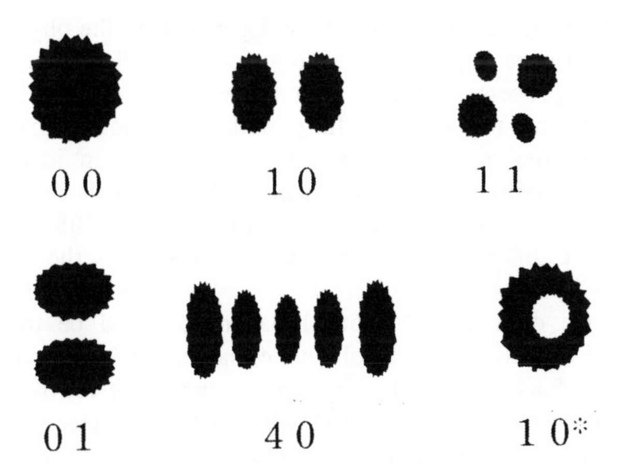

Figure 2.4. A schematic representation of the beam cross sections of lasers oscillating in different transverse modes.

which occasionally derives from a particle of dust on a mirror. It is actually the first of an additional set of cylindrically symmetric modes which do not normally oscillate, since most lasers lack perfect axial symmetry. In some lasers, under the right conditions, many different transverse modes can oscillate simultaneously, each at a slightly different frequency. For the laser systems to be discussed here, the higher order transverse modes are simply to be avoided. Perhaps the greatest usefulness of the TEM nomenclature is to impress one's erudition on someone lacking experience with lasers. For this purpose, any of the conventions for specifying the different modes would be equally acceptable.

For the synchronously pumped systems, it is important that the pump and dye lasers both operate in TEM_{00} mode. Restriction to the TEM_{00} mode is accomplished by simply inserting an aperture within the resonator, thus preventing oscillation of the spatially wider modes. This actually somewhat reduces the overall intensity of light in the laser, since the higher order modes have wider spatial distribution and can make use of more of the population inversion in the lasing medium. Our applications require the TEM_{00} mode, because the intensity across the center of the TEM_{00} mode is Gaussian in shape, permitting the focusing of higher intensities into smaller volumes. Any small loss in overall intracavity power resulting from restriction of the laser to TEM_{00} mode is compensated by the greater efficiencies of processes such as mode locking, frequency doubling, dye laser pumping, and cavity dumping, which depend on highly localized beams.

2.2.1.6. Homogeneous versus Inhomogeneous Broadening

Many of the properties of a laser are dependent on the physical processes which give rise to the broadening, or frequency dependence, of the lineshape function. These processes fall into two classes, homogeneous and inhomogeneous broadening mechanisms. In pure homogeneous broadening, each excited atom of the population inversion behaves in an identical manner and contributes equally to the broadness of the lineshape function. In pure inhomogeneous broadening, each excited atom behaves differently, and it is the sum of these different contributions which generates the breadth of the function.

An example of a homogeneous broadening would be the "collisional mechanism" of excited ions in a solid-state matrix, such as the Nd^{3+} ions in a YAG crystal. Each ion has kinetic energy, causing it to vibrate about its equilibrium position. This motion causes it to have frequent close encounters with neighboring ions, and these encounters continually alter its environment. The frequency of the light that can be emitted by the ion is dependent on the environment at the moment of the stimulated emission event. The changes in environment are fast relative to the length of time during which the ion can participate in lasing, and therefore any particular ion can emit radiation over a range of different frequencies and can participate in the lasing of more than one of the axial modes. Pure homogeneous broadening gives rise to a lineshape function which is Lorentzian in shape.

An example of an inhomogeneous broadening mechanism would be Doppler shifting in the argon ion laser. Different argon ions may have substantially different kinetic energies in the high-temperature plasma in which they are formed. The energy of the Doppler-shifted radiation that they emit is given by the sum of the energy of the transition and the kinetic energy of the ion in the direction of emission. Lineshape broadening derives from the fact that different ions with different kinetic energies contribute different frequencies. Since the ions are likely to retain the different kinetic energies for the entire excited state lifetime, a particular ion can therefore take part in only one axial mode. Pure inhomogeneous broadening gives rise to a lineshape function which is Gaussian in shape.

As illustrated in Figure 2.5, lasers operating with these different broadening mechanisms should behave quite differently. The top of the figure (panels A and B) shows small-signal gain curves [$\beta(\nu)$ versus ν] derived from the Lorentzian lineshape function of an ideal homogeneously broadened laser. Panel A shows the gain curve at the time the laser is turned on. The frequency positions of the possible resonant cavity modes are indicated under the curve by vertical lines. When the laser is first turned on, any cavity mode whose gain exceeds the threshold value, β_{th}, is free to begin lasing. As mentioned earlier, however, in a cw laser an equilibrium is soon reached at which the steady-state gain becomes equal to the threshold value. Since any excited atom or member of the population inversion can contribute equally to any of the resonant cavity modes, the entire gain function is uniformly decreased. Ideally, only that mode corresponding to the highest position on the curve is left to resonate.

In an inhomogeneously broadened laser, on the other hand, different atoms cannot necessarily contribute to different modes, and the gain curve does not uniformly decrease after lasing begins. The depletion of the popula-

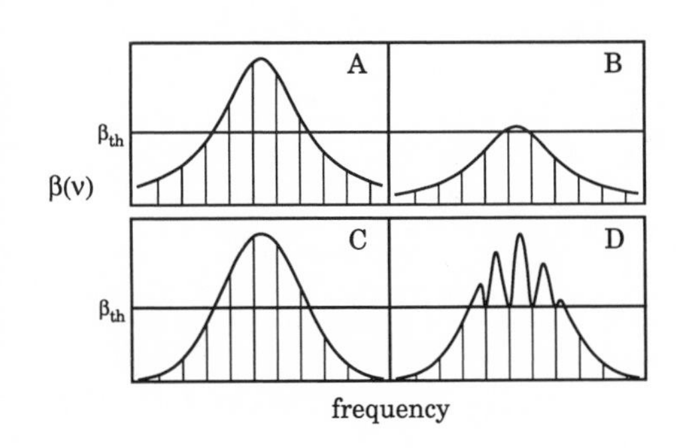

Figure 2.5. Gain curves illustrating the effects of different broadening mechanisms. A and B show gain curves for a homogeneous broadening; C and D show gain curves for inhomogeneous broadening. A and C show the curves at the time lasing begins, and B and D show the curves after equilibrium has been reached. In this figure, $\beta(\nu)$ represents the small-signal gain coefficient, while β_{th} is the threshold value of $\beta(\nu)$ required for lasing.

tion inversion can only occur at those frequencies which are resonant in the cavity. This is illustrated in the bottom half of Figure 2.5 (panels C and D). When the laser is turned on, the gain drops to the threshold value at each frequency position, and the corresponding axial modes continue to resonate. Different axial modes oscillate simultaneously. For obvious reasons, this phenomenon is called *hole burning*.

These, of course, are ideal situations which do not occur in nature. Any given laser will have contributions from both homogeneous and inhomogeneous mechanisms. For example, the Nd:YAG laser has homogeneous broadening from collisional mechanisms but has significant inhomogeneous mechanisms as well.[19] All three lasers to be discussed here, the argon ion, the Nd:YAG, and the dye laser, possess sufficient inhomogeneous broadening for lasing to occur in many modes simultaneously. As already mentioned, this feature is required for pulse generation by mode locking.

2.2.1.7. Brewster's Angle

The output beams of the lasers discussed in this chapter are highly polarized, and this polarization derives from the use of surfaces within the lasers oriented at *Brewster's angle*. When a light beam is incident on the surface of a transmitting plate, a certain fraction of the beam is reflected and a certain fraction transmitted, depending on the angle of incidence and the polarization of the beam. This is illustrated in Figure 2.6, which shows the reflectivity of incident light polarized with the direction of its electric vector either parallel to the surface (curve labeled h) or in a plane perpendicular to the surface (curve labeled v).

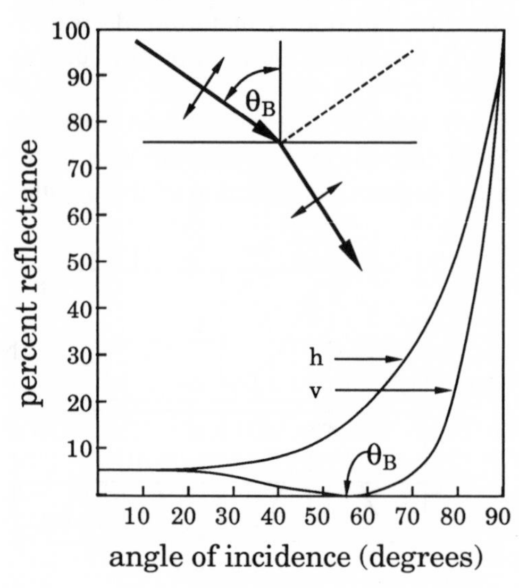

Figure 2.6. External reflection at a glass surface, illustrating Brewster's angle. The curve labeled v gives the reflectance for an incident beam with a polarization direction lying in a plane perpendicular to the surface; h gives the reflectance for light polarized parallel to the surface. In a laser which emits vertically polarized light, v and h would correspond to the reflectances of the vertical and horizontal components of the intracavity beam, respectively.

When the direction of a light beam is normal to the surface (0° angle of incidence in the figure), about 4% of either polarization is reflected. At high angles of incidence, nearly all of either polarization is reflected. At an intermediate angle, equal to about 56° for an air–glass interface, a point is reached where no reflection at all is observed for light which is polarized in a plane perpendicular to the surface. This angle is called Brewster's angle, θ_B, and can be calculated from

$$\theta_B = \tan^{-1}(n_2/n_1) \tag{2.15}$$

where n_2 and n_1 are the refractive indices inside the material and outside the surface, respectively. Brewster's angle is the angle at which the electric vectors of the reflected and transmitted beams would be orthogonal. Surfaces which are cut or placed at Brewster's angle within the lasers discussed here minimize losses for light which is vertically polarized, while introducing losses for other polarizations. The polarized light which experiences the minimum loss competes most effectively for the energy of the population inversion.

2.2.1.8. *Frequency Doubling*

Nonlinear optical effects can be used to generate new light frequencies from laser beams. In frequency doubling, or *second-harmonic generation* (the terms are synonymous), a laser beam entering a crystal can give rise to a new coherent light beam at twice the frequency (half the wavelength) of the original beam. It is required for generating the green laser beam necessary for pumping a dye laser from the infrared output of the Nd:YAG and is often used for extending the dye laser output to the UV. It is difficult to make intuitive arguments about the behavior of light, and I believe that this is particularly true about nonlinear optical effects. An attempt will be made here to do so, nevertheless. A rigorous treatment of nonlinear optical effects can be found in the book by Bloembergen.[20]

Light travels through a dielectric medium, such as glass or a crystal, as a periodic displacement of the electronic charge. The electrons in the material respond to the incoming oscillating field of the light and are accelerated laterally. Their slightly delayed movements then contribute to the oscillating electric field, and the light radiates forward through the material. Since it takes time to accelerate charges, the light travels more slowly than it would in a vacuum. This response of a dielectric material to a perturbing radiation field is described by a quantity called the polarization. In all of classical optics, this charge polarization is assumed to be linearly proportional to the electric field strength of the radiation. Although this is an excellent approximation for ordinary light beams in traditional optics, the high optical power densities available from lasers can readily exceed this limit, especially in some materials. Under these conditions, the simple proportionality breaks down. If one uses special crystalline materials which are particularly susceptible to nonlinear effects, and the proper conditions are maintained, new light frequencies can be produced, with surprising efficiency.

The polarization depends in complex ways on the direction of the beam in the crystal and is therefore really a tensor, but for simplicity we can consider the polarization, P, and the electric field strength, E, in only one dimension[21]:

$$P = \chi E(1 + a_2 E + a_3 E^2 + \cdots) \tag{2.16}$$

In this expression, χ is called the polarizability, and the a_i are called the nonlinear coefficients. At normal light levels, only the first term is required to accurately describe the behavior of the light. Very high values of E, on the order of megawatts per square centimer, are typically required for the nonlinear terms to become significant. These power densities, however, are easily reached by most focused laser beams, particularly those in which the energy has been localized in time by pulsing techniques.

For a travelling light wave, $E = E_0 \sin \omega t$, the second-order term becomes

$$a_2 \chi E_0^2 \sin^2 \omega t = \tfrac{1}{2} a_2 \chi E_0^2 (1 - \cos 2\omega t) \tag{2.17}$$

The first term in the parentheses gives rise to a simple displacement of charge independent of time. The second term gives rise to an oscillating displacement at twice the frequency of the original light, which generates the second harmonic or frequency-doubled light.

To understand intuitively how the second harmonic is generated, one might imagine that the perturbed electrons are attached to the atoms with springs.[3] For relatively low light intensities, their degree of displacement is directly proportional to the strength of the electric field. When they are displaced by the light, a linear force restores them to their equilibrium positions, and the light passes through slower, but unaltered in frequency. At much higher light intensities, the amplitude of the oscillating electric field is much greater and the electrons no longer respond in a linear manner. Whereas small displacements result in a linear force to restore the electrons to their initial positions, the force is no longer linear for such large displacements. The stronger force begins to restore the electrons to their initial positions much too quickly, and an oscillation is initiated at twice the frequency of the incident light. This is the second harmonic. Although certainly not a rigorous explanation, this does give an intuitive feeling for the origin of the frequency-doubled beam.

Two beams pass out of the crystal superimposed on each other, the *fundamental* beam of unaltered frequency and the *second harmonic*. In order for the second harmonic to be reinforced within the crystal, its phase must align with the phase of the fundamental beam. Otherwise, each oscillating charge would be out of step with the next, and the second harmonic would be eliminated by destructive interference. In general, the index of refraction of the crystal for the two different wavelengths (the fundamental and second harmonic will be different, and the phase conditions will not be met for frequency doubling. It is the birefringence of the crystal that can be used to overcome this difficulty. When light enters the crystal, it is divided into ordinary and extraordinary rays which have mutually perpendicular polarizations and experience different refractive indices. A birefringent crystal can often be aligned with the incoming beam in such a way as to properly match the phases of the two internal beams. Two different techniques, called type I and type II phase matching, are used.

In a type I frequency doubler, the crystal is oriented such that the fundamental beam enters as the ordinary ray. The second harmonic is generated at 90° to the fundamental as the extraordinary ray. Therefore, one must match the refractive index of the fundamental frequency in the ordinary ray to that of the second harmonic in the extraordinary ray. For many crystals it is possible to choose an angle of incidence for this to occur. The crystal can then be tuned to other nearby wavelengths by changing its orientation relative to the beam. This is called an angle-tuned frequency doubler. The tuning can also be done by changing the temperature at a fixed angle of incidence. This is called a temperature-tuned frequency doubler.

In a type II frequency doubler, the crystal is oriented such that the fundamental strikes the crystal at a 45° angle between the polarization directions of the ordinary and extraordinary rays. Thus, the beam enters the crystal with equal amplitudes in both rays. The second harmonic is generated in the extraordinary polarization. Therefore, the second harmonic exits the crystal with a polarization 45° from that of the fundamental. It is difficult to formulate a simple intuitive description of how type II phase matching is accomplished, although both temperature- and angle-tuned frequency doublers can be made.

Besides frequency doubling, nonlinear effects can be used to combine frequencies in other ways. For example, third or higher harmonic generation can lead to multiples of the incident frequency by making use of the a_3 and higher terms of Eq. (2.16), although the efficiencies of such processes are generally very low due to difficulties with phase matching. With appropriate phase matching, two coherent beams can also be mixed within a crystal to produce either a sum or difference frequency. For example, a sum of the second-harmonic and initial beams can be used in this way to produce the third harmonic. Also, by using a device called an *optical parametric oscillator*, a single coherent beam can be used to generate two beams whose frequencies sum to the original frequency. A process called *four-wave-sum mixing* uses a metal vapor as a nonlinear optical medium. One can obtain far-UV frequencies which are the sum of two times a visible frequency plus a different UV frequency.[22] Further frequency shifting is possible by making use of stimulated Raman scattering. This technique, called *Raman shifting*, can be used to generate a new coherent beam whose frequency is lower by the energy of a vibrational mode.[23] Descriptions of these various means for obtaining new frequencies of coherent light are outside the scope of this article.

The conversion efficiency of a second-harmonic generator can be summarized by the following proportionality:

$$P_{sh} \propto \frac{l^2 P_f^2}{w_0^2} \left[\frac{\sin^2 \Delta\phi}{(\Delta\phi)^2} \right] \tag{2.18}$$

The second-harmonic power, P_{sh}, is proportional to the square of the fundamental power, P_f. It is also proportional to the square of the interaction length within the crystal, l, and inversely proportional to the square of the beam waist, w_0, a measure of the focused diameter of the beam. The quantity in the brackets is a phase match factor which can vary from zero to one.

To maximize the efficiency of frequency doubling, one would like to focus the incident beam as tightly as possible for as long a distance in the crystal as possible. Unfortunately, these are mutually contradictory demands. Most TEM_{00} beams (with a Gaussian cross section) can be focused to near a diffraction-limited radius or beam waist, w_0, given by

$$w_0 = \frac{\lambda f}{\pi w_1} \tag{2.19}$$

The beam waist is proportional to the wavelength and the focal length of the lens, f, and inversely proportional to the radius of the spot size, w_1, of the beam striking the lens.† Using a long-focal-length lens to increase the interaction length in the crystal will increase doubling efficiency only up to a point, because at the same time the beam will be limited to a larger beam waist. Longer crystals also cost more. For frequency doubling the output of our dye laser, we have compromised by using a 250-mm focal length lens and a 1.5-cm path length potassium dihydrogen phosphate (KDP) crystal.

The proportionality of Eq. (2.18) has interesting consequences for pulsed lasers. For cw lasers, frequency doubling is a relatively inefficient process because of the power-squared dependence. Doubling is much more efficient when the light energy is localized in time in the form of a pulse. Also, because of the power-squared dependence, frequency doubling will usually narrow the pulse shape of a pulsed laser.

Frequency doubling can actually be quite efficient, even for the relatively

† The radii w_0 and w_1 are usually defined for a Gaussian beam to be the distance from the center to the point at which the field amplitude is reduced by $1/e$ (intensity decreased by $1/e^2$).

low peak power levels of pulsed mode-locked lasers discussed in this chapter. For example, the average output power of our Nd:YAG laser at 1064 nm is about 10 W, and after frequency doubling, as much as 1.8 W can be produced in the green at 532 nm. We actually defocus the fundamental infrared beam in order to limit the pumping power for the dye laser to about 800 mW. Also, the cavity-dumped output of the dye laser, which ranges from about 10 to 100 mW, depending on conditions, can be easily frequency doubled to give adequate intensity in the UV for use as a fluorescence excitation source.

2.2.1.9. Interactions between Light and Sound

Creating and manipulating the light pulses in the laser sources described here requires the use of two *acousto-optic devices* called a mode locker and a cavity dumper. These depend on interactions between light waves resident within the laser cavity and sound waves within optical elements of fused silica. The mode locker depends on a physical process called the *Debye–Sears effect*, in which light is diffracted into multiple beams by an acoustic wave. The cavity dumper depends on a different interaction called a *Bragg reflection*, in which a laser beam is simply deflected. It is called a Bragg reflection, by analogy to the effect familiar to biophysicists, in which X rays appear to be reflected off planes of atoms in crystals. Since these effects are important for understanding laser pulsing, a brief description of them is given here. There are a number of good sources of information on these effects. The interested reader might enjoy the very readable account of these and other interactions between light and sound waves written by Adler.[24] More rigorous approaches can be found in the early papers by Raman and Nath[25, 26] or the classic text by Born and Wolf.[27]

If one uses a transducer to radiate ultrasonic waves of a few megahertz down into a glass of water and then shines a small laser beam through the glass, the light does not travel through unaffected, but splits vertically with multiple images appearing on both sides of the original beam. If the amplitude of the acoustic signal is increased, then more of these images appear and their relative amplitudes vary in a complicated pattern with the acoustic amplitude. If the frequency of the signal is increased, then the images move further apart. This effect was first described in 1932 by Debye and Sears[28] and by Lucas and Biquard[29] and is now commonly called the Debye–Sears effect.

To better understand this phenomenon, imagine a light wave of angular frequency ω incident on a transparent slab of material of length l, as illustrated in Figure 2.7A. Assuming that the refractive index of the material of the slab is higher than that of its surroundings, then the wavelength of the light within the slab will be shortened. Assume also that we have a means of modulating the refractive index of the slab uniformly at frequency Ω (angular frequency in radians per second, as opposed to hertz). At times when the

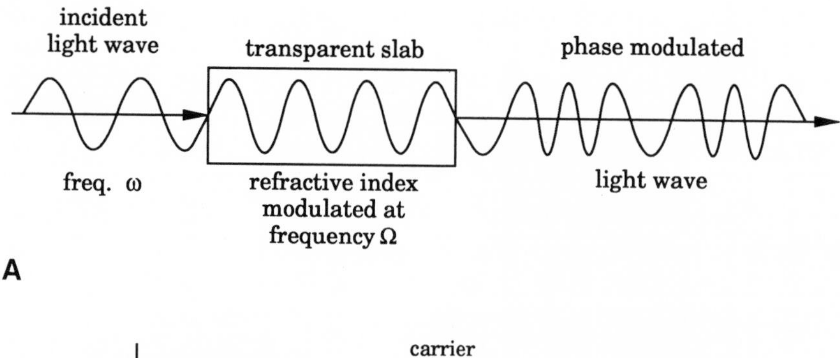

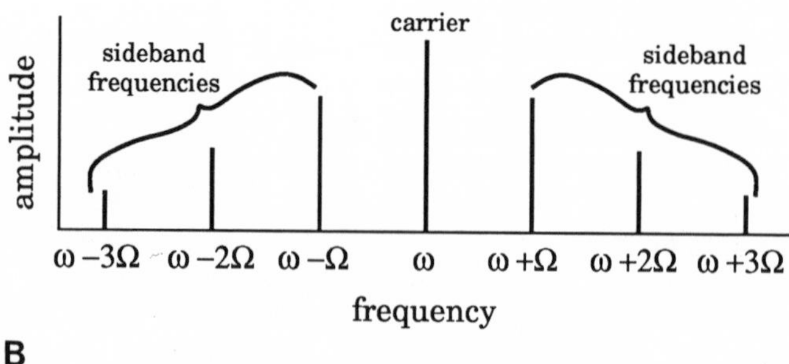

Figure 2.7. Phase modulation of light of frequency ω which travels through a transparent slab whose refractive index is modulated at frequency Ω. The lower section of the figure shows the frequency breakdown of the phase-modulated beam. The output consists of a carrier wave of unaltered frequency superimposed with a series of sideband frequencies shifted either up or down by increments of the modulating frequency. Both ω and Ω are angular frequencies in radians per second.

refractive index is high, the light will travel slower; when it is low, the light will travel a little faster. The light exiting the slab becomes *phase-modulated.* Phase-modulated waves are discussed in texts on frequency modulation (FM) systems and have familiar properties (at least to electrical engineers). The output light wave consists of a carrier wave of frequency ω summed with a series of waves with frequencies which differ from that of the carrier wave by multiples of Ω. This is illustrated in Figure 2.7B. The amplitudes (which can be calculated from Bessel functions) shown in the figure depend on the *phase excursion*, or degree to which the refractive index of the slab is modulated. A greater phase excursion results in increases in the amplitudes of higher order sidebands, at the expense of the amplitudes of the carrier and lower order sidebands.

Now, stack many of these transparent slabs on top of one another to get the column of material shown in Figure 2.8. A direct means of modulating the refractive index of each slab would be to pass a sinusoidal acoustic wave down the column, compressing and dilating individual slabs at the acoustic

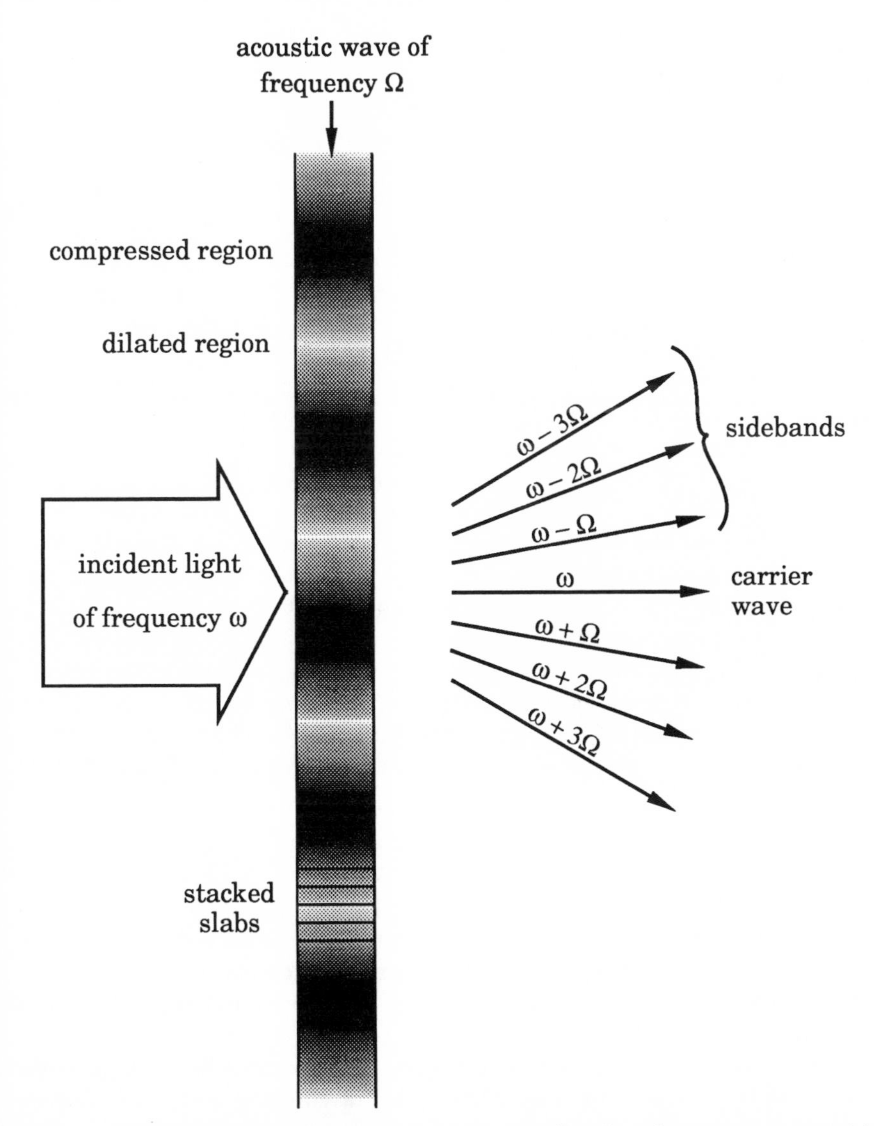

Figure 2.8. The Debye–Sears effect. An acoustic wave traveling through a transparent material diffracts the light into sidebands of differing frequency. The thickness of the column or interaction distance between the light and material must be adequately small for this effect to occur.

frequency Ω. If one directs light of angular frequency ω perpendicular to the column, then each little slab produces phase-modulated light as discussed above. Also, since the refractive index varies vertically throughout the column, the column behaves like a diffraction grating. Geometric arguments can be made[24, 26] that the carrier wave will pass directly through and exit colinear with the incident wave, but the sidebands will be deflected from the carrier by angles, θ_n, given by

$$\theta_n = \tan^{-1}(n\lambda/\Lambda) \qquad n = \text{an integer} \tag{2.20}$$

where λ is the wavelength of the light, and Λ is the wavelength of the acoustic wave.

This is the Debye–Sears effect that we saw in our water glass. Increasing the amplitude of the acoustic wave causes an increase in the number of sidebands with observable amplitudes. Increasing the frequency of the wave, on the other hand, decreases its wavelength, increases the angles of deflection of the sidebands, and results in a spread of the pattern. The acoustic wave acts very much like a diffraction grating, except that the diffracted beams are shifted in frequency from the incident light. This shift in frequency can also be explained as a Doppler shift in the light as it reflects off the moving acoustic wave front.

Another method for increasing the sideband amplitudes would be to make the column thicker, forcing the light to traverse more material. This cannot, however, be done indefinitely. Light will not travel in a perfectly straight line through a particular slab but will be diffracted or bent into the adjacent slabs. Such diffracted light will no longer contribute to the pattern but will give rise to sidebands which destructively interfere with those generated at the original location in the column of material. One can define a thickness, l, of material, which is required for the Debye–Sears effect to effectively occur[24]:

$$l \ll \Lambda^2/2\pi\lambda \tag{2.21}$$

This is sometimes called the *critical length*, although it is obviously not precisely defined. Acousto-optic modulators which work below this length are said to operate within the *Raman–Nath regime*, after the authors who first presented a successful mathematical theory for the Debye–Sears effect and described its limits.[25, 26]

The acousto-optic mode locker that we will describe later (Section 2.2.3.2) actually uses a standing, rather than traveling, acoustic wave. Such a wave is illustrated in Figure 2.9. Instead of moving downward, the waves remain in place. At any particular location in the column (for example, the dashed line of the figure), the refractive index varies sinusoidally from a maximum value to a minimum and back at frequency Ω. Twice during the cycle the refractive index variations are at a maximum; twice they vanish. At times when the refractive index is uniform, the beam passes through unaffected, but at other times diffraction occurs. What results is a sinusoidally varying Debye–Sears efect at twice the frequency of the acoustic wave. Since the acoustic frequency remains the same, the locations and individual frequencies

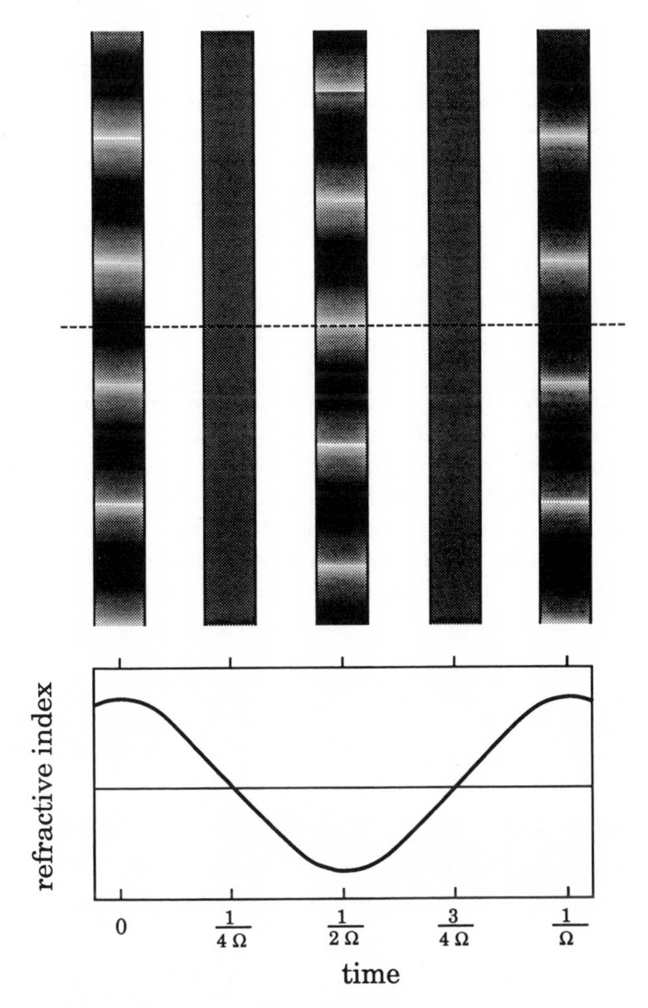

Figure 2.9. A standing acoustic wave shown at different times. The wave remains in place and the refractive index varies sinusoidally as a function of time. In the bottom section of the figure this variation is shown for the position of the dashed line. A standing acoustic wave acts like a sinusoidally varying diffraction grating and produces a varying Debye–Sears effect at twice Ω, the frequency of the acoustic wave.

of the diffracted sidebands do not vary, but since the phase excursion varies, their relative amplitudes do change. As we will see, an acousto-optic modulator utilizing a standing acoustic wave is used to modulate the intracavity losses in the mode-locked argon ion and Nd:YAG lasers, at twice the frequency of the acoustic signal sent through it.

The Debye–Sears effect requires that the interaction distance be very short relative to the acoustic wavelength. Another important effect, the Bragg reflection, can occur when the relative interaction length becomes much longer. As mentioned above, sidebands do not normally appear under these conditions, because they are eliminated by destructive interference. However, one can observe a single strong sideband if the incident light beam is rotated relative to the plane of the acoustic waves by a specific angle. This is illustrated in Figure 2.10. At the proper angle, scattered light reinforces itself and appears to reflect off the acoustic wave front leaving at the same angle. The Bragg angle, α, is given by

$$\alpha = \sin^{-1}(\lambda/2\Lambda) \tag{2.22}$$

A macroscopic view of this phenomenon is shown in the schematic view of a Bragg cell in Figure 2.11. The cell usually consists of a piece of fused silica with a piezoelectric transducer attached to it. The transducer radiates an acoustic wave through the cell. Depending on whether the incident light is oriented properly to strike an oncoming or departing acoustic wave, the light frequency, ω, is either shifted up or down in frequency by Ω, the acoustic wave frequency. Part of the light passes through unaffected.

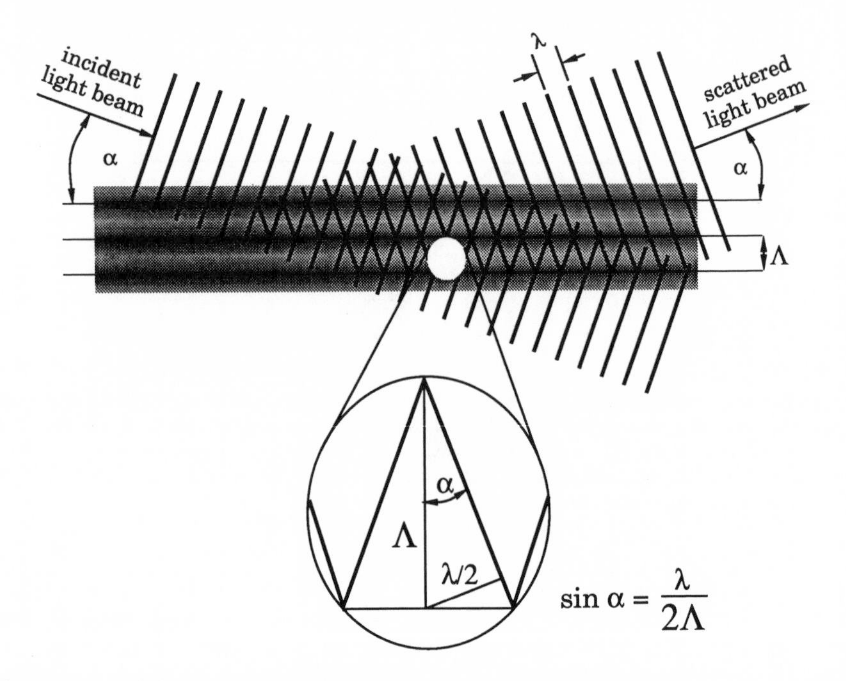

Figure 2.10. A microscopic view of the Bragg reflection. If a light beam strikes an acoustic wave at the Bragg angle, α, a first-order sideband becomes reinforced to form a scattered beam. The scattered beam also forms an angle of α with the acoustic wave front.

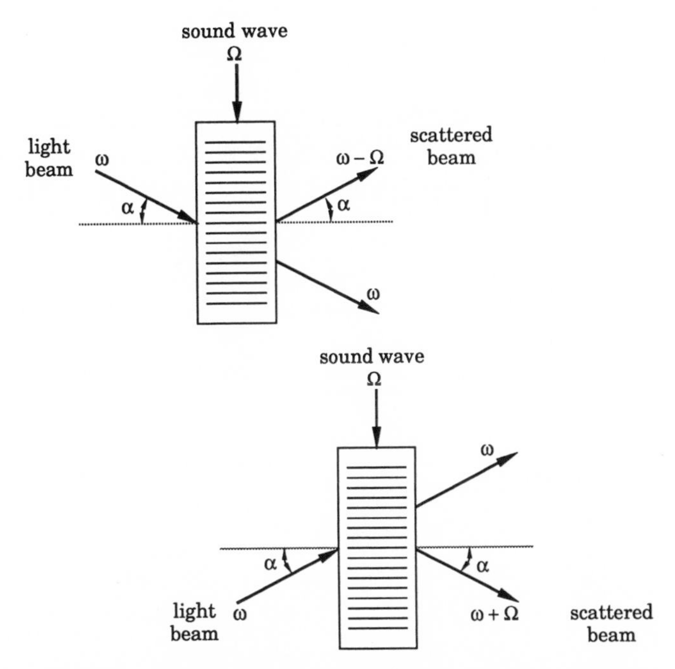

Figure 2.11. A macroscopic view of a Bragg cell. Depending on whether the incident light beam strikes an oncoming or departing acoustic wave, the scattered beam will be shifted either up or down in frequency. In this figure, ω and Ω are the angular frequencies of the incident light and the acoustic waves, respectively, while α is the Bragg angle.

An acousto-optic modulator operating under conditions where

$$l >> \Lambda^2/2\pi\lambda \tag{2.23}$$

and with incident light at the Bragg angle is said to operate in the *Bragg regime*. Typically, very high acoustic frequencies of the order of nearly 1 GHz are required to satisfy Eq. (2.23). As we will see, for our laser this effect is used by the cavity dumper to deflect a pulse of light out of the dye laser.

2.2.2. The Argon Ion, Nd:YAG, and Dye Lasers—Energy Levels and Physical Layout

2.2.2.1. The Argon Ion Laser

A simplified diagram of an argon ion laser is shown in Figure 2.12A. The active medium consists of argon gas at a low pressure of about 0.7 Torr, enclosed in a discharge tube. A high current (about 50 A for the Spectra-

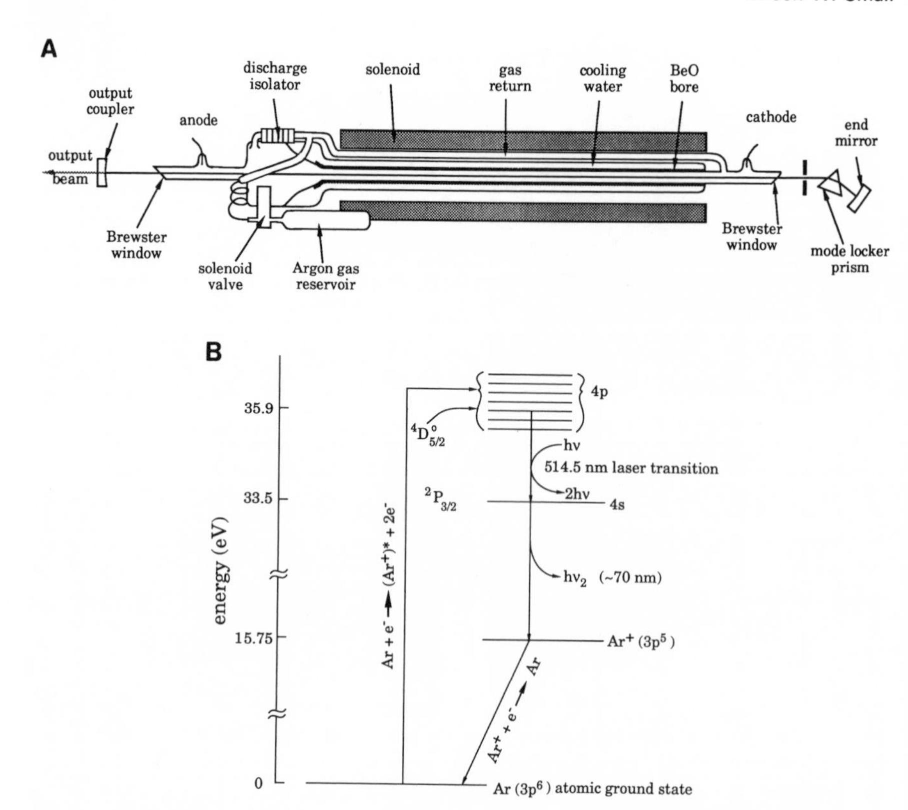

Figure 2.12. The argon ion laser. (A) A schematic drawing illustrating the main components of the laser. (B) An energy level diagram showing how a cycle of ionization and neutralization of argon atoms is used to supply the energy for lasing. When viewing the energy diagrams shown in this section, it might help to recall the conversion: energy in eV = 1240/wavelength in nm.

Physics model 171-09 laser that we have used) is passed through the tube, and collisions between electrons and argon atoms generate Ar^+ ions and impart the excitation energy necessary for lasing. Although a number of different materials can be used to line the center or *bore* of the plasma tube, Spectra-Physics chooses the ceramic material BeO because of its high thermal conductivity, good strength at high temperatures, and low rate of argon adsorption. The BeO lining is also highly resistant to erosion by the plasma and is contoured to match the shape of the discharge within the tube. The plasma tube is conduction-cooled through direct contact with circulating water and is surrounded by a solenoid which produces a magnetic field of about 1000 G, greatly increasing the electron density within the center of the bore. Since the positive Ar^+ ions are accelerated down the tube toward the anode, a gas return is provided outside the tube and fitted with a discharge isolator to prevent a discharge from striking down this secondary path. The resonant cavity is bounded by a highly reflective mirror at one end and another mirror with partial transmittance at the other end. Both mirrors have long-focal-length spherical surfaces to minimize refraction losses in the cavity. An adjustable aperture is provided to maintain the laser in TEM_{00} mode. A gas supply, operated using a solenoid valve, is used to replace Ar that is adsorbed by surfaces during operation. Brewster's angle windows are used to seal the ends of the plasma tube to polarize the beam vertically, and an acousto-optic mode locker is used to obtain pulsed output. The mode locker will be described in Section 2.2.3.2.

An energy level diagram for the argon ion laser is shown in Figure 2.12B.[2] A collision between an argon atom and an electron ionizes the atom and supplies an energy of about 36 eV. Lasing from stimulated emission at 514.5 nm occurs as an electron drops from a $4p$ to a $4s$ level. (The levels are described by the term symbols ${}^4D^0_{5/2}$ and ${}^2P_{3/2}$, respectively.) The ion then quickly drops further in energy by the release of radiation at ~70 nm (far UV) to the ion ground state. Subsequent recombination with an electron results in the neutral atom. This path is similar to that in the four-level laser in that the lower level of the lasing transition is rapidly depleted but differs in that ionization is involved and excitation occurs directly into the upper level of the lasing transition and not into pumping bands.

Because there are many different excited states produced in the argon plasma, argon ion lasers can lase at a number of different wavelengths simultaneously. A prism (in our case the mode-locking prism) is inserted in the cavity in order to get the laser to produce the single 514.5-nm beam. Since limiting the laser to 514.5 nm does not lead to a significant increase in the power level at this wavelength compared to that in the multiline laser, it can be concluded that there are few nonradiative transitions between the excited states of the argon ion.

In our laboratory, we used a Spectra-Physics model 171-09 argon ion laser for approximately nine months with excellent results. The laser was extremely dependable and gave very good long-term stability at power levels which exceeded specifications. After nine months of use, in which we put over 2000 h on the laser, finally the plasma tube failed when a crack developed and it filled with water. This tube lasted approximately twice its expected lifetime. A disadvantage of an argon ion laser is that the tube does have a relatively short lifetime, and the replacement cost of the tube (~$16,000) is a great deal of money for most grant budgets. Another disadvantage is that an argon ion laser requires a large amount of electricity. Our laser required three-phase electrical service rated at 460 V and 75 A per phase. It can be difficult to convince a Physical Plant to supply such service at the same time as they are reducing lighting over work areas in order to save money. It was partly for these reasons that we ultimately chose to purchase a frequency-doubled Nd:YAG laser as a pumping source.

2.2.2.2. The Nd: YAG Laser

The physical layout of a Spectra-Physics Series 3000 Nd:YAG laser is shown schematically in Figure 2.13A. The active component is a rod of neodymium-doped yttrium aluminum garnet crystal ($Y_3AL_5O_{12}$ with <1% Nd), 4 mm in diameter and about 80 mm long. The rod is placed at one focus of a gold-plated elliptical reflector. The active Nd^{3+} ions in the rod are optically pumped using a krypton arc lamp located at the other focus of the reflector. The pumping lamp dissipates approximately 2 kW, mostly in the

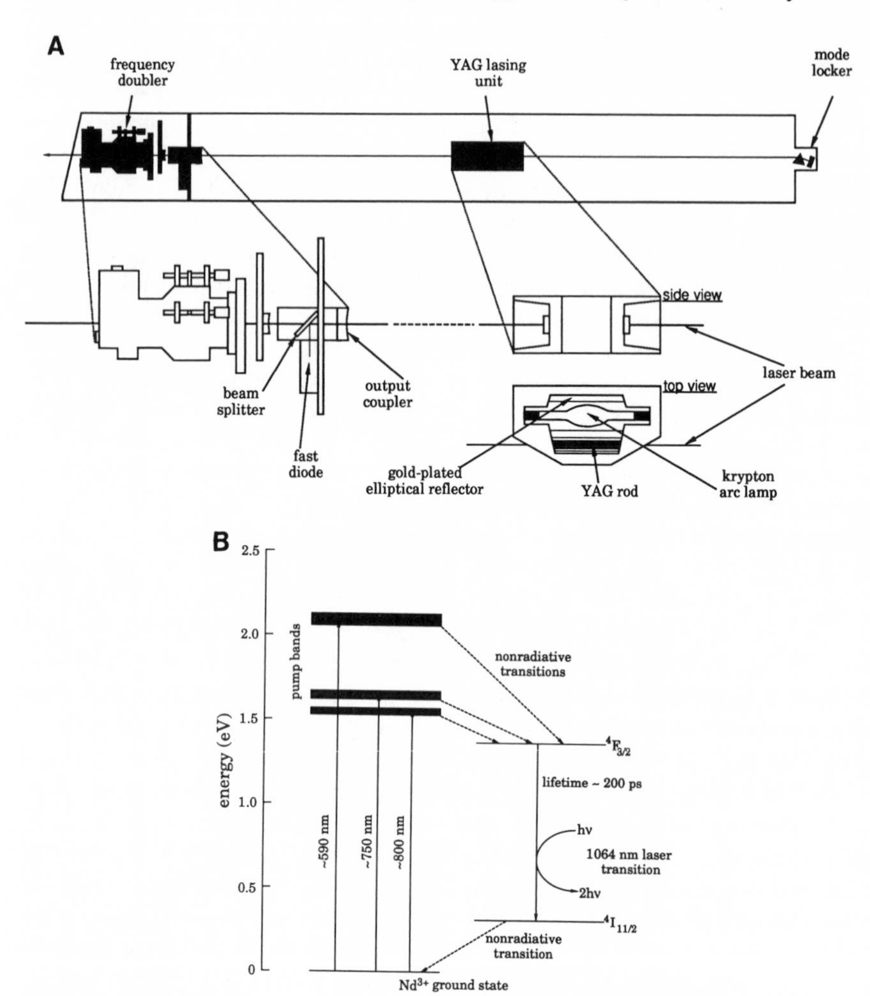

Figure 2.13. The Nd:YAG laser. (A) A schematic drawing illustrating the main components of the laser. (B) An energy level diagram showing the classic four-level structure of the laser transitions.

form of heat. Optical absorption and subsequent nonradiative transitions heat the rod considerably, and therefore both the rod and arc lamp must be efficiently cooled. This is done using a closed-loop deionized water circulation system in which water is in direct contact with the rod and lamp and fills the elliptical chamber. Pyrex flow tubes surround the rod and lamp to control turbulence for even cooling. The rear mirror of the resonant cavity is highly reflective at the 1064-nm wavelength of the laser and is placed behind a mode locker as in the argon ion laser.

An energy diagram for Nd^{3+} is shown in Figure 2.13B.[19] The Nd:YAG laser is a classical four-level laser. Light from the krypton arc lamp is used to optically pump the Nd^{3+} ions into absorption bands at about 590, 750, and 800 nm. (These are actually highly degenerate single levels of the neodymium ion, dispersed by crystal field splitting.) The energy is rapidly depleted by nonradiative processes, leaving the ions in a metastable state ($^4F_{3/2}$) with a lifetime of about 200 μs. Lasing occurs by stimulated emission at 1064 nm down to a lower state ($^4I_{11/2}$) which is rapidly depleted by nonradiative processes to the Nd^{3+} ground state.

Outside the resonator, the 1064-nm beam passes through a beam splitter which directs a small fraction of the energy to a very fast photodiode for monitoring the mode-locked pulse shape. Next, the beam passes a manually operated shutter made of heavy-gauge aluminum. The shutter is necessary because it permits the laser to warm up without exposure of the frequency-doubling crystal to the IR beam. Instabilities in the laser during the warm-up process can cause an effect called "self-Q switching," which produces random, extremely high intensity pulses that could destroy the doubling crystal. A KTP crystal ($KTiOPO_4$) is used to frequency double the 1064-nm output to the green at 532 nm. KTP crystals have very high nonlinear coefficients [a_i in Eq. (2.16)] and a high damage threshold and can be used as either type I or type II second-harmonic generators.[30] In the Spectra-Physics system, the KTP crystal is used as a type II doubler, which means that the second harmonic is produced with a polarization at 45° from vertical. A polarization rotator (crystalline quartz with the optic axis perpendicular to the surface of the plate) is used to rotate the plane of polarization back to the vertical for pumping the dye laser.

In the Nd:YAG laser, the heating of the rod affects the focusing of the optical system. Optical absorption heats the rod fairly uniformly, but the water cooling along its cylindrical surface establishes a radial temperature gradient, and consequently a change in refractive index across the rod. The rod therefore acts as a thermal lens. The pumping lamp current is adjusted until a compromise is reached between the need for a high population inversion in the rod and the destabilization of lasing due to strong focusing within the rod. Thus, the output power of the Nd:YAG laser is essentially set by its design, and this has important consequences for its use.

In the argon laser, the intensity can be easily controlled by varying the current through the plasma tube, but such control is not readily available for the Nd:YAG. In order to control the intensity for pumping the dye laser, we typically decrease output power by defocusing the 1064-nm light incident on the doubling crystal. This changes the efficiency of frequency doubling and probably extends the life of the crystal. There is an additional consequence of the lack of intensity control which affects fluorescence decay measurements. The argon ion laser has a feedback control, which maintains the long-term stability of the light output. A photodiode monitors the intensity of the laser output, and its signal is used to control the current. This is not done with the Series 3000 Nd:YAG, and therefore the output intensity can drift in time. We have not found this to be problem if adequate time is taken to warm up the system before critical experiments are begun, but it must be regarded as a disadvantage of this system. Such a feedback control, however, has been added to more recent Spectra-Physics models.

In spite of possible problems with long-term stability, there are substantial advantages of the Nd:YAG system in terms of dependability. As opposed to argon ion plasma tubes, Nd:YAG crystals appear to last indefinitely. After about 300 h of use, one must replace the inexpensive (~$250) krypton pumping lamp (a fast, relatively easy process). The only expensive item ($5000–6000) that must occasionally be replaced is the doubling crystal. We have had to do this once in the past seven years. Two advantages of the mode-locked and frequency-doubled Nd:YAG laser over the argon ion laser are the narrower pulse width (~70 ps vs. ~140 ps for the argon ion laser) and the considerably higher peak power of the Nd:YAG laser. It can thus be a better pump source for the dye laser, if narrow pulse widths are desired.

2.2.2.3. The Dye Laser

Figures 2.14A and 2.14B are schematic drawings of the Spectra-Physics model 375 dye laser, set up in a configuration for synchronous pumping (discussed in Section 2.2.3.3) and cavity dumping (discussed in Section 2.2.3.4). The combined system consists of two main components, the dye laser and the cavity dumper, connected by a quartz tube. Each unit contains an end mirror and forms one end of the dye laser resonator. The dye laser unit contains the active lasing medium, and the cavity dumper is an acousto-optic device which provides a means of extracting light pulses from the laser. Precise maintenance of the length of the resonator is necessary for synchronous pumping, and therefore the length is determined using a quartz tube with a very low coefficient of thermal expansion. The dye laser unit is fixed to the table, and the cavity dumper optics are forced, using springs, up against the quartz tube, permitting the cavity dumper to "float" relative to the optics in the dye laser.

The lasing medium consists of a flat stream of concentrated dye solution

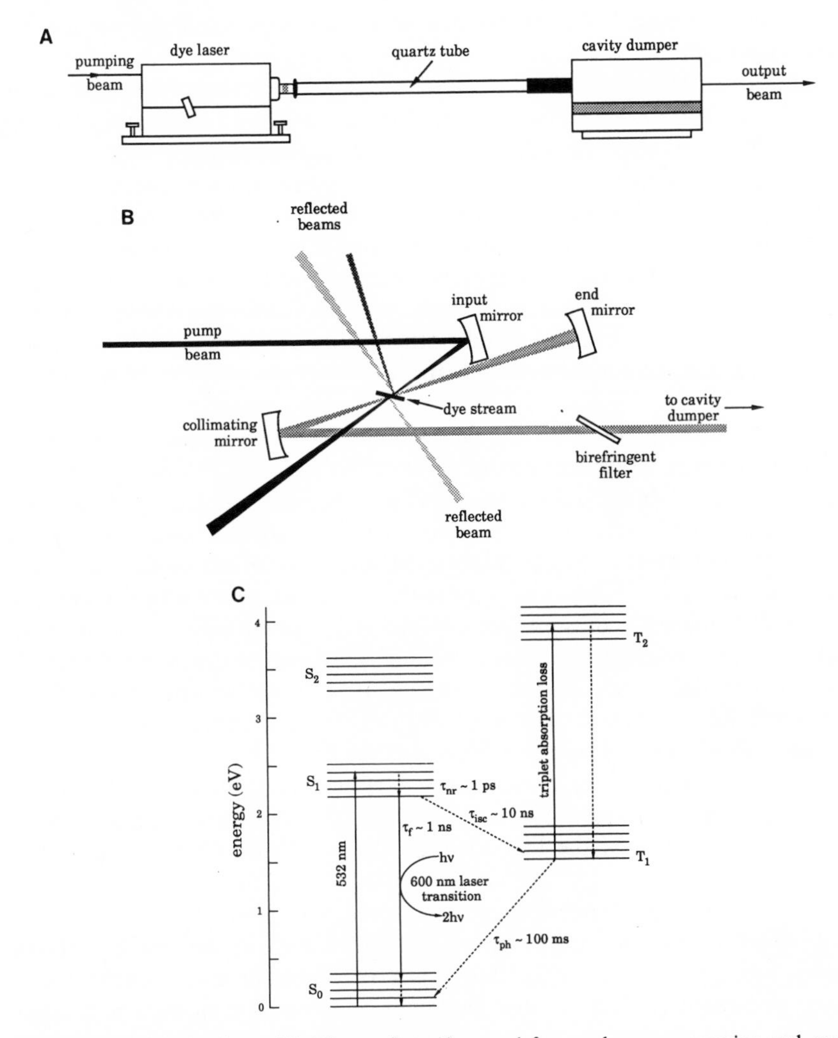

Figure 2.14. The dye laser. (A) The configuration used for synchronous pumping and cavity dumping. (B) A schematic drawing of the optical components within the dye laser housing. (C) A Jablonski diagram showing the energy levels of the dye molecules.

(for rhodamine 6G, about 2 m*M* in ethylene glycol), 5 mm wide and ~300 μm thick, forced at about 60–80 psi through a carefully polished nozzle. The stream is collected in a piece of large tubing and returned to a reservoir for reuse. Ethylene glycol is used as a solvent because it is nonvolatile and its high viscosity greatly reduces turbulence, resulting in an optically smooth stream. A focused laser beam (usually from an argon or Nd:YAG laser) is used to optically pump the dye to obtain a population inversion. Lasing occurs through the dye stream, located between a highly reflective end mirror and a collimating mirror.

The dye stream is oriented at Brewster's angle relative to the lasing beam in order to minimize reflective losses and favor vertically polarized light in the laser. The dye laser contains as a wavelength-tuning element a Lyot (birefringent) filter, which consists of one or more (in our case, two) crystalline quartz plates oriented at Brewster's angle relative to the beam. The plates are cut parallel to their optic axes, and a rotation of the plates about a line normal to the surface changes the range of wavelengths of vertically polarized light which can pass through with minimal loss. Using a double Lyot filter reduces the lasing bandwidth to about 60 GHz ($\sim$0.07 nm), and by rotating the filter one can tune the dye laser over about 60 nm of the fluorescent band.

At the risk of insulting my fellow fluorescence spectroscopists, I have shown the energy levels of rhodamine 6G in the familiar Jablonski diagram of Figure 2.14C. Each excited electronic state contains a number of vibrational sublevels, each of which contains many closer rotational levels, which, for clarity, are not shown. Singlet levels are shown on the left of the figure and triplet levels on the right. For dyes such as this at room temperature in a hydrophilic solvent, collisions and electrostatic perturbations broaden the multiple levels, causing the absorption and emission spectra to be essentially smooth. For the discussion here, however, it is easiest to draw the different vibronic levels as though they were distinct, to illustrate that the dye laser operates on what is essentially a four-level scheme.

The green light of the pumping laser (either 514.5 or 532 nm) excites the molecule from the lowest vibronic level of the lowest electronic singlet into an upper vibronic level of the first excited singlet. (Some other dyes are often pumped into higher singlet levels.) Vibrational relaxation on a time scale of about a picosecond populates the lowest vibronic level of the first singlet, from which lasing will occur. The fluorescent lifetime of this state is a little over a nanosecond, and while this is not very long, it does qualify it as a metastable state relative to that into which pumping occurred. Stimulated emission returns the molecule to an upper vibronic level of the ground state, which again decays in about a picosecond by vibrational relaxation to the lowest vibronic level, thus depopulating the lower laser level. Note that lasing cannot occur directly to the lowest vibronic level of the ground state, because self-absorption could then compete with stimulated emission and prevent lasing. This can be seen in Figure 2.15, which shows the absorption, emission, and wavelength lasing region of the dye rhodamine 6G (R6G). The lasing region corresponds only to that region of the dye emission which does not overlap with the absorption band.

The two possible pump wavelengths of 514.5 and 532 nm are indicated on Figure 2.15. For optimum lasing, an R6G concentration of 2×10^{-3} *M* in

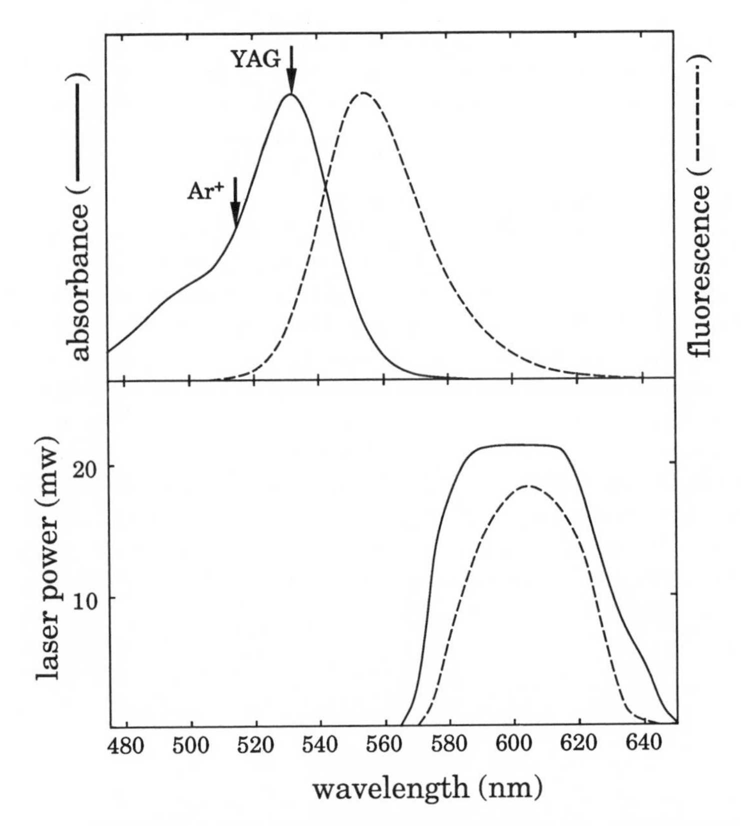

Figure 2.15. Properties of the laser dye rhodamine 6G. The top section of the figure shows the absorption and fluorescence emission. The two pumping wavelengths of 514.5 nm (argon ion laser) and 532 nm (frequency-doubled output of the Nd:YAG laser) are indicated on the figure. The lower section of the figure shows wavelength tuning curves for the dye when pumped at 532 nm. The dashed curve was obtained at the concentration recommended by Spectra-Physics (0.96 g/liter), and the solid curve was obtained after dilution to 0.55 g/liter.

ethylene glycol is recommended by Spectra-Physics for both pumping sources, the argon ion laser and the frequency-doubled Nd:YAG laser. Obviously, the 532-nm light will be absorbed much more effectively, and therefore one could use a lower concentration of dye. We have found this to be the case. Decreasing the concentration of dye for the Nd:YAG-pumped system results in substantially higher output powers as well as a greater wavelength tuning range.[31] This occurs with little degradation of the pulse width as far as our fluorescence decay measurements are concerned. The lasing regions for both the recommended dye concentration and diluted dye are shown in the figure.

Laser dyes are chosen to have very high quantum yields for singlet emission. However, the population of the excited state must be very high for lasing to occur, and, consequently, spin-forbidden intersystem crossing to the lowest triplet can begin to compete for the energy (Figure 2.14C). This occurs because of the very long relative lifetime of the lowest triplet, which must return via slow spin-forbidden phosphorescence to the ground state. Unless something is done to remove molecules in the triplet state, their population can build up significantly. This not only reduces the needed singlet population inversion, but can also reduce the efficiency of the dye laser by introducing an absorption loss due to allowed transitions to higher triplet levels.

Removal of triplets is the main reason for the use of a fast-moving dye stream in the dye laser. In the Spectra-Physics system, the stream velocity approaches 25 m/s. Since the active area in the stream (where the focused pump beam strikes) has a diameter of about 15 μm, the flight time of a dye molecule through the area is less than a microsecond. A particular molecule can participate in the absorption and stimulated emission cycle repeatedly, but the flight time is sufficiently short to minimize the number of triplet states formed. This procedure is quite effective for the rhodamine dyes, but for some others it is necessary to introduce triplet quenching agents such as 9-methylanthracene or cyclooctatetraene to further reduce the concentration of dye molecules in the triplet state.

2.2.3. Laser Pulse Generation

Perhaps the simplest way to get a pulse of light out of a laser is to physically block the resonance of light inside the laser cavity while the population inversion is generated and then quickly remove the restraint. If the laser has sufficient initial gain [see Eq. (2.9)], then lasing begins and the intracavity power quickly rises. The process is usually repeated to create a train of light pulses. This is called Q-switching, where Q refers to the quality of the resonator, a measure of its ability to support the oscillation of light. Many different techniques can be used for Q-switching a laser, including rotating mirrors, electro-optic light switches such as Pockels cells and Kerr cells, and dye solutions called saturable absorbers. The giant Q-switched pulses are relatively broad in time ($\sim$5 to 50 ns), and their repetition rates are low, usually ranging from about 1 to 100 pulses/s. For a pulse fluorometer, we would like a much higher rate of pulsing but do not require very high intensities. Obviously, a Q-switched laser is not an appropriate source for us. Instead, we rely on another process called mode locking, which produces very narrow pulses of lower intensity but much higher rate. In order to produce mode-locked pulses, one must begin with a laser which can be run in cw mode. After mode locking, such a laser is generally referred to by the oxymoron *cw mode-locked laser*.

2.2.3.1. Mode Locking

Mode locking is another way of creating pulsed light from a laser. In Section 2.2.1.5, we described how a typical laser resonator will support the simultaneous oscillation of many different axial modes, each at a different frequency. The different frequencies derive from the fact that the resonant cavity will support oscillations of lights at an integral number of half-wavelengths between the mirrors. Normally in a cw laser there would be no particular phase relationship between these different colinear light waves. If one were to examine the electric vector of the light coming out of the laser as a function of time, one would find that it would represent a fraction of the sum of these axial modes. It would therefore fluctuate in an apparently random manner depending on the frequencies, amplitudes, and phases of the axial modes active in the laser.

This is shown schematically in Figure 2.16A. The time axis is marked off in units of L/c, the length of time required for light to travel the length of the laser cavity (about 6.1 ns for the Spectra-Physics argon ion or Nd:YAG laser). In this hypothetical laser, seven different axial modes are oscillating, each at a slightly different frequency, ν_n, satisfying the condition $\nu_n = nc/2L$, where n is an integer [see Eq. (2.13)]. The curves shown are meant to represent the fraction of each mode which leaks out of the laser through the output coupler. Their phases have been chosen randomly. For this example, the amplitudes of the waves have a Gaussian distribution meant to approximate an inhomogeneously broadened Gaussian gain curve. The figure is unrealistic in that in a real laser there would be many more oscillations of the electric vector of the light within the time period of the figure, but our conclusions will be quite independent of the particular frequencies chosen so long as they show the proper relationship to one another. We sum the amplitudes of the axial modes to get the output of the laser shown at the bottom of the figure. Since we chose random phases for the different modes, the output amplitude of this cw laser fluctuates randomly.

For a mode-locked laser, illustrated in Figure 2.16B, one uses an acousto-optic modulator near one of the mirrors to introduce a periodic loss [α_{loss} in Eq. (2.9)] in the cavity. The modulator acts like a shutter, opening once for every round trip time for the light in the resonator. When it is fully open, the light can pass through unimpeded, and this tends to align the phases of the different axial modes with one another. In Figure 2.16B, this alignment has occurred at L/c and $3L/c$. Where the amplitudes align, constructive interference reinforces the intensity to form a pulse. Between the pulses, destructive interference eliminates the light. The rate at which pulses are produced by the mode-locked laser, or its *repetition rate* ("rep rate"), is thus

$$\text{repetition rate} = c/2L \tag{2.24}$$

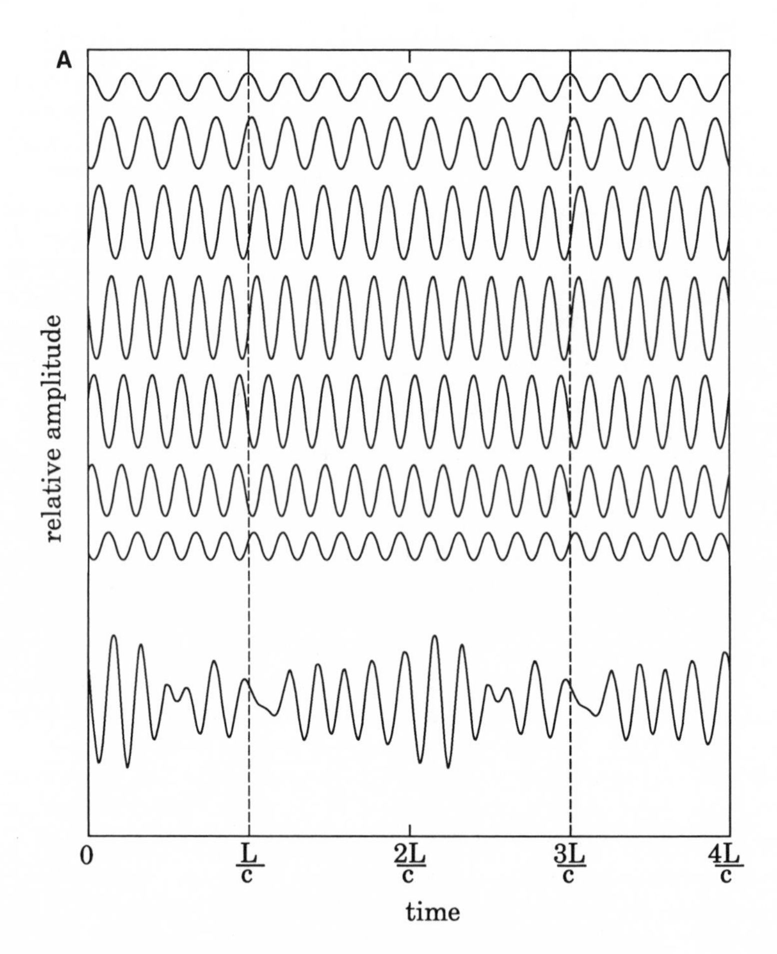

Figure 2.16. Simulation of the output of a mode-locked laser. (A) This section of the figure illustrates the output of a cw laser without mode locking. Time is marked in units of L/c, the length of time it takes light to travel the length of the laser. The seven curves in the upper section of the figure indicate the amplitude fluctuations of seven axial modes as they leak out through the output coupler. They have been chosen to have random phases. The actual output of the laser, the sum of these amplitude fluctuations, is shown at the bottom of the figure. Because the components are not related in phase, the sum varies in a complex manner in time. (B) In a mode-locked laser, the phases are aligned with one another. This alignment is shown at times of L/c and $3L/c$, which are indicated by dashed lines on the plot. When the contributions from the different axial modes align, they reinforce each other to form a pulse. The output of this mode-locked laser therefore has a repetition rate of $c/2L$. In a real laser there would be many more amplitude fluctuations over this range of time. Also, the apparent repeat in the pattern of the output shown in (A) would tend not to occur, since individual axial modes will occasionally change phase.

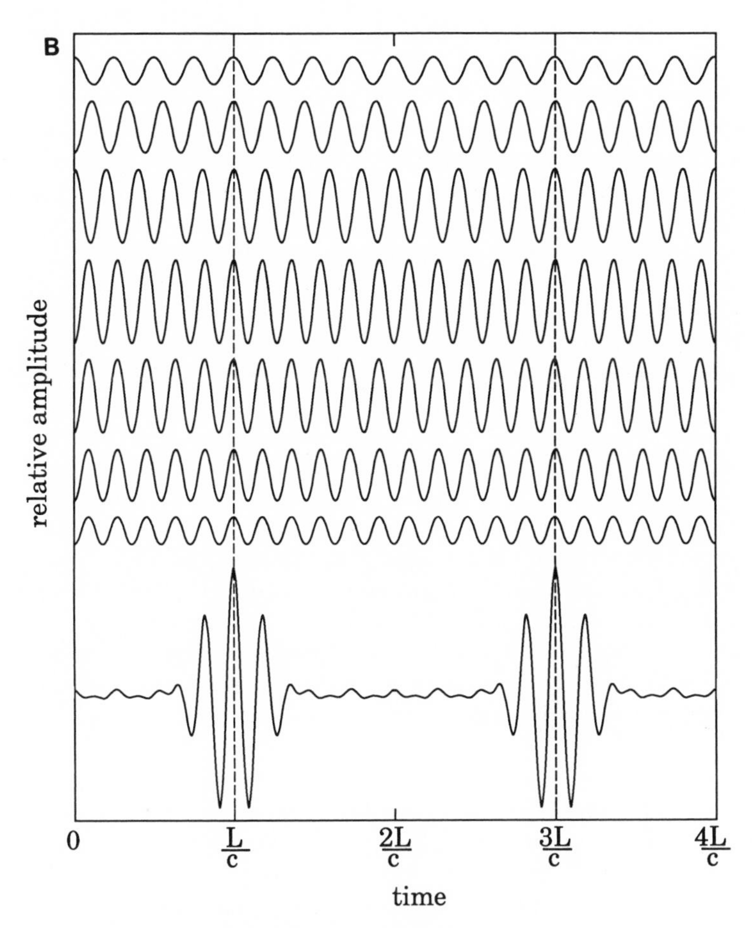

Figure 2.16. (*Continued*).

In the mode-locked laser, a single pulse of light oscillates back and forth between the end mirrors. (Mode locking can also be done using higher multiples of the axial mode frequency, resulting in more than one pulse, but this will not concern us here.)

Expressed more formally, the output amplitude of a laser, $E(t)$, can be described in the time domain by the sum of the mode amplitudes. In complex notation:

$$E(t) = \sum_j E_j \exp[i\omega_j t + \phi_j(t)] \tag{2.25}$$

where the E_j are the individual axial mode amplitudes, the ω_j are their frequencies, and the ϕ_j are the corresponding phases. In the case of a cw laser, the ϕ_j differ from mode to mode, but for a perfectly mode-locked laser they will be equal. Equation (2.25) shows that the amplitude in the time domain is the Fourier transform of the amplitude in the frequency domain. From the properties of Fourier transforms, we therefore know that the bandwidth of the laser or the frequency spread of the light is inversely related to the pulse width in time. The wider the bandwidth, the narrower is the pulse width that can be obtained. Looking back at Figure 2.16B, the way to get narrower pulses would be to sum more different frequencies together. Changing the frequencies to be summed (as long as they retain their relationship to one another) will not change the shape of the pulse envelope in time.

This inverse relationship between bandwidth and pulse width derives directly from Heisenberg's uncertainty principle. We usually see this principle in the form

$$\Delta E \, \Delta t \approx \hbar \tag{2.26}$$

The greater the uncertainty in the energy of the light (bandwidth of the laser), the less is the uncertainty in the time of the pulse (pulse width). As we learn in quantum mechanics, a more rigorous expression of Heisenberg's uncertainty principle requires the use of Fourier analysis as in the example presented here (see, for example, Ref. 32). The narrowest pulse that may be obtained from a perfectly mode-locked laser is limited by this Fourier transform relationship, and a pulse which approaches this fundamental limit is called *transform-limited.*

This Fourier relationship can be used to diagnose how well a laser is mode-locked. A commonly used rule of thumb is that a pulse is nearly transform-limited if its pulse width [full width at half-maximum (fwhm)] times its bandwidth (fwhm) is about 0.4. (Take the pulse width in nanoseconds and multiply it by the bandwidth in gigahertz.) This product will actually depend on the shapes of the pulse width and bandwidth functions. A number of these products have been tabulated[33] and are shown in Table 2.1.

The pulses from the Spectra-Physics lasers can be reasonably close to

Table 2.1. Pulse Width–Bandwidth Products for Several Pulse Width Functions

Pulse width function	Pulse width–bandwidth product[a]
Gaussian	0.441
Squared hyperbolic secant	0.315
Lorentzian	0.221
Symmetric two-sided exponential	0.142

[a] These values were taken from Ref. 33.

transform-limited. For example, a typical pulse width from the mode-locked model 171-09 argon ion laser is about 140 ps, which when multiplied by the bandwidth of 6 GHz gives a product of 0.8. Similarly, the series 3000 Nd:YAG laser has a pulse width of about 80 ps, which when multiplied by its bandwidth of 9 GHz yields a product of 0.7. Both of these lasers have pulse shapes which approximate Gaussians. When mode-locked through synchronous pumping, the model 375 dye laser is even closer to the transform limit. It can produce pulse widths of less than 2.5 ps, which when multiplied by a bandwidth of 60 GHz (using a birefringent filter as a wavelength-tuning element) gives a product of 0.15. Since the pulse shape is described as a two-sided exponential, this is essentially the transform limit.

We do not operate our dye laser near the transform limit, because for our studies there is no need for such a narrow pulse. Instead, we tune our dye laser for higher average power and obtain a pulse width in the range of about 10–25 ps, still more than adequate for measuring fluorescence lifetimes which are greater than 20 ps or so. A laser tuned in this way gives better long-term stability as well as increased output.

2.2.3.2. The Acousto-Optic Mode Locker

The Spectra-Physics mode locker on our Nd:YAG laser consists of a small fused-silica prism placed horizontally near the rear end mirror of the laser, as illustrated in Figure 2.17. The beam passes through the upper edge of the prism and reflects off the back mirror. The two upper sides of the prism

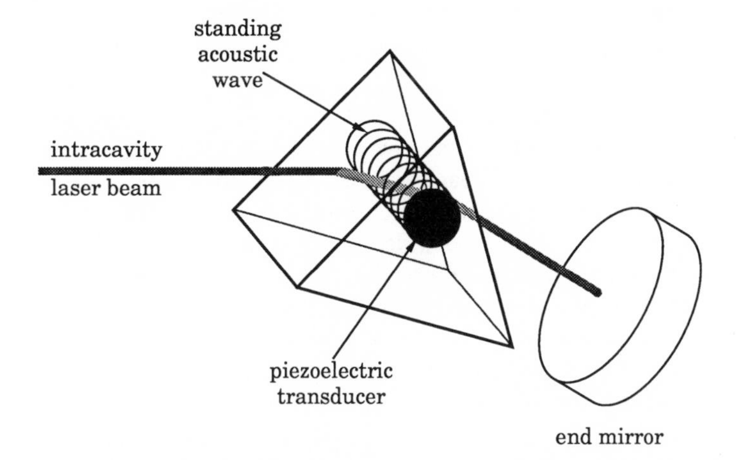

Figure 2.17. A schematic picture of the acousto-optic mode locker used in the Spectra-Physics argon ion and Nd:YAG lasers. A piezoelectric transducer produces a standing acoustic wave across a small fused-silica prism. The acoustic wave acts as a periodically changing diffraction grating, introducing optical losses in the laser cavity at twice the acoustic frequency.

are polished, and the angles are chosen such that the beam can both enter and exit at Brewster's angle to minimize reflective losses and vertically polarize the light in the cavity. The two triangular ends of the prism are polished as well to permit acoustic resonance between them. A small piezoelectric transducer at one end of the prism radiates acoustic waves inward at 41 MHz. At first, these waves enter with spherical wave fronts, but soon resonance occurs between the two polished ends, and standing waves become established. Since the velocity of sound in fused silica is 5.90×10^5 cm/s, the acoustic wavelength for the 41-MHz frequency is about 0.041 cm. The critical length under these conditions [from Eq. (2.21)] is about 0.64 cm for the 514.5-nm argon ion laser or about 0.31 cm for the 1064-nm Nd:YAG laser. The actual dimensions of the standing wave in the prism are determined by the size of the transducer, and this is proprietary information. However, one must conclude that the path the beam must travel is smaller than the critical length, since the mode locker operates near the upper limit of the Raman–Nath regime. Thus, the mode-locker prism acts as a periodic diffraction grating, with the Debye–Sears effect appearing and disappearing at twice the frequency used to drive the transducer. When the effect appears, light energy in the resonant cavity is diffracted out of the beam, creating a loss which reduces the gain of the resonator (Eq. 2.9). When the effect disappears, the light is free to pass. The resulting pulse train has a repetition rate of 82 MHz with pulses separated by about 12.2 ns, the round trip transit time for light in the laser. A single pulse resonates back and forth between the mirrors.

2.2.3.3. *Mode Locking the Dye Laser by Synchronous Pumping*

We have now discussed the pumping laser, either a mode-locked argon ion laser or a frequency-doubled mode-locked Nd:YAG laser (see Figure 2.1), in significant detail. In the Spectra-Physics systems, either pump laser provides a train of reasonably narrow pulses at 82 MHz in the green region of the spectrum. The pump beam is focused onto the dye stream of the dye laser, which absorbs a significant fraction of the incident light, generating a population inversion. The pulse width of the pump laser is of the order of 100 ps or so, but the dye laser becomes mode-locked with much narrower pulses. As mentioned in Section 2.2.3.1, pulse widths can be nearly transform-limited and as narrow as 1.5 to 2.5 ps depending on the wavelength-tuning option chosen. With a much higher frequency bandwidth than the pump laser, it is not surprising that the mode-locked pulse could be narrower, but how is this achieved?

The process is called synchronous pumping. To synchronously pump, one must match the dye laser cavity optical length almost exactly to that of the pumping laser. Matching is done such that the round trip transit time of a pulse in each laser is almost identical, so that each laser will have the same mode-locked repetition rate. This requires stable alignment of the lengths of both lasers to within about a micron of each other. It ensures that the pump pulse and the dye laser pulse reflecting off the far end mirror of the laser will arrive at the dye stream precisely aligned with one another in time. The pump pulse creates the population inversion at exactly the right time for the dye laser pulse to induce stimulated emission. Thus, the dye laser, unlike the pumping laser, is mode-locked by periodically increasing the gain in the cavity rather than modulating a loss.

The events at the dye stream are shown schematically in Figure 2.18. First, the relatively long pump pulse arrives. Some residual excitation remains from the previous event, and therefore the gain of the laser does not begin at

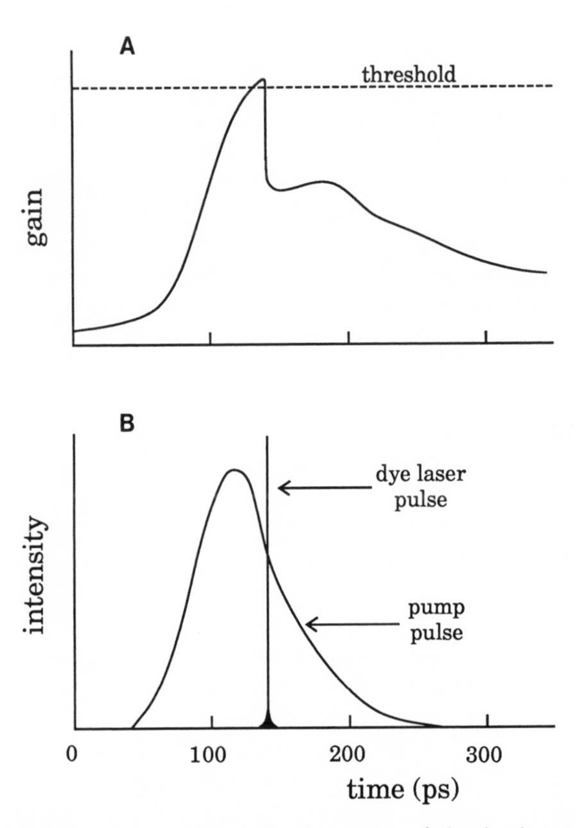

Figure 2.18. A diagram of the events at the dye stream of the dye laser mode-locked by synchronous pumping. The arrival of the pump pulse causes an increase in the gain of the dye laser. The circulating dye laser pulse is timed to arrive at the dye stream just after the gain rises above the threshold for lasing in the dye laser. The intensities of two pulses in the lower section of the figure are not drawn to scale. The resulting peak power of the dye laser is about an order of magnitude higher than that of the pumping laser.

zero. The arriving pump pulse pushes the gain up quickly, and as soon as it passes the threshold value, lasing can occur. Because of the careful match of the repetition rates of the pumping and dye lasers, this is the moment of the arrival of the circulating mode-locked pulse in the dye laser. Stimulated emission, excited by the circulating pulse, very rapidly depletes the population inversion, suddenly dropping the gain well below the threshold value. This fast depletion is greatly favored by an unusual property of dye, an extremely high stimulated emission cross section. This means that a particular excited dye molecule has a very high probability of interacting with the circulating beam and undergoing stimulated emission. The sudden drop in gain extinguishes the stimulated emission, and further absorption of the current pump pulse is insufficient to again raise the gain above threshold.

As mentioned earlier, since our applications do not require the very narrow, nearly transform-limited pulses, we do not optimize the synchronous pumping process but instead opt for higher output intensities. To do this, we take the output of the dye laser, frequency double it to the UV, scatter it with a dilute Ludox solution (an aqueous suspension of silica available from Dupont), and measure its intensity using our decay fluorometer. The dye laser is then adjusted or *tweaked* to maximize the UV intensity. Since the UV intensity depends on the square of the dye laser pulse intensity, efficient doubling requires a reasonably narrow pulse shape. We can therefore increase overall power without drastically degrading the pulse shape.

The resulting dye laser pulses are tunable in wavelength over a range of about 60 nm (see Figure 2.15), depending on the dye used and its concentration. The repetition rate of our dye laser is, of course, 82 MHz, the same as that of the pump laser. This is too fast for pulsed fluorescence decay measurements in the nanosecond time range, and something must be done to limit this high rate. The cavity dumper accomplishes this for us, as well as providing a very efficient extraction of the pulses from the dye laser cavity.

2.2.3.4. Cavity Dumping

So far, for each of the lasers we have discussed, a fraction of the light energy has been passed out from the resonator through a partially transmitting mirror. The cavity dumper, an acousto-optic device, instead extracts a fraction of the number of dye laser pulses by acting like a very fast mirror with direct access to the relatively high intracavity power. The active component of the cavity dumper consists of a small block of fused silica with a piezoelectric transducer attached to one edge. The transducer sends a very high frequency acoustic pulse (779 MHz) into the block. A pulse of this frequency has a wavelength of 7.6×10^{-4} cm, and from Eq. (2.21) (for 600-nm light in the dye laser), the critical length is easily calculated to be 2.4×10^{-3} cm. This is much less than the pathlength of the light traversing the acoustic wave within the cell. Thus, this acousto-optic device works in the Bragg regime and is called the *Bragg cell.* It is mounted between two spherical mirrors at the far end of the dye laser in the "double pass" configuration[34] shown in Figure 2.19. M1 is the end mirror. M1 and M2 focus the beam to a small point in the cell. The prism deflects the output pulses below M2 and out of the cavity dumper housing. Because of the three-dimensional nature of the process, two views are shown.

The Bragg cell is oriented so that the intracavity beam strikes it in the vertical dimension at Brewster's angle to minimize reflective losses and favor the lasing of vertically polarized light. In the horizontal dimension, the intracavity beam strikes the cell at Bragg's angle [$\sim 2.3^\circ$ from the normal—from Eq. (2.22)]. Since the acoustic wave is directed sideways through the cell, the first Bragg reflection of the intracavity beam is shifted down in frequency by Ω, the angular frequency of the acoustic wave. The deflected beam reflects off M1 and back to the Bragg cell, where the process is reversed. A fraction of the beam is deflected back into the cavity and the remainder passes through to M2, the prism, and out of the housing. The part of the original beam which passes through without being deflected is reduced in intensity. It reflects off M1 and back to the same point on the Bragg cell, where a fraction of it undergoes a Bragg reflection, up in frequency by Ω. The remainder passes back into the cavity. The two frequency-shifted reflections which exit the resonator, one up in frequency and one down, are superimposed on each other. The up- and down-frequency-shifted beams have essentially the same amplitudes and can interfere with each other either constructively or destructively depending on the phase relationship between the acoustic wave and the light beam in the laser. As we will see, this double-pass arrangement allows one to adjust the phase of the acoustic wave to reinforce and efficiently transfer a pulse out of the laser, or to eliminate it almost entirely.

In the figure, we assume that the amplitude of the light in the laser is

$$E(t, s) = E_0(t, s) \cos \omega t \tag{2.27}$$

This is the same standing wave shown in Eq. (2.11), except that now the position dependence (described by the variable s) and the pulse shape information have all been included in the amplitude $E_0(t, s)$. If γ is the single-pass diffraction efficiency of the acoustic cell (fraction of the light intensity reflected on each pass—remember that the intensity is proportional to amplitude squared), then one can easily calculate the output amplitude, $E_{\text{out}}(s, t)$, of the cavity dumper.[34] A little reflection shows that $E_{\text{out}}(s, t)$ has the form

$$E_{\text{out}}(s, t) = (1-\gamma)^{1/2}\, \gamma^{1/2} E_0(s, t)[\cos(\omega - \Omega)t + \cos(\omega + \Omega)t] \tag{2.28}$$

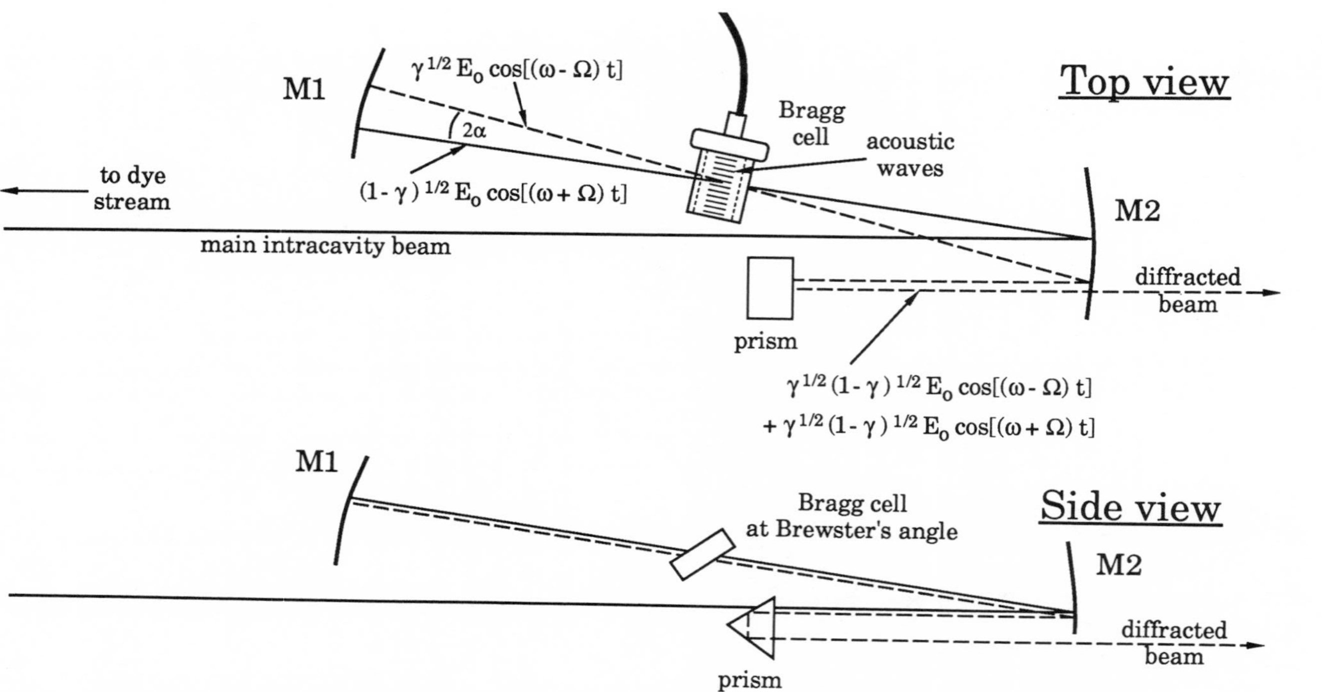

Figure 2.19. The optical components of the cavity dumper. A Bragg cell is arranged in a double-pass configuration to deflect the mode-locked pulses out of the dye laser. The intracavity beam is shown as a solid line, and deflected beams are shown as dashed lines. When a light pulse from the intracavity beam strikes the Bragg cell, part of it passes straight through and part is deflected and shifted down in frequency by Ω, the acoustic frequency. This is the first pass. The deflected portion reflects off M1 and back through the Bragg cell. A fraction of this light pulse is deflected again back into the cavity, but most passes through to be deflected out of the laser by the prism. The undeflected portion of the intracavity pulse also reflects from M1 back toward the Bragg cell, where it passes through a second time. Again, a fraction of the pulse is deflected, but this time it is shifted up in frequency by Ω. This deflected pulse from the second pass is superimposed on that from the first pass. The diffracted beam exiting the laser thus consists of pulses which contain equal amplitudes of light shifted both up and down in frequency by Ω.

Squaring the amplitude to find the output intensity, $I(t)$:

$$I(t) = \gamma(1-\gamma)\,E_0^2(s,t)(1+\cos 2\Omega t) \tag{2.29}$$

We find that the intensity of the output pulse varies as a cosine function of time at *twice* the frequency of the acoustic wave. It is therefore possible to time the acoustic pulse in the Bragg cell relative to the phase of the laser pulse such that either a maximum intensity is deflected out of the beam or essentially no intensity at all. Spectra-Physics calls this method "integer plus $\frac{1}{2}$ timing." It is designed to specifically extract a particular pulse, while suppressing both the previous mode-locked pulse and the trailing one.[34] The acoustic frequency has been chosen so that, when divided by the mode-locking frequency, it yields an integer plus $\frac{1}{2}$. In this case, 779 MHz ÷ 82 MHz is equal to $9\frac{1}{2}$. If one pulse is timed to arrive at time equal to 0, then it will be efficiently deflected out of the cavity. The next pulse will then arrive $9\frac{1}{2}$ cycles later (when $1+\cos 2\Omega t$ is equal to zero), and most of it will not be deflected because of destructive interference between the up- and down-shifted reflections. The timing of the pulses relative to the function $1+\cos 2\Omega t$ is shown in Figure 2.20.

The radio frequency (rf) driver for the cavity dumper produces an envelope of high-amplitude oscillation at 779 MHz, at a repetition rate that can be set on the front of the instrument. It is synchronized with the circulating dye laser pulse using a diode. This pulse, which produces a similarly shaped acoustic wave in the Bragg cell, is illustrated at the top of Figure 2.20. The center and maximum-amplitude point of the acoustic wave corresponds to the point which diffracts the narrow dye laser pulse out of the cavity. At a time 12.2 ns before and after the passage of the highest point where the adjacent mode-locked pulses might be deflected, destructive interference, as described above, prevents the light from being deflected. The acoustic wave is short enough so that mode-locked pulses even further away cannot be significantly deflected. A combination of the destructive interference and diminishing intensity of the acoustic wave provides an *extinction ratio* (ratio of the intensity of the main deflected pulse to that of either of the two adjacent pulses) which is specified by Spectra-Physics to be at least 500:1. Our lasers have consistently exceeded this specification.

There are four adjustments on the cavity dumper driver: repetition rate, power, phase, and timing. The repetition rate we generally set to 800 kHz. The power level allows us to set the amplitude of the acoustic wave to maximize the efficiency of the Bragg reflection. The phase control permits us to move the phase of the rf signal within the overall envelope of the pulse, and thereby maximize the extinction ratio. The timing moves the envelope without changing the phase, thus increasing one adjacent pulse while decreasing the other.

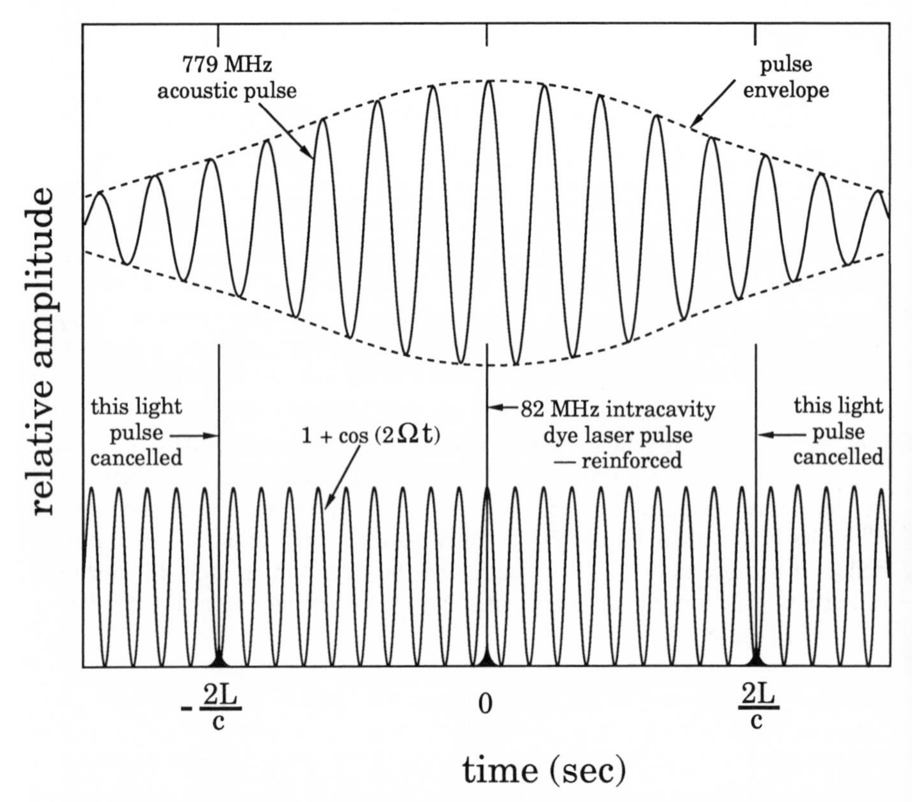

Figure 2.20. A diagram illustrating integer plus $\frac{1}{2}$ timing in the cavity dumper. Three adjacent mode-locked dye laser pulses are illustrated in the bottom part of the figure. Since the repetition rate of the laser is 82 MHz, they are separated in time by $2L/c$ or about 12.2 ns, where L is the optical cavity length of the dye laser. The acoustic pulse, which is sent down through the Bragg cell by the transducer, is illustrated in the top part of the figure. Its frequency is chosen to be 779 MHz, which when divided by the repetition rate yields $9\frac{1}{2}$, an integer plus $\frac{1}{2}$. Due to the double-pass configuration in the cavity dumper, the dye laser output consists of two light beams, one shifted up in frequency by the acoustic frequency and one shifted down. They thus interfere with one another in a manner which modulates the output at a frequency twice that of the acoustic wave. This modulation function is shown in the bottom part of the figure. It shows how the output of the laser would be modulated, if the laser were operated with continuous light rather than mode locked. Depending on the time of arrival of the intracavity dye laser pulse relative to that of the acoustic pulse, the output pulse if either reinforced or cancelled.

2.2.3.5. Subsequent Manipulations of the Light Pulse Train

The laser systems described so far produce a train of picosecond light pulses with tunable wavelengths at a repetition rate that we usually choose to be 800 kHz. The wavelength range that we can obtain with our Nd:YAG system is from about 550 nm to about 640 nm, depending on the dye we use.[31] This range can be greatly extended toward the red by the use of other dyes and a different set of reflective optics for the dye laser. Much of our work requires the use of UV, and we therefore usually frequency double the output.

A schematic diagram of the optical components that we place between the laser system and the sample cuvette is shown in Figure 2.21. After the beam passes out of the cavity dumper, it is reflected with a front surface mirror (Newport Corporation enhanced dielectric-coated surface ER.1) toward the center of the table. A long-focal-length (250 nm) glass lens focuses the light through an optional type I KDP angle-tuned doubling crystal. The doubling cyrstal (purchased from Cleveland Crystals) is mounted in a cell filled with an oil with good UV light transmittance and a refractive index roughly matching that of the crystal and the windows on the cell. Also, the entrance window is antireflection-coated for 600-nm light, and the exit window is coated for 300-nm light. This greatly minimizes reflective losses as well as reflections which might contribute to the beam and be measured by the instrument. The cell is mounted on a stable holder which permits smooth and easy angle adjustments. The fundamental and second-harmonic beams exit the frequency doubler superimposed on one another and pass through an

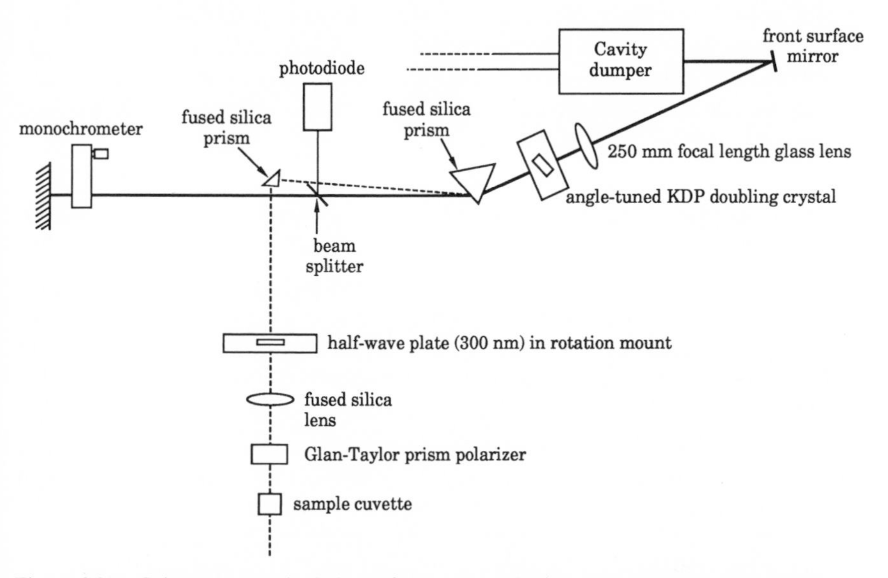

Figure 2.21. Subsequent manipulations of the output of the laser system. The dye laser output is focused through a frequency doubler, and the second harmonic is split from the fundamental using a prism. A fraction of the fundamental is used to generate a pulse to time the laser flash, and the remainder is directed to a monochromator to calibrate the wavelength of the laser. The UV beam is deflected toward the fluorometer and passed through a half-wave plate which rotates the plane of polarization and is used for intensity control. The beam is then focused through a prism polarizer onto the sample.

apex of an equilateral prism, which separates them. The fundamental then passes through a beam splitter, and the nondiverted portion is directed to a Schoeffel Instrument Corporation GM100 miniature monochromator. This is used to calibrate the wavelength of the laser. The fraction of the beam which is diverted by the beam splitter is directed toward a fast photodiode which is used to supply a timing pulse for the decay fluorometer. The UV beam is reflected 90° toward the sample using a small fused-silica 45° prism. (A prism is used rather than a mirror, as good stable reflective surfaces for UV light are difficult to obtain.) In order to use the laser output directly without frequency doubling, one simply moves the 45° prism into the fundamental beam.

The dye laser output itself is vertically polarized. As we have discussed, the type I frequency doubler produces the second harmonic at right angles to the fundamental. In order to rotate the polarization of the incoming excitation beam, we pass it through a half-wave plate. We then focus it with a fused-silica lens through a calcite crystal polarizer and onto the sample. A half-wave plate cut to the proper thickness for a particular wavelength will rotate plane-polarized light to any desired orientation. Our half-wave plate is cut for 300-nm light. At other wavelengths, this plate converts the incident plane-polarized light to an elliptically polarized beam,[†] but the final polarizer will convert it back to plane-polarized light of the desired orientation. One advantage of using the half-wave plate in this way is that it can be rotated to change the intensity of light incident of the sample. It will be straightforward to add a servo mechanism for controlling light intensity on the sample by rotating the plate.

2.3. The Microchannel Plate (MCP) Photomultiplier

In our pulse fluorometer, we excite a fluorescent sample with a brief flash of the polarized laser light. Some of the light is absorbed by the sample; some of the excited molecules subsequently emit fluorescence; and a small fraction of the emitted light is collected by the remaining optics and directed to a detector. The low levels of light reaching the detector are best treated as individual photons. The detector and associated electronics must be able to detect single photons and, as precisely as possible, time their arrival relative to the laser flash. Detectors which have been used in single-photon pulse fluorometry include photomultipliers with traditional dynode structures, microchannel plate (MCP) photomultipliers, and, very recently, avalanche photodiodes. While there are reasons why one might choose any of these options, the MCP photomultiplier best fulfills the needs for our instrument. It is the only device now available which can supply extremely fast response

[†] One frequently used but relatively expensive solution to this problem is to use a variable retarder called a Soleil–Babinet compensator.

at high quantum efficiencies throughout the UV and visible regions of the spectrum. The purpose of this section is to explain in relatively simple terms how an MCP photomultiplier works and to give some examples of the performance of such tubes.

2.3.1. How an MCP Photomultiplier Works

To detect a photon using a photomultiplier, one depends on the photoelectric effect. A single photon of light strikes a photocathode within an evacuated tube. Its energy ionizes a single atom and imparts kinetic energy to the released electron. If the electron succeeds in escaping from the photocathode material into the vacuum surrounding it, the electron is accelerated in an electric field in the direction of the anode of the tube, which is positively charged relative to the cathode. (Sometimes the anode is held at ground potential and the cathode is given a negative charge; sometimes the cathode is held at ground potential and the anode is given a positive charge.) Since a single photoelectron would not give rise to a measurable current pulse at the anode, an electron multiplier is inserted in its path. A single photoelectron thus gives rise to a burst of electrons at the anode, which is detected electronically as a small pulse of current.

The two different kinds of photomultipliers, illustrated in Figure 2.22, differ in how they amplify the current in the electron multiplier. A traditional photomultiplier uses a series of dynodes, charged metal structures coated with a secondary emitting surface of a semiconductor material. Several different designs are used, but the one that is diagramed in Figure 2.22A is the linear focused design in the RCA 8850 type of tube that we have used. Each dynode in the chain is given a progressively higher positive charge relative to the cathode. The photoelectron strikes the first dynode, displacing a few electrons, which are accelerated to the second dynode. At the second dynode, each of these electrons again displaces a few electrons, to be accelerated and multiplied as they travel further down the chain until they arrive as a pulse of current at the anode.

An MCP photomultiplier (Figure 2.22B), on the other hand, uses thin plates of glass which have many microscopic channels through them. The surfaces of each plate between the channels are coated with a thin conducting layer, and a large voltage (~ 1000 V) is placed across the thickness of the plate. Each channel is lined with a secondary emitting surface and functions as an individual electron multiplier. The photomultiplier illustrated in Figure 2.22B is of the *proximity-focus* design, in which the photocathode is placed very close (~ 3 mm) to the first MCP. As illustrated in the inset to the figure, the photoelectron travels the short distance to the first MCP, enters one of the channels, and strikes the wall producing secondary electrons. These electrons

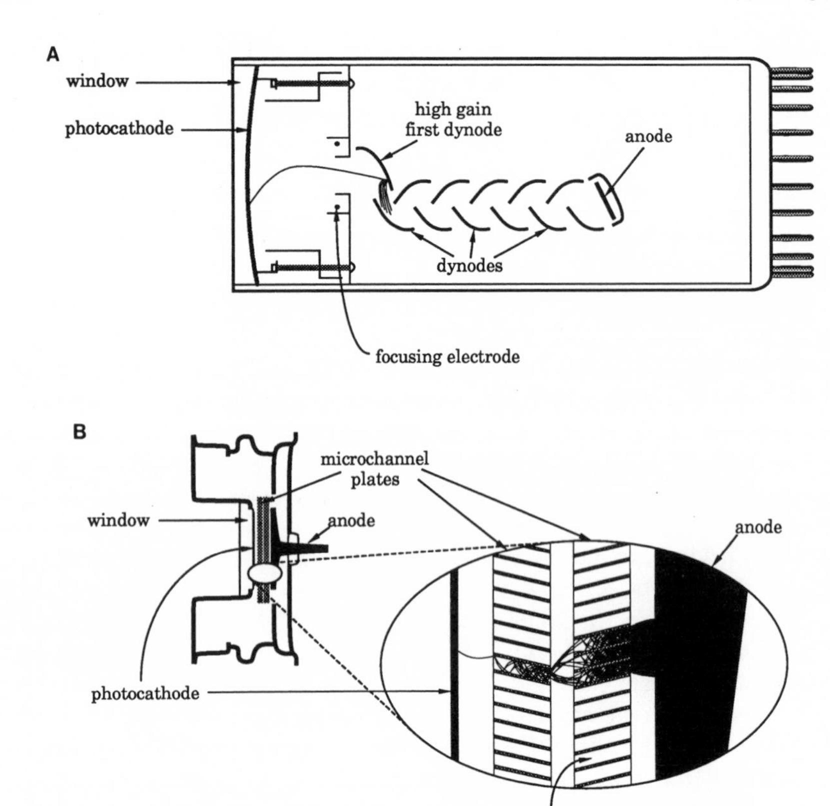

Figure 2.22. A comparison between a traditional photomultiplier with dynodes and a microchannel plate photomultiplier. (A) The RCA 8850 (or C31000M) is a traditional photomultiplier of the linear focused design. A single photoelectron is shown ejected from the photocathode. It is accelerated in an electric field which focuses it onto a high-gain, gallium phosphide-coated first dynode. Between 30 and 50 secondary electrons are accelerated through a series of lower gain dynodes (5–8 electrons emitted per collision) to become a pulse of current at the anode. (B) In a proximity-focus MCP photomultiplier, the photoelectron is accelerated only a short distance before it strikes the first microchannel plate. If it passes into one of the channels, it strikes the semiconducting surface and causes the ejection of an average of between one and two secondary electrons. The secondary electrons are accelerated further down the channel, colliding with the walls and causing the release of more electrons. Between the plates the electrons spread out and pass into about seven channels of the next MCP, where they are again multiplied to give a cloud of electrons which strike the anode to produce a pulse of current.

in turn induce more secondary events, cascading through the channel and exiting from the far end of the channel. The electrons then spread out and enter a number of adjacent channels in the next MCP, where they again increase greatly in numbers. MCP photomultipliers typically have two or three plates. After the cascade passes through the last plate, it is collected at the anode as a pulse of current. Since the electrons have much shorter overall distances to travel in the MCP as opposed to the dynode chain, there is much less fluctuation in the lengths of time between the arrival of photons and the subsequent anode pulses, and there is much less spread in time in the burst of electrons as it travels down the tube. For these reasons, the MCP tubes can time events much more accurately than the dynode-based tubes. For a pulse fluorometer, this translates into a narrower instrument response time which is very insensitive to the wavelength of the light it is measuring.

Two important properties of photomultipliers are the *quantum efficiency* and gain of the electron multiplier. The quantum efficiency is the ratio of anode pulses produced per photon striking the cathode. It depends on the ability of the light to penetrate the window of the tube, the efficiency of the photocathode material, and the probability that a released photoelectron will give rise to an anode pulse. The quantum efficiency is a function of the wavelength of the light striking the tube and ranges from 0 up to about 0.3 for some photomultipliers (the maximum is somewhat lower for MCP tubes). The gain, or the number of electrons which arrive at the anode for each primary photoelectron, is typically (for a double Hamamatsu MCP photomultiplier) in the range of about 5×10^5. It shows very little dependence on the wavelength of the light.

2.3.1.1. The Photocathode

Any material will emit electrons if it is struck with an appropriate wavelength of electromagnetic radiation, but to make a photocathode which will do so when struck by light in the visible or near-UV regions of the spectrum, one must choose elements which can be readily ionized. Metals can serve this function and have been used for photocathodes, but metals are very inefficient. They reflect most of the light which strikes them and do not readily release the electrons which become freed from the atoms by ionization. For very efficient photocathodes, one must use compounds which act as semiconductors. It is worth very briefly reviewing the properties of such materials.

If one calculates the allowed energy levels for electrons in the periodic lattice of a solid, one finds that they are grouped into distinct bands of energy. The lowest band is called the filled band. It does not participate directly in the interactions with surrounding atoms. The next level, which is involved with the bonding between atoms, is called the valence band. The highest energy band contains electrons which are relatively free to move about in the solid and is called the conduction band. The energy bands are separated by gaps, or ranges of energy that no electron can possess in the solid. The energy separation between the top of the valence band and the bottom of the conduction band is commonly called the *band gap*. The energy levels are filled according to the Pauli exclusion principle, beginning at the lowest filled band

and distributed upward to higher levels with populations in different levels filled according to a Boltzmann distribution. In an insulator, the conduction band remains empty and is separated by a large band gap from the valence band. Thermal energy is insufficient to promote an electron from the valence band to the conduction band, and electrons therefore are unable to freely move throughout an insulator. In a metal, on the other hand, the conduction band may be partially filled or the valence band may overlap in energy with the conduction band. The application of an electric field can move the free electrons from one state in a band to another of nearly identical energy, and current flows. A semiconductor is intermediate between the insulator and conductor in that the band gap is very small. Thermal energy alone is sufficient to promote some electrons across the gap into the conduction band. The application of an electric field causes the movement of free charge carriers (either negative electrons or positive *holes*) from one state in a band to another, and current will flow, but not as freely as in a conductor.

The energy levels of a semiconductor photocathode are shown schematically in Figure 2.23. The band gap energy is labeled E_g. In the photocathode, a single photon promotes an electron from the valence band to the conduction band, imparting kinetic energy. The amount of total free energy imparted to the electron is equal to the energy of the photon of light. For example, at 400 nm the electron would receive 3.1 eV. To be counted, the electron must then diffuse through the photocathode material to the surface and escape into the surrounding vacuum, overcoming the surface potential or electron affinity of the material (labeled E_a in the figure). The electron may suffer collisions within the material on the way to the surface, losing sufficient energy so that it no longer possesses enough to leave the photocathode. Therefore, minimizing these collisions will increase the probability of the successful release of a photoelectron and consequently increase the quantum efficiency of the photomultiplier.

There are two kinds of collisions that are significant. The first is called *lattice scattering*, in which the electron collides with an atom and introduces vibrational energy or a phonon into the crystalline lattice. The energy loss for such a scattering event is quite small, typically of the order of about 0.01 eV, and the mean free path through the material is typically 30–40 Å. In the absence of other scattering events, electrons produced at depths of several hundred angstroms can have sufficient energy to escape from the material. The second kind of scattering event is called *electron–hole pair production* or *impact ionization*. In this case, an electron with energy over a threshold value (labeled E_{th} in the figure; generally more than $2 \times E_g$) can collide with a valence electron, knocking it free and leaving a positive "hole." Such a collision costs the electron a relatively large amount of energy. This energy loss is obviously greater than or equal to the band gap energy, since that is the amount of energy required to release the second electron from its atom.

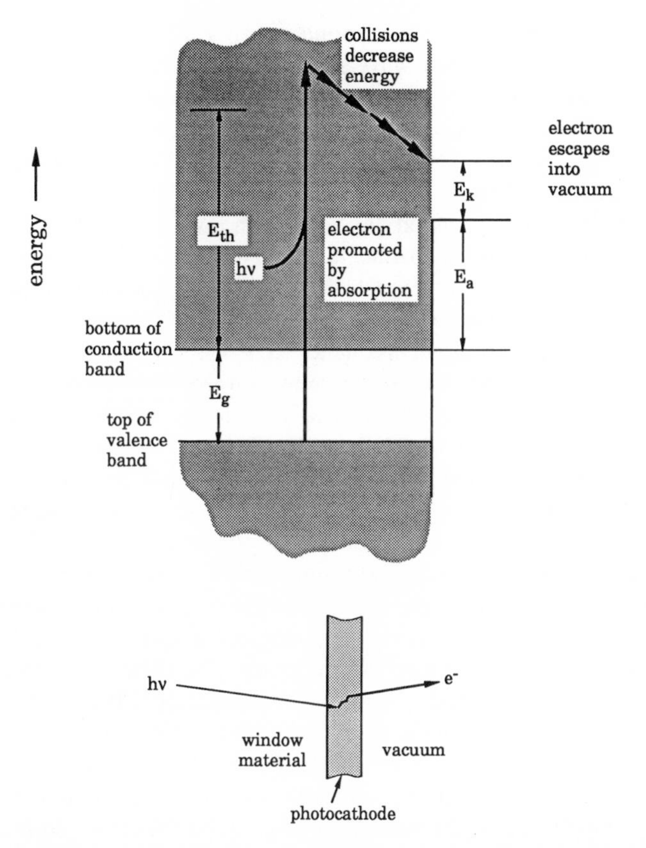

Figure 2.23. Energy level diagrams for the emission of a photoelectron from the photocathode. The event itself is illustrated in the lower section of the figure. A photon penetrates the window and enters the photocathode coating on the inside of the tube. An electron released on the absorption of the photon diffuses to the surface and is released into the vacuum of the tube. The left side of the upper part of the figure shows the energy levels within the material, and the right side shows the relative energy levels for the electron which is released. When a photon is absorbed by the material, an electron is promoted from the top of the valence band into the conduction band. Various collisional events decrease the free energy of the electron. If the electron reaches the surface of the photocathode with sufficient free energy to overcome E_a, the electron affinity of the material, it leaves with a kinetic energy of E_k.

Photocathode materials are chosen to have high values of E_{th} to minimize the number of these collisions. In most photocathodes, the mean free path for electron–hole pair production will generally be of the order of 10–30 Å. Finally, if the photoelectron reaches the surface with enough kinetic energy to overcome the electron affinity of the material, it is then released from the surface. It leaves with the remaining kinetic energy, E_k. (The interested reader will find a discussion of these processes in a paper by Ghosh.[35])

The photocathode cannot respond at all unless the wavelength of the light is less than a threshold wavelength, because an electron must be given sufficient energy $(E_g + E_a)$ to escape. The quantum efficiency of the cathode will then increase with increasing frequencies of incident light, because the higher energy imparted to the photoelectron will permit its escape from deeper inside the material. In the high-frequency range, the photocathode response is generally limited by the window material on which the photocathode is coated. Borosilicate glass, for example, has a wavelength cutoff of about 300 nm, whereas UV-transmitting glasses generally transmit light well below 250 nm and fused silica well below 200 nm. It is important to note, however, that the upper frequency end of the photomultiplier response curve may not be a usable range for pulse fluorometric measurements, since window materials can introduce artifacts into decays because of luminescence, when used at wavelengths where they absorb significantly.[36]

The same photocathode materials can be used for either traditional dynode or MCP photomultipliers. Those that we have used, for both kinds of tubes, are called bialkali and multialkali photocathodes. Each consists of polycrystalline coatings of alkali antimonides prepared by exposing antimony layers on the window material to alkali vapors at elevated temperatures in different sequences. The maximum quantum efficiency for the release of photoelectrons is about 0.3 for both materials. The bialkali photocathode has the formula Na_2KSb, and the multialkali consists of a Na_2KSb layer with a thin layer of K_2CsSb over it. Both of these photocathode materials have band gap energies of about 1.1 eV, but they differ in electron affinity. The bialkali material has an electron affinity of 1.0 eV and therefore should have a long-wavelength threshold of about 600 nm[35] (1.1 + 1.0 eV is equivalent to a photon of ~600 nm). Apparently, special photocathode preparations for the bialkali photocathode can also lower the electron affinity somewhat, because the photomultipliers that we have used with bialkali photocathodes show sensitivity out to about 660 nm. The multialkali photocathode can have a much lower electron affinity of about 0.24 eV and a response curve which extends out to about 925 nm in the near UV.†

Most background noise in single-photon fluorescence measurements derives from thermal excitation of electrons in the photocathode. Thermal electrons released from the cathode pass through the electron multiplier and become full-sized pulses at the anode. Due to its relatively high electron affinity barrier, the bialkali photocathode typically produces very low noise levels, in the range of a little less than a hundred to a few hundred counts per

† For those with some familiarity with solid-state physics, it should be mentioned that the bialkali layer is a *p*-type material and the K_2CsSb layer is nearly intrinsic material. Thus, the multialkali photocathode contains a heterojunction, and this construction is responsible for the very low apparent electron affinity and extended red response.

second depending on the particular tube. This can translate into just a few counts per second when collecting a typical fluorescence decay. For example, if one were to collect data over a total range of 50 ns using a laser repetition rate of 800 kHz, only $\frac{1}{25}$ of the noise would actually be counted. Although this very low noise level could be reduced by operating the tube at lower temperatures, doing so would also slightly effect the efficiency of the tube. Therefore, generally nothing is gained by operating these tubes below room temperature. The multialkali tubes, on the other hand, have a lower electron affinity and a much higher noise level and consequently must be cooled for single-photon work.

The photocathode material can be easily destroyed. Exposure to significant levels of light, even in the absence of an applied voltage, can bleach it. The photocathode material is quickly destroyed by contact with air and easily damaged by ion bombardment from within the tube. Also, the uniform application of photocathode materials to different substrates inside electron tubes is a significant accomplishment. Differences in photocathodes provide a source of variation in the photomultiplier tubes that we use.

2.3.1.2. The Microchannel Plate

The manufacture of microchannel plates was made possible by the development of fiber-optic techniques for precise drawing of glass into fibers. A plate is typically made in the manner illustrated in Figure 2.24. A rod of chemically etchable core glass is placed inside a tube of cladding glass, typically of a lead glass composition. The core is ultimately etched out, and the cladding glass forms the MCP matrix. It is important, therefore, for the two glasses to be chosen such that they do not diffuse into one another when partially melted. The rod and cladding glass are drawn into single fibers which are stacked into a hexagonal array, fused together, and then drawn again. These hexagonal multifibers are then stacked and fused together within a glass envelope to form a *boule*. The boule is cut in slices about 1 mm thick at an angle of between 5 and 15° of the original fibers and polished. The soluble core glass is then chemically etched out of the plate, forming channels (typically 12 to 25 μm in diameter), and the entire plate is reduced in an oven in the presence of hydrogen gas. The reduction process produces a semiconducting layer (involving the formation of elemental lead from the lead oxide in the glass) on the inside of the channels for the secondary emission of electrons, described below. Nichrome or other metallic electrodes are then vacuum-deposited on both surfaces of the microchannel plate, and the plate is baked again at about 300 °C to facilitate outgassing of impurities.

When the plate is placed in a vacuum and a voltage applied across it, each channel can act as an electron multiplier. In the photomultiplier, a photoelectron emitted from the photocathode is accelerated toward the MCP

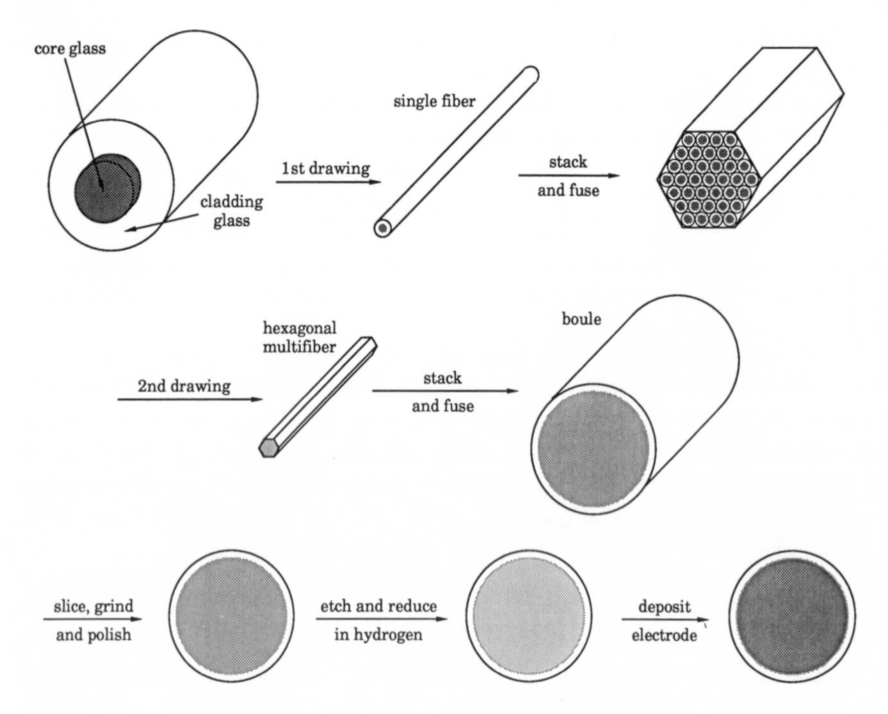

Figure 2.24. An illustration of one commonly used method for making microchannel plates.

and has a certain probability of striking the inside of a channel. If it does so and strikes with sufficient energy, then it causes secondary emission of electrons. Typically, between one and two electrons are released into the channel, where they are accelerated until they strike the wall of the channel further down. If they acquire sufficient energy (20–50 eV), they each in return cause the secondary emission of electrons and so forth down the channel. A large pulse of electrons exits at the far end of the channel.

If one would like to increase the gain of the MCP, then one could do so by increasing the length of the channels, but very long channels would become saturated in their ability to release secondary electrons. Increasing the voltage over the plate will increase the gain by increasing the energy of the secondary electrons as they strike the walls of the channel, increasing the secondary emission ratio (number of electrons released for each collision of an electron with the wall). Too high a voltage, however, will decrease the number of possible collisions with the wall, because the trajectory of each electron will cause it to travel further down the channel before it collides. Thus, compromises must be reached to maximize the gain of the MCP. In practice, a well-designed plate will have a maximum gain of about 10^4.

The highest gain is actually limited by a process called positive ion feedback. With positive ion feedback, secondary electrons can strike residual gas molecules in the tube, either free in the channels or adsorbed to the channel walls. If the energy of the collision is sufficient to ionize the molecule, it becomes a positive ion which travels back toward the photocathode in the opposite direction to the electrons. When the positive ion strikes the photocathode, it can release electrons which can act very much like photoelectrons. They are accelerated toward the electron multiplier, amplified, and ultimately counted as an event. In a standard photomultiplier tube such as the RCA 8850 with a dynode chain, this process requires about 20 ns and results in an "afterpulse" which appears in the instrument response function at about 20 ns after the main peak and with an intensity about three orders of magnitude lower than that of the main peak. In an MCP photomultiplier with much shorter path lengths, the process is much faster, on the order of 0.5 ns. Obviously, one would like to eliminate the positive ion feedback from the instrument response, because it distorts the function and also because bombardment with positive ions is one of the main causes for the aging of a photocathode.

There are a number of measures that can be taken to increase the gain of an MCP photomultiplier while minimizing positive ion feedback. First, microchannel plates are usually used in tandem, with the channels in one plate at angles relative to those of the other, as illustrated in Figure 2.22. For two plates or three plates, this is referred to as the "chevron" or "Z" configuration, respectively. Other manufacturers have developed plates with curved channels.† These are generally used in conjunction with a second straight channel plate. All of these configurations, the chevron, Z, and curved, reduce positive ion feedback because they make it difficult for the ions to travel all the way back to the photocathode without first colliding with a channel wall. Another feature of the Hamamatsu tubes that we have worked with is the use of an aluminum film on the entrance surface of the first MCP in the photomultiplier. This feature is apparently responsible for a great reduction of positive ion feedback in these tubes.

2.3.2. Photomultiplier Performance

For a single-photon fluorometer, we would like to be able to time the arrival of a single photon at the photocathode as accurately as possible, but statistical fluctuations of the events within the photomultiplier, as well as other artifacts, will place limits on the accuracy with which we can do this.

† Photomultipliers which use microchannel plates with curved channels were developed by the Laboratories d'Electronique et de Physique at Limeil-Brevannes, near Paris. Lo and Leskovar have evaluated the performance of these tubes.[37]

Ideally, the fluorescence decay, $F(t)$, is given by the convolution between the profile of the light flash, $E(t)$, and the impulse response function, $f(t)$. The function $f(t)$ is the fluorescence that would be observed if the light flash were infinitely narrow in time and the instrument could respond infinitely fast to the fluorescence. Expressed mathematically,

$$F(t) = E(t) \otimes f(t) \tag{2.30}$$

One would like to measure $E(t)$ and $F(t)$, deconvolute $F(t)$ using $E(t)$, and solve for the decay parameters of $f(t)$.

On a real fluorescence instrument, one instead measures the functions $E_m(t)$ and $F_m(t)$. These functions are the convolutions of the ideal functions, $E(t)$ and $F(t)$, with instrumental distortion functions, $D_E(t)$ and $D_F(t)$, respectively.[38]

$$E_m(t) = D_E(t) \otimes E(t) \tag{2.31}$$

$$F_m(t) = D_F(t) \otimes F(t) \tag{2.32}$$

If $D_E(t) = D_F(t)$, then it can be shown using the associative property of convolutions that

$$F_m(t) = E_m(t) \otimes f(t) \tag{2.33}$$

and one can use the measured functions to deconvolute and solve for the parameters of interest in $f(t)$.[38, 39]

For a good decay fluorometer under the right conditions, D_E will be much like D_F, and accurate deconvolutions can be performed. At some level, however, analyses of the data will become limited by the differences between these functions. This limit could be easily reached if one is attempting to resolve two closely spaced decays, or perhaps if one is attempting to obtain information on distributions of lifetimes in a complex decay.

The differences between the distortion functions result from many sources. These include the use of different wavelengths, optical paths, or counting rates for measuring $E_m(t)$ and $F_m(t)$. They also can result from certain subtle errors in the electronics. Many of these errors derive directly from properties of the photomultiplier, and the advantage of the MCP photomultiplier is that it greatly minimizes several of these problems. In this section, I will discuss properties of photomultiplier tubes and compare the performance of a microchannel plate photomultiplier to that of a more traditional design of photomultiplier, the RCA C31000M, a variant of the RCA 8850 with a fused-silica rather than a Pyrex window.

2.3.2.1. *Timing*

Some of the most important sources of error in the photomultiplier are illustrated in the schematic anode pulse outputs shown in Figure 2.25. A typical pulse resulting from a single photoelectron is shown in Figure 2.25A. In order to use the anode pulse for timing, it is important for its rise time to be short. The rise time is usually defined as the length of time required for the pulse to increase from 10% to 90% of its maximum (in this case, negative) value. Typical rise times for fast photomultipliers such as the RCA 8850 are a little less than a nanosecond. In the MCP photomultiplier, the electrons have a much shorter path and therefore less time to spread out, and typical rise times for a 12-μm-channel-diameter double-plate tube are about 220 ps. Adding a third MCP increases the rise time to about 280 ps. Hamamatsu now sells MCP tubes which contain extremely narrow 6 μm channels, and for such tubes, in the two-plate configuration, the rise time can be as low as 150 ps.

Timing the single-photon event is usually done using a constant-fraction

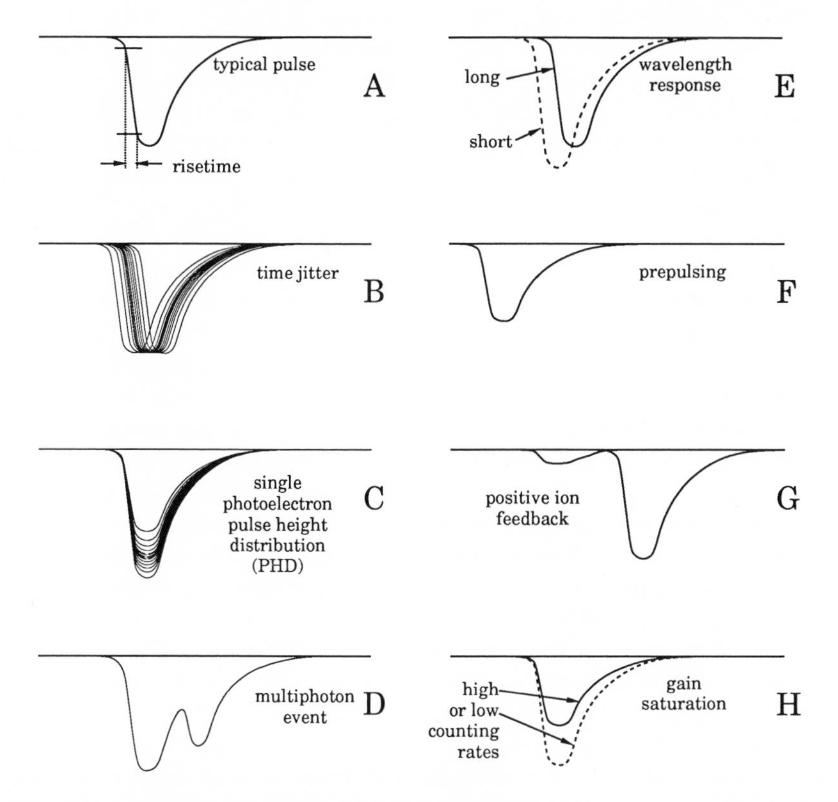

Figure 2.25. A schematic drawing of photomultiplier output pulses. Several features which can affect the performance of a pulse fluorometer are illustrated.

timing discriminator, and presumably a faster pulse rise time will increase the accuracy of the timing. Unfortunately, at this time, constant-fraction timing discriminators are manufactured primarily for the nuclear industry, which does not generally require accurate timing of pulses with such very fast rise times. Therefore, to my knowledge it is not yet possible to purchase a discriminator which can really take advantage of the fast rise times available with MCP tubes. There are two problems with the commercially available constant-fraction discriminators. First, pulse shaping circuitry encountered by the pulse as it enters the discriminator is too slow to respond effectively to the fast rise time. Second, constant-fraction timing discrimination requires a user-selected external delay. A faster rise time requires a shorter delay, but even the minimum external delay possible is much too long for correct timing. Much of our work has been done with an EG&G Ortec model 934 Quad Constant Fraction Discriminator for timing, but we had to modify the unit internally to improve its timing. We now use a Phillips Scientific model 6915 discriminator. Although internal modifications were not necessary, its input circuitry is slow enough to degrade its performance. Tennelec will modify their model 454 or 455 for use with MCP photomultipliers,† but my laboratory has no direct experience with how well these units function. All three of the above manufacturers are aware of the deficiencies of their constant-fraction timing discriminators when used with MCP photomultipliers. However, no schedule for improvements has been advised by any of them. Tuning of the timing discriminator is necessary for good instrumental performance and is discussed in Section 2.4.2.1.

All photomultipliers have some degree of time jitter, or variations in the arrival time of the anode pulse relative to the primary photoelectron event (Figure 2.25B). There are two main sources of this jitter. First, the photoelectron must pass out of varying depths within the photocathode to arrive at the vacuum interface with a variable amount of kinetic energy in a variable direction. The electron can emerge from any location on the illuminated portion of the photocathode. An electric field accelerates the electron toward the electron multiplier. Different electrons therefore have different trajectories and require different lengths of time to arrive. The second source of time jitter is the electron multiplier. Simple statistical variations in the electron trajectories within the multiplier will also give rise to time jitter. Some electron cascades will pass through the multiplier faster than others.

The probability that an anode pulse will arrive at a given time after the primary photoelectric event is called the *transit time spread.* It cannot be directly measured. Instead, one measures E_m [see Eq. (2.33)], the *measured excitation.* It is the response of the instrument to scattered light. Its width contains contributions from the temporal width of the laser flash, broadening

† Contact Mr. Bruce Coyne, Tennelec, P.O. Box 2560, Oak Ridge, Tennessee 37831-2560.

due to optical components, transit time spread, and further broadening from the timing electronics.

It is obvious from the designs of the two photomultipliers shown in Figure 2.22 that the MCP photomultiplier will have greatly reduced transit time spread from both of the sources mentioned above, because all of the path lengths traveled by the electrons are much shorter. We have measured the fwhm of the measured excitation for an 8850 photomultiplier to be 290 ps at 600 nm.[36] As will be discussed in Section 2.3.2.3, it is strongly dependent on the wavelength. For the Hamamatsu R1564U and R2809U MCP photomultipliers, much lower fwhm values of 42 and 19.1 ps, respectively, have been reported.[40]

2.3.2.2. Pulse Height Distribution (PHD) and Energy Windowing

Anode pulses vary considerably in their amplitude, because of statistical variations in electron trajectories and collision energies within the semiconductor layer of the electron multiplier. Constant-fraction timing will not work perfectly under all circumstances, and the fluctuation in the pulse heights will result in time jitter in the electronics of the fluorometer. These anode pulse fluctuations (Figure 2.25C) are greatly reduced in the 8850 type of photomultiplier, relative to those for more standard designs, by the use of a very high gain gallium phosphide semiconductor layer on the first dynode and a very high voltage drop between the photocathode and the first dynode. Under these circumstances, a single photoelectron dislodges about 30–50 secondary electrons for subsequent amplification. The remaining 11 dynodes use beryllium oxide for a coating and produce only about 5–8 electrons per electron striking them. This very high first amplification stage greatly reduces the pulse height fluctuations at the anode.

There are two reasons why pulse height resolution is important. First, it will result in decreased timing errors, since all the pulses will be about the same size. The second is that it will allow the distinction of single- from multiple-photon events. A single laser flash will excite many molecules in our sample which will subsequently emit fluorescence. One must limit the detector to responding to only single photons to avoid serious timing artifacts. If two photoelectrons are emitted from the photocathode at about the same time, then the total pulse charge delivered to the anode will be about twice as large as that resulting from one photoelectron. Similarly, three or more photons arriving at nearly the same time can be distinguished from a single event based on the total charge delivered to the anode. One can electronically integrate the anode charge over a time period ($\sim 1\ \mu s$) sufficient to capture all possible events derived from a single laser flash. The electronics which perform this integration produce a pulse whose voltage is a linear function of the total charge. A single-channel analyzer is then used, with upper and lower threshold limits, to produce a logic pulse only when the anode current corresponds to a single-photon event. The logic pulse is then used to gate the data collection process to accept only the single-photon timing events. This was first done using an RCA 8850 tube by Schuyler and Isenberg,[(41)] who called the process *energy windowing*. The RCA 8850 was mounted in a base which permitted the simultaneous use of both the anode pulse for timing the event and a pulse from the 11th dynode for energy windowing.

The Hamamatsu MCP photomultipliers have only a single output pulse for a 50-Ω load, and it is therefore necessary to develop a new means of accomplishing energy windowing. We use a special-order model 6955 amplifier available from Phillips Scientific. This unit accepts the 50-Ω input,[†] splits it, and puts out a current pulse for the charge-sensitive preamplifier used for energy windowing as well as a 10 times amplified negative pulse into a 50-Ω load for timing.

A pulse height distribution (PHD) can be measured by setting the decay instrument to collect fluorescence and, instead of feeding the linear voltage pulse into the single-channel analyzer, putting it into the input of a pulse height analyzer. One obtains distributions of peak heights such as those shown in Figure 2.26. In the figure, two photomultipliers, an RCA C31000M and a Hamamatsu R1564U triple-MCP photomultiplier are compared. On the left-hand side of the figure, PHD curves for each tube are shown for single-photoelectron events collected at a rate of 20,000 fluorescence events per second using a laser repetition rate of 800 kHz. Thus, one out of 40 laser flashes will result in a single-photon event, and very few double events will occur. These are the standard conditions under which we run our instrument. The energy windowing positions are shown on the plot. Single-photoelectron events, each deriving from a different flash of the laser, give rise to anode pulses which show a distribution of amplitudes. This is illustrated schematically in Figure 2.25C. One often judges the quality of the single-photon PHD by expressing the peak-to-valley ratio indicated on the upper plot, and both tubes yield ratios which would be considered to be excellent.

On the right-hand side of Figure 2.26, the fluorescence intensities incident on the photocathodes have been increased to very high values at the same time as the repetition rate of the laser has been reduced to 40 kHz. Now, most laser flashes result in multiple events. Several new peaks with larger pulse heights appear in the plot, and the scale on the x axis has been compressed in order to view them. The single-photoelectron peak (labeled 1) is on the left.

[†] It is important to terminate the input of this preamplifier. Due to the design of the Hamamatsu tubes, charge can build up and destroy the sensitive electronics of the 6955 as well as other models of preamplifiers. We have found a 4-MΩ resistor to ground to be adequate for bleeding off the excess charge without deteriorating the input pulse.

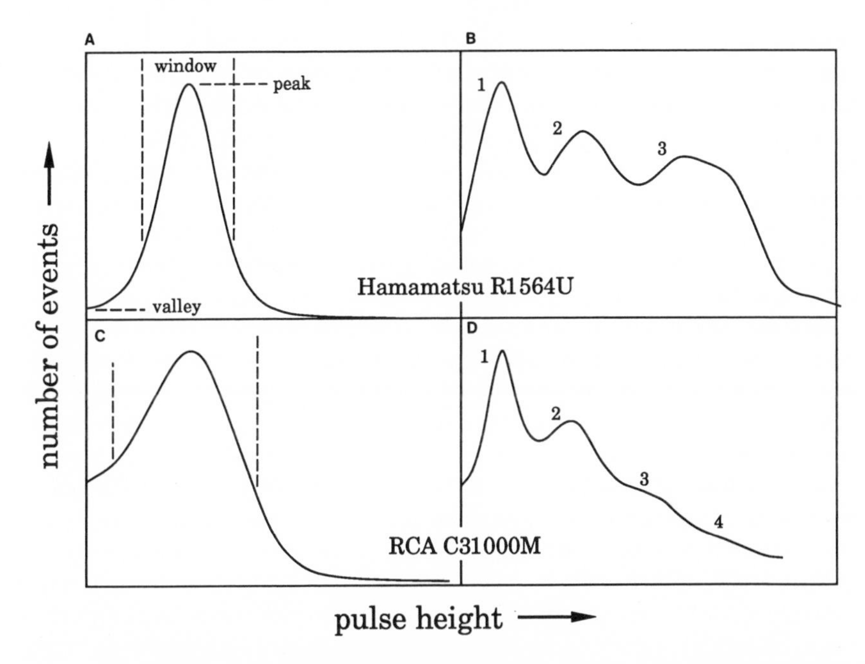

Figure 2.26. A comparison of PHD curves for an MCP photomultiplier (Hamamatsu R1564U variant with three plates in a Z configuration and a UV-transmitting window) and an RCA C31000M. The left-hand side of the plot shows the single photoelectron distribution obtained when the counting rate is maintained at 20 kHz, the standard counting rate used on our instrument. Typical energy windows chosen for these photomultipliers are indicated on the plots. The right-hand side of the plot shows PHD curves obtained at much higher counting rates. The number of photoelectrons per laser flash is indicated above each peak.

Other peaks are labeled with the number of photoelectrons which gave rise to them. The peaks fall off suddenly for the MCP tube because of the saturation of one of the amplifiers used in the measurement. Also, only four peaks can be seen for the C31000M, but at much higher counts rates, up to seven can actually be distinguished. We interpret the curves of Figure 2.26 to indicate the PHD for this particular MCP to be as good as or better than the PHD of this particular C31000M. We have found considerably varying PHD curves with different Hamamatsu MCP photomultipliers. Some function very well and some not at all. If one wishes to use energy windowing, then this requirement should be discussed with the manufacturer.

Figure 2.25 illustrates how the anode pulse might appear for a double-photon event resulting from a single flash of the laser. Timing will occur from the first event, neglecting the arrival of the later photon, thus biasing data to shorter times. Multiple-photon events give rise to a particularly insidious form of data distortion, that of the introduction of an apparent contaminant fluorescence to the measured decay. This is easy to show for a simplified case, that of a single-exponential decay in the absence of a convolution (and making the assumption that all anode pulses can be distinguished from one another). Double-photon events will appear as a simple exponential decay with half the lifetime of that of the sample. Thus, a one-component analysis becomes a two-component analysis. Relaxing these restrictions by, for example, examining the effects of double-photon events on multiexponential decays, one greatly adds to the complexity of the apparent contaminant decay that is summed with the data.[42]

All biological samples will contain impurities, often at levels which will prevent correct data resolutions. Also, contamination of the scatter sample with a trace fluorescent impurity, or luminescence from optical components such as photomultiplier windows, will appear as apparent fluorescence components. These contaminations, when not resolved, can readily prevent correct resolutions of decays[36] and can lead to other errors such as the recovery of a relatively broad distribution of lifetimes for a sample which really consists of a single decay and an apparent contaminant representing less than 0.5% of the data.[43] These errors are particularly troublesome since, although one can minimize some of them, one would certainly not want to design an instrument or use an analysis method which is insensitive to them.†

2.3.2.3. *Wavelength Response*

At the long-wavelength edge of the photocathode response, a photon will barely have the energy to overcome the band gap energy and the surface potential of the photocathode material. All photoelectrons which succeed in escaping must come from very near the surface and will possess very little kinetic energy as they leave. Higher energy light, on the other hand, will release electrons from the surface, as well as deep within the photocathode, and the escaping electrons will possess a much wider range of initial kinetic energies. As the electrons escape from the surface, they will have kinetic energies in different directions and will therefore have different trajectories toward the electron multiplier. Photoelectrons derived from shorter wavelengths will, on the average, have slightly higher energies when they arrive at the electron multiplier and will tend to arrive somewhat sooner. The effects of these phenomena on the anode pulse are illustrated schematically in Figure 2.25E. Shorter wavelengths give rise to an instrument response which is shifted slightly to shorter times relative to one which is measured using longer wavelengths.

† Some laboratories solve this and other problems by shortening the collection time. This greatly improves the data fit as judged by χ^2 and prevents resolution of the unwanted components. I personally think of this approach as looking at the fluorescence with dark glasses in the hope that what you do not see will not hurt you.

This wavelength effect is probably the main source of the "time origin shift" encountered in fluorescence decay data, since the fluorescence is generally measured at longer wavelengths than the excitation profile. Also, photoelectrons from shorter wavelengths of light, because they have more divergent trajectories, have a wider distribution of arrival times at the electron multiplier and therefore give rise to more photomultiplier time jitter. This translates into a broader instrument response for light at shorter wavelengths and is one source of instrument response width variations (sometimes called "lamp drift").

Time origin shifts for a typical fluorescence measurement using an 8850 type of photomultiplier range from about 30 to 100 ps and are easily corrected by data analysis methods. Instrument response width distortions can be considerably more troublesome and have led to more complex methods for correcting them. A plot of the half-width of the measured excitation for the RCA C31000M photomultiplier as a function of the wavelength of the light is shown in Figure 2.27. The measured half-width of the instrument response function varies from about 750 ps at 280 nm to about 350 ps at 640 nm. Of course, a normal fluorescence measurement would require only a small fraction of this wavelength range and would result in a correspondingly smaller width error.

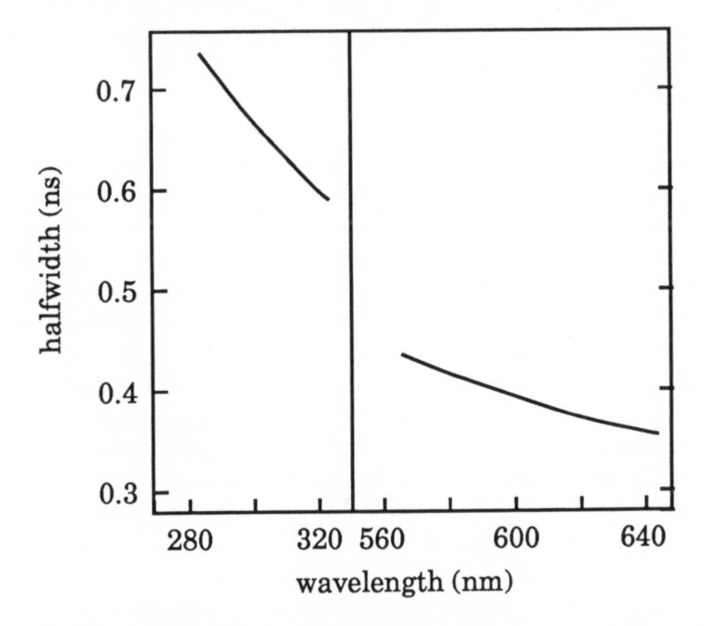

Figure 2.27. Half-width of the measured excitation as a function of wavelength, obtained using an RCA C31000M. The decreasing half-width with increasing wavelength results from timing properties of the photomultiplier.

It is apparent from the design of the proximity-focus MCP photomultiplier shown in Figure 2.22B that wavelength effects will be negligible in this kind of tube, since the photoelectrons have a much shorter distance to travel before they reach the first MCP. We have measured a 20-ps decrease in half-width in the measured excitation on changing the wavelength from 280 to 560 nm. Similarly, Visser and van Hoek observe a 15-ps change over the range of 300 to 600 nm.[44] Probably very little of this width change can be attributed to the MCP photomultiplier. The photomultiplier contribution is believed to be of the order of about 6 ps.† With such a small wavelength dependence, one can routinely measure fluorescence samples without regard to a wavelength effect, even exciting them in the UV and measuring their emission far out in the red region of the spectrum.

2.3.2.4. *Extraneous Pulses*

Using a scatter sample and laser excitation, one can obtain measured excitations such as those shown in panels A and B of Figure 2.28 for the C31000M and MCP photomultipliers, respectively. These are logarithmic plots which tend to emphasize extraneous features of the instrumental response. (Generally, in the literature, those who advocate the use of particular photomultipliers for use in decay fluorometers do not show enough data to reveal all of the artifacts visible in this figure.) Note that the time scales are different. The response of the MCP photomultiplier is substantially faster than that of the dynode-based tube.

Our first experience with the laser on our decay fluorometer taught us that, although our measured excitation with the old flashlamp technology appeared to have been of high quality, it contained a number of serious artifacts which had been hidden by the broadness of the light flash. Two of these, luminescence from the photomultiplier window[36] (an increase in the measured excitation background at long times) and optical reflections (small duplicates of the measured excitation delayed and superimposed on it), required immediate changes to the instrument to remove them. The luminescence problem was solved by switching from a standard RCA 8850 photomultiplier to a C31000M, which is equivalent except for its window material. Reflections were minimized by carefully tracing through the optical system looking for parallel surfaces separated by a distance which could give rise to the observed delay (1 ns corresponds to ~30 cm). It was necessary to slightly cant several surfaces to redirect certain reflections away from the collection optics. Note that optical reflections could give rise to very serious errors in fluorescence decay data if the measured fluorescence contained a reflection not present in the measured excitation [$D_E(t) \neq D_F(t)$ in Eqs. (2.31)

† David Fatlowitz, Hamamatsu Corporation, personal communication.

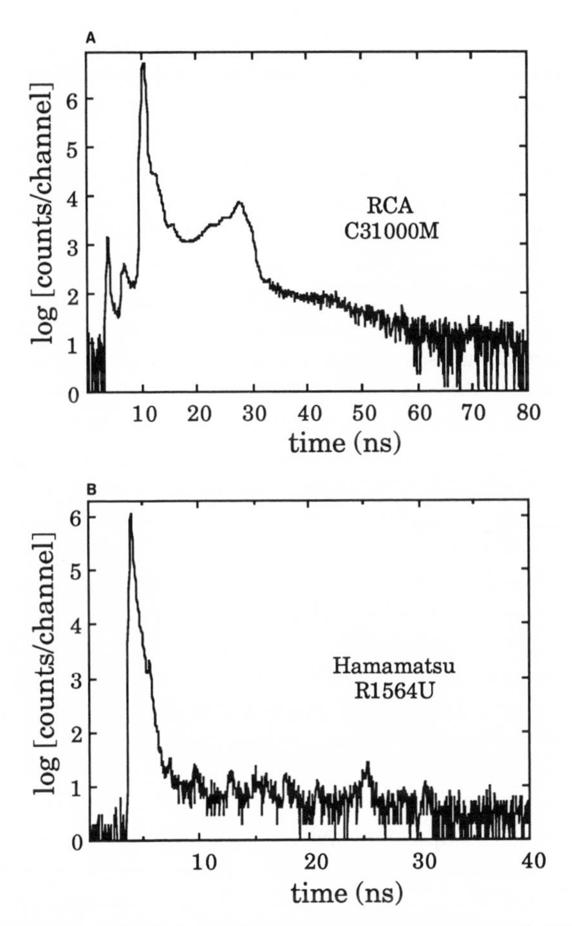

Figure 2.28. Measured excitations obtained using an RCA C31000M photomultiplier with a linear-focused dynode structure and a Hamamatsu R1564U MCP photomultiplier (a variant with a UV-transmitting window). Wavelengths of light used for the two curves were 300 nm for the C31000M and 310 nm for the R1564U.

and (2.32)]. This could occur, for example, if one used a different interference filter for measuring each function (as one is likely to do) and one of the filters participated in a collected reflection.

After initial changes in the instrument, a level of refinement can be reached at which the quality of data becomes limited by the intrinsic properties of the photomultiplier. The response of the RCA C31000M photomultiplier in Figure 2.28 contains two residual peaks before the main response. Their relative sizes can be changed by changing the geometry of the focusing of the fluorescent light on the photocathode. They can be minimized by spreading the light as evenly as possible across the photocathode surface.[45] The photocathode is nearly transparent, and the gallium phosphide on the first dynode is a semiconductor, like the photocathode, capable of emitting photoelectrons. The prepulsing in Figure 2.28A results from photons passing through the photocathode and initiating electron cascades by striking the first dynode. It is not clear why there are two such peaks, but perhaps one derives from a reflection from one of the shiny surfaces inside, near the face of the tube. Again, because the photoelectric effect is sensitive to wavelength, $D_E(t)$ will differ from $D_F(t)$.

The peak located about 20 ns after the main peak, and about three orders of magnitude lower in intensity, derives from positive ion feedback. A number of different kinds of positive ions are involved in forming this peak (such as H^+, H_2^+, He^+, O^+, O_2^+; see Ref. 46). Since heavier ions require longer times to be accelerated, the most probable arrival times for the different ions vary, and therefore the peak is relatively broad and shows some structure. As far as we have been able to determine, the afterpulse is well convoluted with the corresponding fluorescence and thus does not contribute to significant differences between $D_E(t)$ and $D_F(t)$.

By comparison to the response of the C31000M, the MCP photomultiplier response is extremely clean (Figure 2.28B). The response rises suddenly in just a few channels without any prepulsing. It decays within about 3 ns to a low background level about five orders of magnitude below the peak intensity. A positive ion feedback peak is not defined in the output of this particular tube, but the feedback probably contributes to the tailing of the pulse. Another contribution to the tailing of the pulse shape probably derives from reflections from the aluminum surface of the first MCP.[47] The main reflection should arrive approximately 20 ps after the peak, because this is the length of time required for the light to travel 6 mm, or twice the distance between the photocathode and the first MCP. Often, a small peak or shoulder can be resolved, which probably derives from this reflection. With three reflective surfaces, two on the window and one on the MCP, multiple reflections must contribute as well. Further contributions to the tail of the measured excitation could possibly be made by dye laser afterpulsing caused by misalignment of the dye laser cavity length. The small peak which follows the main peak in Figure 2.28 is due to a remaining reflection in the optics of the fluorometer.

The photomultiplier outputs resulting from prepulsing and positive ion feedback are illustrated schematically in Figures 2.25F and 2.25G, respectively.

2.3.2.5. *Gain Saturation*

After an electron cascade travels down through a dynode chain in a traditional photomultiplier tube, electrical charge on the dynode surfaces is depleted. Some time is required for recovery of the dynodes, since time is required for the charge to leak back into the semiconductor layer. This phenomenon gives rise to time-dependent changes in the gain of the tube as illustrated in Figure 2.25H, since any electron cascade which proceeds down the channels before they have adequately recovered from the previous cascade would be reduced in intensity. This gain saturation has a peculiar, long-term time dependence in some kinds of photomultipliers[48] and is a measurable property of the RCA 8850 type of photomultiplier, but it has not been a serious problem for our measurements because of the low light levels that are used for the fluorescence experiment. It becomes important to ask, however, whether such phenomena occur in MCP photomultipliers as well. One might expect a more drastic effect in the latter case, since the diffusion of the electrons back into the channel wall is a very slow process, requiring milliseconds. If one were measuring an anisotropy, for example, and determining the parallel and perpendicular polarized fluorescence components at different counting rates, then the average photomultiplier pulse size for each determination could be different. This could lead to serious and unpredictable timing artifacts and could also prevent correct energy windowing to eliminate multiple-photon events from one of the curves.

The single-photoelectron PHD curves shown in Figure 2.29 for three counting rates on an MCP photomultiplier indicate that gain saturation can indeed be a problem for these tubes. For this reason, we are always careful to collect scatter and fluorescence data at the same standard rate, the rate at which our energy window has been set. We also apply the same rule to the

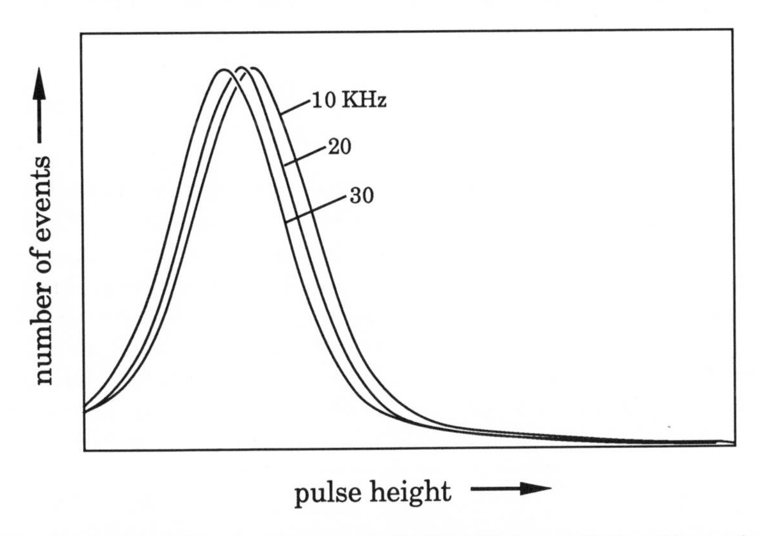

Figure 2.29. Gain variation of a Hamamatsu R1564U MCP photomultiplier with counting rate. The three PHD curves, from left to right, were measured using counting rates of 10, 20, and 30 kHz. Photomultiplier gain decreases with increasing counting rate.

collection of the different polarized components of the anisotropy. This requires special methods for determining instrumental sensitivity corrections, which have been described.[49]

2.4. Tuning a Monophoton Decay Fluorometer for High Performance

Using the synchronously pumped laser system as an excitation source permits the collection of data at very high rates. Our standard rate of collection is 20,000 single-photon events per second, and at this rate over 10^7 counts of data may be obtained in just 10 min. One quickly finds that the quality of the data, and the power to analyze the data to extract decay parameters, becomes limited, not by statistical errors in the data, but by nonrandom instrumental errors. This occurs even though the use of an MCP photomultiplier leads intrinsically to low error. This section will examine some other sources of error within the electronics and how they can be removed. We call this "tuning" the decay fluorometer.[36, 50]

2.4.1. Judging the Quality of Data

If one is attempting to remove instrumental errors, one needs a measure of whether the data have, in fact, been improved by a particular operation. Adequacy of a data fit is often judged by the value of the familiar reduced chi squared:

$$\chi^2 = \frac{1}{N+n} \sum_{i=1}^{N} \frac{(F_{ic} - F_{io})^2}{F_{io}} \tag{2.34}$$

where N is the number of channels of data, n is the number of independent decay parameters, and F_{ic} and F_{io} are values of the calculated and observed fluorescence intensities in the ith channels, respectively. Lower values of χ^2 are generally said to indicate a better fit. $F_{io}(t)$ is used as a weighting function in Eq. (2.34) because in normally distributed noise the standard deviation is proportional to the square root of the number of counts of data in the channel. If data contain only random errors, then χ^2 should remain near a value of 1 regardless of how many counts of data are collected. This is not the case for the real fluorescence data collected at high counting rates, because nonrandom instrumental errors, which generally show a linear relationship with the amount of data in the channel, quickly dominate. We find that even with a well-tuned instrument, χ^2 will often be quite large, reflecting errors which are mostly in the rising edge of the data. One is faced with the situation that the longer the data collection time, the better the data will become in terms of statistical reliability, and also the higher the value of χ^2. Thus, in a sense, a high χ^2 actually indicates the reverse of what it is often used to indicate.

In tuning our instrument, our first step is to use the fluorescence of a compound which will give us a clean single exponential decay. We measure a reasonably large amount ($\sim 10^7$ monophoton counts) of fluorescence in each data set and adjust the instrument to improve the fit to the data as judged by the deviation function. The ith channel of the deviation function, D_i, is defined as

$$D_i = \sum_{i=1}^{N} \frac{F_{ic} - F_{io}}{F_{io}^{1/2}} \tag{2.35}$$

The final step we use in judging the quality of the data is to perform a method of moments analysis with λ invariance.[50] Stated briefly, λ-invariance testing examines whether the recovered decay parameters are invariant with respect to the data transformation called *exponential depression.* If $F(t)$ is given by the convolution of $E(t)$ with the impulse response function $f(t)$ [see Eq. (2.30)], then it is easy to show[51] that

$$F_\lambda(t) = E_\lambda(t) \otimes f_\lambda(t) \tag{2.36}$$

where

$$E_\lambda(t) = e^{-\lambda t} E(t) \tag{2.37}$$

$$F_\lambda(t) = e^{-\lambda t} F(t) \tag{2.38}$$

$$f_\lambda(t) = e^{-\lambda t} f(t) \tag{2.39}$$

The raw data are multiplied by $e^{-\lambda t}$, deconvolution is performed, and the resulting decay parameters are "undepressed" or corrected for the transformation. Typically, λ is varied over a range of a factor of 3 in value, and one looks for the invariance of the recovered lifetime or other decay parameter as a function of λ. One effectively obtains a varied weighting of the data to shorter times. Flat regions in the λ-invariance plots verify that the correct model was chosen for the decay and that one has achieved all of the resolution possible for the particular data set. Although we use the method of moments for our analyses, λ invariance is general and can be used with any deconvolution method. For example, it has been used with least-squares fitting.[52]

Problems with the data that are not visible in the fit will often lead to significant curvature in the λ-invariance plots. This is the most sensitive test we have for determining whether the instrument is properly tuned.

2.4.2. Tuning

Many adjustments must be made on a decay fluorometer for it to properly take advantage of the fast laser source as well as the MCP photomultiplier. Two of the most dramatic improvements that we were able to make in the data quality obtained from our instrument are discussed here. They involve the proper tuning of the timing discriminators and the proper tuning to avoid time base nonlinearities. A more subtle, but nevertheless important, improvement resulted from the development of a method to eliminate multiple-photon events. The use of λ-invariance testing to detect this distortion will be illustrated.

2.4.2.1. Tuning the Timing Discriminators

Since the photomultiplier pulses vary in amplitude, it is essential to use constant-fraction timing discrimination (CFTD) in order to accurately time their arrival. This is also true, to a lesser degree, for timing the arrival of the laser pulse, because even the most stable laser source will have an intensity which will drift slowly in time. The CFTD is perhaps the most critical unit in the single-photon fluorometer, and yet its tuning must be done essentially by trial and error.

CFTDs work by splitting the incoming timing pulse unevenly so that most of it is inverted in polarity and passes through a delay circuit. The remaining fraction (usually about 0.2 or 0.3) is then summed with the delayed pulse to create a zero crossover signal, which is used to trigger a circuit to produce a fast timing pulse. This is shown schematically in Figure 2.30. Discriminators usually permit several adjustments. These include the adjustment of a threshold value below which incoming pulses will not trigger an output; a *walk* adjustment to make small changes in the voltage taken as the zero crossover; and the adjustment of the delay time, t_d, illustrated in Figure 2.30. The delay time is usually selected by adding an external cable to terminals on the outside of the unit. As mentioned earlier in Section 2.3.2.1, the fast pulses from an MCP or a photodiode monitoring the laser flash require that one bypass the external terminals and insert very short delay cables internally. This requires that for each tested delay the unit must be disassembled and the new decay cable carefully soldered in. One makes the necessary adjustments to the cable length and other external adjustments while observing the crossover signal on an oscilloscope, triggered by the output timing pulse. Unfortunately, in our experience, very good-looking crossover signals can lead to either good or bad measured excitations. After one establishes the rough ranges for the delays and adjustments, it is necessary to use trial and error to refine them.

Even after amplification, the anode pulse from the photomultiplier can be

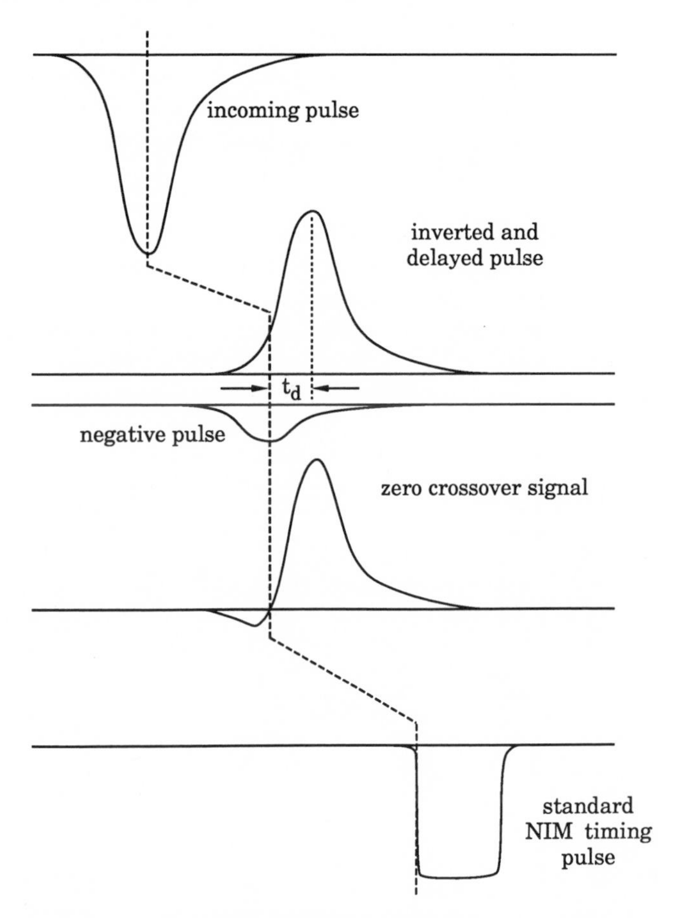

Figure 2.30. Schematic drawing which shows the functioning of a constant-fraction timing discriminator (CFTD). An incoming amplified pulse is split. The larger part of the signal is inverted and delayed and summed back with the negative portion. In theory, the delay must be chosen so that the peak of the negative pulse cancels the appropriate fraction of the delayed positive pulse to form a zero crossover. In other words, the delay, t_d, should be chosen such that the negative peak of the negative signal occurs at the time the positive pulse has reached the desired fraction of its total pulse height. The zero crossover signal is used to trigger a circuit which generates a standard Nuclear Instrument Modules (NIM) timing pulse. The timing pulse is not shown to the same scale as the other pulses, because it is much larger and broader in time.

quite small relative to the noise on the signal. Some of the noise is correlated with the laser flash and thus can have a dramatic effect on the instrument response. The CFTD of the photodiode circuit which detects the laser pulse can be adjusted while collecting a measured excitation with a scatter sample. The adjustments are correct if manually changing the size of the diode pulse does not significantly change the channel number of the peak of the measured

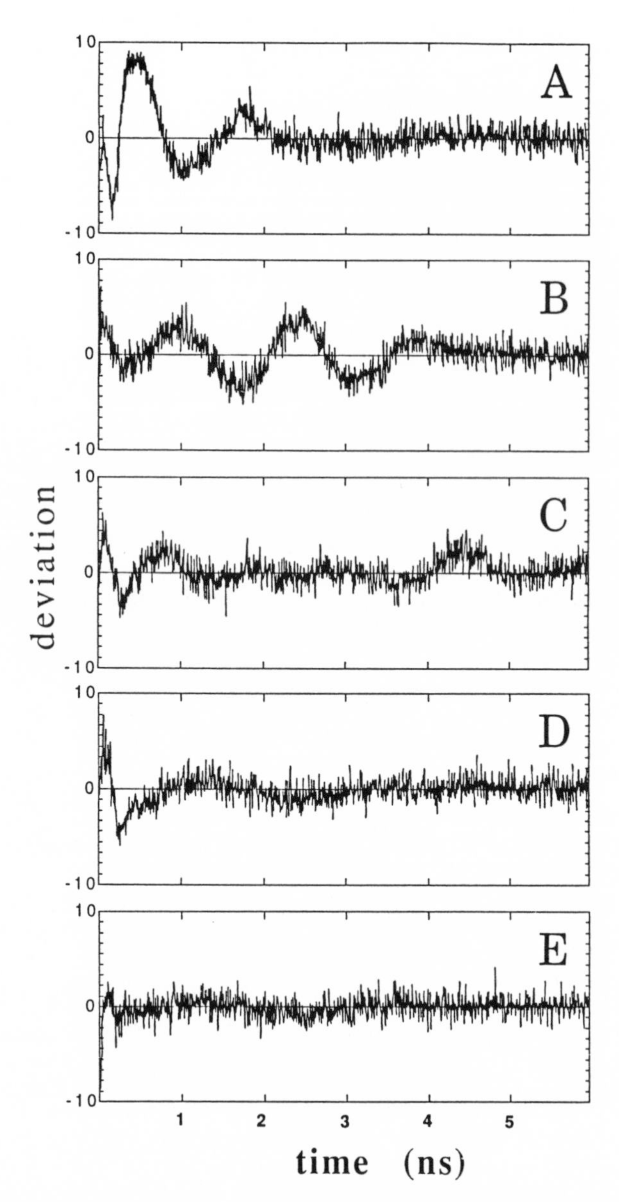

Figure 2.31. An example of "bump" removal. Time base nonlinearities in the instrument can become apparent as "bumps" on the data. Supplying delays to walk across the time range of the TAC, and making compensating changes in the pulse height analyzer, can almost eliminate this distortion. From top to bottom, these illustrations represent delay times of A: 30, B: 32, C: 34, D: 38, and E: 40 ns. The sample was *p*-terphenyl in ethanol excited at 280 nm with the fluorescence observed at 360 nm. A total of 1024 channels were used per data set with a channel width of 0.0058 ns/channel.

excitation. The CFTD of the photomultiplier circuit is more difficult to tune. We have found it easiest to simply tune it to give the narrowest measured excitation without the introduction of spurious peaks at other times.

Some of the problems we have observed with our instrument, which could be traced directly to poorly tuned CFTDs, include double pulsing, bad rise time on the leading edge of the measured excitation, irregular background levels after the pulse, and "bumps" on the data. Many of these errors would have been hidden if we had been using a slower instrument or if we had only collected 10^6 total counts of data, but they quickly dominated the deviation and autocorrelation functions when 10^7 total counts were collected.

2.4.2.2. Time Base Nonlinearities

In our instrument, the time-to-amplitude converter (TAC) is used to measure the time between the arrival of the timing pulse resulting from the fluorescent photon and the arrival of the delayed pulse resulting from the diode detection of the laser flash. Ideally, the TAC puts out a pulse whose amplitude is a perfect linear function of the time interval between these two timing pulses. The pulse height analyzer (PHA) then uses an analog-to-digital converter to measure the amplitude of the TAC output in order to record a single event in a channel of its memory, indicating the time of arrival of the fluorescence photon relative to the laser flash. We refer to a distortion in the timing as a *time base nonlinearity*.

We have found that our TAC (an Ortec 467 time-to-pulse height converter) has very small but significant nonlinearities near the extremes of its range. These nonlinearities appear as "bumps" on the fluorescence and can be seen in the deviation and autocorrelation plots if a sufficient number of data

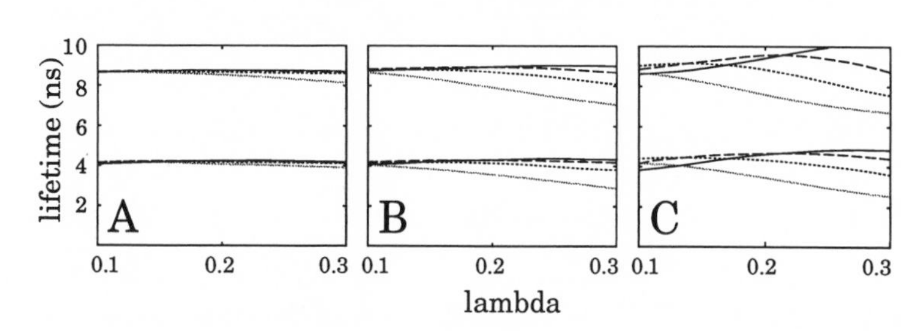

Figure 2.32. Effect of multiple-photon events and excessive count rate on a two-component resolution. The decays were collected from a mixture of 3-aminopyridine ($\tau = 4.28$ ns) and 3-aminonaphthalene-6-sulfonic acid ($\tau = 8.76$ ns) in ethanol. Excitation was at 300 nm and emission was through a Corning CS 0-52 cutoff filter. (A) The windowed count rate was 20 kHz. (B) The unwindowed count rate was ~30 kHz. These curves are distorted by the effects of multiple-photon events. (C) The windowed count rate was near 40 kHz. These curves are distorted because of excessive count rate. In each section of the figure, results are shown for MD values of 1 (·····), 2 (-----), 3 (– – –), and 4 (——).

counts are measured. An even more serious source of "bumps" on the data is the CFTD. Such errors in our instrument probably derive from cross-talk between the two timing discriminators. However, any noise which is correlated in time with the laser flash may give rise to such bumps or nonlinearities in the time base.

To switch to a different region of the TAC range, one changes the delay on the photodiode timing pulse to either increase or decrease the time interval to be measured. A compensating change must then be made in the region of the PHA. This adjustment on our PHA is called the digital zero offset. Figure 2.31 shows the removal of "bumps" by making these adjustments.

2.4.2.3. *Removal of Multiple-Photon Events*

The use of energy windowing can significantly improve the accuracy of decay resolutions. This is illustrated in Figure 2.32. The top section of the figure shows three λ-variation plots for an analysis of a two-component fluorescence decay generated by measuring the fluorescence of a mixture of two dyes with lifetimes (τ) of 4.28 and 8.76 ns. Each plot shows analyses done at three different values of moment index displacement (MD).[53–55] (Use of MD is an integral part of the method of moments analysis.[56]) In a method of moments analysis, one looks for flatness of the plots (invariance of the recovered lifetimes versus λ) and agreement between different values of MD. The curvature and spread of the lines is regarded as a measure of the uncertainty of the recovered decay parameters. Panel A of Figure 2.32 shows the λ-variation plots for the standard measuring conditions that we use. Panel B and C show the effects of multiple photons and the use of too high a counting rate, respectively. Both the inclusion of multiple photon events and the measurement at too high a counting rate significantly distorted the results.

2.4.3. Performance of a Tuned Instrument

2.4.3.1. *A Clean Single-Component Fluorescence Decay*

Panels A, B, and C of Figure 2.33 show the fluorescence decay, deviation, and autocorrelation plots, respectively, for a highly purified sample of rhodamine 575 (a laser dye purchased from Exciton Chemical Company, Inc., Dayton, Ohio) dissolved in ethanol. The dye was excited in the UV at 300 nm and the fluorescence observed in the red peaking at about 550 nm. A total of 1.1×10^7 counts of fluorescence data were collected. Note that the data fluctuations appear random even with this very large number of events collected. The data have been corrected for a 19-ps time origin shift. The spike in the deviation function at short times corresponds to the peak of the measured

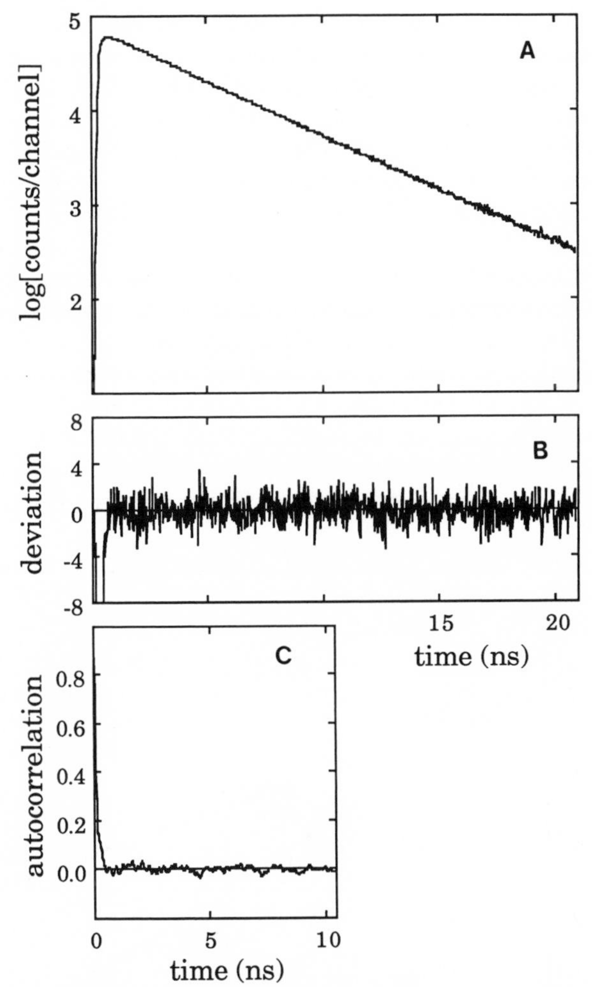

Figure 2.33. A clean single-component fluorescence decay. (A) The measured and calculated fluorescence decays of rhodamine 575 in ethanol. A total of 1.1×10^7 single-photon counts were recorded. The recovered lifetime was 3.72 ns. (B) The deviation function, and (C) the autocorrelation function for the above analysis.

excitation and results from the width error due to the wavelength dependence of the instrument over this wide wavelength range (discussed in Section 2.3.2.3). It has no significant effect on the recovered lifetime.

2.4.3.2. Four-Component Analyses with and without Deconvolution

Tyrosine was quenched with varying amounts of KI, and a series of fluorescence decays were measured in the range of 0.2 to 3.4 ns. Tyrosine in ethanol was also measured and found to have a lifetime of 4.03 ns. All of the analyses indicated single-exponential decays as evidenced by λ-variation plots, and a Stern–Volmer plot of the quenched lifetimes was linear, indicating dynamic quenching. Four of these decays with lifetimes of 0.38, 1.01, 2.03, and 4.03 ns are shown in Figure 2.34. They were summed together to give the four-component sum shown above them. Note that even the fastest lifetime fluorescence (0.38 ns) shown in the plot appears linear over almost four decades of decay.

A three-component analysis of the four-component sum indicated lack of resolution. A four-component analysis recovered the correct four lifetimes, and a five-component analysis recovered the same four plus a negligible amount of a fifth component.[31] The analyses all satisfied λ-invariance and MD agreement criteria. The expected values and the lifetimes recovered from the deconvolution analysis are listed in Table 2.2. The lifetimes are accurately recovered.

Since the fluorescence decay curves of Figure 2.33 look highly linear, one

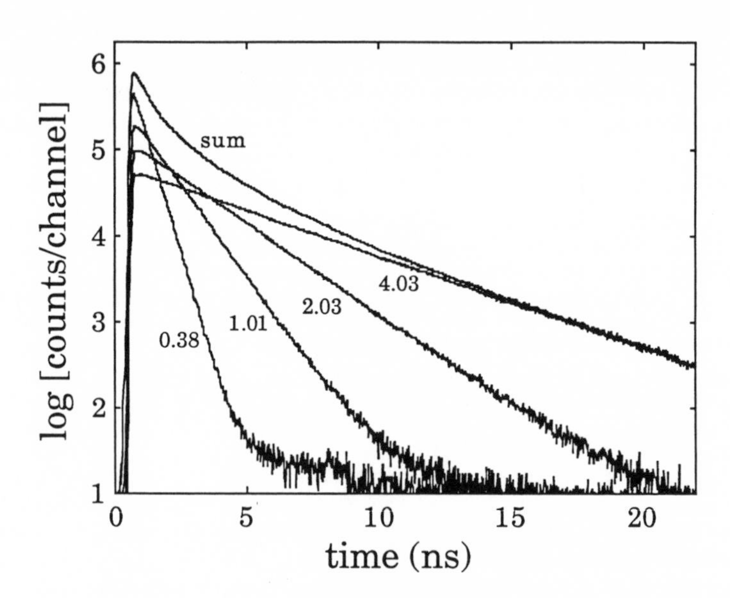

Figure 2.34. An example of a four-component decay. Fluorescence decays were measured for tyrosine solutions quenched with various amounts of KI. Three of these single-exponential curves with lifetimes of 0.38, 1.01, and 2.03 ns are shown in the figure, as well as a fourth decay with a lifetime of 4.03 ns, measured on tyrosine in ethanol. The four decays were added together to give the sum indicated in the figure. The correct four lifetimes, including that of the subnanosecond decay, were accurately recovered by analysis of the data both with and without a deconvolution. The samples were excited with the 280-nm frequency-doubled output of the synchronously pumped dye laser system, and the fluorescence was measured with a Hamamatsu R1564U triple-MCP photomultiplier with a fused-silica window.

might ask whether a deconvolution is actually necessary. In fact, it is not. Also listed in Table 2.2 are the results of a straight method of moments analysis of the data without a deconvolution. In order to avoid the convolution artifact, it was necessary to start the analysis 300 ps after the peak intensity. Again, the lifetimes are accurately recovered.

Another test was done using various aminonaphthalene derivatives quenched with acrylamide. Clean single-exponential fluorescence decays were obtained with lifetimes of approximately 1, 2, 4, 8, and 16 ns, and they were mixed together in various combinations to give four-component data sets. Analyses recovered the correct lifetimes. However, when all five decays were added together, the analysis failed, indicating that the lifetimes had not been resolved. After fixing the central lifetime using a Cheng–Eisenfeld filter,[57] the remaining four lifetimes were easily recovered.

2.4.3.3. *Anisotropy Decays without Deconvolution*

Because of the fast laser source and the fast MCP photomultiplier, we were able to accurately recover a subnanosecond lifetime in the above example of a four-component mixture of quenched tyrosine decays. We were even able to do this without deconvolution. This indicates a lack of instrumental artifacts distorting the data. In this situation, it becomes possible to directly

Table 2.2. Analyses of Four- and Five-Component Data

System	Total counts		Lifetimes (ns)				
			τ_1	τ_2	τ_3	τ_4	τ_5
Iodide-quenched tyrosine	4×10^7	Expected:	0.38	1.01	2.03	4.03	
		Deconvolution:	0.43	1.14	2.38	4.24	
		No deconvolution:	0.41	0.91	2.01	4.16	
Various amino-naphthalenesulfonic acids quenched with acrylamide,[a] analyzed with deconvolution	4×10^7	Expected:	1.06	2.02	4.12	8.02	
		Recovered:	1.03	1.73	3.70	7.87	
	1.6×10^8	Expected:	2.02	4.12	8.02	16.73	
		Recovered:	1.87	3.64	7.38	16.36	
	1.6×10^8	Expected:	1.06	2.02	8.02	16.73	
		Recovered:	1.07	2.08	8.02	16.70	
	2.0×10^8	Expected:	1.06	2.02	4.12	8.02	16.73
		Recovered[b]:	1.11	2.28	4.12	7.60	16.70

[a] The four derivatives used were 1-aminonaphthalene-*x*-sulfonic acid with *x* equal to 2, 4, 5, and 7. They were dissolved in 0.01 *M* Tris and adjusted to pH 4.8 with glacial acetic acid and quenched with various amounts of acrylamide to give the expected lifetimes shown in the table. The approximately 1-, 2-, 4-, 8-, and 16-ns lifetimes were obtained using derivatives with *x* equal to 2, 5, 5, 4, and 7, respectively.

[b] This five-component analysis was only possible when one lifetime (4.12 ns) was fixed using a Cheng–Eisenfeld filter.[57]

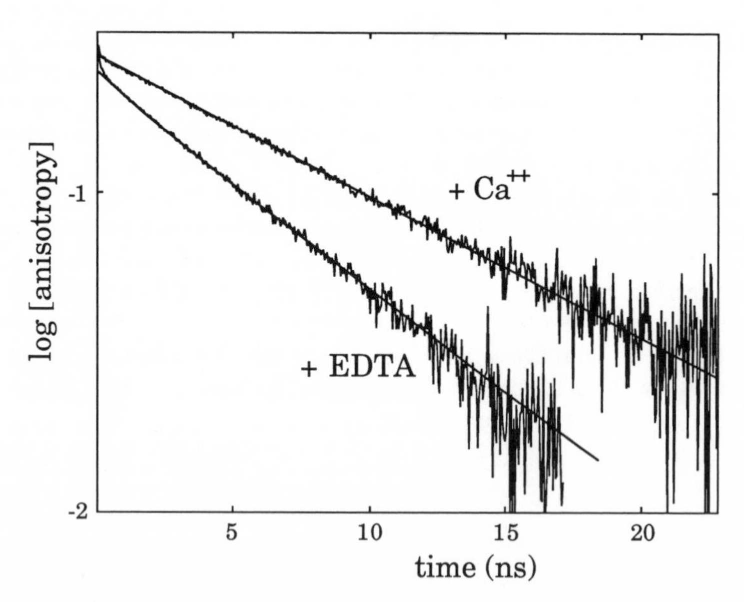

Figure 2.35. Anisotropy decay of the dityrosine fluorescence of photo-cross-linked calmodulin.[48] The fast response of the instrument combined with the high rate of data collection permits the direct plotting of the anisotropy without deconvolution. The two curves shown in the plot, corresponding to the presence and absence of calcium, result from fluorescence decays with average lifetimes of 3.8 and 3.6 ns, respectively.

calculate anisotropy decays without deconvolution, as long as the resolution of very short lifetimes is not required. An example of such an anisotropy is shown in Figure 2.35.

2.5. Conclusion

Combining picosecond laser sources and very fast MCP photomultipliers extends the capabilities of pulse fluorometry. In a fluorescence decay experiment, measured excitations can now readily be obtained with half-widths of less than a hundred picoseconds. The MCP photomultipliers have essentially no measurable wavelength dependence which would distort the measured fluorescence relative to the excitation. Perhaps even more importantly, the measured excitations drop rapidly to background levels several orders of magnitude (~ 5) lower than peak intensities in less than three nanoseconds. This narrow excitation profile represents a vast improvement over that which could be obtained with the older instrumentation, using flashlamps and slower photomultipliers. Much faster fluorescent lifetimes can now be measured accurately.

In addition, for most fluorescence studies involving lifetimes in the nanosecond range, combining picosecond laser sources with microchannel plate photomultipliers results in measured fluorescence curves which approach the ideal Dirac delta function response. Complex decays can even be analyzed without deconvolution. In fact, in Section 2.4.3.2 such an analysis of quenched tyrosine fluorescence data, consisting of four reasonably close lifetime components, was presented. One of the components is subnanosecond. All four components can be recovered with reasonable accuracy either with or without deconvolution. Since deconvolution of such data is easy, one would probably just do it. However, analysis of anisotropy curves without deconvolution permits their direct plotting and analysis. Examples of such curves were presented in Section 2.4.3.3.

Another major advantage of the new instrumentation is that it permits the collection of data at much higher rates than before. A measured fluorescence intensity consisting of over 10^7 counts of data can be collected in about ten minutes. The large amount of data per decay curve reduces the statistical errors in the data. This improves the ability to resolve decays in multicomponent mixtures, and also the ability to measure physical events which occur much slower relative to the fluorescence decay lifetime. For example, anisotropies can be accurately measured out to times manyfold longer than the lifetime of the fluorescence probe. Such analyses are not limited by statistical errors, but by errors in the normalization procedure used for the different polarization components.

With such great reductions in the statistical noise on fluorescence curves, nonrandom instrumental errors become rapidly apparent. Hence, it is important to "tune" a modern decay fluorometer to minimize those errors. For example, timing artifacts, previously invisible under a broad instrument response, can be carefully eliminated by tuning the timing discriminators. Other nonlinear timing responses can be eliminated or avoided as well. For example, small optical reflections within the instrument can be directed away from the detector, and nonlinear regions of gain in the time-to-amplitude converter can be avoided. The integration of fast light sources with fast detectors makes these improvements possible. The reward of this effort is a pulse fluorometer capable of recording much higher quality data than was previously possible.

Acknowledgments

Many thanks are due to those who assisted me in the preparation of this chapter. Dr. Louis Libertini, showing impressive facility with Macintosh computers, prepared all of the final figures using Cricket Draw, Cricket Graph, and MacDraw. He also was responsible for the measurement of most of the data shown in Sections 2.3 and 2.4. Dr. Jeanne Rudzki Small, who is quite knowledgeable herself in the fields of lasers and fluorescence decay techniques, helped to turn my rantings into English. Neeraj Bhatt of Spectra-Physics, whose endless knowledge of the inner workings of Spectra-Physics lasers is nothing short of incredible, answered many questions about the lasers. Over the years, many others at Spectra-Physics have helped me as well. David Fatlowitz and Dr. K. Oba of the Hamamatsu Corporation have been extremely helpful in our evaluation of microchannel plate photomultipliers. Their generosity in loaning us several different tubes for our inspection has made it possible to critically evaluate their use for pulse fluorometry. Special thanks are due to Mr. Fatlowitz for critical comments on the manuscript. Finally, I would like to thank Dr. Warner Peticolas, in whose colorful laboratory I, as a graduate student, had my first introduction into the wonders of laser light.

This work was supported by NIH grant GM25663.

References

1. D. C. O'Shea, W. R. Callen, and W. T. Rhodes, *Introduction to Lasers and Their Applications*, Addison-Wesley, Reading, Massachusetts (1978).
2. M. J. Beesley, *Lasers and Their Applications*, Halsted Press, New York (1976).
3. C. B. Hitz, *Understanding Laser Technology*, Penn Well Publishing Co., Tulsa, Oklahoma (1985).
4. A. E. Siegman, *An Introduction to Lasers and Masers*, McGraw-Hill, New York (1971).
5. M. R. Topp, Pulsed laser spectroscopy, *Appl. Spectrosc. Rev. 14*, 1–100 (1978).
6. G. R. Fleming, Applications of continuously operating, synchronously modelocked lasers, *Adv. Chem. Phys. 49*, 1–45 (1982).
7. A. G. Doukas, J. Buchert, and R. R. Alfano, in: *Biological Events Probed by Ultrafast Laser Spectroscopy* (R. R. Alfano, ed.), pp. 387–416, Academic Press, New York (1982).
8. I. S. McDermid, in: *Comprehensive Chemical Kinetics* (C. H. Bamford and C. F. H. Tipper, eds.), pp. 1–52, Elsevier, Amsterdam (1983).
9. D. V. O'Connor and D. Phillips, *Time-Correlated Single Photon Counting*, Academic Press, London (1984).
10. G. R. Fleming, *Chemical Applications of Ultrafast Spectroscopy*, Oxford University Press, New York (1986).
11. B. Leskovar, Recent advances in high-speed photon detectors, *Laser Focus 20*, 73–81 (1984).
12. B. Leskovar, Microchannel plates, *Physics Today 30*, 42–48 (1977).
13. W. A. Shurcliff and S. S. Ballard, *Polarized Light*, Van Nostrand, Princeton, New Jersey (1964).
14. A. L. Schawlow and C. H. Townes, Infrared and optical masers, *Phys. Rev. 112*, 1940–1949 (1958).
15. A. Einstein, Zur Quantumtheorie der Strahlung, *Physik. Zeitschr. 18*, 121–128 (1917).
16. A. Einstein, in: *The Old Quantum Theory* (D. T. Haar, ed.), pp. 167–183, Pergamon Press, Elmsford, New York (1967).
17. C. Fabry and A. Perot, Méthodes interférentielles pour la mesure des grandes épaisseurs et la comparaison des longueurs d'onde, *Ann. Chim. Phys. 16*, 289–338 (1899).

18. H. Maiman, Stimulated optical radiation in ruby, *Nature 187*, 493–494 (1960).
19. R. B. Chesler and J. E. Geusic, in: *Laser Handbook* (F. T. Arecchi and E. O. Schulz-DuBois, eds.), pp. 325–368, North-Holland, Amsterdam (1972).
20. N. Bloembergen, *Nonlinear Optics*, Benjamin, New York (1965).
21. S. R. Leone and C. B. Moore, in: *Chemical and Biochemical Applications of Lasers* (C. B. Moore, ed.), pp. 1–27, Academic Press, New York (1974).
22. R. T. Hodgson, P. P. Sorokin, and J. J. Wynne, Tunable coherent vacuum-ultraviolet generation in atomic vapors, *Phys. Rev. Lett. 32*, 343–346 (1974).
23. N. Bloembergen, The stimulated Raman effect, *Am. J. Phys. 35*, 989–1023 (1967).
24. R. Adler, Interaction between light and sound, *IEEE Spectrum 4*, 42–53 (1967).
25. C. V. Raman and N. S. N. Nath, The diffraction of light by high frequency sound waves: Part I, *Proc. Indian Acad. Sci. 2*, 406–412 (1935).
26. C. V. Raman and N. S. N. Nath, The diffraction of light by sound waves of high frequency: Part II, *Proc. Indian Acad. Sci. 2*, 413–420 (1935).
27. M. Born and E. Wolf, *Principles of Optics*, Pergamon Press, Oxford (1965).
28. P. Debye and F. W. Sears, On the scattering of light by supersonic waves, *Proc. Natl. Acad. Sci. U.S.A. 18*, 409–414 (1932).
29. R. Lucas and P. Biquard, Propriétés optiques des milieux solides et liquides soumis aux vibrations élastiques ultra sonores, *J. Phys. Rad. 3*, 464–477 (1932).
30. F. C. Zumsteg, J. D. Bierlien, and T. E. Gier, $K_xRb_{1-x}TiOPO_4$: A new non-linear optical material, *J. Appl. Phys. 47*, 4980–4985 (1976).
31. L. J. Libertini and E. W. Small, On the choice of laser dyes for use in exciting tyrosine fluorescence decays, *Anal. Biochem. 163*, 500–505 (1987).
32. J. L. Powell and B. Crasemann, *Quantum Mechanics*, Addison-Wesley, Reading, Massachusetts (1965).
33. G. J. Blanchard and M. J. Wirth, Transform-limited behavior from the synchronously pumped cw dye laser, *Opt. Comm. 53*, 394–400 (1985).
34. R. H. Johnson, Characteristics of acoustooptic cavity dumping in a mode-locked laser, *IEEE J. Quantum Electron. QE-9*, 255–257 (1973).
35. C. Ghosh, in: *Thin Film Technologies and Special Applications, Proc. SPIE 346*, 62–68 (1982).
36. L. J. Libertini and E. W. Small, Resolution of closely spaced fluorescence decays—the luminescence background of the RCA 8850 photomultiplier and other sources of error, *Rev. Sci. Instrum. 54*, 1458–1466 (1983).
37. C. C. Lo and B. Leskovar, Studies of prototype high-gain microchannel plate photomultipliers, *IEEE Trans. Nucl. Sci. NS-26*, 388–394 (1979).
38. I. Isenberg, in: *Biochemical Fluorescence: Concepts*, Vol. 1 (R. F. Chen and H. Edelhoch, eds.), pp. 43–77, Marcel Dekker, New York (1975).
39. I. Isenberg and E. W. Small, Exponential depression as a test of estimated decay parameters, *J. Chem. Phys. 77*, 2799–2805 (1982).
40. H. Kume, K. Koyama, K. Nakatsugawa, S. Suzuki, and D. Fatlowitz, Ultrafast microchannel plate photomultipliers, *Appl. Opt. 27*, 1170–1178 (1988).
41. R. Schuyler and I. Isenberg, A monophoton fluorometer with energy discrimination, *Rev. Sci. Instrum. 42*, 813–817 (1971).
42. J. J. Hutchings and E. W. Small, Energy windowing with Hamamatsu microchannel plate photomultipliers (J. E. Lakowicz, ed.), *Proc. SPIE 1204*, 184–191 (1990).
43. L. J. Libertini and E. W. Small, Application of Method of Moments analysis to fluorescence decay lifetime distributions, *Biophys. Chem. 34*, 269–282 (1989).
44. A. J. W. G. Visser and A. van Hoek, in: *Time-Resolved Laser Spectroscopy in Biochemistry* (J. R. Lakowicz, ed.), *Proc. SPIE 909*, 61–68 (1988).
45. P. R. Hartig, K. Sauer, C. C. Lo, and B. Leskovar, Measurement of very short fluorescence lifetimes by single photon counting, *Rev. Sci. Instrum. 47*, 1122–1129 (1976).

46. S. Torre, T. Antonioli, and P. Benetti, Study of afterpulse effects in photomultipliers, *Rev. Sci. Instrum. 54*, 1777–1780 (1983).
47. D. Bebelaar, Time response of various types of photomultipliers and its wavelength dependence in time-correlated single-photon counting with an ultimate resolution of 47 ps fwhm, *Rev. Sci. Instrum. 57*, 1116–1125 (1986).
48. M. Yamashita, Time dependence of rate-dependent photomultiplier gain and its implications, *Rev. Sci. Instrum. 51*, 768–775 (1980).
49. E. W. Small and S. R. Anderson, Fluorescence anisotropy decay demonstrates calcium-dependent shape changes in photo-cross-linked calmodulin, *Biochemistry 27*, 419–428 (1988).
50. E. W. Small, L. J. Libertini, and I. Isenberg, Construction and tuning of a monophoton decay fluorometer with high-resolution capabilities, *Rev. Sci. Instrum. 55*, 879–885 (1984).
51. I. Isenberg, R. D. Dyson, and R. Hanson, Studies on the analysis of fluorescence decay data by the method of moments, *Biophys. J. 13*, 1090–1115 (1973).
52. S. Hirayama, Application of the lambda invariance test on non-exponential decays, *J. Photochem. 27*, 171–178 (1984).
53. I. Isenberg, On the theory of fluorescence decay experiments. I. Nonrandom distortions, *J. Chem. Phys. 59*, 5696–5707 (1973).
54 E. W. Small and I. Isenberg, The use of moment index displacement in analyzing fluorescence time-decay data, *Biopolymers 15*, 1093–1101 (1976).
55. E. W. Small and I. Isenberg, On moment index displacement, *J. Chem. Phys. 66*, 3347–3351 (1977).
56. E. W. Small, The method of moments and the treatment of nonrandom error, in: *Numerical Computer Methods, Methods in Enzymology*, Vol. 210 (L. Brand and M. L. Johnson, eds.), in press (1991).
57. S. W. Cheng and J. Eisenfeld, in: *Applied Nonlinear Analysis* (V. Lakshmikantham, ed.), pp. 485–497, Academic Press, New York (1979).

3

Streak Cameras for Time-Domain Fluorescence

Thomas M. Nordlund

3.1. Introduction

3.1.1. What Is a Streak Camera?

A streak camera is a device designed to measure the time dependence of electromagnetic radiation emitted from a source on very fast time scales. In today's context, the radiation can range in frequency from the X-ray to the near-infrared region and in time scale from about 10^{-13} to 10^{-7} s. Streak cameras of the photoelectronic type[1-3] (see Sections 3.1.2 and 3.1.3 below) have been employed in numerous laboratories since the early 1970s to measure the time dependence of events occurring in laser, chemical, biological, solid-state, and gaseous systems on time scales of picoseconds. Streak cameras were originally developed because of a need to measure motion of objects or physical processes which occur very rapidly after some initiating impulse. Typical examples are projectile travel (ballistics), shock waves caused by explosions, and flow dynamics.[4] Modern applications include these original ones but more and more are being dominated by laser diagnostics, spectroscopic applications, time-resolved X-ray imaging and diagnostics, and optoelectronics.

Three basic physical processes underlie the operation of modern photoelectronic streak cameras:

(i) conversion of photons to photoelectrons, which are accelerated to high velocity

---

Thomas M. Nordlund • Departments of Biophysics, Physics and Astronomy, and Laboratory for Laser Energetics, University of Rochester, Rochester, New York 14642. The author's present address is Department of Physics, University of Alabama at Birmingham, Birmingham, Alabama 35294.

Topics in Fluorescence Spectroscopy, Volume 1: Techniques, edited by Joseph R. Lakowicz. Plenum Press, New York, 1991.

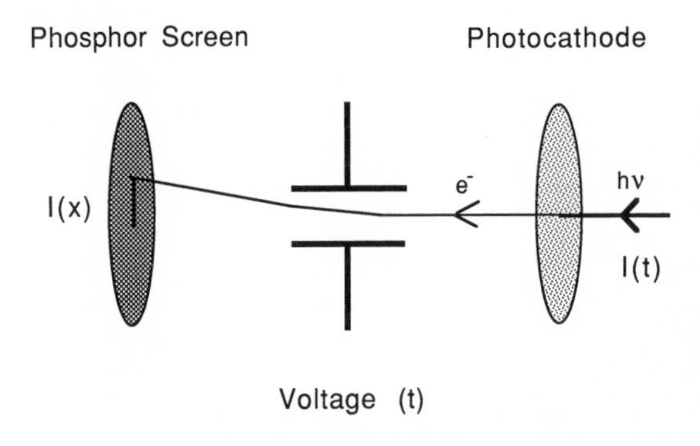

Figure 3.1. Conversion of time dependence of light emission $I(t)$ to position dependence of phosphor screen emission $I(x)$. Light ($h\nu$) strikes the semitransparent photocathode, from which are emitted photoelectrons which are accelerated toward the phosphor screen and deflected by an electric field increasing linearly with time. A "streak" of phosphor glow is created on the screen.

(ii) deflection of the photoelectrons by a transverse electric field which rises in strength as time increases

(iii) detection of the transverse position of impact of the photoelectrons on a phosphor screen

The result of these processes is that the time dependence of the radiation intensity is translated into the distance dependence of electron impact on the phosphor screen (Figure 3.1). The relation between time and distance is governed by the shape of the voltage ramp (rising electric field) applied to the plates which deflect the photoelectrons.

In its simplest form, a photoelectronic streak camera (mechanical streak cameras will only be briefly considered in Section 3.1.4) consists of an image converter tube (ICT; Figure 3.2) and a device to record the emission from the

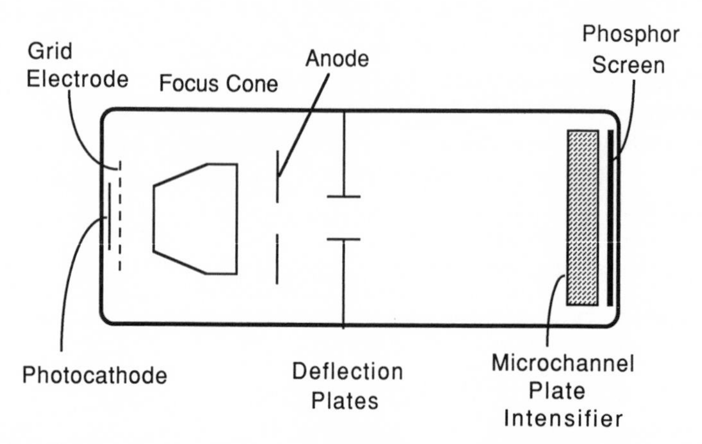

Figure 3.2. Diagram of an image converter tube (ICT), or streak tube. Electrons emitted by the photocathode are extracted by the grid, kept in focus by the electric field of the focus cone, and accelerated toward the anode. After passing through a hole in the anode, the electrons are deflected by the voltage ramp applied to the deflection plates. In some tubes a microchannel plate intensifier multiplies the number of electrons, and the electrons then strike the phosphor.

phosphor screen. The ICT contains a photocathode to convert photons to electrons, plates to accelerate, focus, and deflect the electron beam, and the phosphor screen which luminesces where the photoelectrons hit. The recording device could be as simple as a camera with high-speed film. In its commonly implemented form, however, a streak camera system usually includes the ICT, an image intensifier, an optical multichannel analyzer (OMA), a computer, and a laser for excitation of the sample and triggering of the streak camera. A block diagram of such a system is shown in Figure 3.3.

3.1.2. Scope of the Chapter

In this chapter, the basic components, capabilities, and uses of photoelectronic streak cameras (termed simply streak cameras from this point on) are described, with special emphasis on applications in fluorescence spectroscopy. Considerable space will be devoted to description of the components of a streak camera system, as streak cameras are still rather mysterious to most fluorescence spectroscopists. Among the methods in ultrafast spectroscopy, streak cameras occupy rather well-defined niches, including the time regime 0.5 to 500 ps and multidimensional spectroscopy. In this chapter, I will not describe streak cameras from an electrical engineer's point of view, will not

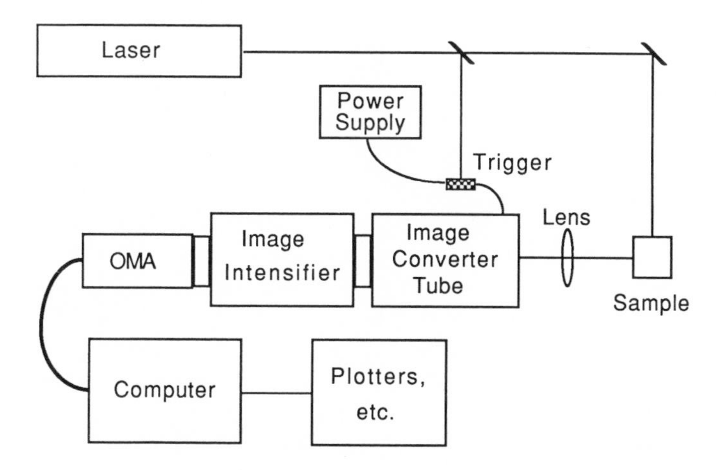

Figure 3.3. Schematic of a typical streak camera system. The trigger synchronizes the image converter tube (ICT) with the laser excitation pulse. The output of the ICT, emission from its phosphor screen, is amplified by the image intensifier, producing a brighter replica on the intensifier phosphor. The optical multichannel analyzer (OMA) records the position-dependent emission from the image intensifier.

address the question of choosing a streak camera, and will not attempt to review the current literature on characteristics and uses of streak cameras. Indeed, a rather restricted set of examples of the use of streak cameras is given, each example serving to illustrate a specific feature of the camera. Liberal use of diagrams is intended to give the reader a feel for how a streak camera is built and how it operates.

There exist at present four primary centers in the world for the development of streak camera technology: the Soviet Union, the United States, England, and Japan, each of which has taken a somewhat different approach to their cameras. In this chapter, I will not attempt to give equal time and space consideration to each of the various approaches researchers have taken in their development of measurement systems. Comparisons of the commercially available systems can be found in the literature.[5, 6] Some descriptions are made of particular manufacturers' streak cameras either because of my familiarity with them or because of the particular features of the cameras. Research directions of many of the active research groups can be found in Proceedings of the International Congresses on High Speed Photography and Photonics, the latest published of which is the 18th.[7] Research and commercial development of streak cameras is presently very rapid, with new models appearing yearly. It should be expected that new and especially more user-friendly cameras will be appearing.

3.1.3. Basic Features and Abilities of a Photoelectronic Streak Camera

The most familiar feature of streak cameras is their ability to time-resolve events on extremely short time scales. The limiting short-time resolution of streak cameras has historically followed the ability of technology to controllably produce events on short time scales. Although it was shown very early that these devices could have time resolutions of 10 fs,[8] time resolution approaching this limit did not come about until the early 1980s, after subpicosecond dye lasers were developed. Since then, several groups have demonstrated ICTs with time resolution below a picosecond.[9, 10, 62] The limiting time resolution of commercially available streak cameras is presently about 1 ps. At the long end of the time scale, photoelectronic streak cameras are in principle not limited, but for practical applications it is not useful to use them for times longer than about 100 ns. This will be discussed more fully in Section 3.2.2.2.

The second most familiar characteristic of streak cameras in their intrinsic ability to do two-dimensional spectroscopy. Streak cameras have always been designed as "cameras," that is, devices which can record events in two dimensions. Photoelectronic cameras maintain their imaging capabilities after photons are converted to photoelectrons inside the tube by the use of focusing electrodes. The two dimensions available can be used in a number of ways. Traditionally, one dimension is used to record the time dependence of a signal. (It need not be, however.) The most common use of the second dimension is to record the signal as a function of a geometrical coordinate along the sample under study or, if a frequency-dispersive element is added to the detection system, to record the signal as a function of wavelength. Modifications of streak cameras, called framing cameras, have an external shutter or an additional set of deflection electrodes inside the tube and provide one additional dimension for recording.[11, 12] Two-dimensional images of objects can be recorded with such devices with a frame time of hundreds of picoseconds or less. These have been used to time-resolve the implosion of laser-fusion targets[13] by measuring X-ray emission and can also be used for fluorescence emission at visible wavelengths. Methods for 10-ps resolution in framing cameras have been devised,[14] using frequency "chirping" of an optical (or X-ray) probe pulse to encode time-dependent information. An example of the use of framing cameras can be found in Section 3.8.5.

3.1.4. History

The development of picosecond streak cameras has been from the beginning a rather international affair. Scientists from England, the United States, the Soviet Union, Germany, and France had well-developed and active programs in photoelectronic streak camera research in the 1950s. Much of the interaction between the various groups took place at the International Congresses on High Speed Photography, normally held every two years. These congresses in turn had their beginnings in meetings of the (American) Society of Motion Picture and Television Engineers (SMPTE) and, even earlier, before the invention of television, the Society of Motion Picture Engineers, whose initial interests centered around recording motions of rapidly moving objects. In fact, the development of streak cameras was quite closely tied to early cinematography. The first SMPTE committee on High Speed Photography was formed in 1948, and the International Congresses began in 1952.

The first types of "streak cameras" were not of the photoelectronic type but stemmed from mechanical and "flash" cameras. The drum streak camera, which transported film fixed to a rotating drum past a slit, was first demonstrated in 1882 by Mallard and Le Chatelier. Hubert Schardin, John Waddell, and Harold Edgerton were responsible for three types of high-speed cameras: the Cranz–Schardin system of multiple light sources and lenses, the rotating prism camera, and the electronic flash camera, respectively. The high-speed photograph that is most well known to the general public is perhaps that of a bullet in flight. (One still sees advertisements of mechanical streak cameras containing this photo.) Such an image was first obtained by Ernst Mach, the greatest ballistics expert of his day. Cranz was Mach's successor, and Schardin was Cranz's student, whose thesis project was the development of the Cranz–Schardin photosystem. The first commercially available rotating prism camera, a mechanical type of streak camera, was used at the 1932 Olympic Games.

A key figure responsible for the formation and nurture of the field of high-speed photography was C. Francis Jenkens, first president of the SMPE. Jenkins was an inventor who began his own business at the age of 16 in 1883. At 26, Jenkins received his first patent, for image-compensation cameras. He worked until his death in 1934 on high-speed cameras, though this work was overshadowed by Kodak and Fastex cameras in terms of development of commercial systems. Jenkins was one of the pioneers who created the marriage between short-duration spark illumination systems and dry-plate photoemulsions.

High-speed photographic technology continued along two general lines after the 1930s. High-speed motion picture and video cameras are the present-day result of one of those lines. The other technological direction led to what we now term streak cameras. Two main types of streak cameras were produced: mechanical cameras and deflecting image converter tubes, or photoelectronic streak cameras. The mechanical cameras, which move film past a slit at high speed or deflect an image onto a fixed piece of film with a rapidly rotating prism or mirror (or a combination of the two), had time resolutions of 10–100 ns by 1957,[(15)] limited primarily by spool rupture and motion noncompensation.[(16)] The primary efforts for improvement of these cameras were directed toward improving the mechanical properties of the moving parts, primary problems being film spool rupture and mirror deformation.[(16)] In 1949, a photoelectronic streak camera system built by Courtney-Pratt could resolve processes down to 10–100 ps.[(17)] Figure 3.4 charts the time resolution of streak cameras from 1957 to the present. A further noteworthy difference between the ICTs and mechanical streak cameras is the better spatial resolution, better by a factor of 3–10, of the mechanical variety.[(16)]

The precursors to modern photoelectronic streak cameras or image converter tubes began to take shape following the demonstration by Zavoiskii and Fanchenko[(8)] that the conversion of light pulses to photoelectrons, followed by a sweep of the electrons by an increasing electromagnetic field, could result in time resolution of 10^{-14} s. Within a few years, groups in a number of countries had prototypes of electromagnetic deflection-type ICTs. These developments were watched with interest by the governments of those countries. The Fifth International Congress on High Speed Photography, held in Washington, D.C., in 1960 was opened by a greeting from President

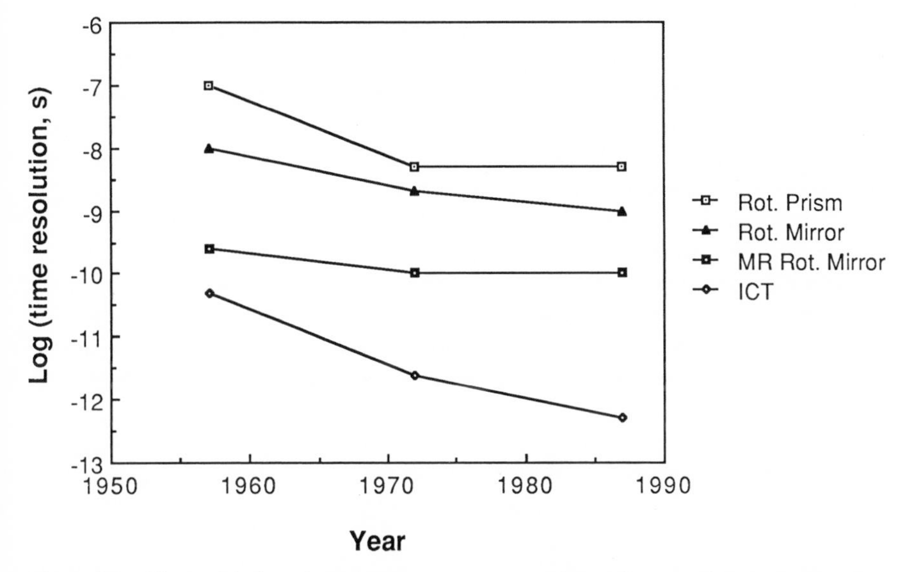

Figure 3.4. Time resolution of streak cameras versus year. Rot. Prism, rotating prims, mechanical type streak camera; MR, multiple reflection; ICT, image converter tube (photoelectronic streak camera).

Dwight Eisenhower and a Resolution of the Congress of the United States. The Congressional resolution stated in part

> Whereas photographic techniques which can magnify the time scale of scientific phenomena are extremely important to the research and engineering activities of every nation; and ...
> Whereas it is the belief of the Congress that:
> (1) the democratic environment of the free world is the best environment for achievement in science; and
> (2) scientists and engineers have special advantages and opportunities to assist in achieving international understanding ...
> Whereas the Congres is interested in (1) promoting international understanding and goodwill; (2) enhancing the excellence of American science, both basic and applied; and (3) furthering international cooperation in science and technology by creating the necessary climate for effective interchange of ideas: Now, therefore, be it *Resolved by the Senate* (*the House of Representatives concurring*). That it is the sense of the Congress that all interested agencies of the Federal Government should participate actively to the greatest practicable extent in the Fifth International Congress on High Speed Photography ...[18]

Governmental interest has perhaps flagged slightly, as the 16th Congress, held in Strasbourg, France, in 1984 was opened by the West German Minister for Research and Technology.

The period from 1960 to the present has been marked by a steady improvement in the time resolution and sensitivity of ICTs, though the basic design has not changed. The transition from 10^{-10}-s time resolution to true picosecond resolution occurred in the early 1970s and was primarily due to the realization by Schelev, Bradley, and others that the application of a large photoelectron extraction field to the photocathode would minimize spreading effects which limited time resolution. [See Section 3.3.1.1, especially the discussion following Eq. (3.10)]. Perhaps the most significant debate over the basic design of ICTs concerned whether to use electric or magnetic fields to deflect the photoelectrons. In 1960, the editor of the Proceedings of the Fifth International Congress on High-Speed Photography debated this point with an author,[19] stating that magnetic deflection was preferable because it gave better spatial resolution in the image. This question was again addressed at the 15th Congress in 1982.[16] ICTs employing magnetic fields for improved spatial resolution but electric field deflection have appeared on the market. The ITT F4127 is an example. Kinoshita *et al.* introduced a 0.5-ps ICT in 1984 also employing electric field deflection but magnetic field focusing.[20]

3.2. System Building Blocks

3.2.1. Modes of Operation

Present-day photoelectronic streak camera systems can be configured to operate in three general modes: single-pulse, synchroscan, and steady-state. The mode of operation is generally determined not by the design of the streak tube but rather by the excitation source and/or the detector which reads and stores the signal from the ICT. These modes are normally correlated with the repetition rate of the signal:

- Single-pulse mode: <10 Hz
- Synchroscan mode: ≈ 10 kHz–100 MHz
- Steady-state mode: ≈ 10 Hz–10 kHz

3.2.1.1. Single-Pulse Mode (<10 Hz Repetition Rate)

The single-pulse or low-repetition-rate mode of operation was the first to be developed and is the most straightforward to understand. In this mode, a sample is excited by one excitation pulse from a laser or other excitation source, and the streak camera is triggered to record the signal created by this pulse. The light emitted from the sample is focused onto the photocathode, the photoelectrons are swept across the ICT phosphor screen, and a recording device reads and stores the signal from the phosphor. There are two sets of events which must be synchronized. First, the ICT must be triggered to record when the sample is emitting. Second, the signal recording system must be synchronized to the ICT phosphor screen signal. The first synchronization must be accurate to a picosecond, if time resolution of about one picosecond is desired. The synchrony between the phosphor screen emission and the recording thereof must only be precise to about ten microseconds, since the lifetime of the screen phosphorescence is hundreds of microseconds. In some cases, this may be the end of the experiment—a single pulse and a single recorded signal. If, however, the signal is weak, one may want to sum several or many shots together to improve the signal-to-noise ratio. In this case, the whole procedure is repeated and the signal from the next excitation is either stored separately or added onto the previous signal. The choice between addition and separate storage is usually a question of data storage space: 100 shots can require a large amount of storage space if each signal is stored separately (see Section 3.2.5).

The number of repetitions and the repetition rate for an experiment are governed by the photodynamic properties of the sample. There are two aspects to consider: (i) the size of the signal, which includes effects of the absorbing species' optical properties and the desired signal-to-noise ratio, and (ii) the recovery of the sample from an excitation pulse. Consider, as an example, the case of fluorescence emission from a dipole-allowed transition. In single-shot mode, the laser energy needed to create large enough signals typically ranges from about 0.1 to 100 μJ. (Note that in this mode the excitation energy, rather than the intensity or power, governs the signal size.) If a single 10-μJ pulse is needed to produce an acceptable signal-to-noise ratio for one shot, then if one desires or is forced to work with pulse energies to 0.1 μJ, roughly 10^4 such shots must be summed. This may or may not be practical, depending upon the system repetition rate, the data storage available, the system stability over time, and the operator's persistence. A common reason to keep excitation energies low is to avoid high-intensity nonlinearities.[21] A 10-μJ, 1-ps laser pulse at 600 nm, focused to a 5×10^{-4}-cm^2 area, for example, produces an average intensity of 2×10^{10} W/cm^2 in the sample during the pulse, corresponding to a fluence of 6×10^{16} photons/cm^2. If the absorption cross section of the sample is about 5×10^{-18} cm^2 or higher, then high-intensity effects will probably be encountered. An estimated criterion for the onset of one type of nonlinear effect, ground state depletion, is when the product of the absorption cross section and the fluence exceeds 0.1:

$$f = \sigma F > 0.1 \tag{3.1}$$

where f is the fraction of absorbing molecules excited by the pulse, σ is the absorption cross section in cm^2, and F is the fluence in photons/cm^2. The onset of ground state depletion is also a good marker of the beginning of excited state absorption, another effect to be considered in single-pulse excitation. The relation between cross section and extinction coefficient is

$$\sigma\ (\mathrm{cm}^2) = 3.824 \times 10^{-21}\ \varepsilon\ (M^{-1}\ \mathrm{cm}^{-1}) \tag{3.2}$$

A cross section of $5 \times 10^{-18}\ \mathrm{cm}^2$ corresponds to an extinction coefficient of $1300\ M^{-1}\ \mathrm{cm}^{-1}$. Single-shot mode is the mode of choice if one wants to study effects of high excitation intensities or fluences. The second factor to be considered when choosing repetition rates and number of repetitions is whether the signal may be expected to change from shot to shot. If, for example, one laser pulse causes a 1% permanent (relative to the time between repetitions) change in the sample, then after 70 laser shots, the sample is more than 50% altered. The most common cause of sample change during a series of optical excitations is photochemistry. The usual solution when the sample is a liquid is to flow the liquid to present fresh sample to each excitation pulse. Problems of electronic excited state nonrecovery between shots (nonrelaxation of triplets, for example) are not normally encountered because of the low repetition rate in single-shot mode.

3.2.1.2. Synchroscan Mode (10 kHz–200 MHz Repetition Rate)

The synchroscan mode of operation uses a relatively high repetition rate source, such as a continuous-wave (cw) mode-locked laser, for sample excitation. Synchrony between excitation and detection is enforced by driving the excitation source and the streak camera with a common oscillator. Figure 3.5 shows the layout of a such a setup. The oscillator normally drives a power amplifier or resonant circuit which produces voltages high enough to drive the ICT (see Section 3.2.2.2). The synchrony between excitation and ICT scan must be accurate to about a picosecond if 1-ps time resolution is desired. This

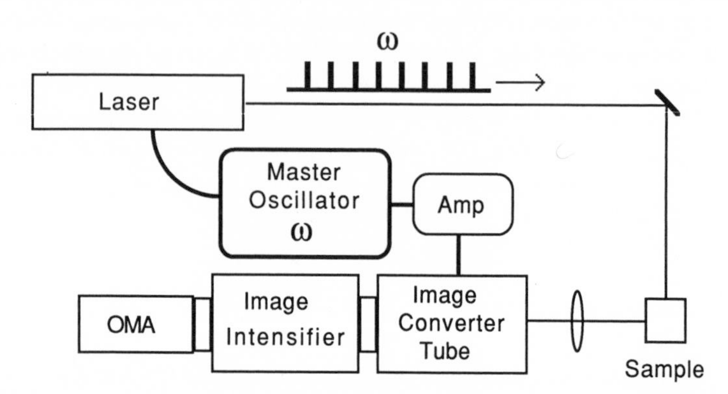

Figure 3.5. Schematic of a synchroscan streak camera system. A master oscillator, usually operating at 1 to 150 MHz, synchronizes the streak camera sweep with the sample excitation.

can be achieved, although commercial units typically display about ±4-ps jitter (see Section 3.2.2.2b). At frequencies used for synchroscan operation, the readout and data storage system (the OMA in Figure 3.5) cannot read the ICT signal after every excitation pulse. In fact, there is no use in doing so, as the phosphor decay time of the ICT, being relatively long compared to the time between excitation pulses, will produce a signal which is constant as a function of time. Thus, the OMA or other recording device can scan the signal of the ICT continuously and average until the signal is acceptable.

Operation of a streak system with a cw mode-locked laser typically involves frequencies of 50–150 MHz or about 1 MHz if the laser is cavity-dumped. At the former rate, significant problems with excited state non-recovery and photochemistry can be encountered if states with lifetimes of about 10 ns or longer are produced. The extent of the problem depends upon laser intensity and excited state decay times. Since the size of the signal scales with the average laser power, a compromise must often be made between signal-to-noise ratio and sample integrity. Flowing liquid samples or operation of the laser in cavity-dumped mode can also alleviate sample recovery problems.

A potential problem similar to excited state nonrecovery in synchroscan mode is that of flyback. In synchroscan mode, a sine wave is usually applied to the deflection plates. If the rising part of the sine wave causes the photoelectron beam to be swept in the forward direction across the phosphor screen, then the falling part will cause the beam to be swept back across the screen in the opposite direction (flyback) if the sample is still emitting after one-half the period of the sine wave. This makes the signal almost impossible to analyze. In some models, the flyback problem has been eliminated by incorporating a second set of electrodes, oriented perpendicular to the deflection electrodes, which deflect the electron beam off to the side of the phosphor screen during sweepback.[57] Note that this does not eliminate the problem of sample nonrecovery mentioned in the previous paragraph.

Synchroscan mode normally places one in a low-excitation-intensity/fluence regime when doing optical fluorescence measurements. The energy of individual excitation pulses of cw mode-locked lasers ranges typically from 0.1 to 10 nJ. From Eq. (3.1), one sees that ground state depletion and excited state absorption effects are not normally encountered when using high-repetition-rate, low-pulse-energy lasers. On the other hand, if one wishes to study nonlinear effects, this type of excitation source is not optimal. Cavity dumping can increase the pulse energy by up to a factor of 100, so that under some circumstances significant ground state depletion can occur.

A major advantage of synchroscan operation is that the repetition rate allows for easy adjustment of sample or system to achieve optimal signal quality. Typical problems encountered in doing laser spectroscopy are laser instability and sample or optics positioning problems. Remember that in doing time-resolved spectroscopy with <10-ps resolution, geometry and alterations of the light path become crucial. The time it takes for light to travel through 1 mm of a liquid or solid sample, for example, is about 5 ps. The ability to see a continuous display of the signal as the laser or sample is adjusted is a great convenience.

3.2.1.3. *Steady-State Mode (10 Hz–10 kHz Repetition Rate)*

The steady-state mode spans the repetition rate region between that of single-shot and synchroscan modes. The upper and lower frequency limits of this mode are not universal but rather depend upon the characteristics of the readout system and the excitation source. The most common reasons for using this mode are to avoid accumulation of long-lived excited states or to accommodate the repetition rate of the excitation source or the scan rate of the readout system. There is no fundamental characteristic of this mode of operation which distinguishes it from synchroscan mode. The primary difference between single-shot and synchroscan modes, besides the repetition rate difference, is that in single-shot mode the signal from the ICT is read out after each excitation pulse, in synchrony with the ICT phosphor signal, whereas in synchroscan operation the ICT signal is read out continuously. In this sense, steady-state mode is like synchroscan. However, the repetition rate of the former is far enough below the usual implementations of synchroscan systems (1–100 Mhz) that it is not usually referred to as synchroscan. As the scan rates of readout systems for ICTs improve, "steady-state mode" will become "single-pulse mode." Some modern femtosecond, amplified dye laser systems are designed to operate at about 1 kHz.[22] OMA systems may take as long as 500 ms to record and store the signal from the ICT phosphor. Under such circumstances, the OMA cannot read out the individual signal associated with each excitation pulse even though the ICT phosphor decay time would allow it. The ICT can be triggered so that it presents a new record for each excitation pulse, but the signal would be accumulated continuously, as in synchroscan mode.

3.2.2. Image Converter Tube

The image converter tube is the heart of the photoelectronic streak camera system. Its purpose is to convert the time dependence of a light input on a picosecond time scale to an easily measurable distance dependence on a detector screen. Figure 3.6 illustrates this conversion. The ICT does this by using the photoelectric effect to eject electrons from a photocathode when the light impinges (Figures 3.1 and 3.2). If the light source is imaged onto the photocathode, an electron replica of this image will be produced on the other

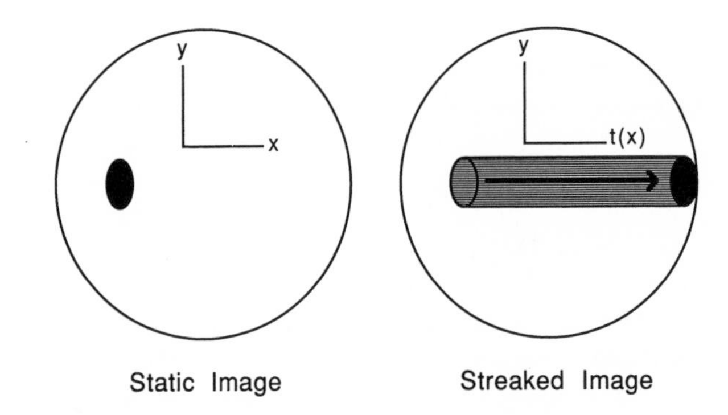

Figure 3.6. Phosphor screen output image in focus mode (statis image) and in streak mode (streaked image). The two-dimensional image is swept across the phosphor screen as a function of time, transforming the position axis, x, into a time axis, $t(x)$.

side of the cathode. The number of photoelectrons in this replica will be proportional to the number of photons which hit the cathode. If the light intensity varies as a function of time, the photoelectron density will also vary with time. An extraction grid or mesh electrode draws the photoelectrons from the cathode, and they accelerate toward the anode. If the light is pulsed, then a three-dimensional electron bunch will travel toward the anode. The transverse cross section of the electron bunch will reflect the light beam cross section. A longitudinal cross section of the traveling electron bunch reflects the time dependence of the light intensity. A focusing electrode is typically needed to keep the electron bunch properly in focus as it travels down the tube axis. The electron bunch then passes through a hole in the anode and enters the region between the deflection plates. In synchrony with the arrival of the electrons, a linearly rising voltage ramp is applied across the deflection plates. The front part of the electron bunch is deflected very little, since the field is low. Electrons which arrive later are deflected farther. If the voltage rises linearly with time, then the deflection angle is directly proportional to time. In a typical ICT, the voltage would rise a few kilovolts per nanosecond. The electrons impact on a phosphorescent coating on the end of the ICT.

3.2.2.1. Photocathodes

Photocathodes of commercial ICTs are generally of type S-1, S-11, S-20, or S-25. S-1 is a near-infrared-sensitive cathode made of Ag–O–Cs, used generally in the wavelength range 800–1100 nm. Its peak sensitivity is typically a factor of 10–20 less than those of the other cathodes, though it is the most sensitive beyond about 900 nm. S-1 photocathodes built for streak tubes have been noted for their limited sensitivity lifetime.[5] Tubes manufactured in the 1970s sometimes deteriorated markedly in response over periods of months, though their photomultiplier counterparts seemed not to do so. Presently, they are much improved, but sensitivity decrease is still more of a concern than for the other types of photocathodes. Gex *et al.*[23] have discussed the mechanism of this sensitivity loss in terms of a mechanism proposed by Wu Quan De. They believe that the loss of response of S-1 photocathodes is caused by free cesium atoms which leave the surface of the cathode and deposit on the ICT phosphor screen. The response loss is partially reversible by rebaking the tubes. Because of the response extending toward the infrared, S-1 ICTs will also exhibit more thermal noise than the others.

S-20 and S-25 cathodes are most commonly employed for spectroscopy in the visible. The former responds approximately from 200 to 750 nm. The lower limit is variable, depending upon the absorption properties of the glass, fused silica, or sapphire used for the input window and the photocathode thickness. The photosensitive material in these cathodes is a layer of Sb–Na–K which is doped on the surface with Cs. They are sometimes termed multialkali, and designations such as MA may be used. S-25 photocathodes are similar to S-20 photocathodes, but the response extends farther to the red. Some manufacturers use the description S-20.⟨wavelength⟩, where wavelength indicates the wavelength of maximum response, or ERMA (extended red multialkali) rather then S-25 in labeling of cathode response, since increased red response is obtained by increasing the thickness of the Sb–Na–K–Cs layer and monitoring the quality of the material, rather than by using different materials. Cathode response typically peaks between 400 and 500 nm in S-20 tubes, but the peak can be shifted possibly as far as 700 nm. (One does not yet find S-20.⟨700 nm⟩ tubes one the market.) The payment for the extended red response is a reduced response in the bluer part of the spectrum.[24] Though the manufacture of S-20 tubes is still somewhat of an art, stable and relatively reproducible responses can be obtained. The S-11, or monoalkali, photocathode was one of the first types to be developed for use in image converter tubes. The photosensitive material is Cs_3Sb. S-11 photocathodes display a narrower wavelength sensitivity bandwidth, responding from about 200 to 650 nm.

Image converter tubes can also be used to image and time-resolve X-ray emission. In this case, the photocathode is usually a layer of gold about 100 Å thick evaporated onto a beryllium faceplate. This would respond to X rays of about 1 to 10 keV.

The type of photocathode affects not only the color response of the ICT but also the time resolution of the tube for a given wavelength, due to the dependence of photoelectron velocity on the work function of the cathode material. This is discussed further in Section 3.3.1.1.

3.2.2.2. Deflection and Trigger Systems

Two types of systems can be used to electrically drive the converter tube deflection plates, one appropriate for synchroscan operation and one for single-shot or steady-state mode.

Synchroscan employs a high-quality oscillator/amplifier combination to provide the needed high-voltage sine wave (see Figure 3.5). If the synchroscan frequency of operation is lower than about 10 MHz, however, a sine wave is not appropriate, because dV/dt cannot be large enough unless sine waves of several tens of kilovolts peak height are used (see Section 3.2.2.2a). The oscillator must be the same one as that which drives the excitation of the sample in order that there be as little jitter and drift as possible. A typical example is the driver for the mode locker for a cw mode-locked laser. The mode-locked oscillator, operating at 50–150 MHz, will have its output split to drive both the mode locker and the amplifier for the streak camera sweep. Alternatively, the output of the mode-locked laser can be monitored with a PIN photodiode, and the diode's output used to drive an oscillator which is then amplified. The amplifier must be capable of producing outputs of several kilovolts, though little current is needed.

Single-shot mode provides a kilovolt voltage ramp of fast (about 1 ns) rise time to drive the photoelectron beam across the screen. After the sweep, the waveform should remain at a voltage high enough to keep the beam off screen until the signal has disappeared. The fall time of the voltage is typically microseconds or longer, so that samples with emission lifetimes of up to microseconds can be examined without the flyback problem discussed in Section 3.2.1.2. The generation of these kilovolt voltage ramps is not a simple engineering problem. Examples of devices designed to provide appropriate voltage drives are avalanche transistor stacks, spark gaps, krytrons, microwave triodes, and photoconductive switches. Triggering pulses for the kilovolt devices are derived directly from the excitation pulse itself. A high-speed PIN photodiode monitoring the laser pulse is appropriate for triggering an avalanche transistor stack. The photon energy needed to turn the PIN photodiode on sufficiently is relatively small, less than a microjoule. The diode must have a rise time of less than 1 ns in order to be effective. Spark gaps can provide the high-voltage ramp, but they must be triggered directly by a relatively large laser pulse, millijoules or more, since the laser must break down the gas within the gap. Photoconductive switches are the simplest electrical devices which can be used for single-shot streak camera triggering[25] and do not demand too much energy from the laser, roughly 100 μJ. Figure 3.7 shows how the voltage ramp is applied to the deflection plates. The circuit is passive, consisting only of a dc high-voltage supply, resistors, capacitors, inductors, and the semiconducting switch. The inductor is often just the stray inductance of the wires. The switch can be chromium-doped

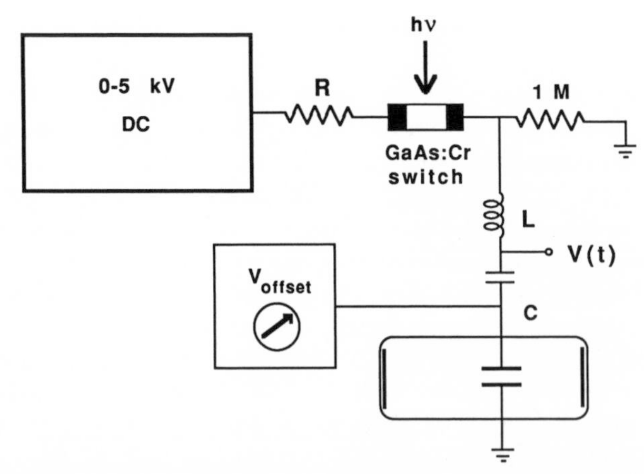

Figure 3.7. A photoconductive switch used to create the ICT deflection voltage ramp and synchronize it with the sample excitation. The chromium-doped gallium arsenide switch has high resistance in the absence of light. A part of the laser pulse, *hv*, is used to create electron–hole pairs in the semiconducting switch, making the switch conductive for the lifetime of the pairs. A high-voltage ramp is then applied to the deflection plates. The shape of the ramp is determined by the RLC circuit. A dc offset voltage is applied to the plates in streak mode to position the photoelectron beam properly on the phosphor screen.

gallium arsenide, which has high resistivity until the laser pulse arrives, creating electron–hole pairs and making the switch a good conductor.

3.2.2.2a. Sweep Speed and Time Linearity. A typical voltage pulse applied to the deflection plates of a streak camera in single-shot mode is shown in Figure 3.8. The pulse shown has a relatively fast rise time, a long fall time, and ringing following the peak. When the streak camera is placed in operate mode, a dc offset voltage is first applied to the plates to keep the electron beam off screen to the left when no other voltage is present. When the voltage ramp arrives, the beam will be at the left edge of the screen when the voltage reaches V_{min}. Voltage V_{max} will bring the beam to the right side

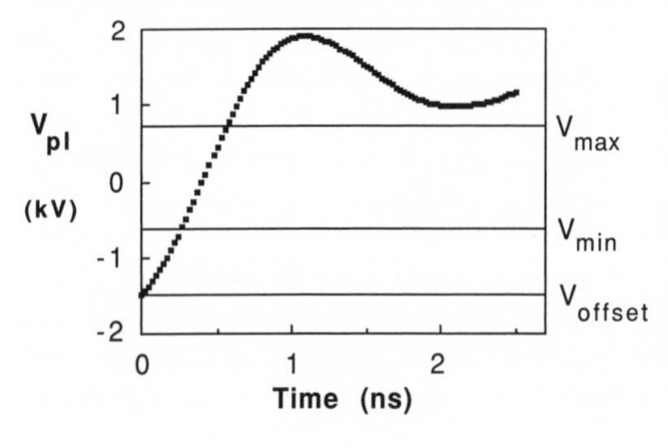

Figure 3.8. Typical voltage ramp applied to deflection plates in single-shot streak mode. V_{pl} is the total voltage applied to the deflection plates. The on-screen detector region corresponds to voltages between V_{min} and V_{max}. The offset voltage determines which part of the voltage ramp waveform is included in the on-screen region.

of the screen. Between V_{min} and V_{max} is a region in which the voltage is approximately linearly rising with time, and the photoelectron beam will be somewhere in the middle of the phosphor screen. The portion of the rising curve comprising the "on-screen" region can be adjusted by adjusting V_{offset}, effectively translating the voltage curve up or down in the figure. The most linear part of the ramp voltage can thus be chosen.

Any point that is on screen corresponds to a voltage between V_{min} and V_{max} and a rate of voltage change (slope) dV/dt. Assuming this slope is constant, or taking the average value of the slope over the interval V_{min} to V_{max}, will give the speed at which the electron beam is swept across the phosphor. $(V_{min} - V_{max})/\langle dV/dt \rangle$ gives the time for the electron beam to be swept full screen, where $\langle dV/dt \rangle$ is the average slope. If the phosphor full-screen distance (diameter) is D, then the *streak* or *sweep speed* v_s is

$$v_s = D\langle dV/dt \rangle/(V_{min} - V_{max}) \tag{3.3}$$

The maximum possible value of streak speed for most tubes lies in the range 10–100 μm/ps $= 10^9$–10^{10} cm/s; values of D range from 1.5 to 5.0 cm; $V_{max} - V_{min}$ may be a few kilovolts.

The voltage ramp is never strictly linear with time. The deviation of dV/dt from its mean is usually 1–20%. In some experiments a correction for this nonlinearity must be made. A common method for correcting the data for the nonlinear sweep employs controlled time delays of a laser pulse. The simplest way is to send the laser pulse through an etalon. The etalon is a piece of glass whose sides are ground precisely parallel and coated on each side with a partially reflective coating. If the etalon thickness is d, the reflectivity is R, and the light beam is normal to the etalon, then exiting from the etalon will be a train of pulses, each separated in time from the previous by

$$\Delta t = 2dn/c \tag{3.4}$$

and decreased in height compared to the preceding pulse by a factor of R^2, where $c = 3.00 \times 10^{10}$ cm/s and n is the index of refraction, as shown in Figure 3.9. The thickness should be chosen so that three or more etalon pulses will fit across the full screen time scale. The more closely spaced the etalon pulses are, the more detailed will be the time linearity information. The spacing of the observed pulses on the streak camera phosphor screen can be used to calibrate the time axis, time versus x, where x is the position on the phosphor screen.

Nonlinearity in the time dependence of the sweep affects the recorded signal in two ways. First, time will not be linearly related to the position along the phosphor screen. This can be corrected for as indicated above and involves only the horizontal time (x) axis. The second effect of sweep non-linearity involves the vertical (intensity) axis. The magnitude of the signal

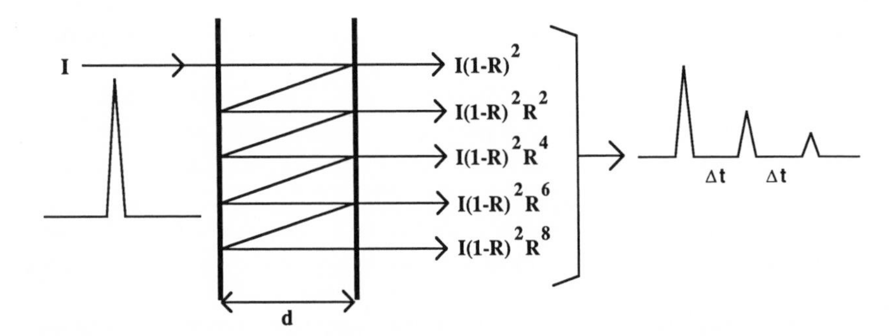

Figure 3.9. Production of a train of optical pulses of calibrated time delays with an etalon. A single incoming pulse is multiply reflected by an etalon of thickness d, coated on each side to produce reflectivity R. Each pulse transmitted through the etalon is delayed by a time $\Delta t = 2dn/c$ with respect to the preceding pulse and reduced in amplitude by a factor of R^2, where n is the index of refraction of light and $c = 3.0 \times 10^{10}$ cm/s. All beams are collinear but shown displaced here for clarity. A train of pulses also is reflected back from the etalon.

observed at any given point on the screen will be inversely proportional to the streak speed at that point (inversely proportional to dV/dt). A slower sweep speed will produce an apparently larger signal. This can be understood most simply by considering a light source which emits constantly in time. The photoelectron beam produced by this source will be of constant intensity in time. As the electron beam is swept across the phosphor, the intensity of the phosphorescence from a given region δx around the point x will be proportional to the number of photoelectrons which hit that region. Since the "dwell" time of the electron beam on the phosphor region is inversely proportional to the streak speed at that point, a larger signal will be recorded when the sweep is slower. Correction for this effect must be made using a light source with well-characterized intensity versus time dependence. Such a source can be the train of etalon pulses described above or can be the emission from a source with a known decay time, preferably single exponential. A convenient decay time is one that is long compared to the full time scale of the streak camera. Many laser dyes give large signals and have decay times which are strictly single exponential and which are long compared to the picosecond time scales of many streak camera experiments. The technique to correct any observed signal for the intensity effect of streak speed nonlinearity is to first correct the time axis as described above and then divide the observed signal by the observed single-exponential-decay calibration signal and multiply the result by $\exp(-t/\tau)$, where τ is the (known) decay time of the calibration dye. This technique will also correct for any response nonuniformities in the phosphor screen or the optical multichannel analyzer (or other detection device). The most accurate correction will be obtained when the emission wavelength of the calibration dye is the same as that of the sample to be measured.

A properly chosen etalon pulse train can be used to correct both the time and intensity axes.[6] Since the intensity of each pulse will be less than that of the previous pulse by a factor of R^2,

$$I_{n+1} = I_n R^2 \tag{3.5}$$

where R is the reflectivity of each etalon surface, the pulse train will be described by an envelope which decays as an exponential:

$$I = I_0 \exp(-t/T) \tag{3.6}$$

where

$$T = 2(\Delta t)^{-1} \ln R \tag{3.7}$$

where Δt is the time between etalon pulses. This exponential decay can be used in place of that of the intensity calibration dye. A disadvantage is that the etalon train provides intensity information only at the points corresponding to the etalon pulses. The dye method, on the other hand, provides a continuous measure of the intensity response across the entire screen.

Figure 3.8 shows a voltage ramp pulse which rings after rising to its maximum. Ringing does not always occur. In well-engineered systems it in general does not. However, in cases where the fastest voltage rises are desired to maximize the rate of photoelectron sweep, ringing may be a necessary part of the fast rise. Ringing indicates that the circuit (see Figure 3.7, for example) formed by the deflection drive electronics is underdamped. Damping can be added to the circuit, but the sweep will then slow. The ringing is not a problem unless a minimum of the ring drops below V_{max}, at which point the photoelectron beam would come back on screen from the right. This, of course, would completely confuse the recorded data unless the light emission being recorded has decayed to zero long before the ringing occurs.

3.2.2.2b. Time Jitter. Jitter can be described as the fluctuation in the delay time between the time of arrival of a δ function (in time) light pulse at the ICT photocathode and the beginning of the streak camera recording scan. If a fixed light pulse is presented many times to the streak camera, the presence of jitter will manifest itself as a variation in the recorded position of the pulse on the ICT phosphor screen. The largest contributions to streak camera jitter are normally associated with the triggering or deflection electronics. Values range from 2 to 100 ps rms, shot-to-shot. If only a single shot is to be recorded, jitter makes no difference, as long as the jitter does not cause the signal to occur off the phosphor screen. In cases where signal averaging is desired or in synchroscan operation, jitter will smear out the time resolution. In the presence of 50 ps of jitter, for example, a 5-ps pulse with a Gaussian shape in time will appear broadened in time by a factor of ten after averaging. The shape of the broadened pulse will be determined by the statistics of the jitter. Clean resolution of a 5-ps signal demands that jitter be reduced below 5 ps. Commercial synchroscan cameras have reduced jitter to less than ± 5 ps. The best commercial single-shot cameras have jitter of about ± 20 ps, shot-to-shot. Use of the photoconductive switch described in Section 3.2.2.2 results in jitter of ± 1 ps. The jitter observed with any given streak camera deflection drive system will depend upon the stability of the laser, if the laser is used to trigger the deflection drive.[25] Shot-to-shot variations in the laser pulse energy or pulse shape can cause the jitter to increase. A minimum in jitter is always obtained when careful attention is paid to the stability of the excitation source. In single-shot mode with a photoconductive switch, for example, laser pulse energies should be stable to about $\pm 5\%$ in order to achieve ± 2 ps shot-to-shot jitter.

3.2.2.3. Maintaining Focus

The photoelectrons traveling down the streak tube must maintain their relative transverse positions if the cross-sectional image of the incident photon beam is to be maintained. However, electron–electron repulsion and stray fields will tend to defocus or spread the electron beam. It is thus required that an element be included in the ICT to maintain focus of the electrons. In most commercial tubes this is done with a focusing cone placed before the anode and deflection plates. The electric field's geometry is determined by the shape of the cone. The voltage applied to the focus cone is set so that it will focus electron beams traveling in the relatively low velocity region just after the grid electrode (see Figure 3.2). Electric field focusing of electron beams is not free of aberrations, as in most optical lenses.[16] One of the problems has been the electron crossover in image-inverting electrostatically focused tubes. At the point of crossover, a high electron density induces space charge effects which defocus the beam. Magnetic lenses can be designed with fewer aberrations[20] to allow for better focusing and thus, in principle, better spatial and temporal resolution.[26]

Proximity focusing is another method which has been used in commercial ICTs. The principle of proximity focusing is that no defocusing occurs if one places the detection surface directly against the emitter. A model produced by General Engineering and Applied Research (now defunct) used a microchannel plate[27] (MCP; see Section 3.2.3) in order to keep the photoelectrons collimated. Figure 3.10 shows a schematic of this tube, adapted from a 1978 advertisement of the manufacturer. The electrons are kept in "focus" by the microchannels, which confine the electrons to travel along small-diameter

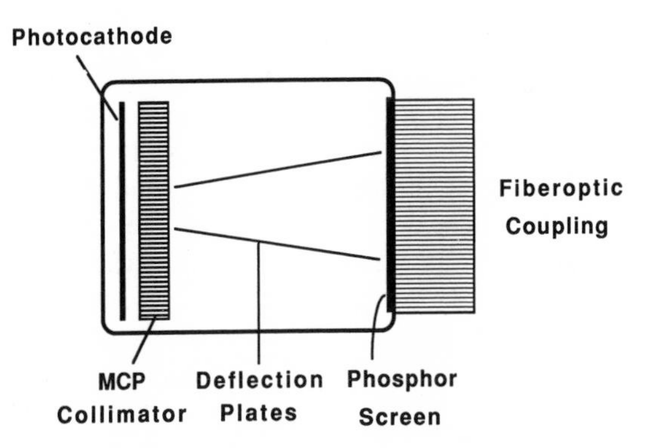

Figure 3.10. Diagram of a proximity-focused streak tube which appeared briefly on the market in the 1970s. The unique feature of this design is the use of a microchannel plate to collimate (focus) the photoelectron beam. This tube was small, approximately 3 cm in length. This is about one-sixth to one-tenth the size of streak tubes of standard design.

cylindrical tracks. This tube is very small compared to other ICTs, measuring only 3 cm in length. Most other tubes range in length from 20 to 40 cm.

The normal way to specify the focusing characteristics of an image converter tube is in terms of ability to resolve lines on a black and white bar chart. The spatial resolution is usually given in terms of the number of line pairs per millimeter (lp/mm) which can be resolved if such a bar chart is imaged on the photocathode and the phosphor screen output is analyzed with no applied deflection voltage ("focus mode"). Note that per millimeter refers to distance on the photocathode unless otherwise specified. In many tubes, the electron optics produces a magnified image of the photocathode image on the phosphor so that lp/mm could refer to the phosphor screen image. The spatial resolution of most streak tubes lies in the range 15–35 lp/mm. This can be compared to the resolution of medium-speed film, which is in the range 50–100 lp/mm. A feeling for spatial resolution can be gotten by comparing these numbers to the spatial resolution of the eye. If one looks at an object, a minimum number of features of that object must be resolved by the eye in order for the observer to detect, orient, recognize, and identify that object. This number of features can be specified in terms of line pairs across the smallest dimension of the object. The line pairs needed in order for various degrees of resolution by the eye are listed in Table 3.1.

Table 3.1. Line Pairs (lp) Needed for Various Degrees of Resolution by the Eye[a]

Task	Resolution needed across object's smallest dimension
Detection	1.0 ± 0.25 lp
Orientation	1.4 ± 0.35 lp
Recognition	4.0 ± 0.8 lp
Identification	6.4 ± 1.5 lp

[a] Ref. 28.

The spatial resolution of the ICT affects the temporal resolution. A tube which has better focus produces a better defined image on the phosphor when the image is a streaked image. Given that all other factors are the same, a tube with better spatial resolution will have better time resolution (see Section 3.3.1.1). It should also be kept in mind that the spatial resolution of a streak camera system depends also upon the resolution of other components of the system, such as the image intensifier (Section 3.2.3) and data readout device (Section 3.2.4). Most streak cameras are limited in spatial resolution by the ICT, however.

3.2.2.4. Phosphor Screens

The most common method for measuring the photoelectron intensity employs a phosphorescent screen deposited on the back end of the streak tube. The (unfortunate) principle is to convert the photoelectron intensity back into a position-dependent light intensity which can be measured at leisure by a two-dimensional photodetector. The phosphors are usually type P-20, P-11, or P-4, aluminized on the front to prevent phosphorescence from going back to the photocathode. The choice between the phosphors is made by matching phosphor spectral output with the response of the subsequent stage of signal measurement, usually the photocathode of an image intensifier. P-20 emits yellow-green light at 560 nm; P-11 emits in the blue, peaking at about 470 nm; P-4 emits yellow light, peaking at 440 and 570 nm. The intrinsic phosphorescence decay times for P-20, P-11, and P-4 are 60, 34, and 60 μs, respectively,[29] but these times can vary considerably depending upon the manufacturer and the operating conditions of the tube. The decay also apparently has a long tail, so that detectors measuring the phosphorescence typically are gated for hundreds of microseconds.

There is a net gain in signal during the process of conversion of incident light on the photocathode to phosphor light exiting the ICT. (There is also, of course, a net gain in noise.) The photoelectrons which strike the phosphor have energy of 10–15 keV which can be dissipated in the phosphor. Since phosphorescence is in the visible, corresponding to about 2.5 eV, it is clear that many photons could be excited by each photoelectron. One finds in the steady state that one 15-keV photoelectron produces about 600 phosphor photons in a typical ICT phosphor. The quantum efficiencies of visible-wavelength photocathodes are around 10% (number of photoelectrons produced per number of incident photons), so the overall photon gain of the ICT in the steady state (number of phosphor photons per photon incident on the photocathode) is about 60. Under picosecond to nanosecond pulsed excitation, however, the gain appears to be a factor of 100 less.[30] The reason for this large efficiency decrease apparently lies in the complex mechanism by which high-energy electrons convert their energy into phosphorescence. The short-duration, high-current photoelectron pulses produced by picosecond light sources in streak camera tubes favor conversion of the photoelectron energy to forms other than phosphorescence.

3.2.2.5. Direct Charge Readout Systems

In the last several years, prototype ICTs have appeared in which two-dimensional charge-coupled devices (CCDs) have been placed directly in the streak tube in the place of the phosphor. The object is to directly measure the current of the photoelectrons without resorting to conversion back into light. The reason for doing this is clear. In typical streak camera systems, photons are first converted to photoelectrons by the ICT photocathode; the photoelectrons are converted back to photons by the ICT phosphor; the ICT phosphor photons are coupled to an image intensifier and converted to electrons by the image intensifier photocathode; the intensifier photoelectrons are converted to intensifier phosphor photons; the intensifier phosphor photons are coupled into a silicon intensifier target (SIT) vidicon tube and converted to photoelectrons which, finally, are measured by determining the charge induced on the silicon target. The total is five conversions of photons to electrons (or vice versa) and two photon couplings. It is easy to see the great opportunity for loss of signal and gain of noise. There is also the previously mentioned decrease in phosphor signal under short-pulse conditions which reduces sensitivity. The best solution is to minimize the conversions and couplings rather than to try to improve efficiencies of each conversion. Such improvements are not merely academic. Precision fluorescence studies of biological systems at low light excitation intensities or measurements of samples in which radiative transitions are weak are often limited by signal-to-noise considerations. (See Sections 3.3.2, 3.8.2, and 3.8.4; see also Section 3.2.4 for further description of recording systems.)

A description of an ICT with integral CCD has appeared. Weiss *et al.* described a modified Hamamatsu tube and concluded that despite the drawbacks that the integral CCD tends to be damaged by the bakeout processing of the tube, leading to increased noise and decreased dynamic range, that the spatial resolution of the tube is inferior to that of standard tubes by a factor of 2–6, and that the photoelectrons may damage the shift registers and output amplifiers of the CCD, the principle of direct photoelectron readout is viable.[31]

3.2.3. Image Intensifiers

The phosphorescence glow from the phosphor screen is very weak when the intensity of a picosecond light pulse hitting the photocathode is low enough to avoid pulse broadening in the photoelectron beam. (See Sections 3.3.1.2 and 3.2.2.3.) It is in fact too weak for measurement with the normal readout systems: vidicons, photodiode arrays, or film. Some sort of image intensification must therefore be done to produce a usable system. Two placement schemes for image intensifiers have been used: inside the streak tube, between the deflection electrodes and the phosphor screen (integral), and outside the tube, connected to the phosphor screen via lens or fiber-optic connection (external). Different designs of intensifiers must be used, depending upon where the intensifier is to be placed. External intensifiers must have a photocathode in order to convert phosphor photons into electrons. Two classes of intensifiers are commercially available, denoted by the method in which the photoelectrons are focused: proximity-focused intensifiers and electromagnetically focused intensifiers. Modern proximity-focused intensifiers are of the microchannel plate (MCP) type and will differ depending upon

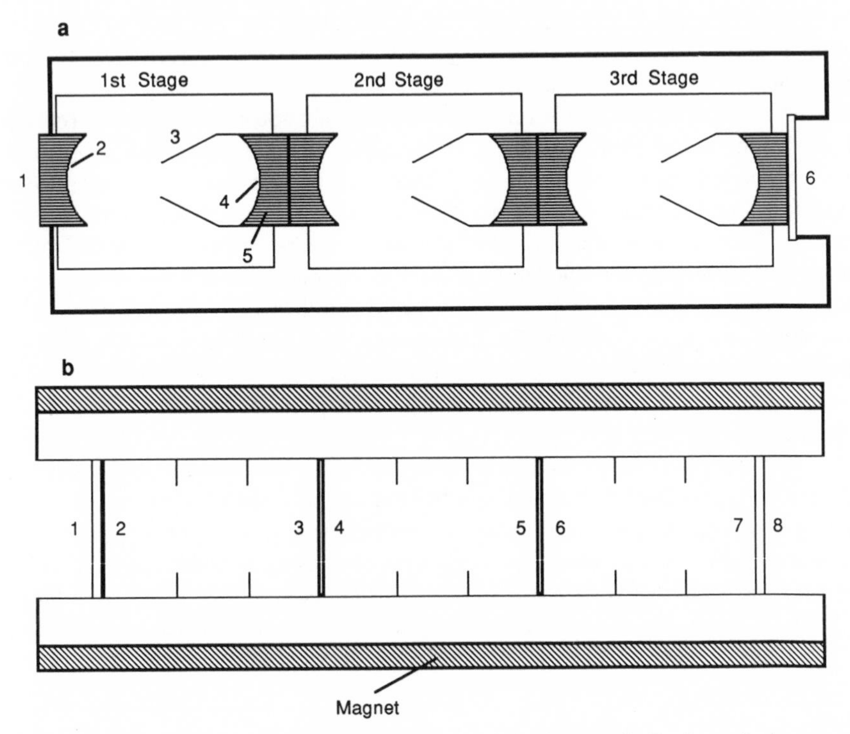

Figure 3.11. (a) Schematic of a fiber-optically coupled, electrostatically focused, three-stage image intensifier. 1, Fiber-optic input window; 2, photocathode; 3, focusing electrode; 4, phosphor screen; 5, fiber-optic output. The second and third stages are similar to the first. (b) Magnetically focused, three-stage image intensifier. 1, Input window; 2, 4, 6, photocathodes; 3, 5, 7, phosphor screens; 8, output window. The phosphor–photocathode combinations 3–4 and 5–6 are separated by a thin plate of mica.

whether they are internal or external to the ICT. The internal MCP intensifier is designed to accept as input the photoelectron beam from the ICT photocathode. The external MCP intensifier must accept as input the photons from the ICT phosphor screen. The electromagnetic intensifier, focused either electrostatically or magnetically, is placed outside the ICT and therefore must input phosphor photons. Figures 3.11 and 3.12 show diagrams of three types of intensifiers.[32] Microchannel plates are also used as the electron gain element in some electrostatically focused intensifiers. In this case the MCP(s) are placed between the electrostatic focusing electrode and the phosphor.

Microchannel plate intensifiers, both proximity and electrostatically focused, are becoming the standard for use in streak camera systems. The magnetically focused tube diagramed in Figure 3.11 can have good spatial resolution and very high gain, but the tube is relatively large, needs high magnet currents and thus a large power supply, and is expensive. The microchannel plate operates on the principle of electron multiplication by secondary emission. Microchannel plates are made from bundles of very small glass tubes coated on the inside with a secondary electron-emitting compound and fused together and cut transversely.[33] Tube diameters are typically 10 to 40 μm. The microchannels are oriented at an angle of about 7° from normal. The thickness and the diameter of the whole plate are about 0.5 to 1.5 mm and 20 to 70 mm, respectively. The surfaces of the plates are coated with an evaporated metal electrode, and 1 to 2 kV are applied across the plate. In an external MCP intensifier, photons strike a photocathode, releasing a stream of electrons which travel toward the MCP, strike the microchannel walls, and knock out secondary photoelectrons, which are accelerated down the tube and produce further secondary electrons. A tube length-to-diameter ratio of 40 has been found to be optimum for production of electron gain. MCPs can

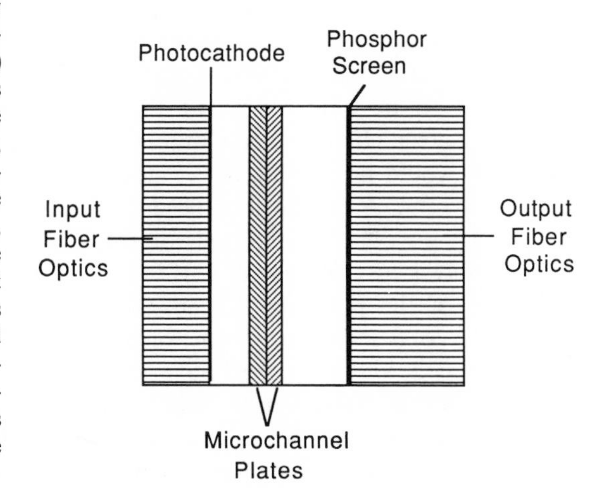

Figure 3.12. A fiber-optically coupled, proximity-focused, two-stage microchannel plate (MCP) image intensifier. Photoelectrons emitted by the photocathode are accelerated toward the first MCP, where they enter cylindrical channels of 10–40 μm diameter. The channels are oriented at about 7°, causing the electrons to strike the channel sides and knock out secondary electrons. The channels of the second MCP are oriented 7° in the other direction (Z configuration). The multiplied electron beam exiting the MCPs are further accelerated to the phosphor.

be stacked to increase gain, as in Figure 3.12, where two MCPs are butted together. The secondary electrons exiting from the first MCP enter the second, and the gain of the microchannel element is squared. The overall luminous gain (luminous flux of phosphor screen light divided by luminous flux of light striking the photocathode) of MCP intensifiers depends upon the voltage applied across the plate. For a single plate this gain varies between about 5×10^3 and 2×10^4. A two-plate intensifier will have gain between 1×10^3 and 1×10^6. Note that the gain in an MCP intensifier which has a photocathode also includes the gain from the photocathode–phosphor combination. An integral MCP intensifier is much like an external one, except that the incident photoelectrons come directly from the streak tube's photocathode, and any focusing is done by the streak tube's electron optics. The phosphor screen of an external intensifier has properties very similar to those of the phosphor of the ICT; in fact, the materials are identical. The primary differences are that the photoelectron energy in intensifier tubes is usually less, especially in MCP intensifiers, and that intensifier tubes are not operated in short-pulse mode. The intensifier amplifies the light output from the ICT phosphor, which has a decay time on the order of 10^{-4} s, so that the efficiency decrease caused by picosecond or nanosecond pulse excitation does not apply.

3.2.4. Signal Readout and Storage

The next step in the recording of time-dependent fluorescence signals in a streak camera system is to measure and store the two-dimensional, position-dependent emission from the intensifier phosphor screen in the form of an electrical signal which can eventually be stored in a computer. The ideal device to do this should have (i) a high quantum yield for conversion of phosphor photons to electrons, (ii) two-dimensional recording ability, (iii) low noise, (iv) a linear response, and (v) a fast scan rate. In principle, one would like to be able to record the entire two-dimensional view of the phosphor screen. If the screen were to be divided into a 500×200 grid of pixels this would necessitate the readout of 10^5 data points for each shot of the laser system. A typical time needed to read out one pixel is 20–80 μs, so that 2 to 8 s would be required for the entire readout. This time is often unacceptably long. Because of this, data scans are usually made along one dimension of the phosphor screen, for example, along the time axis. This reduces the total scan time to tens or hundreds of milliseconds.

Intensified vidicon tubes have most often been used to record the phosphor output, though new camera systems now are mostly offered with cooled CCD readouts. An intensified vidicon tube is simply a TV camera which has an extra stage of image intensification. The structure of such an SIT tube is shown in Figure 3.13. The light from the intensifier phosphor is coupled via lens or fiber optics to the faceplate of the SIT tube. The photons then strike a photocathode, emitting electrons which are accelerated by a 10-kV potential toward a silicon target. The target provides the gain, for when the high-energy electrons strike, they excite on the order of 10^3 electron–hole pairs. The silicon target is a thin silicon wafer on which has been placed a two-dimensional matrix of *p–n* junction diodes. The center-to-center spacing

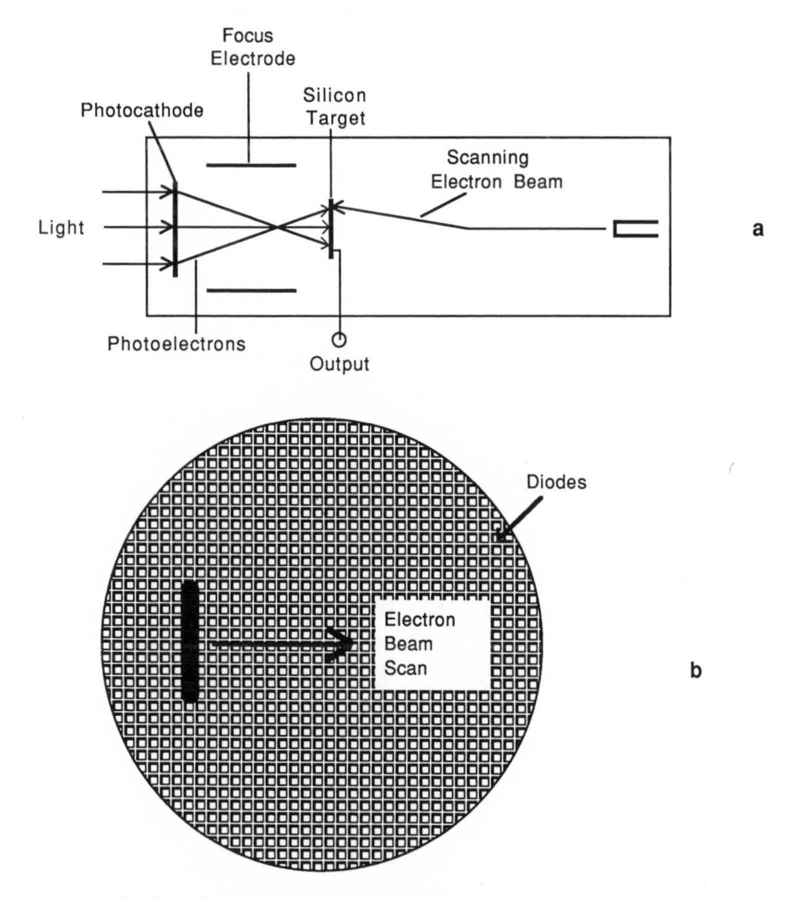

Figure 3.13. (a) Silicon intensifier target (SIT) tube. Light from the image intensifier strikes the SIT tube photocathode. The photoelectron beam is kept in focus and accelerated into the silicon target, where each photoelectron strikes a small diode, knocking out multiple electron–hole pairs. The holes are swept to the output side of the target, and their charge is read out via a neutralizing scanning electron beam. The SIT tube controller records the scanning beam position and charge for each diode. These two numbers give the position and intensity of the image intensifier phosphor screen emission. (b) Diagram of part of the surface of the SIT tube target, showing the diodes which define the "pixels" at which data can be recorded. The electron beam is scanning a vertical group of about 10 pixels along the horizontal axis.

of the diodes is 15–25 μm. If the optical imaging of the phosphor glow is a 1:1 magnification, then the streak camera system will have a potential spatial resolution of 15–25 μm. The readout of the SIT tube is accomplished by scanning an electron beam across the *p*-side of the diodes, where the holes have been collected. At the position of each scanned diode, the electron beam charge necessary to neutralize the charge of the holes is measured, giving a number proportional (ideally) to the number of photoelectrons which hit that diode. The recording electronics must record both the position of the scanning electron beam and the neutralization charge at each position.

SIT tubes have low noise relative to the other components of the streak camera system.[32] The biggest problem caused by the use of SIT tubes is nonlinearity with respect to input light level. The effect is termed lag. Lag in a SIT tube is capacitive. It occurs because of the finite time it takes for the scanning electron beam to remove the charge deposited on the silicon target. Capacitive lag, or the time necessary to completely discharge the target, increases as the light level decreases. Thus, if the electron beam is scanned at a fixed rate, it will more completely (measured by percent) read out high light levels than low light levels. This will give an apparent super-exponential response of the SIT tube with respect to light intensity. The brute-force way to improve linearity is to repetitively scan the target, so that the low light levels are more accurately read out. There is a tradeoff with noise, however, as the noise accumulated is proportional to the number of scans. Most systems scan the target 5 to 20 times. This gives a total scan time of $(5\text{–}20 \text{ scans}) \times (500 \text{ pixels/scan}) \times (40\ \mu\text{s/pixel}) = 0.1\text{–}0.4$ s. Internal electrical gating methods have been devised to improve the linearity of these devices at a lower number of scans.[34, 35] Manufacturers have not universally applied these methods, however. Even with the improvements, it is necessary to determine the linearity. Usually, this is done for the streak camera system as a whole, including the linearity of the ICT and intensifier tubes. The replacement of SIT with CCD readouts[62] can reduce the nonlinearity problems. CCD readouts are now made by a variety of manufacturers for use with streak cameras.

SIT tubes do contribute significant dark current to signals. Usually, this dark current can effectively be subtracted from the total signal. The dark current can be reduced by cooling the tube, but this adds to the lag problem. It takes even longer to read off the charge on the silicon target at low temperatures. For this reason, SIT tubes are often operated at room temperature.

A gating pulse is often applied to a grid in the SIT tube in order to pass signal only when the image intensifier phosphor is emitting. The gating time is typically about a millisecond. If the total scan time of the SIT tube is 0.1 s, then background or dark noise coming from the intensifier and streak camera is reduced by a factor of about 100. The total signal-to-noise ratio is not increased by a factor of 100, however, because the SIT tube noise is not reduced by this gating and the signal is also reduced by the gating. Under high-signal conditions it may be preferable not to gate the SIT tube in order to avoid this attenuation of the signal due to gating, which may be a factor of 2.

Controllers for the readout of SIT tubes generally are capable of recording the signal from each pixel (diode) on a 500×200 pixel target. The actual scans are normally in one direction (horizontal), however, grouping 5–200 pixels in the vertical dimension. Each vertical group, scanned along the horizontal axis, is termed a track, illustrated in Figure 3.14. The horizontal start and stop positions of the scan can be varied by the SIT tube controller, allowing a pseudo-two-dimensional scanning ability. A given horizontal position on the SIT tube array is termed a channel. Controllers can "simultaneously" scan and store many tracks, though each track is in fact scanned in sequence. It must again be kept in mind that the scan time will increase with the number of tracks. Ten tracks scanned five times at 40 μs/pixel each would take 1 s. This would limit the repetition rate of a single-shot experiment to less than 1 Hz.

More elaborate scanning control systems have been devised, all of which demand considerably enhanced computerization and expense. The objective is often to give real two-dimensional scans, so that the usual 2D image can be stored. Scanning such as this is not generally available in commercial systems.

SIT controllers are generally not put together by the user, in contrast to

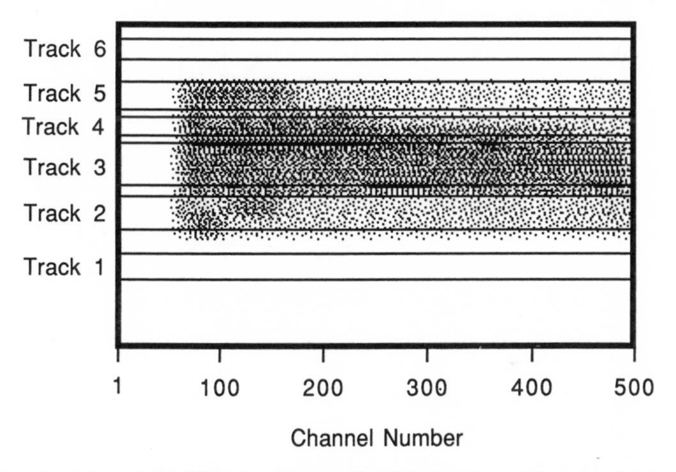

Figure 3.14. Scanning of the SIT tube detector (OMA) together with a simulated streak. The horizontal axis is the time axis of the streak camera sweep. The vertical axis may be one physical dimension of an emitting object or may be the wavelength of the sample emission, dispersed by a diffraction grating. The SIT tube scans one or more tracks along the horizontal axis, recording the signal at each of the channels (diode positions).

the computer which may be used to store and manipulate data once they are acquired. The scan control involves a rapid and rather complex series of timing, gating, and read/write procedures which are beyond the capabilities or interests of most spectroscopists.

3.2.5. Computer and Data Manipulation

Micro- or minicomputers of varying capability are used to command the SIT-tube controller. The computer is often part of the OMA package which is commercially available. Typical computers used in the 1980s have been based on the D.E.C. LSI-11, IBM PC-XT or -AT, Macintosh II, or other computers of similar size. As a rule, camera manufacturers now make readouts compatible with any IBM PC-AT type computer, though additional boards must also be purchased. The SIT tube controller is made to be compatible with standard computers, so in principle the user could purchase a computer separately to control the overall operation of the streak camera system. A number of simple but useful built-in functions are usually included with the commercial systems, such as external and event triggering, data display, integration, simple math functions ($\pm$, $-$, $\times$, $\div$, log, ln, exp, powers, and roots, performed on the 500-channel data array), background subtraction, windowing, smoothing, differentiation, and area and peak detection. These functions are intended to be for general data characterization, not for fitting of particular kinetic or spectral models to data. Modeling must be done by user-supplied programs or routines which may be purchased commercially. The built-in functions may seem simple enough for the casual user to write, but the speed at which the user-written functions are performed often forces the user to spend considerably more time in programming. In particular, the rate at which 500-channel data sets are displayed on a CRT screen may be 5–10 times slower in a casually written routine. When real-time data display after every data shot is desired, the display time may be crucial. One rarely wants to wait 15 minutes for the averaging to be completed before seeing what the signal looks like.

Specifications for the digital section of the OMA system, which includes both the OMA controller and the computer, are typically as follows:

- Signal A/D conversion: 8 to 14 bits (16-bit computer, with double-precision capability)
- RAM data storage: 4 kB to 1 MB
- Disk data storage: $3\frac{1}{2}''$ or $5\frac{1}{4}''$ floppy disk or 20–80-MB hard disk

The use of computers in recording and analyzing streak camera data will be addressed again briefly in Section 3.5.2.3, but the topics of computer control of instrumentation, computer acquisition of data, and computer techniques of data evaluation are in general beyond the scope and intent of this article.

3.3. Streak Camera Capabilities

Streak cameras are used by relatively few research groups interested in UV or visible fluorescence spectroscopy. Even among those interested in picosecond events, the streak camera has been regarded with mistrust and skepticism. Three of the reasons for this are the historical difficulty of operation of streak cameras, the number of possible artifacts in streak camera systems (time and intensity nonlinearity, flyback artifacts), and restricted dynamic range. Systems which are presently appearing on the market have effectively reduced or eliminated these problems. Most modern systems are modular, with interchangeable and well-engineered parts. Plug-ins with various sweep speeds are available, for example, in which the sweep nonlinearities have been reduced, sometimes to negligible levels. The operation of these streak cameras has become very similar to the operation of oscilloscopes. The dynamic range of streak cameras is also increasing as averaging becomes more and more the norm. Even the size of the entire streak camera system is now often governed more by the size of the computer and its peripherals. In spite of this, it is questionable whether streak camera use will increase dramatically among spectroscopists in chemistry, biology, biochemistry, condensed matter physics, and biophysics. The primary reasons for this lie in the areas of signal-to-noise ratios, dynamic range, and capabilities of alternate technologies. In particular, time-resolved single-photon counting is viewed as having much superior noise and dynamic range characteristics for fluorescence spectroscopy and time resolution which now approaches that of streak cameras. Because of this, we briefly consider at the end of this section a comparison of streak cameras and photon-counting systems for time-domain fluorescence spectroscopy. Other alternate methods, such as fast photomultipliers coupled with boxcar or lock-in amplifiers or phase fluorometry, should in principle also be considered, but we do not here.

3.3.1. Detector Characteristics

Many of the properties and characteristics of streak cameras, photon counting devices, and other photomultiplier-based systems are similar, since the basic light detectors operate on the same physical effects: the photoelectric effect and electron multiplication through secondary emission. In a photon-counting system, photons strike a photocathode, producing photoelectrons which are accelerated, multiplied, discriminated, and counted. In a streak camera, photons strike a photocathode, producing photoelectrons which are accelerated, multiplied, deflected, and counted. The only new element which affects time resolution in a streak camera as opposed to photomultiplier-based devices is the deflection step. The instrumentation associated with streak camera and photon-counting systems may also be very similar, especially if the streak camera is operated in synchroscan mode at high repetition rate. Picosecond photon-counting systems normally also use a cw mode-locked laser for excitation, operating at repetition rates of 1–150 MHz.

3.3.1.1. Time Resolution

Many sorts of time resolution must be considered for any system which measures the time dependence of a transient signal. The most general and inclusive sort of time resolution is that which is observed when the real excitation source is used under real experimental conditions to excite a signal from a real sample which responds instantaneously. Usually this sample is a scatterer which approximates the geometry of real samples to be used in experiments. We call this the *system response time*, $\Delta\tau_{\mathrm{sr}}$. A more fundamental response time is termed the *detector response time*, $\Delta\tau_{\mathrm{dr}}$, and is usually defined by the detector's response to an infinitely short impulse. We will speak of the detector response time as the full width at half-maximum (fwhm) of the response observed after a δ-function excitation of the system. The time resolution of the detector will depend to a lesser extent on the shape of the detector response function. The *fwhm of the actual source of excitation*, $\Delta\tau_p$, often does not approximate a δ function in time and imposes a limitation on the overall system response time.

The finite speed of light imposes another restriction on the time resolution of a picosecond detection system. Since light travels at 3.00×10^{10} cm/s divided by the index of refraction, signals coming from various parts of a sample of finite dimensions will arrive at the detector at different times. This can be a complicated effect even when the geometry is rather simple. A rough approximation is to consider the additional *broadening of the fwhm of an excitation pulse caused by sample geometry* $\Delta\tau_g$. The extent of broadening due to geometry can be determined by the signal observed from scattering off an object which approximates the geometry and optical properties of the sample of interest, assuming all of the other broadening effects are known.

Jitter may limit the time resolution when the signal is averaged over many excitation pulses. The *jitter distribution width*, $\Delta\tau_j$, can be defined by the fwhm of the signal observed when a series of δ-function pulses are applied and the signals summed. The jitter width applies strictly when the number of shots summed is large, such that the observed shape of the distribution does not change if more shots are summed. It also is only useful if the response function jitter distribution can usefully be characterized by the fwhm of the distribution. If, for example, the jitter is not random and the resulting distribution is very asymmetric or multipeaked, the effects of jitter must be treated explicitly and separately from the other effects.

The last time resolution we consider is much more difficult to define precisely. It is the shortest time which can be extracted from "ideal" data by deconvolution or other data analysis techniques. It usually can be no shorter than about one-tenth of the system fwhm. We will call it the "*deconvolution time*," $\Delta\tau_{dv}$. The reason the deconvolution time is so slippery a number is that there is no universal agreement on the protocol for determining its value. The value depends partly upon the apparatus, partly upon the signal-to-noise ratio, partly upon the data analysis, partly upon the experimenter's judgment of allowable error and degree of certainty, and partly upon the sample which is being measured and its mechanism for light emission. A person could quite safely ignore all claims about deconvolution times and assume that under favorable circumstances (a strong emitter with simple exponential decay, for example) a decay time of about one-tenth the system fwhm time can be extracted. We discuss convolution time resolution further in Section 3.4.

The fwhm response time $\Delta\tau_{sr}$ of a system can be limited by any of the previously mentioned time factors (excluding $\Delta\tau_{dv}$). In general, the system response time depends upon the square root of the sum of the squares of the response times:

$$\Delta\tau_{sr}^2 = \Delta\tau_{dr}^2 + \Delta\tau_p^2 + \Delta\tau_g^2 + \Delta\tau_j^2 \tag{3.8}$$

The time resolution discussion to this point applies equally well to any time-resolving detection system. The detector response time, on the other hand, is specific for specific detectors. The detector response time of a streak camera depends upon three factors. The first is the difference in velocities which photoelectrons acquire when they leave the photocathode. Light striking the photocathode knocks out photoelectrons which statistically vary in velocity (direction and magnitude). As a result, the time needed for the electrons to traverse the streak tube will be statistically distributed. This so-called *transit time spread*, $\Delta\tau_{tr}$, depends upon the wavelength of the photons, since more energetic photons will knock out a different distribution of electron velocities. If two photoelectrons differ in initial velocity along the axis of the tube by an amount Δv_z, then the difference in transit time will be

$$\Delta t = (m\,\Delta v_z)/(eE) \tag{3.9}$$

where m and e are the mass and charge of the electron, respectively, and E is the magnitude of the electric field near the photocathode. This will translate into a transit time distribution width (fwhm) in seconds of[2]

$$\Delta\tau_{tr} = 2.34 \times 10^{-8}\,\Delta\varepsilon^{1/2}/E \tag{3.10}$$

where $\Delta\varepsilon$ is the photoelectron energy spread in eV, and E is the electric field in volts/cm. Note that $\Delta\tau_{tr}$ may be reduced by increasing the electric field. Typical values of $\Delta\varepsilon$ and E are 0.5 eV (530-nm light incident on an S-20 photocathode) and 2×10^4 V/cm, giving $\Delta\tau_{tr} = 0.83$ ps. The photoelectron energy spread is proportional to the most probable value energy, which in turn is proportional to the difference between the incident photon energy and the photocathode work function. If light incident on the photocathode is polychromatic or if emission at two different wavelengths is being compared, then an additional wavelength dependence of the transit time enters. The effect on transit time will depend upon the photocathode and the wavelengths being measured. Knox, however, has demonstrated in the case of an S-20 photocathode that transit time wavelength dependence in the streak tube is offset by travel time through the lens which is used to collect light from the sample.[36] In the streak tube, blue light excites more energetic photoelectrons which arrive at the phosphor before those excited by red light. On the other hand, red light travels faster than blue light in glass, offsetting the photoelectric effect. For an *f*/1.9 double Gauss lens, the total time delay between the arrival of 350-nm-excited photoelectrons and 800-nm-excited photoelectrons is about 10 ps in the streak tube employed by Knox. Use of a single 1-cm-thick fused-silica lens decreases this time delay to about 2–3 ps. If an integral MCP is in the streak tube, an additional contribution to the transit time spread must be included.[6] The second limitation on time resolution is sometimes termed the "technical time resolution" and is the spread due to the finite minimum width of an image on the photocathode (equal to the width on the phosphor screen if the electron magnification of the ICT is 1) and the finite sweep speed of the electrons across the phosphor. If v_s is the sweep speed in cm/s and a is the size of the minimum resolvable distance element, then the technical time resolution is

$$\Delta\tau_{tt} = a/v_s \tag{3.11}$$

The minimum element which can be resolved on the phosphor screen depends on several factors, including imperfections in the electron beam focusing and, especially for very short light pulses, electron–electron repulsion, which tends to broaden a tightly packed bunch of electrons.[37] The repulsion effects are sometimes considered separately. If $a = 50\ \mu$m and $v_s = 1 \times 10^{10}$ cm/s, then $\Delta\tau_{tt} = 0.5$ ps. If there are N resolvable elements along the streak direction and the voltage needed to sweep the beam over this total distance is ΔV, then the voltage ramp applied to the deflection plates must rise at a rate

$$dV/dt = \Delta V/(\Delta\tau_{tt}\mathrm{N}) \tag{3.12}$$

to achieve the technical time resolution $\Delta\tau_{tt}$. If $N = 1000$ and $\Delta V = 2.0$ kV, then the voltage must rise at 4.0×10^{12} V/s = 4.0 kV/ns to achieve 0.5-ps technical time resolution. The total detector response time can be written

$$\Delta\tau_{\rm dr}^2 = \Delta\tau_{\rm tr}^2 + \Delta\tau_{\rm tt}^2 \tag{3.13}$$

Common values of $\Delta\tau_{\rm dr}$ for modern cameras are of the order 0.5–10 ps.

3.3.1.2. Dynamic Range and Signal-to-Noise Ratio

The dynamic range and signal-to-noise characteristics of streak cameras are intimately related. Dynamic range can be defined as the ratio of the largest undistorted signal which can be measured to the smallest discernible signal. The system for which the dynamic range is specified must also be described. The presence of an image intensifier of gain 1000, for example, will increase the dynamic range of a streak camera by an order of magnitude or more because it decreases the lower limit of signal detectability. (It also adds noise.) In a streak camera system it is the noise of the ICT–intensifier combination which normally places the lower limit on the smallest measurable signal. The upper signal level at which distortion sets in is usually determined by charge effects at the photocathode. When too much charge is demanded of the photocathode by an incident pulse of light, the photoelectron charge packet will be broadened. The dynamic range of a streak camera is normally determined by measuring the apparent pulse width of an incident laser pulse versus intensity; see Figure 3.15. The ratio taken is that of the peak amplitudes of an upper and a lower usable signal, with the upper usable signal defined as that at which the pulse width is broadened 20% as referenced to the width of a pulse near the middle of the dynamic range.

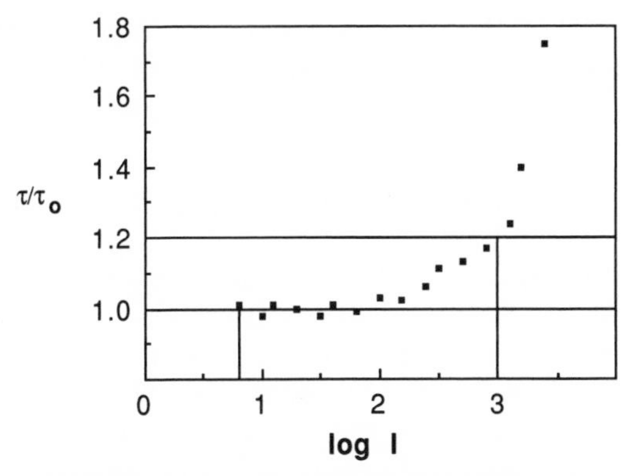

Figure 3.15. Dynamic range determination of a streak camera system. The ratio of the intensity at which an excitation pulse is 20 % broadened to the lowest intensity at which the signal is discernible is the dynamic range for the streak camera system at the time resolution corresponding to the pulse width.

Dynamic range must be specified for a given time resolution or width of the incident light pulse. This is partly because the photocathode will be able to provide more total charge before distortion sets in if it is given longer to do so. In Section 3.2.2.4, the reduction in streak camera phosphor response as pulse width is shortened was also discussed: as the photoelectron beam pulse gets shorter and shorter, the number of phosphor photons produced per electron decreases. Both of these effects conspire to restrict the dynamic range of streak camera systems. There is at present no generally accepted theory to describe the dynamic range of streak camera systems. Majumdar[38] has proposed the following equation for the dynamic range of a streak camera system:

$$\text{D.R.} = C\sigma(\lambda)\tau^2/Q \tag{3.14}$$

where $\sigma(\lambda)$ is the photocathode quantum efficiency, τ is the pulse width, Q is the total charge required per cm^2 from the cathode, and C is a constant which depends somewhat on the emission characteristics of the cathode as a function of wavelength. This equation predicts that the dynamic range is proportional to the pulse width squared. Majumdar presents data supporting this dependence, but other data point to a linear dependence on the pulse width.[39, 40]

Signal averaging or summation affects dynamic range in an obvious way, since the lower limit of detectability can be extended through averaging. In comparing various streak tube/intensifier combinations with regard to their intrinsic dynamic range, comparison must be made under similar conditions, for example, for a single shot. Dynamic range comparisons of streak cameras operating in different modes must be made with care. It is not reasonable, for example, to compare the dynamic range of a synchroscan system after averaging 4 h at 100 MHz to that of a single-shot mode streak camera after one shot. Assuming that the lower limit of detectability is determined by random noise, then the signal-to-noise ratio S/N will depend on the square root of the number of shots averaged together:

$$\text{S/N} \propto (\text{number of shots})^{1/2} \tag{3.15}$$

This implies that the dynamic range will also depend upon the square root of the number of shots. It is always observed that averaging increases the dynamic range of streak cameras. The square root dependence is not always observed, however. The noise characteristics of complete streak camera systems are often not well understood because of the number of components and the fact that various components (especially the image intensifier) are operated at different gain settings depending upon the size of the signal observed. The effectiveness of averaging also depends upon the time characteristics of the "random" noise. In simple terms, the noise must vary rapidly compared to the signal averaging time or the noise will not be averaged out. It is generally said that the streak tube is the source of most of the noise in a streak camera system. However, if very small signals are being observed and the intensifier is at high gain, noise from the intensifier may dominate. Depending upon the quality of the intensifier tube, some of this noise may be in the form of spikes which occur in the signal at the rate of 10^{-2} to 1 Hz. Averaging over periods of minutes or even hours will not effectively reduce noise in this case.

The smallest signal discernible with a streak camera system depends upon characteristics of all components of the system: the streak tube, the image intensifier, and the SIT tube or other device used to record the signal. It also depends upon the wavelength of the incident light and the duration of the light pulse. Noise from the streak tube and intensifier imposes the normal limitations on the smallest signal. Accordingly, we examine below the factors which affect the signal-to-noise ratio, S/N, of a streak tube–intensifier combination.

Kinoshita and Suzuki[41] calculated the noise characteristics of a streak tube with integral MCP intensifier. We follow their discussion. There are at least six factors which determine noise at the output of a streak tube:

(i) the statistical fluctuation in the input light
(ii) fluctuation in the conversion and amplification process from light at the photocathode to light emitted by the output phosphor
(iii) dark current (thermionic emission, field emission from high voltages, and background light)
(iv) fluctuation due to nonuniform streak speed
(v) fluctuation of the readout system (e.g., the OMA)
(vi) apparent fluctuation due to nonuniformity of spatial response across the sweep axis of the MCP and phosphor

Factors (iv) and (v) can usually be neglected and (vi) can be corrected for, so we consider only the first three effects.

The intensity of light as a function of time is obtained by integrating the charge introduced in a channel of the OMA by the streak. A channel may consist of several or many pixels grouped together in the direction perpendicular to the direction of sweep, as in Figures 3.13 and 3.14. A photon has probability P of causing a photoelectron to successfully enter the MCP:

$$P = \eta \beta_1 \beta_2 \tag{3.16}$$

where η is the quantum efficiency of photocathode conversion of photons to electrons, β_1 is the transmittance of the grid electrode, and β_2 is the fraction of the surface area of the MCP which is open. The distribution of the number of secondary electrons produced in the MCP by the first collisions of the incident photoelectrons can be described by a Poisson distribution of mean δ. The distribution of the number of secondary electrons produced by successive collisions in the MCP can be described by a Furry distribution, F, of mean m:

$$F(x) = m^x/(1+m)^{1+x} \tag{3.17}$$

where x is the output intensity of electrons. If the number of dark electrons emitted from the photocathode and scattered from electrodes is N_d electrons/(cm$^2 \cdot$ s), a fraction α of these fall into a spot on the phosphor corresponding to an MCP microchannel being considered, and the time over which the OMA integrates signal from the phosphor is T, then the mean number of dark electrons which are counted is $\alpha N_d T$. A Poisson distribution of these dark electrons is also assumed. The distribution of output electrons from the MCP will be given by the compound distribution of the Poisson and Furry distributions. The mean and standard deviation of this output distribution are

$$\mu = \text{mean} = (N + \alpha N_d T/\eta) Pm\delta \tag{3.18}$$

$$\mu_s = \text{mean signal} = NPm\delta \tag{3.19}$$

$$\sigma = \text{standard deviation} = [(N + \alpha N_d T/\eta)\, Pm^2\delta(2+\delta)]^{1/2} \tag{3.20}$$

The signal-to-noise ratio of the output is

$$\begin{aligned}(\text{S/N})_{\text{out}} &= \mu_s/\sigma \\ &= \{N^2 P\delta/[(N + \alpha N_d T/\eta)(2+\delta)]\}^{1/2}\end{aligned} \tag{3.21}$$

The input S/N is just $N^{1/2}$, since the photon statistics are Poisson. Therefore, the noise figure NF of the streak camera, given by the ratio of the input S/N to the output S/N, is

$$\begin{aligned}\text{NF} &= (\text{S/N})_{\text{in}}/(\text{S/N})_{\text{out}} \\ &= \{[(1 + \alpha N_d T/N\eta)(1 + 2/\delta)]/\eta\beta_1\beta_2\}^{1/2}\end{aligned} \tag{3.22}$$

A larger noise figure means that the streak camera adds a larger amount of noise to the signal. The noise can be reduced by decreasing the fraction of dark electrons which enter the MCP directly from the photocathode or indirectly through scattering, by decreasing the integration time T, by increasing the photocathode quantum efficiency, grid transmittance, and MCP open area ratio, and by increasing δ, the mean of the Poisson distribution of first-collision secondary electrons. The integration time can be reduced to about two times the phosphor decay time. The quantum efficiency varies considerably depending upon the type of cathode. S-20 cathodes presently have the highest peak efficiency (at about 450 nm), η of about 0.1–0.2.

Improvements in cathode fabrication may increase this efficiency somewhat. Aging of photocathodes, resulting in loss of conversion efficiency, obviously increases the noise of the tube. Grid transmittance depends upon the size of the grid wires relative to the open area. Smaller and fewer wires give better transmittance, but field uniformity considerations impose a minimum number of wires per cm^2. The open area ratio of the MCP depends upon the wall thickness and microchannel diameter. Reduction in wall thickness is limited by manufacturing techniques.

The lower limit of the dynamic range can be defined as the number of input photons necessary to give an output S/N of 1. The output S/N is proportional to $N^{1/2}$ except at very low values of N. The minimum number of photons needed can be found by plotting $(S/N)_{out}$ versus number of photons input (in a pulse) on a log–log plot, as in Figure 3.16. The best tubes tested by Kinoshita and Suzuki had a minimum number of photons of 50–100, corresponding to 5–10 photoelectrons entering each microchannel. Urakami *et al.* subsequently presented data demonstrating that such tubes can be used for photon counting.[42] They estimated the dynamic range of such a tube at 10^5 for a 100-ms incident light pulse (see Section 3.9.3).

3.3.1.3. *Polarization Dependence*

The sensitivity of streak cameras is almost independent of the polarization of the incident light. If all light is incident normal to the ICT entrance window, no polarization dependence is observed. In the case where a lens is used to collect light and image it onto the photocathode, only the dependence

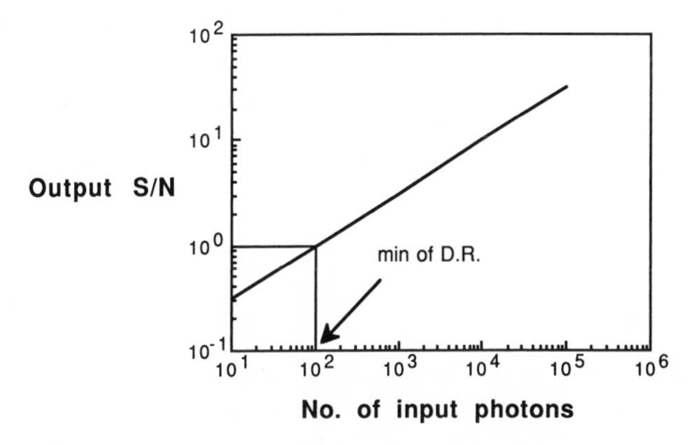

Figure 3.16. Dependence of output S/N on number of input photons. The minimum of the dynamic range (D.R.) is the number of input photons that produces an output S/N of 1. The maximum of the D.R. would be the number of photons corresponding to the 20 % pulse broadening intensity of Figure 3.15. This figure applies to the internally intensified ICT described in the text.

of the entrance window reflectivity on angle of incidence is important. If the f number of the lens is large, then polarization sensitivity can usually be ignored ($<1\%$). See also Section 3.7.1 below.

The independence of streak camera sensitivity from polarization is a very useful detector characteristic, as light emission from many sources has polarization which is changing with time on picosecond to nanosecond time scales. Often this time-dependent polarization measures an electronic process in the sample which is of great interest. Section 3.7 explores the use of streak cameras for fluorescence polarization spectroscopy in more detail.

3.3.1.4. Angle Dependence

Streak camera detection is practically independent of the angle of light incidence. Besides the weak polarization dependence of entrance window reflectivity, one observes only a very small dependence of streak camera time and spatial resolution on angle of incidence. This behavior occurs because the velocity of the photoelectrons immediately after the photocathode in a direction transverse to the acceleration direction depends upon the angle of photon incidence.[37] If a low-f-number lens is used for light collection, then the angle of incidence is distributed and the photoelectron pulse will be broadened.

3.3.1.5. Wavelength Dependence

The photocathode is the major source of wavelength dependence of streak camera characteristics. The sensitivity dependence is determined by the photocathode material, as discussed in Section 3.2.2.1. These photocathodes are the same types as are used in photomultipliers, though sensitivity versus wavelength curves may differ slightly. Gallium arsenide photocathodes which show a broad region of high and almost constant sensitivity (200–850 nm) are not yet available in commercial streak tubes. The time resolution of the streak tube depends to a small extent on the incident photon wavelength and the photocathode material, as mentioned in Section 3.3.1.1.

3.3.2. Signal Averaging

One of the great advantages of single-shot streak cameras is their ability to unambiguously record the time dependence of a single event as, for example, a single shot of a laser pulse whose characteristics may change from shot to shot. Methods such as autocorrelation produce, on the other hand, pulse characteristics averaged over many pulses. Other applications, such as fluorescence spectroscopy, often demand averaging over several or many excitation pulses in order to get the signal-to-noise ratio to an acceptable level or to increase the dynamic range. In these cases one cannot or does not need to take advantage of the single-shot recording ability.

The effectiveness of signal averaging was discussed in Section 3.3.1.2. Averaging is essential in most fluorescence experiments because the number of photons striking the streak camera photocathode is often small. If a short 10^{-6}-J, 530-nm laser pulse excites a sample with absorbance 0.1, then approximately 5×10^{11} photons will be absorbed. If the sample fluorescence quantum yield is 0.1, then 5×10^{10} photons will be emitted. If optics collect emission over 1 millisteradian, then 4×10^{6} photons are collected. Typical reflection losses and color filter bandpasses remove 90% of the photons, leaving about 4×10^{5}. If the fluorescence lifetime is 100 ps, then 4000 photons/ps strike the photocathode. The input signal-to-noise ratio is thus $(40{,}000)^{1/2} = 200$ for a 10-ps interval. If the noise figure of the streak camera is 10, then the output signal-to-noise ratio will be 20. Further optical couplings to the OMA may lose 80% more of the signal, reducing the S/N to 4 per 10-ps interval. This figure is too low for detailed modeling of fluorescence kinetics. Averaging together 100 shots can improve the S/N by a factor of 10. This allows for much more information to be extracted from the fluorescence data: double-exponential time decays can be clearly distinguished from single-exponential ones, 10-ps rise times can clearly be distinguished from instantaneous rise times, etc.

Synchroscan cameras are the extreme example of signal averaging in streak camera systems. An individual laser pulse, typical energy being a few hundred picojoules, will generate a signal which is indistinguishable from noise. Integration of the signal on the OMA for 1 s at a laser repetition rate of 100 MHz will increase the S/N by about a factor of 10^{4}, to levels comparable to that of the single-shot camera with a 1-μJ excitation pulse. A direct comparison of signal-to-noise ratios for single-shot and synchroscan streak cameras should be done carefully. However, the signal accumulation time needed for achieving a given S/N in a 100-MHz synchroscan system with 100-mW source is roughly the same as for a single-shot system operating at 1 Hz with 10-μJ pulses.

3.3.3. Streak Cameras and Time-Resolved Single-Photon Counting

The method of time-resolved single-photon counting is described elsewhere in this volume and in many other review articles. The principle of photon counting is that the intensity versus time curve of a fluorescing sample can be obtained by counting the rate at which photons are emitted from a sample at various times after an excitation of the sample such that one photon or less is produced per excitation. Since only one photon is counted per pulse, time-resolved photon counting must be a high-repetition-rate measurement in order to accumulate enough counts for a good signal-to-noise ratio. Streak camera methods, as we have seen, can be either high or low repetition rate. By the time we reach Section 3.9 we will see, in fact, that streak cameras can now be used in a single-shot, high-signal per pulse mode, in photon-counting mode, or anywhere in between.

The primary characteristics we need to consider in the comparison of streak camera and photon-counting methods are time resolution, dynamic range, sensitivity and signal-to-noise ratio, and wavelength dependence. System cost is another important consideration, but variables such as cost of the excitation source, the electronic expertise of the user, the time and spatial resolution desired, and the wavelength range make precise comparisons difficult. Suffice it to say that either system will likely cost more than $300,000, including the cost of lasers, computers, and accessories. A brief comparison of streak cameras and photon-counting systems is presented below, continuing into Section 3.4 (see also Section 3.9).

The primary advantage of time-resolved single-photon counting (TRPC) is generally thought to be the high dynamic range and high signal-to-noise ratio achievable. Dynamic ranges of 10^4 or more for full-scale time intervals of a few nanoseconds are not out of the ordinary. Furthermore, the time necessary to develop a given signal-to-noise ratio is relatively independent of sample emission yield, except in the limit of low yield. (This simply means that the laser is attenuated for each sample to the point at which only one photon is counted per excitation pulse. If the sample yield is less, then the laser is attenuated less. The "low yield" limit mentioned here is encountered for samples which emit one counted photon or less per laser pulse at full laser power.) This has encouraged the use of many-parameter equations for data fitting, which traditionally has not been done with streak camera data. As we have seen, however, dynamic range and signal-to-noise ratios are defined only for a given time interval. For a given total integration time, the dynamic range of TRPC systems will generally be less as the full-scale time range is decreased below 1 ns. A more fundamental problem is encountered, however, as time intervals decrease below 100 ps. The fwhm response time of the fastest TRPC systems is about 40 ps,[43] even with excitation pulses below 5 ps. This is because the system response time is dominated by the response time of the photomultiplier (photoelectron transit time spread) and associated electronics. If signal decay times comparable to or below 40 ps are to be measured, then improved dynamic range does not automatically translate into improved time resolution.

The second advantage of TRPC is said to be its wide time range. Full scale on the time axis for almost any TRPC system can easily range from about 1 ns to more than 100 ns, with 500 to 1000 time bins per full scale. Commercial streak cameras now are available with time axis plug-ins or variable time settings. In one such camera, the Hadland 675/II, full scale is variable from 1.5 to 500 ns, with 500 time bins. Hamamatsu manufactures plug-ins which can vary full scale from about 0.5 to 100 ns. Many of the older streak camera models have time sweeps that are not so flexible.

The third characteristic of TRPC systems which is put to good use in precision fitting of mathematical models to data is the well-understood Poisson statistical properties of photon counting. When doing a precise fit, it is desirable to weight the data points according to the statistical nature of the data recording process. For photon counting this is straightforward. If a data point in time is n counts, then the weighting of this data point is $n^{-1/2}$. For streak camera data, the statistical weighting factors depend upon the particular streak camera and how its individual components are operated, in particular, the gain. In many cases the system statistics have not been determined. This does not mean that data fits to streak camera data cannot be done. It simply means that the precision with which fitting parameters can be determined is restricted. Unless the streak camera statistics are known, however, fitting of one decay curve with more than three parameters (a sum of more than two exponentials, for example) will not in general be possible.

The advantage that streak cameras have always enjoyed over photon-counting methods has been time resolution. Streak camera time resolution is now about 0.5 ps (fwhm) while the best TRPC systems, employing fast microchannel plate photomultipliers, are about two orders of magnitude longer in fwhm.

3.4. Time Dependence of Fluorescence

The primary purpose of a streak camera is to measure the time dependence of light emitted by a sample. In this section we address the application of streak camera techniques to measurement of fluorescence decays. We assume that the measurement is done at a single wavelength or that the emission spectrum is being collected.

3.4.1. Decay Times

Three time regimes may be identified in which different aspects of detector response become important. These are the short-time, intermediate-time, and long-time regimes. The short-time regime is that in which the time scale of interest is comparable to or shorter than the system response time or the excitation pulse width, whichever is longer. The long-time regime is that in which the sample decay times are much longer than the limiting short-time resolution of the system and comparable to the full-scale time sweep of the streak camera. Intermediate times are defined by default as those between short and long.

3.4.1.1. *Short-Time Regime*

Short times are defined by the system response time, whether it is limited by the excitation pulse or by the detector response, and by the shortest possible full-scale time sweep of the camera. The shortest full-scale time for most streak cameras is in the range 200 ps to 1 ns, longer than typical detector response times and laser pulse widths. For a 2-ps resolution streak camera with a 20-ps excitation pulse, short times are from about 2 to 50 ps. If the pulse were shortened to 0.1 ps, short times would refer to times from about 0.2 to 5 ps.

The time response of streak camera (or photon-counting) systems can be modeled as follows. Assume that $I(t)$ is the intensity of an excitation pulse [photons/$(\mathrm{s} \cdot \mathrm{cm}^2)$] as a funtion of time. Then the response of the sample being excited is described by the appropriate differential equation:

$$dN/dt = -\Sigma I(t) N_0 - kN(t) \tag{3.23}$$

where Σ is the absorption cross section of the absorbing molecule in cm^2, $N(t)$ is the number of molecules per cm^3 in the excited state, N_0 is the total number of molecules per cm^3, and k (equal to the inverse of the decay time τ) is the decay rate in s^{-1}. This equation assumes that the sample response is single exponential in time and that $N(t) << N_0$. If the excitation pulse is Gaussian in time, of fwhm σ_2, then the solution of the differential equation is

$$s(t) = \int_{-\infty}^{t} ds\, e^{-k(t-s)} e^{-s^2/2\sigma_2^2} \tag{3.24}$$

This equation describes the actual sample response to the excitation pulse. If the measuring apparatus has infinitely short time resolution and otherwise accurately reproduces the sample emission, then Eq. (3.24) also describes the signal which will be recorded. If, however, the apparatus has a finite response time, then the measured signal will be

$$s(t) = \int_{-\infty}^{t} du\, e^{-(t-t_c-u)^2/2\sigma_1^2} \int_{-\infty}^{u} ds\, e^{-k(u-s)} e^{-s^2/2\sigma_2^2} \tag{3.25}$$

where σ_1 is the fwhm of the signal observed if a δ-function excitation pulse is applied to the detector, and t_c reflects the delay or transit time of the signal through the detector. Equations (3.24) and (3.25) are valid for any time range of fluorescence decays, but analysis of very short decay times or rise times requires this convolution to be performed when fitting decay rates to the data. Equation (3.25) can be rewritten in terms of a single integral with the product of two factors as integrand: the first is the kinetic response of the sample (single exponential or otherwise); the second is a function which describes the total system response, including both excitation pulse width and detector response effects, to a sample emission of zero time delay and decay time. This latter function can be directly measured by observing light scattered from an appropriate sample.

Figure 3.17 shows simulated responses of a streak camera with 2-ps time resolution when 20-ps excitation pulses are applied to samples of various decay times. The most obvious characteristic of the signal is the great reduction in amplitude as the decay time of the sample gets shorter than the excitation pulse width. If one can design the experiment so that the streak camera can be amplitude-calibrated, then the signal amplitude will be a sensitive measure of the decay time for decays shorter than the excitation pulse. Figure 3.18 shows the same data but with each curve normalized. A new feature of the recorded signal now becomes evident: the rising edges of the signals are progressively shifted to earlier times as the decay time gets shorter. In the limit of instantaneous decay, the signal directly reflects the excitation pulse. This shift of the observed signal with respect to the excitation pulse is also very sensitive to the sample decay time. Decay times well under the excitation width can be easily resolved. This simply reflects what was termed the "deconvolution time" in Section 3.3.1.1: sample decay times significantly shorter than excitation pulses can be measured with good accuracy. The deconvolution time resolution in this case is roughly 1–2 ps. The resolution will depend upon the signal-to-noise ratio, of course, but the position of the rising edge of a signal can be accurately resolved even when the S/N is quite low. Section 3.8.4.3 describes data where this effect is put to use. Note that the time jitter of the detection system must be small for this technique to work.

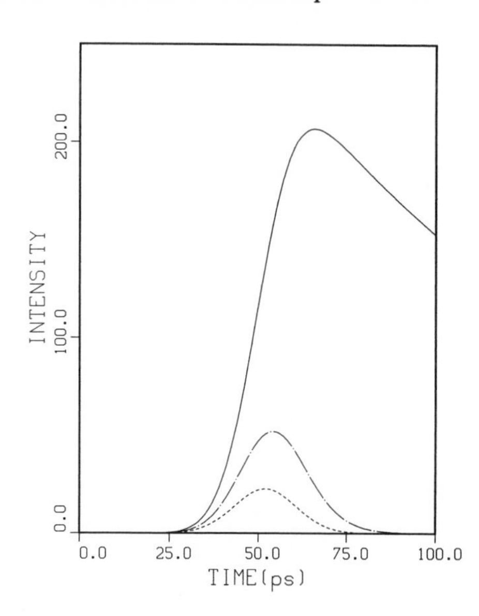

Figure 3.17. Effect of fluorescence decay time on signal amplitude for a finite-width excitation pulse. Decay times are 2, 5, and 100 ps, in order of increasing signal amplitude. The excitation pulse width is 20 ps; the detector response time, 2 ps.

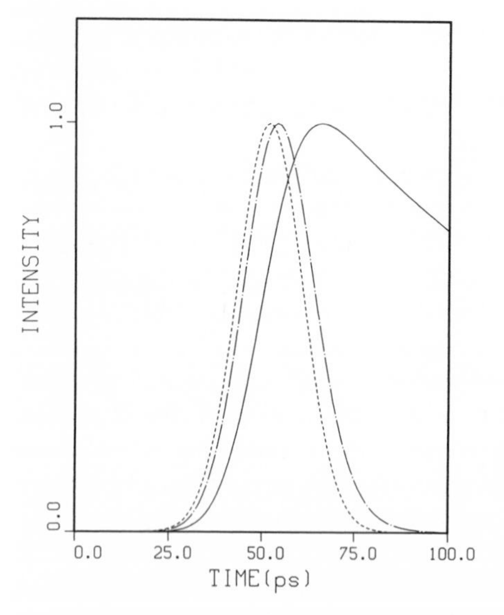

Figure 3.18. Effect of fluorescence decay time on the rising edge of the signal. The data are the same as in Figure 3.17, but each curve is normalized to the same maximum amplitude. Decay times are 2, 5, and 100 ps, from left to right.

Figures 3.17 and 3.18 also reflect the observed signal when a 2-ps excitation pulse is applied to a sample of various decay times when the detector has a response fwhm of 20 ps. This is appropriate for a photon-counting system, though the detector response time has been decreased by a factor of 2 below that presently available in any systems. Finally, Figure 3.19 shows the response of a streak camera with 2-ps time resolution measuring the signal excited by a 2-ps pulse. A very curious effect is evident: the discrimination between the 2- and 5-ps decay curves seems to be greater in

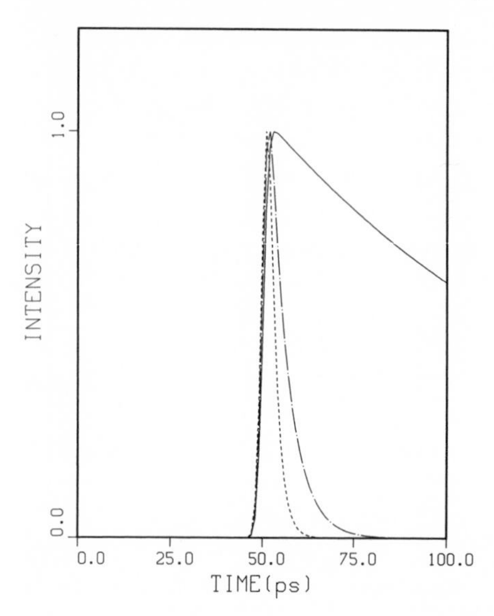

Figure 3.19. Signals observed with excitation pulse width and detector response time of 2 ps. Decay times 2, 5, and 100 ps, from left to right. Note the apparent better distinction between the 2- and 5-ps decays in Figure 3.18.

Figure 3.18 than in Figure 3.19 due to the greater leading edge delay in Figure 3.18. (If the system response is instantaneous, then curves for all decay times will rise instantaneously and at the same time.) The slower response system seems to give better discrimination between the two fast decays than the fast response system does, given the full-scale time sweep!

Sections 3.3.1.1 and 3.3.2 separately discussed streak camera time resolution and signal averaging. At this point it becomes clear that signal averaging increases the deconvolution time resolution. If the shot-to-shot jitter of a streak camera or other detection system is ± 5 ps, then for any given shot the location of a signal (the location of, say, the center of a short pulse) will shift ± 5 ps from shot to shot. Any timing comparison between two separately collected single-shot signals cannot be precise to better than ± 5 ps. However, if two sets of data are collected, each the sum of 100 shots, the timing accuracy will be much better than ± 5 ps if the jitter is random. If the jitter is Gaussian distributed in magnitude and an excitation pulse is 1 ps or less, then the sum of 100 shots will be Gaussian shaped, with Gaussian full width 10 ps (full width to $1/2e$ points: Gaussian full width $= 0.849 \times$ fwhm). The center of this pulse can be determined to much greater accuracy. With good S/N the center can be determined to within 0.5 ps or less. If the individual laser pulses are 20 ps fwhm with ± 5 ps jitter, then the sum of 100 shots will be Gaussian shaped, with width about 22 ps. The center of this pulse can be determined to better than 1 ps.[44]

Fitting fluorescence decay data in the short-time regime can be accomplished by a variety of means, including least-squares analysis and other fitting methods. We will not discuss the relative merits of various fitting routines. All of them will be limited, as noted before, by the (possibly unknown) statistics of the streak camera system. A method of data analysis which is useful when decays are shorter than the system response time is to normalize and calculate the difference between the sample's signal and that produced by an instantaneously responding sample, such as a scatterer. If the excitation pulse is Gaussian in time, the two signals will both be approximately Gaussian, but the former will be shifted to the right by an amount equal to the decay time. The difference will be the derivative of a Gaussian, whose amplitude and width both give the decay time of the sample.[45] The magnitude of the difference in time between the center of the sample decay curve and the half-rise point of a long-decay-time sample also directly gives the decay time of a short-lived fluorescence, since the half-rise point of the long-lived fluorescence is at the center of the excitation pulse. In the time regime of 10 ps or less, the numbers obtained directly from these differences must be corrected for effects of wavelength and solvent refractive index, each of which may amount to a picosecond or so.

Fluorescence Quantum Yields, Lifetimes, and Detectability. The magnitude of a fluorescence signal excited by a continuous source is proportional to the sample's absorption coefficient ε and fluorescence quantum yield Q. The fluorescence quantum yield is in turn proportional to the fluorescence lifetime:

$$Q = \tau/\tau_0 \tag{3.26}$$

where τ is the observed lifetime and τ_0 is the natural or radiative lifetime. In the Strickler–Berg formulation[46] for molecular absorption and emission, τ_0 is inversely proportional to the absorption coefficient, so the intensity of fluorescence is

$$I(\text{steady-state}) \propto \varepsilon^2 \tau c l \tag{3.27}$$

in the limit of low sample absorbance, $A = \varepsilon c l << 1$, where c is sample concentration and l is the light path length. Thus, in steady state, for fixed τ_0, a shorter lifetime means lower signal. With a time-resolving detector the situation is different. If the time resolution is longer than the decay time, then the steady-state description holds. This can be seen in Figure 3.17, where the signal amplitude decreases as the lifetime decreases. In the limit where the system response time is much longer than the fluorescence decay time, the signal amplitude will be directly proportional to the lifetime. If the response time is shorter than the decay time, the maximum amplitude of the signal is independent of the decay time:

$$I(\text{time-resolved}) \propto \varepsilon^2 c I \tag{3.28}$$

This means that a streak camera with 2-ps time resolution can as easily resolve a 5-ps decay as a 500-ps decay in terms of amplitude of signal. In Section 3.8.4.3 the fluorescence decays of nucleic acids in which relatively large-amplitude signals are observed for short-decay-time signals are described.

3.4.1.2. Intermediate-Time Regime

The time range of about 20 ps to 500 ps is ideal for streak cameras, in the sense that very few corrections need be made to obtain kinetic information from fluorescence signals. Convolution is usually unnecessary, unless the excitation pulse is longer than 10 ps. Effects of wavelength, geometry-dependent travel time, and solvent refractive index normally constitute corrections of 5 ps or less. In addition, decay times in this time range easily fit within normal time sweeps of most streak cameras, even those with fixed sweep times. Decay times can be obtained directly from the slope of plots of ln (signal) versus time.

3.4.1.3. Long-Time Regime

Streak camera full-scale time ranges can be adjusted in several of the modern streak cameras from 0.5 ps to hundreds of nanoseconds. Assuming that the excitation pulse is short compared to the full-scale time, long decay times can be defined as those which are comparable to the full scale. This means that the signal will be greater than zero over most of the screen. In this case an important contribution to the shape of the signal can come from the intensity sensitivity as a function of channel number of the OMA. The sensitivity variation with channel number comes from two main sources. First, the diodes constituting individual SIT tube channels may not all have the same response. Diodes typically have sensitivity variations of $\pm 10\%$. Second, the image intensifier may distort the shape of the streak image at high intensification. These two effects together produce variations in sensitivity versus channel of up to $\pm 30\%$. Typically, local variations are about $\pm 5\%$. Differences between the center of the screen and the edge are larger, with edge sensitivity less.

Analysis of decay curves in the long-time regime must take into account the linearity of the time sweep of the streak camera, described in Section 3.2.2.2. When decays are short compared to full scale, the signal is contained in a relatively small part of the time scale. The time sweep is quite linear over small parts of the full scale, often is better than 1%. When the decay curve occupies a major part of the time axis, however, the time linearity becomes important. Correction for the time sweep and for the channel variations in sensitivity can be done as described in Section 3.2.2.2 with a dye of fluorescence lifetime long compared to the full scale of time.

3.4.2. Rise Times

The definition of rise time must be gotten clearly in mind before one speaks of rise times of fluorescence signals. The most common approach is to define the "rise time" as the time it takes for the signal to rise from 10% to 90% of maximum. This is a legitimate way of characterizing curves such as those in Figure 3.18, but the "rise time" obtained in this manner may have no direct connection to physical or chemical processes which are the root cause of the rise. *A "rise time" has physical meaning only in the context of a physical model* for the processes involved. The physical model must be in the form of equations which are used to fit the fluorescence curve. In many cases there are several factors which contribute to an observed 10–90% rise time. One of these factors may, in fact, be the fluorescence decay time.

The first improvement in the analysis of rise times beyond the 10–90% treatment is to insert an exponentially rising component into single-exponential kinetic equations describing the fluorescence:

$$N(t) = N_0[1 - \exp(-k_r t)] \exp(-kt) \tag{3.29}$$

where k_r describes the fluorescence rise and k is the fluorescence decay time. Use of this equation assumes a physical model which may or may not be appropriate for the sample being investigated. The equation assumes that only one species is emitting and that this species obtains its excitation from a single source, with rate k_r. It is applicable, for example, to the case of energy transfer from species 1 to species 2 only in special cases: when 1 does not contribute to the fluorescence, when no other pathways of deexcitation of species 1 exist except transfer to 2, when species 2 cannot be directly excited by the laser pulse, and when transfer directly produces the state in species 2 which fluoresces.

Section 3.8.4.1 will explore an example of the analysis of rise times using a model which is more physically reasonable for complex samples.

Fluorescence rise times are treated much like decay times in the short-

Table 3.2. Processes That Can Be Measured with Streak Cameras

Process	Time scale
Biological systems	
Energy transfer in proteins	<1 ps to 100 ps
Electron (charge) transfer in proteins	1 ps to >1 ns
Photosynthesis: energy transfer	1 ps to 50 ps
Photosynthesis: energy trapping	10 ps to 500 ps
Photosynthesis: exciton annihilation	<1 ps to 1 ns
Protein binding reactions	0.1 ps to 1 ms
Protein conformational motions	0.5 ps to >1 ms
UV damage to DNA	1 ps to ?
Nucleic acid conformational motions	100 ps to >1 ms
Visual processes	1 ps to 10 ms
Chemical systems	
Isomerizations	<1 ps to 1 ns
Rotational diffusion	<1 ps to >1 ns
Vibrational relaxation	10 fs (?) to 50 ps
Electronic relaxation	<1 ps to 100 ns
Exciplex formation	<10 ps to 10 ns
Excited state proton transfer	<10 ps to >100 ps
Geminate recombination	10 ps to 1 ns
Solvation dynamics	0.1 ps to 1 ns
Bond fragmentation	0.1 ps to 1 ps
Transient molecular heating	0.1 ps to 50 ps
Solid-state systems	
Kinetics of melting[a]	1 ps to 1 ns
Semiconductor carrier dynamics	10 fs to 100 ps
Quantum-well optical switches[a]	0.1 ps to 1 ps
Vibrational relaxation[a]	0.1 ps to 10 ps
F-center luminescence	100 ps to 10 ns

[a] Processes are usually measured by nonfluorescence techniques.

time regime. All of the possible sources of time dispersion must be examined for their possible contributions to the observed rise of a fluorescence curve. There are two limiting cases to consider. When the rise time is comparable to the system response time, the convolution equations are critical to obtaining a meaningful determination of the rise time. If the 10–90% rise time is longer than about three times the system fwhm response time, the convolution equations are not necessary. When the convolution equations are used, it is important to use a good model for the kinetics. Equations (3.24) and (3.25), for example, have assumed a single-exponential decay with no rise time, described by the term $\exp[-k(u-s)]$. This term must be replaced by a function $g(u-s)$ which correctly models the fluorescence with its rise and decay characteristics.

3.4.3. Physical Processes and Characteristic Times

The time range for which streak camera techniques are especially useful is from 0.5 to 500 ps. What processes of interest take place in this time range which can be monitored by fluorescence? The list in Table 3.2 is not exhaustive, but it demonstrates that a wide variety of important processes occurring in biological, chemical, solid-state, and other systems can be measured with streak cameras. Most of the processes in Table 3.2 can also be measured with other optical techniques, such as absorption and Raman spectroscopies or electron diffraction. One notes that many processes in modern solid-state systems are becoming too fast for streak camera resolution.

3.5. Wavelength Dependence of Fluorescence

Several methods may be used to record emission spectra with a streak camera. One method is to place narrow-pass filters, such as interference filters, between the sample and streak camera input. This will give the spectrum point by point in wavelength, while at each wavelength a continuous time dependence will be recorded. This is the preferred method if precise knowledge of the time dependence is more important than precision spectra. Generally, interference filters will give points spaced every 10 nm. In addition, to compare data sets collected separately at two different wavelengths, careful normalization of the data must be done if the excitation intensity varied between the two data sets. If the spectrum is broad and any spectral changes are expected to be large, then this method is adequate and simple. Assuming that the filters are all the same thickness, there will be no additional wavelength-dependent time dispersion in the system beyond that of the streak tube itself. In many cases, however, spectral changes may be small and the filter method is inadequate. In this case it is desirable to record entire spectra for each excitation pulse. This method is described below.

3.5.1. Measurement of Time and Wavelength Dependence of Signals: Additional Time Dispersion

Simultaneous measurement of wavelength and time dependence of light emission from a sample is straightforward with a streak camera system: the axis corresponding to the photoelectron sweep direction is used for the time dependence and the orthogonal axis is used for dispersal of spectral information. In Figure 3.14, the horizontal axis could be time, and the vertical, wavelength. Dispersal of UV to near-IR wavelengths can best be accomplished with diffraction gratings or prisms. A schematic of such an apparatus is shown in Figure 3.20. A polychromator (monochromator with exit slit removed) is used to disperse the sample emission spectrum.

Time and wavelength cannot simultaneously be measured to arbitrary resolution. If a highly dispersive diffraction grating is used, for example, different path lengths must be traversed by different parts of the light beam, even for monochromatic light. This has been described by Schiller and Alfano,[47] who found a monochromatic transit time spread due to the grating spectrograph of

$$\Delta\tau_{\lambda} = Dm\lambda/cd \tag{3.30}$$

where D is the diameter of the beam spot on the grating, m is the diffraction order, λ is the wavelength, d is the groove spacing, and c is the speed of light. This transit time corresponds to the difference in transit time between light of a single wavelength which strikes the left and right edges of the grating (see Figure 3.20). The time spread will be less for smaller diameter gratings (larger f-number spectrographs) and for less dispersive (larger groove spacing) gratings. If $D = 2$ cm, $\lambda = 530$ nm, and $m = 1$, then $\Delta\tau = 7$ ps and 42 ps, respectively, for a 200-lines/mm and a 1200-lines/mm grating. The full-screen wavelength ranges in Ref. 47 for the two gratings were 70 nm and 8.5 nm, respectively, for the 200-lines/mm and 1200-lines/mm gratings. The full-screen range depends upon the optical setup. It can be increased to 100–200 nm for the 200-lines/mm grating.

A second transit time spread effect involves different transit times for different wavelengths, even though they initially strike the grating at the same

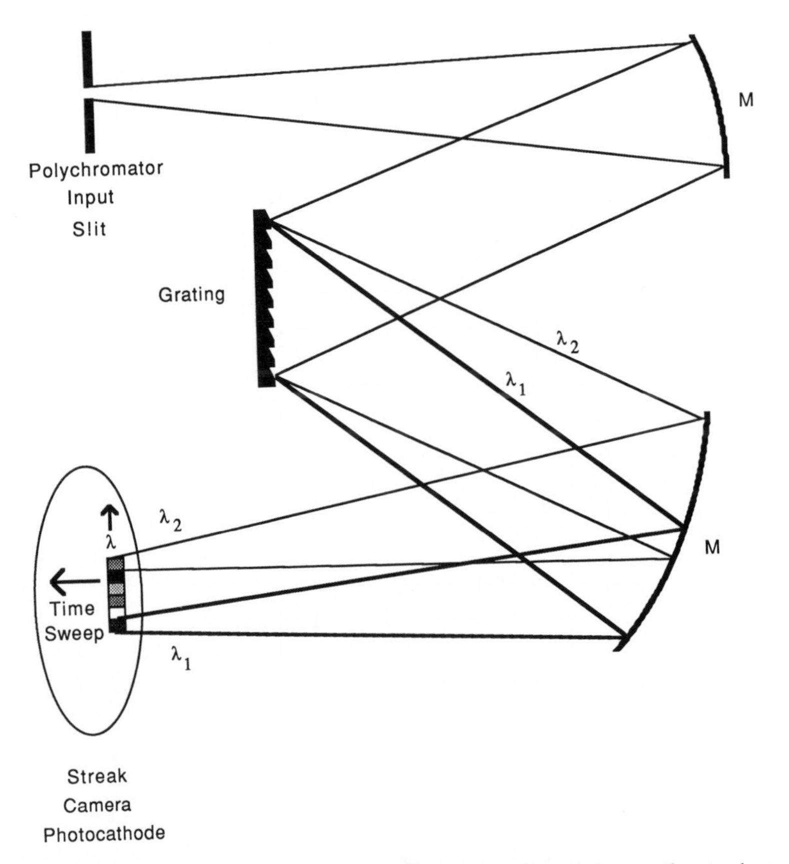

Figure 3.20. Grating polychromator set up to disperse sample emission on the streak camera photocathode. The light coming from the input slit is rendered parallel by the first mirror, M, and directed onto the grating. The dispersed spectrum is focused onto the photocathode by the second mirror, wavelength being dispersed along an axis perpendicular to the direction in which photoelectrons will be swept. The grating is normally on a rotating mount so that different portions of the spectrum can be sent to the photocathode.

spot. The variation of delay with wavelength can be obtained from the grating equation:

$$m\lambda/d = \sin\alpha - \sin\beta \tag{3.31}$$

where α is the angle of incidence and β is the diffraction angle, illustrated in Figure 3.21. If the plane of the next optical component encountered by the diffracted light is set up for normal incidence at wavelength 2, then light at wavelength 1 will travel a longer path, b_1, to get to that plane:

$$b_2/b_1 = \cos(\beta_2 - \beta_1) \tag{3.32}$$

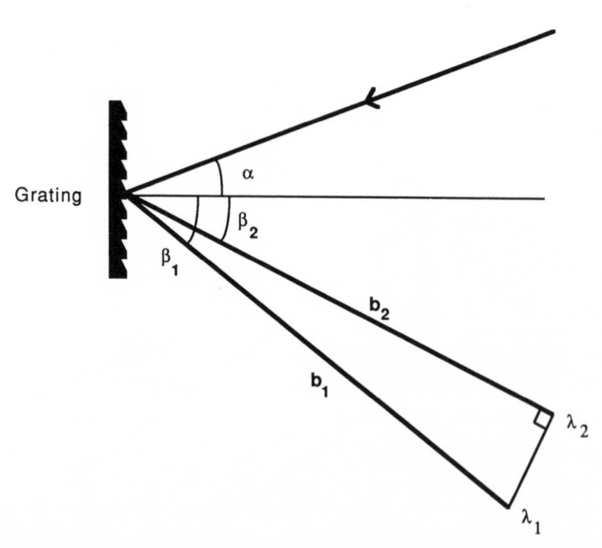

Figure 3.21. Paths traveled by two different wavelengths from the grating to the next optical element in a polychromator. If the element were oriented for normal incidence of λ_2, path b_1 would be longer than b_2.

where the angles and distances are defined in Figure 3.21. The arrival time difference between light at wavelengths λ_1 and λ_2 will be

$$\Delta\tau_\alpha = (b_1 - b_2)/c \tag{3.33}$$

where c is the speed of light and α indicates a fixed angle of incidence. For wavelengths of 530 and 580 nm, the delays for 200- and 1200-lines/mm gratings are, respectively, less than about 5 fs and 60 fs per centimer of path length from the grating to the refocusing element for angles of incidence between 30° and 60°. There can be additional difference in travel time from the refocusing element to the streak camera photocathode, depending upon the optical setup, but this will be even less delay per centimer (with normal focusing setups). The refocusing element itself will contribute to travel time differences, but in a compensatory way for common optical setups. If, for example, a single *f*/5.0, 32-cm focal length lens were used at a distance of 18 cm from the 1200-lines/mm grating and on-axis relative to λ_2, λ_2 would be delayed by about 0.5 ps for visible wavelengths. The orientation and layout of the polychromator output optics affects the relative delay. Overall, the wavelength-dependent time broadening is negligible except for highly dispersive, high *f*-number, long-focal-length spectrographs. For a total grating to photocathode distance of 50 cm, the total differences (530 nm versus 580 nm) for 200- and 1200-lines/mm spectrographs are less than about 0.2 ps and 3 ps, respectively.

3.5.2. Time Dependence of Emission Spectra

A polychromator will disperse spectral information across the streak camera photocathode, and the time sweep will produce temporal information on the other axis. The OMA can therefore be used to scan spectral or temporal information. When it is desirable to determine spectral changes which may be subtle, it is better to scan the axis corresponding to wavelength.

3.5.2.1. Methods of Measurement

The recording of spectra as a function of time is best done by placing a polychromator at the input to the streak camera and by turning the OMA 90° to the normal orientation for time dependence measurements at a single wavelength. This allows the OMA to scan such that channel number corresponds to wavelength. Refer to Figure 3.14. Each track would correspond to a particular time. The position of the center of the track gives the average time; the width of the track gives the width of the time window over which data will be integrated. The time resolution will usually be limited by either the track width or the polychromator input slit height, whichever corresponds to coarser resolution. If the slit height is not limiting, then choosing a narrower track will give better time resolution, but the signal will be smaller. The wavelength resolution will usually be limited by the input slit width. In principle, one could scan N tracks simultaneously and, for each OMA scan, record N spectra. Reference 36 describes the disadvantages in doing this, which include the necessity for correction of each spectrum for local screen sensitivity and for field curvature and distortion in the image intensifier. An alternative method is to scan a single track, corresponding to a single time, store the data, then change the time by varying the delay between the excitation pulse and the streak camera sweep. This can be done with an optical delay line, consisting of a translation stage on which is mounted a retroreflecting mirror or prism. For each successive time, the delay line would be reset and the data recorded on the same location on the OMA screen. No correction for local sensitivity or field curvature is necessary. The timing accuracy in this method is governed by the jitter and drift of the system. In practice, it is necessary to measure the time resolution by putting in a short pulse at various wavelengths of interest. A tunable, short-pulse dye laser is appropriate, or, if only a fixed-wavelength laser is available, dyes of short lifetimes can be used to produce short pulses at a variety of wavelengths. Use of the stimulated Raman effect[48] in liquids is also a very practical method. Focusing a 40-μJ, 30-ps, 530-nm Nd:YAG laser pulse into a 5-cm cell of acetone can produce more than 20 μJ of light in less than 30 ps at about 630 nm. Other common solvents can be used to produce other wavelengths. Further tuning of the wavelength can be done by appropriate deuteration of the solvents. In actual operation, time and wavelength resolution can be on the order of 20 ps and 3 nm (at the same time).

3.5.2.2. Corrected and Uncorrected Spectra

The streak camera record of the emission spectrum will contain distortions due to the wavelength and polarization dependence of polychromator diffraction efficiency, photocathode quantum yield wavelength dependence, electron imaging distortions, and OMA channel sensitivity variations. If the wavelength range measured is in the region of flattest photocathode and grating response, the spectrum will still differ from the real spectrum. Corrections of $\pm 20\%$ are typical. Wavelength regions at the edge of the photocathode response need even larger corrections. Corrected spectra are produced by the standard method of measuring a well-known sample. The OMA can store the measured spectrum and calculate corrections once it is given the true spectrum.

The problem of correction is not as simple as it may seem, however. Tabulated spectra are always recorded in the steady state. Streak camera spectra are normally time-resolved. If the spectrum of the calibration sample is time-independent, then the correction may be done. Another approach is to disable the sweep of the streak camera and record the spectrum undispersed in time (focus mode) or to sweep the camera but integrate the spectrum in time in the OMA. The former method is simplest but least accurate: recorded background light levels will be different in focus mode than in streak mode, and imaging in the ICT will be slightly different when the photoelectrons are deflected by the field. In conclusion, corrected spectra can be produced, but careful attention must be given to the correction process.

The primary purpose of measuring spectra with a streak camera is usually to measure spectral changes with time. In this case it may not be necessary to know the "correct" spectrum but rather only relative changes with time. Use of uncorrected spectra will generally be satisfactory if the spectral region is not too near the edge of response of any of the detector system components.

3.5.2.3. The Mass of Data Problem

The objective of recording time-resolved spectra is to extract the important parameters which describe the spectral changes as a function of time. Any physical theory worth using to fit the measured spectral changes must provide, in the end, a small number of parameters which describe the measurements. Resolution of spectral changes with time requires at least 10–20 spectra, recorded equally spaced over the time interval of interest. The best general approach is to space the spectra equally on a logarithmic time scale. If 20 spectra are recorded, each consisting of 500 to 1000 channels, 10,000 to 20,000 data points will need to be displayed, analyzed, and summarized. Most computers are capable of handling this number of points, but approaches to extracting a small number of the most relevant parameters from these 20,000 data points are not common. An often encountered method is to plot the time dependence of intensity at well-chosen, discrete wavelengths. This, of course, completely eliminates the advantage previously gained by recording the spectrum as a continuous function of wavelength. Another method is to fit the spectrum with Gaussian functions, with each component characterized by its center wavelength, spectral width, and amplitude. Then these Gaussian fitting parameters could be plotted as a function of time. One problem with this method is that each Gaussian component does not necessarily represent a distinct kinetic or chemical species.

A more general approach to the problem of extracting a few relevant parameters to describe 10^4 to 10^5 optical spectra data points has been to use the method of singular value decomposition (SVD).[49, 50] The principle of this method is that one can describe a set of N recorded spectra completely, with no loss of information, by an orthonormal basis set of N spectra. Furthermore, these spectra can be chosen so that the first few basis spectra in the set describe most of the features of the spectra. Later (higher) basis spectra progressively describe more and more of the noise which is contained in the original spectra. The end result is that one is provided with a few basis spectra which completely describe the measured results except for random noise. The basis spectra can then be plotted as a function of time. No initial assumptions need be made about spectral shape or time dependence. (In fact, the method does not care what the spectral variables are. They could equally well be temperature and pressure as time and wavelength.) Once one is provided with this small set of spectral components and their time dependence, one can then proceed to figure out the physical significance of the components—a much easier task than trying to make sense of 20 spectra each consisting of a mix of several spectral species. In Section 3.8.3 an example of this technique applied to an excited state reaction in solution is described.

3.6. Two-Dimensional Detection: Space and Time Coordinates

3.6.1. Position and Time Detection

Time-dependent emission from an object can be resolved as a function of position on the object using the two-dimensional capabilities of streak cameras with two-dimensional OMA readouts. The object of investigation is first imaged onto the photocathode, and then a mask with a slit is placed between the object and the photocathode. The mask can be near the object or near the image on the photocathode. The slit should be oriented so that its long axis is perpendicular to the direction in which the photoelectrons will be swept; that is, if the ICT deflection plates are horizontal, then the slit should be horizontal. A photoelectron image of that part of the object defined by the slit will be swept across the phosphor screen in time. The width of the slit determines the time resolution of the measurement. The spatial resolution is limited by the spatial resolution of the streak camera system but will be less if only a limited number of tracks are scanned on the OMA. In Section 3.8.5 measurements in which the spatial resolution is not limited by track scanning are described.

A recent development has been the commercialization of a picosecond fluorescence microscope system with the ability to construct two-dimensional spatial maps of the lifetime and intensity of fluorescence from samples excited by the 380-, 410-, or 670-nm emission of laser diodes. This system consists of a spectrograph, a synchroscan-type streak camera, CCD camera, video memory, streak camera controller, pulse generator, laser diode head, and microscope stage controller, all controlled by a Macintosh II computer, and is claimed to have time resolution as short as 5 ps and spectral resolution of less than 5 nm.[58] The mapped area is typically 100×100 pixels in size. Applications could include improved inspection of semiconductors, mapping of intracellular fluorescence and locating labeled cell components, and simultaneous fluorescent characterization of large numbers of cells or other microscopic objects.

3.6.2. Two-Dimensional Spatial Detection

A streak camera can be used for real two-dimensional photography if its photoelectron sweep is disabled. In this case a simple photograph of the sample emission will be recorded by the OMA. This photograph can be used, for example, to align the sample and the set the OMA scanning positions. It may also be of use in determining which parts of a complex sample are emitting in which wavelength region, if color filters are placed between the sample and streak camera. The time resolution of these photographs is the time width of the emitted light. If a shuttering mechanism is included in the streak camera, such as, for example, an integral MCP which can be turned on and off by pulsing the biasing electrodes for a time t, then photographs of time resolution t will be obtained.

A device mentioned earlier, the framing camera, can produce picosecond time-resolved photographs in two spatial dimensions with time resolution in the nanosecond to picosecond regime. If a real object were imaged onto the

ICT photocathode, two spatial dimensions of the object would be imaged. The subsequent photoelectron sweep would, however, smear one of the spatial dimensions into the time dimension. To retain the two dimensions of spatial information, the deflected electron beam can be swept across a plate containing one or more apertures along the sweep axis.[12] After passing through the aperture, the electron beam would contain spatial information in a direction along the direction of electron travel. (Different parts of the image pass through the aperture at different times as the beam is swept across the aperture plate.) A second set of deflection electrodes, to which an exactly inverted voltage ramp is applied, removes the transverse sweep of the electrons and restores the spatial information to a transverse position along the phosphor screen. The final record is then a plot of intensity of light emission versus x and y coordinates of the object. Time resolution of 130 ps with spatial resolution of 10 line pairs/mm has been demonstrated.[12]

An alternate method for time-resolved, two-dimensional photography, described in Ref. 11, achieves time resolution of 5 ps. A fast-response external Kerr gate is used to shutter the light emitted from the object. If the shutter time is short compared to the photoelectron sweep time, then the image produced on the phosphor screen of an ordinary streak camera will be an image of the object in two spatial dimensions with shutter time 5 ps. Repetitive shuttering with delay time Δt will produce multiple images at delay times of Δt and shutter times of 5 ps. This method had a demonstrated spatial resolution of 6–16 line pairs/mm and was used to produce time-resolved images of laser-produced plasmas.

3.7. Fluorescence Anisotropy

Fluorescence experiments can be done with polarized light in streak camera systems since the streak camera response itself is almost independent of light polarization (Section 3.3.1.3). The principle of the method is to excite the sample with a polarized pulse light and observe the time dependence of the emitted light, resolved into horizontally and vertically polarized components. The emission anisotropy, $r(t)$, is defined as

$$r(t) = (I_{\parallel} - I_{\perp})/(I_{\parallel} + 2I_{\perp}) \tag{3.34}$$

where $I_{\parallel}$ and $I_{\perp}$ are the intensities of light emitted with polarization parallel and perpendicular, respectively, to the direction of excitation polarization. The denominator of this expression is the total fluorescence intensity. The anisotropy will vary in amplitude between 0.4 and -0.2 for samples in which the molecules are oriented randomly. Normally, the excitation pulse is vertically polarized, and each polarized emission component is individually measured. Alternatively, the excitation polarization can be alternated between vertical and horizontal polarization, and the total emission measured for each excitation polarization direction. Equation (3.34) then becomes

$$r(t) = (I_v - I_h)/[I_v + (I_h/2)] \tag{3.35}$$

where I_v and I_h are the total emission intensities observed with vertical and horizontal excitation polarization, respectively. The advantage to performing streak camera experiments in this latter way is that polarizers must be inserted into the laser beam used for excitation rather than into the emitted light. The laser beam is generally of small diameter and parallel, making the task of ensuring polarization purity easier. High-quality calcite polarizers can be used very effectively. There are a variety of other methods which will work well with a streak camera system.[51]

Time-dependent changes in $r(t)$ on time scales of 10^{-13} to 10^{-7} s can be caused by a number of physical processes. The first is motion of the fluorescing molecules. The anisotropy at short times after excitation is normally nonzero because the polarized laser pulse has selectively excited those molecules whose absorption dipole moments are parallel to the excitation polarization direction. If the molecular emission dipoles rotate randomly and without restriction, the anisotropy will decay to zero at times comparable to the rotational motion. The decay time of the anisotropy in the case of rotational motion of a sphere in a uniform, viscous medium can be described by the Debye–Stokes–Einstein expression:

$$\phi = (6D)^{-1} = 4\pi\eta r^3/3k_B T \tag{3.36}$$

where ϕ, D, η, r, k_B, and T are, respectively, the decay time to $1/e$ of (a single-exponential) $r(t)$, the rotational diffusion coefficient, the viscosity, the sphere radius, the Boltzmann constant, and the absolute temperature. This rotational correlation time can vary from a few picoseconds or less for small molecules to hundreds of nanoseconds for large protein molecules.

A second process which introduces anisotropy changes on fast time scales is energy transfer from the initially excited dipole to a second dipole oriented in a different direction. The time scale can vary from subpicosecond to nanosecond in typical cases. Very fast transfer can take place between molecules which are close together or between different transition dipoles on the same molecule. Anisotropy has not often been used to measure energy transfer processes.

3.7.1. Simultaneous versus Separate Measurements

No matter how a polarization experiment is set up, two signals must be recorded, either $I_{\parallel}$ and $I_{\perp}$ or I_v and I_h. Often these two signals are recorded by exciting the sample two times and measuring one component each time. The drawback to separate measurements is that the excitation source and detector may have changed between measurements. The ideal way would be to record both signals simultaneously, using the two-dimensional capabilities of the streak camera. Each of the signals would be imaged onto a different spot on the streak camera photocathode, and the OMA would scan tracks corresponding to each polarization component. The anisotropy could then easily be calculated by the computer. Since both components are measured for every excitation pulse, no correction need be made for excitation energy variations or detector drift. Care must be taken, however, to ensure that there is no intrinsic optical time delay between the arrival of the two signals at the photocathode. If, for example, a polarization beam splitter is used to divide either the excitation or emitted light beams into two spatially separated beams, the beam splitter will introduce a time delay between the two separated beams, depending upon the design of the beam splitter. This delay, which typically lies between 0 and 20 ps, is intrinsic to the process of polarization beam splitting, since the separation of the two components is accomplished via the refractive index difference (light speed difference) for the two components. When one of the components is delayed relative to the other, a significant apparent anisotropy can be induced for a time somewhat longer than that delay time. Correction for this effect can be done either by the computer or by introducing a compensating optical delay into the beam which arrives too soon (Figure 3.22). In either case, a calibration test with a well-known sample must be done to ensure no polarization delay-time artifacts. Figure 3.22 shows an example of a two-track streak camera transient anisotropy apparatus designed to measure both polarization components simultaneously.

The prerequisite for reasonable anisotropy signals is a good signal-to-noise ratio, since $r(t)$ is proportional to the difference between two signals which may be nearly equal. The two-track anisotropy method has one significant drawback involving the signal-to-noise ratio. If two identical signals are recorded on two different tracks of the OMA, correction for differences in track sensitivity must be done. This is possible, as described earlier, but the correction process contributes to the noise, since the correction factors (the channel-to-channel sensitivity variations) must be measured with a real calibration signal which has some noise of its own. For this reason, streak camera anisotropy experiments which use samples of low fluorescence intensity may achieve better results using a single channel and separately measuring each polarization component. Excitation energy variations must be corrected for in this case by normalizing each signal by the laser energy. A precision photodiode/operational amplifier combination with a peak-reading or integrating voltmeter is appropriate. An error in normalization for energy fluctuations of 4% will produce an offset in anisotropy amplitude of 0.02.

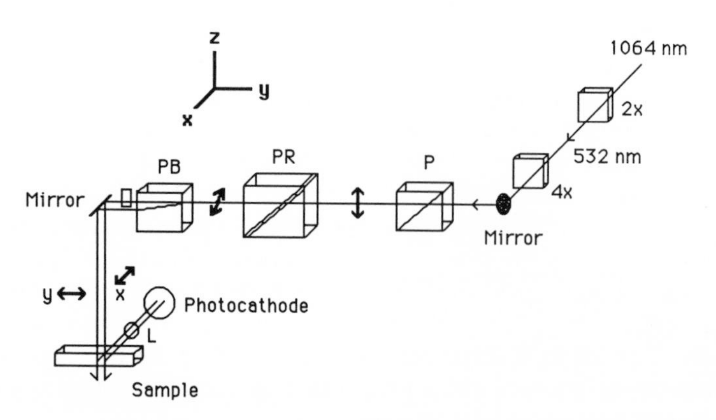

Figure 3.22. Two-track setup for time-resolved anisotropy experiments. The excitation pulse is divided into two beams, polarized orthogonally, and the total emission excited by each beam is collected in each of two spots on the streak camera photocathode. D = Delay, PB = polarization beamsplitter, P = polarizer, L = lens, PR = polarization rotator, 2x = 2nd harmonic generator, 4x = 4th harmonic generator.

There is very little effect on most decay times. A short-time anisotropy error can also be induced if there is jitter or timing drift between when the two polarization components are recorded in the single-channel method. Low-jitter operation is crucial if anisotropy decay times of tens of picoseconds or less are to be measured, especially if the system response time (streak camera response or excitation pulse) is longer than about 5 ps. An offset of 5 ps between the position of time zero for the two polarization components will induce an anisotropy artifact which lasts for a time about equal to the system response time or 5 ps, whichever is longer.

3.7.2. Signal-to-Noise Ratio

Signal-to-noise ratio determines the design of anisotropy experiments with samples of low fluorescence, as described in the previous paragraph. Even with the best of designs, signal averaging is essential to achieving resolution of 0.01 or better in anisotropy amplitude. A single-shot S/N of 5:1 will give about 0.07 units of noise in anisotropy, while the sum of 100 such shots will give 0.007. The averaging process would be crucial, for example, to the determination of the exponentially of an anisotropy decay. Streak cameras are capable of resolving double-exponential decays, but as in any system, there are limits as to how close in decay time and amplitude the two components can be. Signal averaging is also essential to reducing timing jitter effects.

Averaging together 100 shots can reduce timing errors of 5 ps to well below 1 ps (see Section 3.4.1.1).

In summary, picosecond time-resolved fluorescence anisotropy measurements can be effectively performed with streak camera systems, but requirements for noise levels and timing are strict. A rule of thumb is that the signal-to-noise ratio and timing accuracy must be about three times better for anisotropy than for simple fluorescence intensity measurements.

3.8. Applications

Streak camera systems have traditionally been located in larger research centers with either a technological or interdisciplinary bent. Centers for laser research, inertial-confinement fusion research, and fast kinetics research have been prime users of streak cameras. Even in the early 1970s, however, some of the first experiments performed with streak cameras were of a biological or biophysical nature. Campillo and Shapiro,[21] for example, made picosecond energy transfer measurements on photosynthetic systems. Below we consider a few other examples of streak camera applications, with an emphasis on fluorescence experiments of the basic research sort.

3.8.1. Laser Diagnostics

Measurement of the time dependence of laser pulses was one of the earliest applications of the streak camera. There was generally no shortage of photons, and the pulses could be easily handled. Development of picosecond streak cameras in the early 1970s was in fact closely tied to the invention of short-pulse, mode-locked lasers. The relationship was mutually stimulating, for lasers were one of the few reproducible sources for picosecond light pulses suitable for testing the time resolution of streak cameras, and streak cameras were one of the few devices suitable for making certain types of measurements of laser pulses. In particular, the temporal profile of individual laser pulses could be measured by few other techniques. Autocorrelation of laser pulses can be used, and today it is indeed the most widely used method for characterizing high-repetition-rate, subpicosecond pulsed lasers, but it has the disadvantages of averaging pulse characteristics over many shots and not unambiguously determining the pulse shape. Experiments with lasers of pulse width 1 ns or less and repetition rate 10 Hz or less are candidates for streak camera methods. If the experiment needs to simultaneously measure two or more electromagnetic signals on a subnanosecond time scale at low repetition rate, streak cameras are one of the best choices.

Characterization of the shape of a laser pulse in time is also perhaps the most trivial application of streak cameras. It does not make use of the camera's two-dimensional capabilities, it does not severely test the camera's dynamic range or signal-to-noise ratio, it does not assess the camera's ability to deal with processes extending over several orders of magnitude in time, and it does not usually demand a great deal of precision. The recent emphasis on signal-to-noise ratio, dynamic range, and signal averaging has come in large part from experiments demanding more from streak cameras than laser characterization did. Consequently, we quickly pass on from laser pulses to effects produced in matter by laser pulses.

3.8.2. Solid-State Systems

Streak camera measurements on vibrational relaxation of defects in alkali halides is described in detail by Knox in Ref. 36. This reference contains the most inclusive published collection of streak camera methods applied to a single problem. The methods include signal averaging, rise time spectroscopy, wavelength-resolved spectroscopy (both continuous wavelength dispersion with a polychromator and discrete wavelength dispersion with interference filters), and polarization spectroscopy. Molecular ions such as O_2^- in alkali halide crystals are examples of defects whose vibrational states are weakly coupled to the crystal lattice because of the small ion radius and the mismatch of O_2^- vibrational energies (300–1000 cm^{-1}) with lattice phonon frequencies (100–200 cm^{-1}). Because of this weak coupling, vibrational structure in the O_2^- optical spectra is expected to be well resolved, and relaxation of excited O_2^- states is expected to be relatively slow. Both of these expectations are confirmed: O_2^- in KBr shows clearly resolved line spectra; O_2^- in KCl, KBr, and RbI shows vibrational-relaxation-limited fluorescence rise times of 15 ± 4, 41 ± 6, and 79 ± 11 ps, respectively. Also evident in the time-resolved emission spectra are the appearance and decay of two well-resolved hot luminescence bands. In KBr the rise and fall of these bands are completed in about 30 and 80 ps, respectively. The fall times of these bands are approximately equal to the rise time of the ordinary band.

A noteworthy feature of the time-resolved O_2^- fluorescence is the lack of dependence of signal amplitude on emission quantum efficiencies. The quantum efficiency is relatively high, but this is because the fluorescence lifetime of the ion is on the order of 100 ns. In picosecond time-resolved experiments done over a 300-ps time scale, only about one out of 1000 photons is emitted in the time frame of interest, so the signals are relatively small. Steady-state or long-time-scale experiments, on the other hand, cannot resolve the fast vibrational relaxation processes but will collect many more photons.

3.8.3. Solutions: Molecular Relaxation in Solvents

The dynamics of molecules in solvents has been a subject of great interest for the last several decades. Many chemical reactions are controlled by interactions between reacting molecules and the solvent. Diffusion-controlled reactions are the simplest example of the influence of solvent on chemical reactions. Another sort of reaction influenced by solvent is molecular isomerization, such as *cis–trans* isomerization in stilbene. One important question is the relationship between isomerization rate and solvent viscosity. The reaction has been traditionally imagined as an activated process which proceeds by passage over an activation barrier which contains both an energy and a friction term. The friction is determined by the viscosity. The effect of solvent-controlled friction can depend on the time scale of molecular motions involved in the isomerization reaction, so considerable work has gone into the study of dynamic frictional forces (see the reference list of Ref. 52).

Isomerizations are often accompanied by spectral changes if the molecule is fluorescent. The spectral change may be due to solvent relaxation around the chromophore or relaxation of the fluorescent molecule itself. Sommer *et al.*[52] measured a time-dependent spectral shift of ethidium bromide in glycerol using a streak camera system in two-dimensional mode to measure the entire emission spectrum from time 0 to about 2 ns after excitation by a 30-ps, 532-nm Nd:YAG laser pulse, with 20-ps time resolution. The streak camera was of the single-shot, signal-averaging type, used with a photoconductive triggering switch.[44] Figure 3.23 shows the raw spectra which they measured, plotted as emission intensity versus wavelength and log time. This data set consists of twenty 500-point spectra, recorded approximately equally spaced on a logarithmic time scale. It is apparent that not much sense can be made of this data set as it stands. Figure 3.24 shows an SVD analysis of the data set, with the two spectral components found to be significant plotted in panel a and their time dependence in panel b. The two components consist of

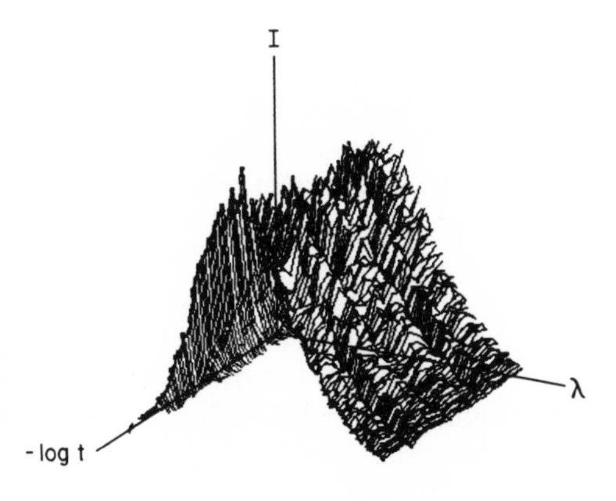

Figure 3.23. Unprocessed emission spectra of ethidium bromide in glycerol as a function of time. Twenty spectra are displayed, spaced equally on the logarithmic time scale, from 0 to 1.8 ns. The wavelength range is approximately 540–700 nm. Reproduced from Ref. 52, with permission.

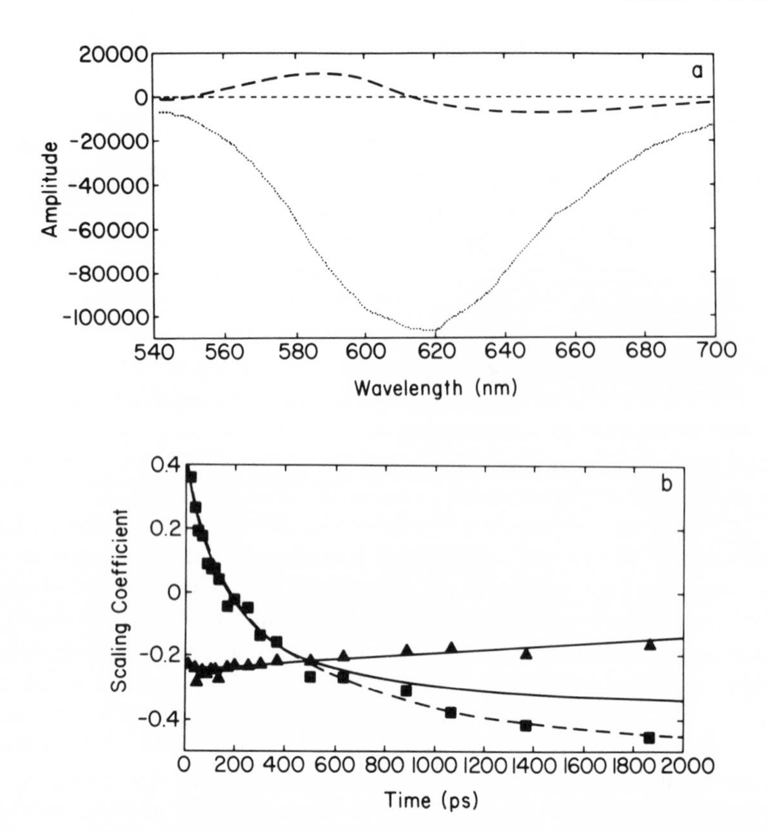

Figure 3.24. Analysis of spectra of Figure 3.23 by SVD. (a) The spectra and relative amplitudes of the first two basis spectra extracted by the SVD algorithm are shown. The dotted line is the first component and is shown inverted. (b) The scaling coefficients of the two components as a function of time. Triangles are the first basis spectrum; squares the second. The solid lines are single-exponential fits to the time dependence of the spectra; the dashed line is the double-exponential fit. The negative scaling coefficient of the first component inverts its sense, restoring a positive fluorescence intensity which decays slowly to zero. Reproduced from Ref. 52, with permission.

a spectrum which describes the total quantity of dye in solution excited by the laser pulse and a spectral shift component which has a multiexponential decay in time over hundreds of picoseconds.

SVD analysis does not automatically indicate the physical meaning of spectral components. That is the responsibility of the investigator. The first SVD basis spectrum of this data set was calculated as the best least-squares fit to the entire set of observed spectra. It was assigned as a measure of the total amount of excited dye because (i) the spectrum closely resembles the static ethidium spectrum, (ii) the magnitude of this spectral component is ten times that of the second component, and (iii) the amplitude of this component decayed slowly, with a decay time of about 4 ns, similar to the 5.9-ns lifetime previously observed for ethidium in glycerol. The second SVD component was calculated, orthogonal to the first, and, together with the first, gives the best two-component least-squares fit to the data. It was assigned as a spectral shift component because of its spectral shape and relatively small amplitude.

The noise in Figure 3.24 is small compared to that in the raw data of Figure 3.23. This demonstrates another characteristic of SVD analysis: noise in the spectra is contained in the higher SVD components, not in the lower, larger amplitude components. SVD is thus both an effective way of summarizing and parameterizing data and an effective smoothing routine.

3.8.4. Biological Systems

3.8.4.1. Photosynthesis

Bruce *et al.*[53] have recently analyzed wavelength-dependent fluorescence decay and rise times from streak camera data of emission from the photosynthetic algae *Porphyridium cruentum*. These data were recorded with the low-jitter, signal-averaging streak camera[44] at discrete wavelengths, using interference filters. The cells contain large protein complexes with several chromophores: phycoerythrin (PE), phycocyanin (PC), allophycocyanin (APC), and chlorophyll *a*. Figure 3.25 shows the kinetics of the fluorescence from PE (570 nm), PC (640 nm), and APC (660 nm), determined by the wavelength of emission. The PE signal rises with the laser pulse (zero rise time), whereas the PC and APC components show rise times of 25 ± 8 and 60 ± 15 ps. These rise times were defined by a kinetic model in which transfer between three kinetic species was allowed at particular rates. The inverses of

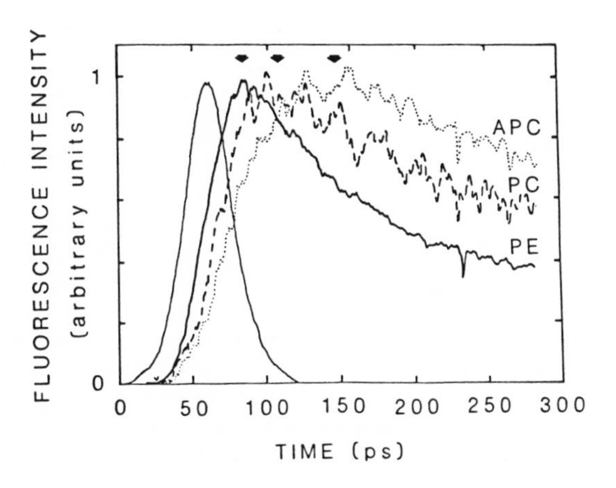

Figure 3.25. Fluorescence emission kinetics of phycoerythrin (PE), phycocyanin (PC), and allophycocyanin (APC). The laser excitation pulse is shown at the left. Reproduced from Ref. 53, with permission.

the transfer rates are the cited rise times and were taken to indicate energy transfer from PE to PC to APC, consistent with previous measurements.

3.8.4.2. Proteins: Structural Dynamics

Protein molecules reside in a conformation which is usually characterized by the X-ray crystal structure. This structure is the time average of the positions of all the atoms in the molecule. In solution, however, protein structure fluctuates on time scales ranging from picoseconds or less to hours or more. The fast structural motions are motions of individual atoms or small groups of atoms, for example, amino acid side chains. Tryptophan and tyrosine amino acids contain aromatic side chains which absorb and emit light in the 200- to 400-nm spectral region. Motions of these side chains have been studied both theoretically and experimentally using optical methods (see reference list in Ref. 54). Figure 3.26 shows an example of a streak camera record of the time-resolved anisotropy of fluorescence from a tyrosine amino acid in the protein lima bean trypsin inhibitor.[54] The figure also contains a computer fit showing a fast (about 40 ps) and a slow (longer than about 3 ns) anisotropy decay process, suggesting fast, restricted rotations of the tyrosine ring. Tyrosine is a chromophore with a fairly low absorption coefficient, about $10^3\ M^{-1}\ cm^{-1}$, so it was necessary to integrate the entire emission spectrum

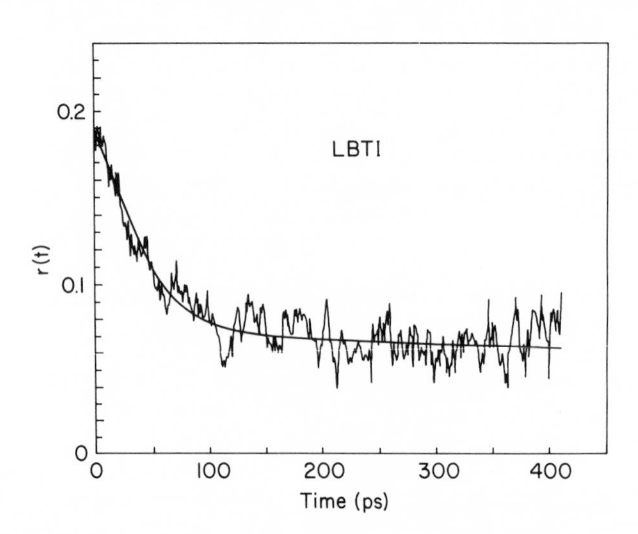

Figure 3.26. Fluorescence anisotropy decay versus time for the single tyrosine residue in lima bean trypsin inhibitor. The smooth curve is a double-exponential fit to the data with components of amplitudes and decay times of about [0.12, 40 ps] and [0.07, >3 ns], respectively. A total of 300 shots were summed for each polarization component in single-shot, signal-averaging mode. Reproduced from Ref. 54, with permission.

from 300 to 380 nm. These data were recorded with the low-jitter, signal-averaging streak camera.

3.8.4.3. *Nucleic Acids*

DNA was considered to be nonfluorescent at room temperature until the early 1980s. Steady-state, low-temperature measurements and extrapolations of quantum yields to room temperature suggested that the fluorescence decay times of common nucleic acid bases were in the range 10^{-13} to 10^{-12} s. Short decay times produce signal amplitudes which are very small if the response time of the system is longer than the decay time (Section 3.4.1.1). The amplitude of a signal recorded by a faster response time system, on the other hand, depends upon the absorption coefficient of the chromophore but not on the decay time. Consequently, streak camera techniques can quite easily measure the emission from nucleic acid bases, whose absorption coefficients are about 10^4 M^{-1} cm^{-1}.[59] Figure 3.27 shows the time-resolved fluorescence from calf-thymus DNA, excited with a 30-ps, 266-nm Nd:YAG laser pulse and detected with the low-jitter, signal-averaging streak camera.[55] The signal is dominated by a decay component of lifetime 10 ps, though inspection of decay curves as a function of emission wavelength reveals wavelength-dependent components ranging in decay time from 4 ps to 65 ps or more.

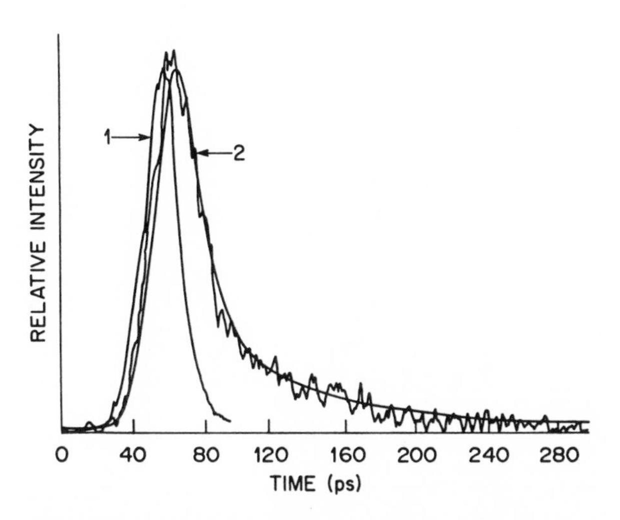

Figure 3.27. Time resolved fluorescence from calf thymus DNA, excited by 30-ps, 266-nm, 1-μJ pulses from a Nd:YAG laser. A total of 400 shots were summed in single-shot, signal-averaging mode. Reproduced from Ref. 55, with permission.

3.8.5. X-Ray Streak Cameras: Fusion Reactions

Photocathodes of streak cameras can be designed to detect X-ray emission as well as emission at longer wavelengths. X-ray streak cameras have long been used to study nuclear processes, both fission and fusion. With the recent advent of X-ray laser research, the streak camera has become a primary device for identifying lasing lines in the presence of a large background of soft X-ray emission.[60, 61] This background can be separated from the lasing signal by the streak camera because the lasing is short-lived and narrow-band, unlike the background. Below we consider in more detail an application of streak cameras to inertial-confinement fusion studies.

Fusion of nuclei of hydrogen isotopes produces large amounts of energy which, if controllable, can be converted into heat to drive electrical power generators. The products of interest in nuclear fusion are neutrons, but copious amounts of X rays are also emitted. These X rays can be used to characterize physical properties of the fusing matter. At the University of Rochester's Laboratory for Laser Energetics and at other laboratories, fusion is being induced by focusing multiple laser beams onto a target microballoon filled with a deuterium–tritium mixture. The laser beams heat and implode the target, and if a sufficient temperature and density are maintained for a long enough time, more energy could be released than is input by the laser ("breakeven").

The process of target implosion creates a plasma at the surface of a microballoon of initial diameter about 500–1000 μm. This plasma emits X rays, and as the implosion proceeds, the X-ray emission follows the imploding outer diameter of the target until at maximum compression a bright flash occurs from the central compressed region. The process takes about 1 ns. Streak cameras coupled with X-ray pinhole cameras can be used to measure this implosion process as a function of time and spatial coordinate.[56] The pinhole camera images X rays from the target onto a mask over the streak camera photocathode. The mask contains a slit about 100 μm wide which

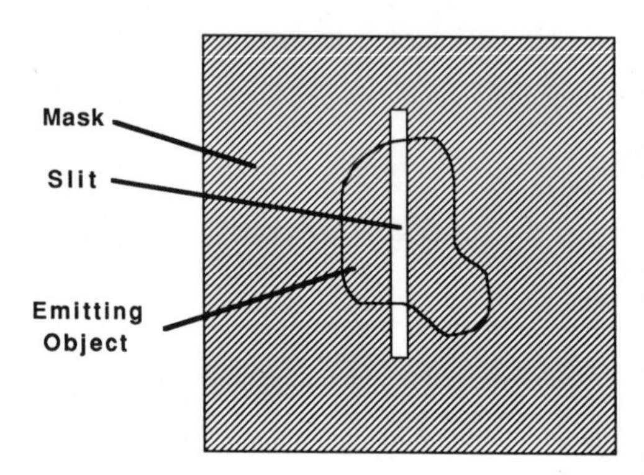

Figure 3.28. Schematic for measurement of emission intensity versus radius and time in Figure 3.29. The 100-μm-wide slit selects a portion of the object for measurement.

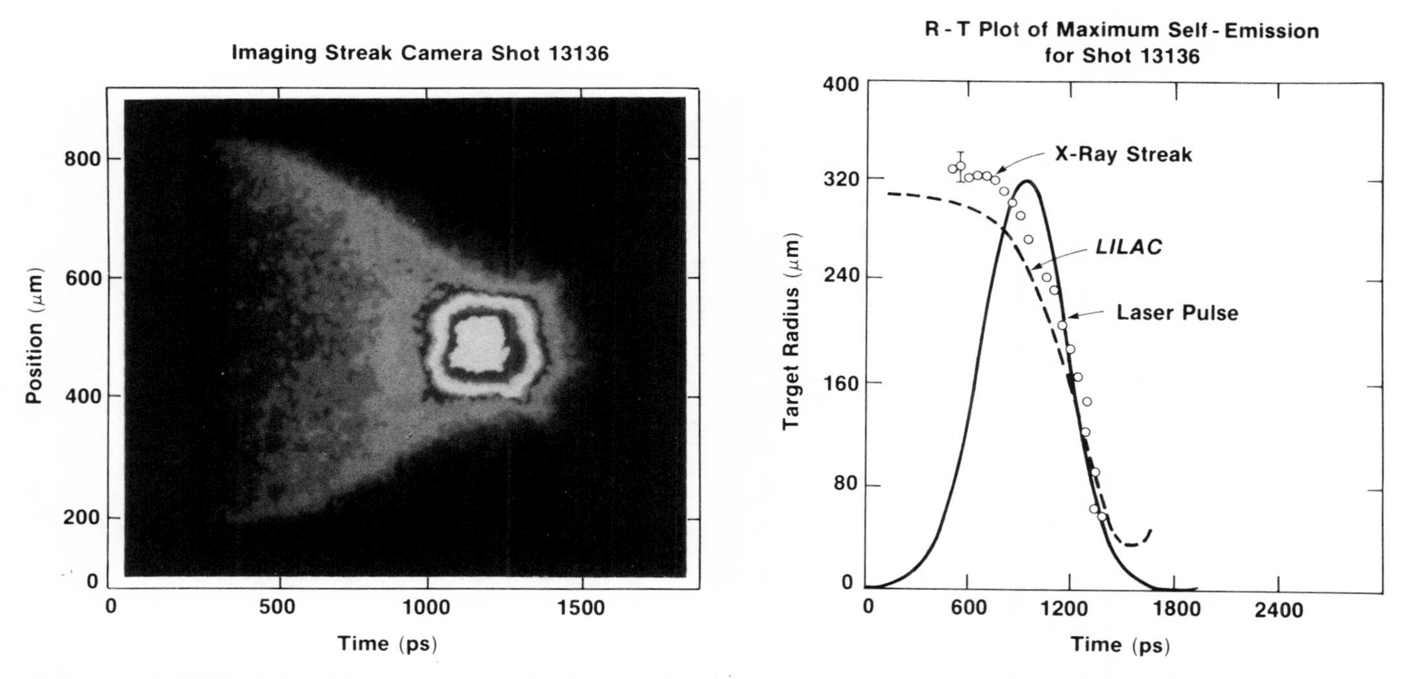

Figure 3.29. Time-resolved X-ray imaging (E ~ 1 keV). *Left*: Intensity of X-ray emission from a target imploded symmetrically by 24 equal beams of total energy 2 kJ, pulse width 600 ps, and wavelength 351 nm. The outer diameter of the glass shell containing the deuterium–tritium mixture appears at early times at positions 200 and 800 μm. At time 1200 ps the target has reached its point of maximum compression and X-ray emission. *Right*: Plot of target radius versus time, along with laser pulse and a hydrodynamic computer simulation of the implosion process. See Ref. 13. Figure courtesy of University of Rochester Laboratory for Laser Energetics.

selects a portion of the target for measurement (see Figure 3.28). The slit is oriented so that the streak camera sweeps the photoelectron image in a direction perpendicular to the slit's long axis. The streak sweep is synchronized with the laser pulse arrival at the target, and as time proceeds, a time- and space-resolved image of the imploding target is created. Figure 3.29 shows an example of target implosion recorded on laser shot 13136 at the University of Rochester. The 24 beams of the OMEGA laser system were focused symmetrically onto a 600-μm glass balloon filled to about 10 atm pressure with a deuterium–tritium mixture. The laser, a frequency-tripled Nd:glass laser emitting at 351 nm, provided a total energy of about 2 kJ in 600 ps. The radius of the target is shown to collapse at a speed of about 3×10^5 m/s. The plot provides the intensity of X-ray emission as a function of position along the imaged target portion and time. The streak record was recorded on film rather than with an OMA in order to retain complete two-dimensional information.

3.9. The Future

It is perhaps dangerous for anyone to predict the future direction and end result of streak camera usage and development. The scientific marketplace of the 1980s made a habit of being rather unpredictable when it came to devices and methods of high technology, for example, nuclear power and genetic engineering. Since the streak camera has long ago lost the interest of politicians (see Section 3.1.4), it may be somewhat safer to guess what is ahead, or at least guess what researchers would like to see, in streak camera development. Below I discuss three aspects of streak camera technology which I consider to be the most important factors to be considered in the development of streak cameras for general use in ultrafast kinetics. These aspects are time range, simplicity of design and operation, and precision.

3.9.1. The 0.1–20-ps Time Regime

Streak cameras remain one of the premier techniques for resolving luminescent phenomena on time scales of 0.1 to 20 ps with good time and amplitude precision. The advantages of streak cameras over techniques such as fluorescence upconversion or Kerr shuttering, which can have even better time resolution, include relatively unrestricted spectral bandwidth, polarization insensitivity, and acceptance solid angle. The streak camera retains its wide lead over photon-counting methods in time resolution, though presently it lags in signal-to-noise ratio on longer time scales. In view of the ease of setup of streak camera detectors and the ease of operation of the systems presently being marketed, the streak camera will continue to serve as an all-purpose detector for radiation processes occurring on times scales of 10^{-13} to 10^{-11} s.

3.9.2. Integrated Detection Systems

The repetitive conversion of photons to electrons back to photons to electrons, and so on, which takes place in streak cameras has long been recognized as a limitation to recording precision data. A few attempts to directly measure the ICT photoelectron current produced results which were not extremely promising (Section 3.2.2.5). Nevertheless, the elimination of several stages in the process of measurement remains a very worthwhile goal. Direct counting of the photoelectrons as a function of position at the end of the streak tube would eliminate noise, image distortion, and intensity non-linearities added by subsequent stages in present cameras. In addition, a price benefit may ultimately accrue by the replacement or elimination of the rather expensive image intensifier stage.

The trend toward systems integrated in the sense of having multiple settings for various purposes is also expected to continue. Commercial streak cameras presently have time bases which are easily variable by changing switch settings or changing time base plug-ins, much like oscilloscopes. (As an aside, Hamamatsu presently markets a streak camera tube which is configured to operate as the world's fastest sampling oscilloscope.) Some tubes are designed with variable electron image magnification, so that time-resolved imaging spectroscopy can in principle be done with variable magnification. Since the forte of streak camera detection is two-dimensional spectroscopy, it may also be expected that easily exchangeable input optical components will become available. This would greatly facilitate the setting up of time- and wavelength-resolved experiments. Presently, alignment and characterization of the time and wavelength response of streak cameras with input poly-chromators can take a day or more, with hours required for modifications such as exchange of gratings and wavelength recalibration.

3.9.3. Time-Resolved Photon-Counting Streak Cameras (TRPCSC)

Fluorescence spectroscopy in the time range of about 50 ps to 10 ns is presently dominated by time-resolved single-photon counting (TRPC), due to the advent of fast-response microchannel plate photomultipliers and fast amplifiers. The time range of TRPC and other methods such as phase fluorometry has been slowly extending downward into the realm previously accessible only to streak cameras. Since TRPC has traditionally had better dynamic range, it has been thought that streak camera methods may gradually disappear for applications such as time-resolved visible fluorescence spectroscopy. TRPC, however, is presently approaching time response limitations caused by the amplification and transport of electrical signals over distances of a meter or so. These limitations can only be eliminated by miniaturizing and integrating the electronic system, that is, placing amplifiers, time-to-amplitude converters, pulse height discriminators, etc., on an integrated circuit ship. Image converter tubes, on the other hand, avoid this electronic speed limitation by converting time dependence to position dependence, which can then be read out at leisure. The intrinsic response time of this method is 10^{-14} s, and there is no obvious reason why the response time of streak tubes will not extend to times below 0.1 ps.[62]

Streak cameras may not disappear in the face of competition from other devices, but they will borrow techniques used with these other devices. Streak camera tubes and photomultiplier tubes are both based on the photoelectric effect and subsequent amplification of the photoelectronic signal. In principle, individual photoelectrons can be counted in either device, but only photomultipliers have had a response (pulse height distribution function) appropriate for photon counting. New streak tubes have now been developed, however, which have the necessary response for photon counting with a streak tube.[42] Together with high-repetition-rate sources and synchroscan technology, this opens the possibility of time-resolved single-photon counting on time scales of 0.1 ps to 100 ns. Very short time response can now be achieved via the streak method of time-to-position conversion of the photoelectron sweep; precision limited only by photon statistics can be achieved simultaneously via the method of single-photon counting.

Acknowledgments

I am grateful to W. Knox and G. Mourou, who taught me the use of streak cameras, and to J. Cuny, D. Floryan, K. Kauffman, P. Jaanimagi, R. Knox, S. Letzring, M. Richardson, and P. Roehrenbeck for helpful discussions. H. Huang provided the simulations in Figures 3.17–3.19. This work was supported in part by NSF grant PCM 83-02601, NIH grant 1 R01 CA-41368, and Biomedical Research Support Grant Program, Division of Research Resources, Public Health Service (2 S07 RR-05403) and by the sponsors of the Laser Fusion Feasibility Project at the Laboratory for Lase Energetics of the University of Rochester.

References

1. D. J. Bradley, B. Liddy, and W. E. Sleat, Direct linear measurement of ultrafast light pulses with a picosecond streak camera, *Opt. Commun. 2*, 391–395 (1971).

2. V. V. Korobkin, A. A. Maljutin, and M. Ya. Schelev, Time resolution of an image converter camera in streak operation, *J. Photogr. Sci. 17*, 179–182 (1969).
3. M. Ya. Shelev, M. C. Richardson, and A. J. Alcock, Image-converter streak camera with picosecond resolution, *Appl. Phys. Lett. 18*, 354–357 (1971).
4. J. S. Courtney-Pratt (ed.), *Proceedings of the Fifth International Congress on High-Speed Photography*, pp. 369–530. Society of Motion Picture and Television Engineers, Inc., New York (1962).
5. S. Letzring, Buying and using a streak camera, *Lasers and Applications II*, 49–52 (1983).
6. N. H. Schiller, Y. Tsuchiya, E. Inuzuka, Y. Suzuki, K. Kinoshita, K. Kamiya, H. Ida, and R. R. Alfano, An ultrafast streak camera system: Temporal disperser and analyzer, *Optical Spectra* (now *Photonics Spectra*) *14*, 55–62 (1980).
7. W. Daheng (ed.), *Proceedings of the 18th International Congress on High Speed Photography and Photonics, Proc. SPIE 1032* (1989).
8. E. K. Zavoiskii and S. D. Fanchenko, Physical fundamentals of electron-optical chronography, *Sov. Phys. Dokl. 1*, 285–288 (1956).
9. W. Sibbett, H. Niu, and M. R. Baggs, Femtosecond streak image tube, in: *Proceedings of the 15th International Congress on High Speed Photography and Photonics* (L. L. Endelman, ed.), *Proc. SPIE 348*, 271–275 (1983).
10. Y. Takiguchi, K. Kinoshita, E. Inuzuka, and Y. Tsuchiya, Development of a subpicosecond streak camera, in: *Proceedings of the 16th International Congress on High Speed Photography and Photonics* (M. Andre and M. Hugenschmidt, eds.), *Proc. SPIE 491*, 46–50 (1985).
11. M. C. Richardson and K. Sala, Picosecond framing photography of a laser-produced plasma, *Appl. Phys. Lett. 23*, 420–422 (1973).
12. M. R. Baggs, R. T. Eagles, W. Sibbett, and W. E. Sleat, Recent developments in the Picoframe I framing camera, in: *Proceedings of the 16th International Congress on High Speed Photography and Photonics* (M. Andre and M. Hugenschmidt, eds.), *Proc. SPIE 491*, 40–45 (1985).
13. R. C. Craxton, R. L. McCrory, and J. M. Soures, Progress in laser fusion, *Sci. Am. 255*, 68–79 (1986).
14. R. S. Marjoribanks, P. A. Jaanimagi, and M. C. Richardson, Principles of streak and framing photography by frequency-encoding on a chirped pulse, in: *High Speed Photography, Videography and Photonics IV, Proc. SPIE 693*, 134–146 (1986).
15. J. S. Courtney-Pratt, Advances in high speed photography 1972 to 1982, in: *Proceedings of the 15th International Congress on High Speed Photography and Photonics* (L. L. Endelman, ed.), *Proc. SPIE 348*, 2–7 (1983).
16. H. F. Swift, Current and future activities in high speed photography and photonics, in: *Proceedings of the 15th International Congress on High Speed Photography and Photonics* (L. L. Endelman, ed.), *Proc. SPIE 348*, 8–14 (1983).
17. J. S. Courtney-Pratt, A new method for the photography study of fast transient phenomena, *Research 2*, 287–294 (1949).
18. Senate Concurrent Resolution 75 of the 86th Congress, 2D Session, June 10, 1960, reproduced in *Proceedings of the Fifth International Congress on High-Speed Photography* (J. S. Courtney-Pratt, ed.), p. V, Society of Motion Picture and Television Engineers, New York (1962).
19. V. S. Komolkov, Y. E. Nesterikhin, and M. I. Pergament, Electron-optical high-speed camera for the investigation of transient processes, in: *Proceedings of the Fifth International Congress on High-Speed Photography* (J. S. Courtney-Pratt, ed.), pp. 118–122, Society of Motion Picture and Television Engineers, New York (1962).
20. K. Kinoshita, N. Hirai, and Y. Suzuki, Femtosecond streak tube, in: *Proceedings of the 16th International Congress on High Speed Photography and Photonics* (M. Andre and M. Hugenschmidt, eds.), *Proc. SPIE 491*, 63–67 (1985).

21. A. J. Campillo and S. L. Shapiro, Picosecond relaxation measurements in biology, in: *Ultrashort Light Pulses* (S. L. Shapiro, ed.), *Topics in Applied Physics*, Vol. 18, pp. 317–376, Springer-Verlag, Berlin (1984).
22. I. N. Duling III, T. Norris, T. Sizer II, P. Bado, and G. A. Mourou, Kilohertz synchronous amplification of 85-femtosecond optical pulses, *J. Opt. Soc. Am. B2*, 616–618 (1985).
23. F. Gex, P. Bauduin, C. Hammes, and P. Horville, S1 photocathode image converter tubes, in: *Proceedings of the 16th International Congress on High Speed Photography and Photonics* (M. Andre and M. Hugenschmidt, eds.), *Proc. SPIE 491*, 276–280 (1985).
24. M. Beghin and G. Eschard, Multialkali photocathode (S20) optimization: Use in streak camera tubes and image intensifiers associated, in: *Proceedings of the 16th International Congress on High Speed Photography and Photonics* (M. Andre and M. Hugenschmidt, eds.), *Proc. SPIE 491*, 281–286 (1985).
25. W. Knox, G. Mourou, and S. Letzring, Jitter-free signal averaging streak camera, in: *Proceedings of the 15th International Congress on High Speed Photography and Photonics* (L. L. Endelman, ed.), *Proc. SPIE 348*, 308–312 (1983).
26. C. B. Johnson, J. M. Abraham, E. H. Eberhardt, and L. W. Coleman, A magnetically focused photoelectronic streak camera tube, in: *Proceedings of the 12th International Congress on High Speed Photography (Photonics)* (M. C. Richardson, ed.), *Proc. SPIE 97*, 56–61 (1977).
27. A. J. Lieber, H. D. Sutphin, C. B. Webb, and A. H. Williams, Subpicosecond x-ray streak camera development for laser-function diagnostics, in: *Proceedings of the 12th International Congress on High Speed Photography (Photonics)* (M. C. Richardson, ed.), *Proc. SPIE 97*, 194–199 (1977).
28. J. Johnson, Analytical description of night vision devices, in: *Proceedings of the Seminar on Direct-Viewing Electro-Optical Aids to Night Vision* (L. Biberman, ed.), Institute for Defense Analyses Study S254 (October 1966).
29. RCA, *Electro-Optics Handbook*, p. 234, RCA Corporation, Lancaster, Pennsylvania (1974).
30. S. Majumdar and S. Majumdar, Reciprocity failure of phosphors used in streak cameras, in: *Proceedings of the 15th International Congress on High Speed Photography and Photonics* (L. L. Endelman, ed.), *Proc. SPIE 348*, 335–337 (1983).
31. P. B. Weiss, P. Black, H. Oona, and L. Sprouse, Characterization of electronic streak tubes including one with internal CCD readout, in: *Proceedings of the 16th International Congress on High Speed Photography and Photonics* (M. Andre and M. Hugenschmidt, eds.), *Proc. SPIE 491*, 679–684 (1985).
32. RCA, *Electro-Optics Handbook*, pp. 173–180, RCA Corporation, Lancaster, Pennsylvania (1974).
33. J. Grad, A. W. Woodland, and C. G. Sluijter, The microchannel plate, in: *Proceedings of the 12th International Congress on High Speed Photography (Photonics)* (M. C. Richardson, ed.), *Proc. SPIE 97*, 223–227 (1977).
34. G. W. Liesegang and P. D. Smith, Improving vidicon linearity in the pulsed illumination mode, *Appl. Opt. 20*, 2604–2605 (1981).
35. H. Staerk, R. Mitzkus, and H. Meyer, Performance of SIT vidicons when exposed to transient light signals, *Appl. Opt. 20*, 471–476 (1981).
36. W. Knox, Picosecond time-resolved spectroscopic study of vibrational relaxation processes in akali halides, Thesis, University of Rochester, Rochester, New York (1983).
37. G. V. Kolesov, I. M. Korzhenevich, V. B. Lebedev, B. M. Stepanov, and A. V. Yudin, The ICT time resolution calculation with allowance for Coulomb interaction of electrons, scatter of their initial vectors and finite image size on the photocathode, in: *Proceedings of the 16th International Congress on High Speed Photography and Photonics* (M. Andre and M. Hugenschmidt, eds.), *Proc. SPIE 491*, 233–238 (1985).
38. S. Majumdar, Dynamic range and recording efficiency of picosecond streak cameras, in:

Proceedings of the 13th International Congress on High Speed Photography and Photonics (Shin-ichi Hyodo, ed.), *Proc. SPIE 189*, 821–824 (1979).
39. E. Inuzuka, Y. Tsuchiya, M. Koishi, and M. Miwa, Dynamic performances and applications of a new two picosecond streak camera system, in: *Proceedings of the 15th International Congress on High Speed Photography and Photonics* (L. L. Endelman, ed.), *Proc. SPIE 348*, 211–216 (1983).
40. W. Friedman, S. Jackel, and W. Seka, Dynamic range and spatial resolution of picosecond streak cameras, in: *Proceedings of the 12th International Congress on High Speed Photography (Photonics)* (M. C. Richardson, ed.), *Proc. SPIE 97*, 544–548 (1977).
41. K. Kinoshita and Y. Suzuki, Noise figure of the streak tube with microchannel plate, in: *Proceedings of the 14th International Congress on High Speed Photography and Photonics* (B. M. Stepanov, ed.), *Proc. SPIE 237*, 159–163, Moscow (1980).
42. T. Urakami, Y. Takiguchi, K. Kinoshita, and Y. Tsuchiya, Characterization of photon-counting streak camera, in: *High Speed Photography, Videoagraphy and Photonics* (B. G. Ponseggi, ed.), *Proc. SPIE 693*, 98–104 (1986).
43. R. Rigler, F. Claesens, and O. Kristensen, Picosecond fluorescence spectroscopy in the analysis of structure and motion of biopolymers, *Anal. Instrum. 14*, 525–546 (1985).
44. W. Knox and G. Mourou, A simple jitter-free picosecond streak camera, *Opt. Commun. 37*, 203–206 (1981).
45. M. Stavola, G. Mourou, and W. Knox, Picosecond time delay fluorimetry using a jitter-free streak camera, *Opt. Commun. 34*, 404–408 (1980).
46. S. J. Strickler and R. A. Berg, Relationship between absorption intensity and fluorescence lifetime of molecules, *J. Chem. Phys. 37*, 814–822 (1962).
47. N. H. Schiller and R. R. Alfano, Picosecond characteristics of a spectrograph measured by a streak camera/video readout system, *Opt. Commun. 35*, 451–454 (1980).
48. M. J. Colles, Efficient stimulated Raman scattering from picosecond pulses, *Opt. Commun. 1*, 169–172 (1969).
49. G. H. Golub and C. Reinisch, Singular value decomposition and least squares solutions, *Numer. Math. 14*, 403–420 (1970).
50. J. H. Sommer, E. R. Henry, and J. Hofrichter, Geminate recombination of *n*-butyl isocyanide to myoglobin, *Biochemistry 24*, 7380–7388 (1985).
51. T. M. Nordlund and D. A. Podolski, Streak camera measurement of tryptophan and rhodamine motions with picosecond time resolution, *Photochem. Photobiol. 38*, 665–669 (1983).
52. J. H. Sommer, T. M. Nordlund, M. McGuire, and G. McLendon, Picosecond time-resolved fluorescence spectra of ethidium bromide: Evidence for a nonactivated reaction, *J. Phys. Chem. 90*, 5173–5178 (1986).
53. D. Bruce, C. A. Hanzlik, L. E. Hancock, J. Biggins, and R. S. Knox, Energy distribution in the photochemical apparatus of *Porphyridium cruentum*: Picosecond fluorescence spectroscopy of cells in state 1 and state 2 at 77K, *Photosynth. Res. 10*, 283–290 (1986).
54. T. M. Nordlund, X.-Y. Liu, and J. H. Sommer, Fluorescence polarization decay of tyrosine in lima bean trypsin inhibitor, *Proc. Natl. Acad. Sci. U.S.A. 83*, 8977–8981 (1986).
55. S. Georghiou, T. M. Nordlund, and A. M. Saim, Picosecond fluorescence decay time measurements of nucleic acids at room temperature in aqueous solution, *Photochem. Photobiol. 41*, 209–212 (1985).
56. G. G. Gregory, S. A. Letzring, M. C. Richardson, and C. D. Kiikka, High time-space resolved photography of laser imploded fusion targets, in: *High Speed Photography Videography and Photonics III* (B. G. Ponseggi and H. C. Johnson, eds.), *Proc. SPIE 569*, 141–148 (1985).
57. T. Minami and S. Hirayama, High quality fluorescence decay curves and lifetime imaging

using an elliptical scan streak camera, *J. Photochem. Photobiol. A: Chem. 53*, 11–22 (library edition) (1990).

58. Hamamatsu technical sheet, Picoseond Fluorescence Microscope System (1989).
59. S. Kobayashi and M. Yamashita, Adenine–adenine interaction in adenyladenosine and polyadenylic acid measure with emission spectroscopy, *Nucleic Acids Symp. Ser. 19*, 199–202 (1988).
60. M. C. Richardson, G. G. Gregory, R. L. Keck, S. A. Letzring, R. S. Majoribanks, F. J. Marshall, G. Pien, J. S. Wark, B. Yaakobi, P. D. Goldstone, A. Hauer, G. S. Stradling, F. Ameduri, B. L. Henke, and P. A. Jaanimagi, Time-resolved x-ray diagnostics for high-density plasma physics studies, in: *Laser interaction and Related Laser Plasma Phenomena*, Vol. 7 (Heinrich Hora and George H. Miley, eds.), Plenum, New York, pp. 179–211 (1986).
61. G. L. Stradling, Streak cameras capture x-ray emissions, *Laser Focus World 1990* (May), 215–223.
62. I. G. Barannik, G. I. Bryukhnevitch, A. M. Prokhorov, and M. Y. Schelev, Pico-femtosecond image converter tubes, in: *Proceedings of the 18th International Congress on High Speed Photography and Photonics* (W. Daheng, ed.), *Proc. SPIE 1032*, 16–29 (1988).

4

Time-Resolved Fluorescence Spectroscopy Using Synchrotron Radiation

Ian H. Munro and Margaret M. Martin

4.1. Introduction

Very often in biochemistry it is the complexity of assembly that confers the biologically significant properties on molecules. Although it may be spectroscopically desirable to reduce the sample to a monomolecular solution, it is not biochemically valid. The concept of breaking down a system into ever smaller constituents in order to ascertain its fundamental properties does not necessarily work for biological systems. Here the essence of the biological property under scrutiny is often the very complexity that is destroyed by a reductive investigation procedure. It is always preferable to attempt to overcome the technical difficulties involved in studying intact biological samples. One of the advantages of using synchrotron radiation is that "wet" biological specimens can often be employed. The intensity of synchrotron radiation at all wavelengths is such that live samples can be probed in a very short time by highly monochromatic light. This minimizes radiation damage or thermal effects that might otherwise disrupt and kill the sample. The use of synchrotron radiation in the X-ray microscope is one example.[1] Although the spatial resolution of this technique is comparable to that of the electron microscope, it has the advantage that biological specimens can be viewed in their native state without metallic staining. Biological electron microscopy is plagued by staining artifacts. In the visible region of the spectrum, lasers can provide a greater power intensity at specific wavelengths, but tunability is difficult even with dye lasers. The availability of continuously tunable synchrotron radiation has opened up the field of UV/visible biochemical

---

Ian H. Munro and Margaret M. Martin • SERC Daresbury Laboratory, Warrington WA4 4AD, England.

Topics in Fluorescence Spectroscopy, Volume 1: Techniques, edited by Joseph R. Lakowicz. Plenum Press, New York, 1991.

spectroscopy. For time-resolved measurements, pulsed lasers can provide up to 10^{12} photons per pulse, whereas synchrotron radiation typically provides 10^{6} photons per pulse at the sample. However, the unique tunability of synchrotron radiation allows the exciting light to be tuned to the absorption maximum. The ability to tune exactly to a specific absorption maximum allows individual species to be resolved from a complex biochemical assembly of different fluorophores, without the need to overpurify the sample. The timing characteristics of a synchrotron radiation source are identical for all emitted wavelengths. This time registration enables us to measure the time profile of the exciting pulses at the fluorescence emission wavelength, thus completely eliminating problems associated with the wavelength dependence of the photoelectron transit time in the detector. Also the repetition rate of the synchrotron radiation source (3.125 MHz in single-bunch mode) is such that data can be accumulated very rapidly (on the order of 10,000 counts/s) while avoiding pileup errors. The inherent polarization of synchrotron radiation can be used to provide geometrical photoselection of molecules, without attenuation of the exciting light by a presample polarizer.

The characteristics of intensity, tunability, time modulation, and polarization make synchrotron radiation an ideal source for biochemical spectroscopy. Many of the technical problems normally encountered can be overcome by the judicial application of its unique combination of properties. Selected examples are used to illustrate recent applications of synchrotron radiation to a range of biochemical investigations. These studies were all undertaken on the high-aperture port of the Synchrotron Radiation Source (SRS) at Daresbury in England.

4.2. Proteins and Peptides

Owing to their fluorescence properties, the aromatic amino acids tryptophan and tyrosine are natural indicators with which to probe a variety of dynamic phenomena in proteins and peptides, via the technique of time-resolved fluorescence depolarization. The time scales of many protein biochemical processes correspond to the fluorescence lifetimes of excited states of these residues, typically in the subnanosecond to few-nanosecond range. This makes time-resolved fluorescence spectroscopy an ideal technique for investigating the molecular dynamics of proteins. These amino acid residues impart intrinsic fluorescence to proteins. Some of these fluorophores, notably single tyrosine residues, have low quantum yields of fluorescence, but the inherent sensitivity of the techniques, combined with the availability of intense light sources, such as lasers and synchrotron radiation, has opened up the field of time-resolved fluorescence studies applied to protein macromolecules and to the study of the excited state behavior of smaller peptide molecules.

4.2.1. Lumazine

Lumazine protein (20 kDa) is the emitter of bioluminescence in the marine photobacteria *Photobacterium phosphoreum* and *P. leiognathi*. The required excitation energy for fluorescence emission is provided by the enzymatic action of luciferase. The highly fluorescent 6,7-dimethyl-8-ribityllumazine prosthetic group and the single tryptophan residue in the polypeptide chain can both be used as intrinsic reporters for intramolecular protein dynamics. Whereas most proteins give rise to complex fluorescence decay kinetics, lumazine proteins can be considered a standard for time-resolved fluorescence depolarization studies of protein dynamics, as both the fluorescence and anisotropies exhibit exponential decay kinetics. Fluorescence from the bound prosthetic group is quite intense (quantum yield = 0.6), and the fluorescence lifetime is of a suitable duration (14 ns) for the investigation of overall protein rotation and of the interaction between lumazine protein and luciferase, which associate under certain conditions to form a protein complex of 100 kDa.

An extensive time-resolved fluorescence study has been carried out on lumazine protein by Visser and co-workers at Wageningen using laser excitation.[2–5] The instrumentation consisted of two parts[6–9]: on the excitation side a mode-locked argon ion laser, a synchronously pumped dye laser, an electro-optic modulator, and a frequency-doubling crystal were available; on the emission side there was the usual arrangement of rotatable polarizer, sensitive, fast photomultiplier, and single-photon counting apparatus. This study was followed up by measurements on the same protein using synchrotron radiation as the excitation source. The tunability of synchrotron radiation is of particular significance in the study of proteins which contain multiple intrinsic fluorophores, since selective excitation of different groun states may be achieved. Förster type energy transfer between residues with overlapping excitation and emission spectra can result in an increased rate of depolarization of emitted fluorescence. Excitation at the red edge of the absorption spectrum decreases the probability of such a transfer event; also the ability to measure the excited state lifetimes and time dependence of emission anisotropy at a range of wavelengths within the excitation spectrum will allow the magnitude and kinetics of these energy migration effects to be assessed.

It is important to compare laser-induced fluorescence results with those from other pulsed excitation sources with regard to data quality, the effect of inclusion of time shifts in the data analysis, and so on. A light source that combines good timing characteristics with continuous wavelength tunability extends the range of dynamic fluorescence properties that can be investigated. For instance, it is very informative to excite lumazine protein over a range of wavelengths, since its absorption spectrum changes on binding. The inherent fluorescence decay processes can be studied from the fluorescence kinetics at

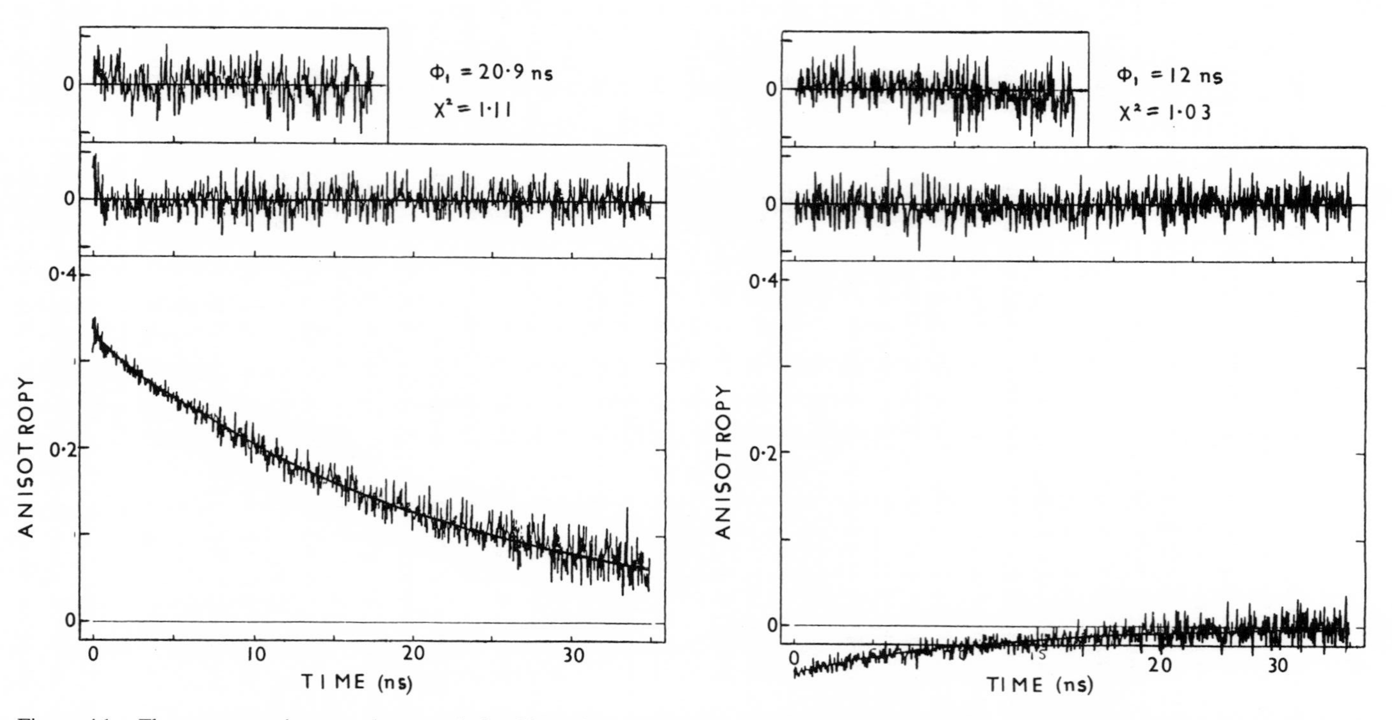

Figure 4.1. Fluorescence anisotropy decay analysis of lumazine protein from *Photobacterium phosphorem*, excited into the first excited singlet state of lumazine (left) and excited into a higher lumazine singlet state (right). Plots of the weighted residuals and autocorrelation function of the residuals illustrate the quality of the fits.

different excitation wavelengths using tunable synchrotron radiation as the light source. If the correlation times differ when excitation wavelength is varied, then conclusions can be drawn about the shape of the protein. If the protein molecule were perfectly spherical, no change in correlation time would be expected. These measurements were carried out using the single-photon counting apparatus on the high-aperture port of the SRS at Daresbury, using microchannel plate photomultipliers in the detection chain to maximize the possible time resolution. Novel dynamic aspects of lumazine protein were demonstrated. Firstly, when lumazine was excited into higher electronic transition, the anisotropy started negatively and decayed to zero with a different correlation time from that obtained after excitation into the first excited state. This indicates the presence of anisotropic protein rotation. Figure 4.1 gives an example of the data analysis showing this negative anisotropy.

Lumazine protein was found to have a correlation time of 20 ns, characteristic of a rigidly associated prosthetic group. If the protein is diluted to submicromolar concentrations, the ligand is dissociated. This was manifested by

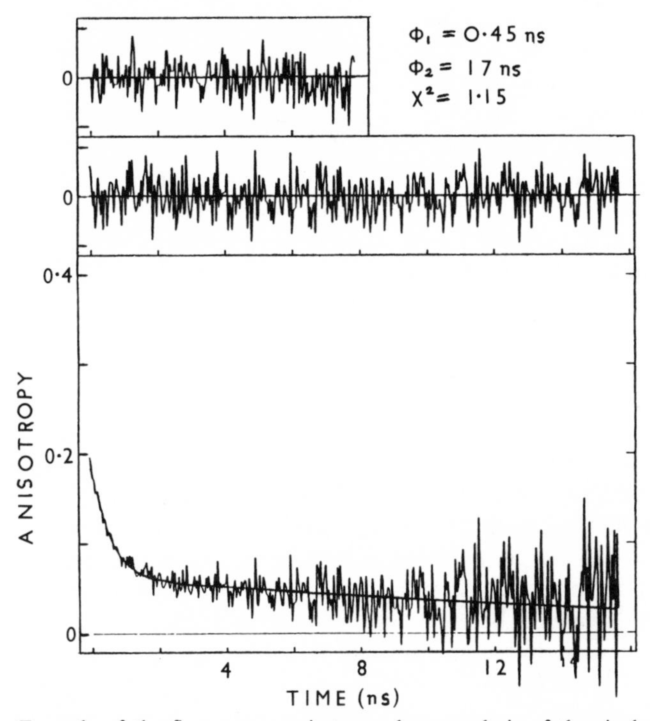

Figure 4.2. Example of the fluorescence anisotropy decay analysis of the single tryptophan residue of lumazine protein, showing rapid mobility (short correlation time ϕ_1) superimposed on the slower protein rotation (ϕ_2).

the presence of a very rapid (100 ps) component in the anisotropy decay. The rapid component could be identified by time-resolved fluorescence depolarization of the free 6,7-dimethylribityllumazine, which results in a correlation time of 160 ps. Such a correlation time is too short for a spherical rotor of equivalent volume. This interpretation of the protein–ligand equilibrium is fully consistent with the observed spectral fluorescence changes.

Lumazine protein was induced to associate with bacterial luciferase to form a protein–protein complex. The anisotropy was found to decay according to a double-exponential function with correlation times of 20 ns and 70 ns, representing free and associated lumazine protein, respectively. The relative amplitudes measure the amounts of free and complexed protein, so that this single experiment directly yields the equilibrium constant.

The fluorescence anisotropy decay of tryptophan was quite different from that of lumazine protein. Analysis revealed a short correlation time (0.46 ns) indicative of a flexible single tryptophan residue (Figure 4.2), which must be located in a very mobile region of the polypeptide chain.

Another line of investigation was to resolve in time the Förster type energy transfer from tryptophan to lumazine. This energy transfer process could be resolved in time since excitation in the absorption band of tryptophan and monitoring lumazine fluorescence reveals a negative preexponential factor after a two-exponential analysis of the fluorescence decay. The short-lifetime component is related to the reciprocal rate constant of energy transfer. Figure 4.3 gives an example of the analysis.

Computer analysis of the synchrotron radiation data required much effort. Because the fitted parameters turned out to be highly correlated, decay curves had to be analyzed simultaneously with linked parameters, requiring a global analysis computer program.[10]

4.2.2. Angiotensin II

Tyrosine has received less attention than tryptophan in the field of time-resolved fluorescence spectroscopy, because of its low absorptivity, low quantum yield of fluorescence in polypeptides, Raman scatter interference resulting from the small emission Stokes shift, masking of its emission by tryptophan, and the complicating factor of excited state proton transfer.[11] Nevertheless, tyrosine has several photophysical properties which are useful in studying the dynamic behavior of proteins, including a short fluorescence lifetime to characterize subnanosecond motions, excited state proton transfer to determine the presence of nearby acid/base groups, and the ability to excite tryptophan residues by resonance energy transfer.

Preliminary studies on simple tyrosine derivatives[12] have shown that a correlation exists between the fluorescence decay kinetics of tyrosine and its

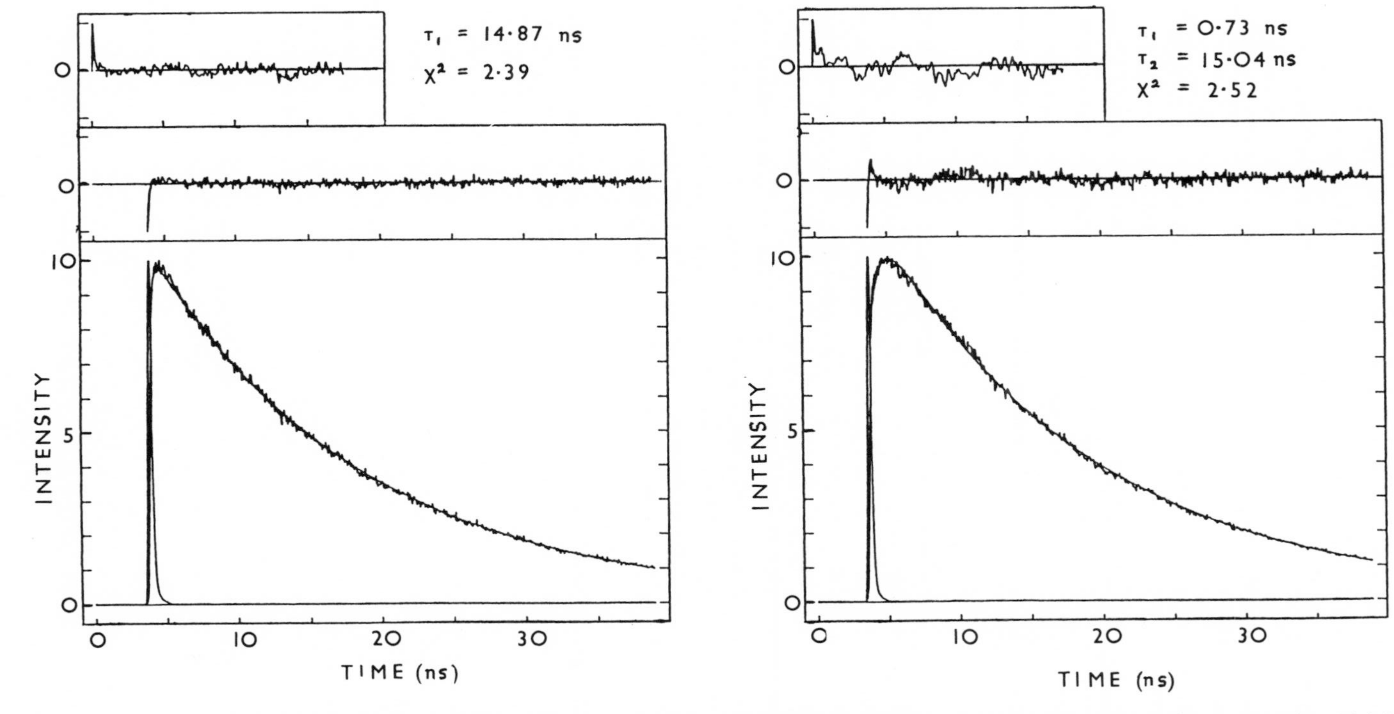

Figure 4.3. Fluorescence lifetime analysis of lumazine protein, directly excited (left) or partially excited via energy transfer from tryptophan (right). The buildup of acceptor fluorescence is manifested by a negative preexponential factor (α_1).

three rotamer populations determined by ^{1}H-NMR. There is rotational isomerism about the $C^\alpha - C^\beta$ bond of tyrosine. The prevalent states are three staggered conformations, or rotamers, of the phenol ring about this bond. Multiexponential decay kinetics of simple tyrosine derivatives can be accounted for by ground state rotamers whose exchanges are on a slower time scale than that of fluorescence decay.

The dynamics of a peptide molecule containing a single tyrosine residue as an intrinsic fluorescent probe can be investigated by time-resolved polarization spectroscopy. One such single-tyrosine peptide is the pressor hormone/ Angiotensin II. The emission anisotropy kinetics of the aromatic amino acid in a peptide molecule incorporates information about the motional freedom of the amino acid residue for intramolecular rotations, as well as the overall molecular reorientation of the peptide.

Angiotensin II solutions were illuminated by 200-ps pulses of 100% polarized light at 280 nm from the high-aperture port of the SRS. The fluorescent light emitted at right angles to the exciting beam was detected by a Hamamatsu microchannel plate photomultiplier. A low-fluorescence LF30 cutoff filter eliminated any scattered 280-nm light. A computer-controlled rotatable polarizer between the sample and the detector was used to select fluorescence in the parallel and perpendicular planes of polarization, and the information was fed into two separate memories of a multichannel analyzer. The scattering from a Ludox suspension was used to give the prompt pulse profile. The profile of the exciting pulses was measured at the emission wavelength 305 nm, to eliminate contributions from the photoelectron transit time spread in the measured data.

Fluorescence decay curves were analyzed by a nonlinear least-squares convolution comparison between the measured and computed data. Table 4.1 lists the fluorescence decay parameters of native angiotensin II and of the peptide denatured by incubation with 7 *M* urea. Both exhibit multiexponential decay kinetics requiring a sum of three exponentials to fit the data. We propose that the three fluorescence lifetime determined for the tyrosine residue in angiotensin II correspond to the three rotameric populations of the phenol ring about the $C^\alpha - C^\beta$ bond. The effect of incubation with 7 *M* urea was to increase the longest fluorescence lifetime from 1.6 ns to 3.33 ns. Urea acts as a denaturant by disrupting hydrogen bonds. Figure 4.4 shows the α-helical

Table 4.1. Fluorescence Decay Parameters for Native and Denatured Human Angiotensin II

Sample	τ_1 (ns)	τ_2 (ns)	τ_3 (ns)
Native angiotensin II	0.12	0.83	1.60
Denatured angiotensin II	0.14	0.98	3.33

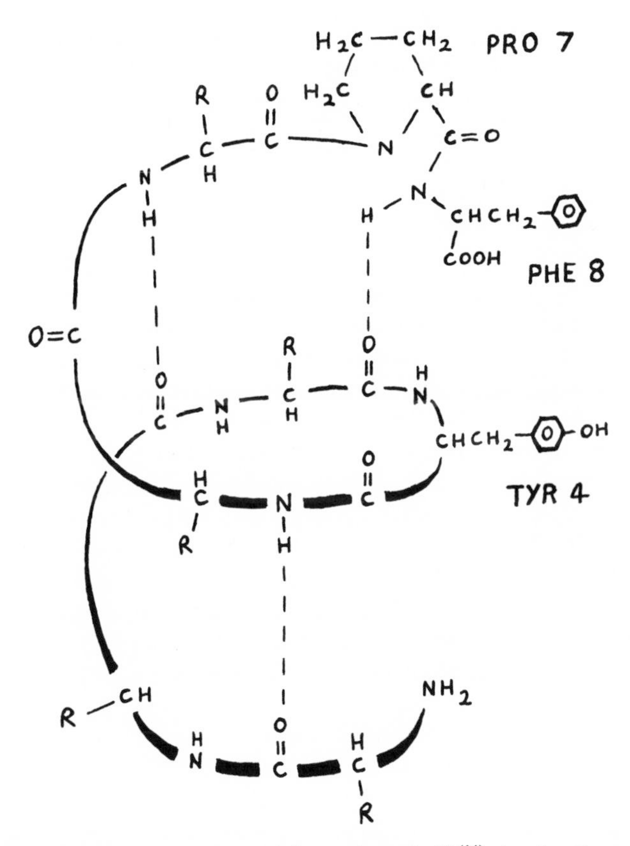

Figure 4.4. The α-helical structural model for angiotensin II,[14] showing the position of the three hydrogen bonds.

structural model for angiotensin II interpreted from NMR and circular dichroism studies.[13] The helix is maintained by three hydrogen bonds. The tyrosine side chain is hydrogen bonded to phenylalanine. In the denatured peptide this bond has been broken, eliminating the effect of excited state proton transfer from the decay kinetics. Since just the longest fluorescence lifetime is affected by denaturation, then this must correspond to the one rotamer involved in the hydrogen bond to phenylalanine.

Rotational correlation times were calculated from the slope of the emission anisotropy decay. Table 4.2 and Figure 4.5 show the emission anisotropy kinetics of native and denatured angiotensin II. The effect of denaturation was to introduce a second, very long (215-ns) rotational correlation time, possibly corresponding to overall rotation of the extended denatured peptide chain along its long axis. The shorter correlation time, which probably corresponds to reorientation of the molecule about the shorter axis, was reduced from

Table 4.2. Rotational Correlation Times for Emission Anisotropy of Native and Denatured Angiotensin II

Sample	ϕ_1 (ns)	ϕ_2 (ns)	χ^2
Native angiotensin II	1.77		1.14
Denatured angiotensin II	0.47	215	1.04

1.77 ns to 0.47 ns. The absence of a very short component indicates that the tyrosine residue does not undergo rapid segmental motion.

The lifetime results lend support to the slow-exchange rotamer model for the tyrosine ring in angiotensin II and demonstrate the quenching effect on the fluorescence decay of the rotamer that is involved in hydrogen bonding in the native helical conformation of the peptide. H-bonding to phenylalanine restricts the motional freedom for intramolecular rotation of the tyrosine residue. The excited state quenching effect may be used in future to indicate the involvement of a tyrosine residue in H-bonding in other peptides and proteins.

The anisotropy results are consistent with the ordered α-helical structural model. That the tyrosine residue does not undergo rapid segmental motion supports the restricted backbone conformation model of angiotensin II.

4.2.3. Lactate Dehydrogenase

The ready tunability of synchrotron radiation, which allows selection of a narrow band anywhere in the excitation spectrum, is particularly useful for studies which use extrinsic fluorescent probes, since many of those which are particularly suitable for formation of dye–protein conjugates have excitation maxima in the 350–500-nm range, a region of the spectrum which is not readily accessible to either nitrogen flashlamps or currently available dye lasers.

Time-resolved anisotropy can reveal the sizes of protein complexes in solution. Such information is important in the elucidation of the molecular mechanisms involved in biochemical phenomena. Holbrook and co-workers have elucidated the role of protein subunit assembly in controlling the enzyme activity of a bacterial lactate dehydrogenase. This provides a molecular explanation for the phenomenon of "glucose repression" in bacteria. Manipulation of this metabolic switch can determine whether a bacterium living on glucose will oxidize it to carbon dioxide and water or will transform it to excreted metabolites such as ethanol and lactate, which are useful fermentation products.

Time-resolved measurements of the fluorescence anisotropy of an

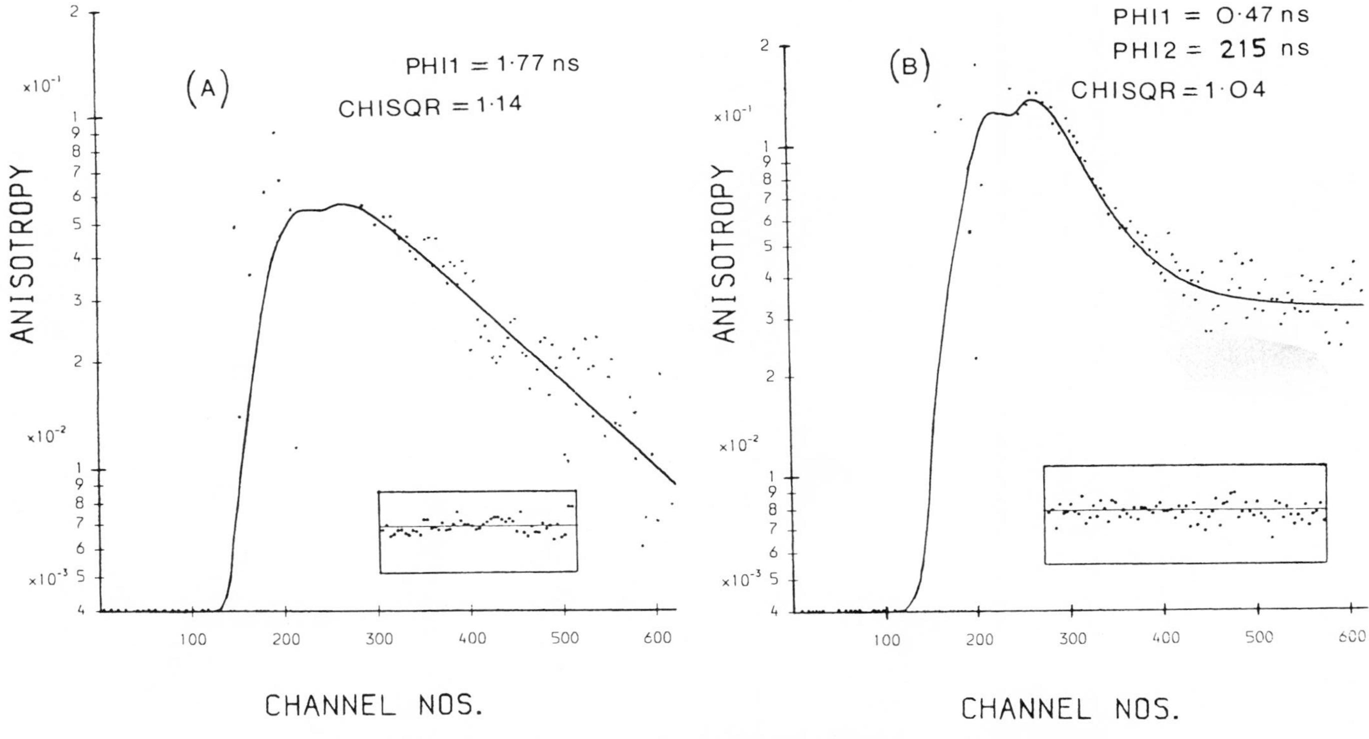

Figure 4.5. Emission anisotropy decays for native (A) and denatured (B) angiotensin II.

extrinsic dye (fluorescamine) attached to lactate dehydrogenase from *Bacillus stearothermophilus* revealed that the rotational correlation time of the enzyme at low concentration ($2\,\mu M$) was 55 ns, while at high concentration ($0.27\,\text{m}M$), or in the presence of the enzyme substrate, fructose-1,6-bisphosphate ($5\,\text{m}M$), the correlation time was increased to 95 ns. During these changes the dye lifetimes remained constant, with half the emitted photons emerging with lifetimes of 5 ns and 11 ns. The two measured correlation times arise from a change in the protein from a dimer (85 ± 12 kDa) to a tetramer (150 ± 22 kDa). Tetramer formation was induced either by an increase in protein concentration or by combination with fructose-1,6-bisphosphate, the metabolic regulator, or effector ligand. In this study the relation of protein assembly state to the expression of biological function has been determined.

4.3. Photosynthesis

Light energy is essential for the process of photosynthesis, whereby electrons produced from the oxidation of water are used to reduce carbon dioxide to carbohydrate. The series of steps by which these electrons are transferred from their source in water to NADPH is commonly represented on a redox scale as the "Z-scheme"[15] shown in Figure 4.6. The photochemical phase of

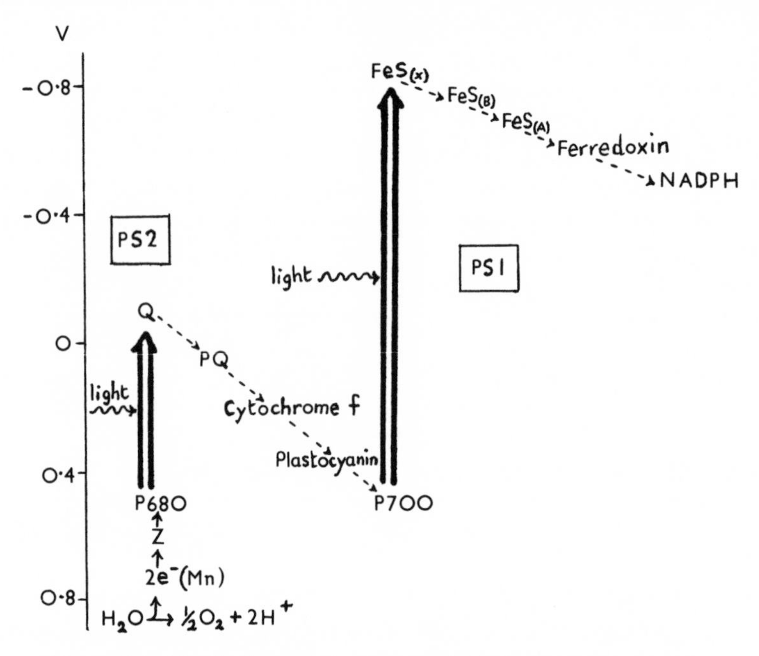

Figure 4.6. The Z-scheme of the photosynthetic electron transport chain.

photosynthesis is localized in two functional assemblies of pigments known as photosystems 1 and 2 (PS1 and PS2). The constituent parts of both of these assemblies are similar in that the light is fed by energy transfer to a reaction center or trap, possibly consisting of a dimeric pair of chlorophyll molecules,[16] by a few hundred light-harvesting pigments (carotenoids, phycobiliproteins, and/or chlorophylls depending on the organism). Once the excitation has arrived at the reaction center, electron transfer can take place from the trap to the primary electron acceptor to result in primary charge separation. This may then be stabilized by the trap becoming re-reduced by electron transfer from the donor side of the photosystem or by the primary electron acceptor passing the electron to the next constituent of the electron transport chain. Alternatively, the primary charge separation may be reversed so that the energy is located one more as an exciton in the reaction center.

These processes involved in energy and primary electron transfer in photosynthetic systems may be routinely probed by fluorescence spectroscopy. However, the complexities of these systems from the point of view of the number of biochemical species present, their organization, and the kinetic relationships between them mean that interpretation of fluorescence signals must be undertaken with caution. Steady-state fluorescence measurements can only yield limited information about the kinetic processes following absorption of a photon, so the technique of time-resolved fluorescence spectroscopy has been applied to this problem to yield valuable information about the rates of the various processes involved.[14]

The initial work on time-resolved fluorescence of photosynthetic material mainly utilized pulsed laser/streak camera systems and has subsequently been found to be complicated by exciton annihilation effects[15–18] due to the high excitation intensities used. The problems can be eliminated by using lower excitation intensities, but streak cameras may not be sensitive enough to detect the correspondingly lower fluorescence intensities. The technique of time-correlated single-photon counting[19] is used to overcome these difficulties, but this generally means poorer time resolution.

Evans and Brown[20, 24] have pioneered the use of synchrotron radiation for these studies. Synchrotron radiation has a longer pulse width than picosecond lasers, although the advantage that lasers have in this respect is diminished when detector response is included in the comparison and when the ease of tunability of synchrotron radiation is taken into account. The high repetition rate of the SRS allows fluorescence decay profiles to be accumulated in less than an hour.

The biggest technical problem with these measurements is the opacity of the samples to be investigated. Whole cyanobacteria cells are about $5 \times 3\ \mu m^2$ in size, and chloroplasts are flattened disks of 5–10-μm diameter. Such preparations strongly scatter light. Even chemical preparations of PS1 and PS2 are suspensions of large supramolecular complexes. None of these samples is anything like optically clean. To separate the photosynthetic systems into optically clean aqueous solutions of their constituent pigment molecules would of course destroy the very energy transfer processes that we are trying to investigate.

Fluorescence decay profiles of various photosynthetic preparations from blue–green algae (cyanobacteria) and higher plants (lettuce) have been measured by the technique of single-photon counting. The pulsed light source can be tuned to an absorption maximum of one of the fluorescent pigments in the photosynthetic system, and the time course of the decay of fluorescence can be observed directly by a fast, red-sensitive photomultiplier. Signals from many exciting pulses can be accumulated in the memory of a multichannel analyzer. This allows the determination of excited state lifetimes of very weakly fluorescent molecules in the complex. By this method, fluorescence decay profiles have been assigned to specific fluorophores, and individual energy transfer species have been resolved from a multicomponent complex assembly of light-harvesting and energy-transferring pigment molecules.

It is well known that the fluorescence decay profiles of photosynthetic organisms and of preparations from them are complex.[17, 20–22] Models based on a sum of three or four exponential components are routinely used to analyze the experimental decay profiles, although it is far from certain whether these are wholly appropriate. What is clear, though, is that the techniques of global analysis,[10] whereby a number of experiments on a sample are analyzed simultaneously using the same decay time parameters, are essential. Also it is proving extremely valuable to undertake fast (picosecond) flash absorption measurements in parallel with the fluorescence spectroscopy.[23, 24] The fluorescence decay profiles of chloroplasts prepared from lettuce leaves, shown in Figure 4.7, have been analyzed with a three-exponential model and found to exhibit lifetimes of approximately 100, 500–600, and 1700–1900 ps.[20] Most of the intensity is invested in the middle lifetime component for chloroplasts, alone or in the presence of an electron acceptor such as potassium ferricyanide, but the addition of the electron donor dichlorophenylmethyl urea (DCMU or diuron) causes a dramatic shift of intensity to the longest component,[20] as shown in Figure 4.7. Here electron transport away from the reaction center has been blocked or at least dramatically reduced. Analysis of similar data by other groups[21, 22] using four exponentials results in the shortest lived component splitting into two (e.g., components of 96 and 254 ps for *Chlorella vulgaris*[21]). The middle, long, and one of the short-lived components are assigned to PS2, and the other short-lived component is assigned to PS1.

Given the complexity of these results from whole organisms and chloroplast preparations, we have attempted to simplify the system under scrutiny by making preparations enriched in one of the two photosystems. It may then be possible to interpret the observed decay profiles on the basis of a physical model.

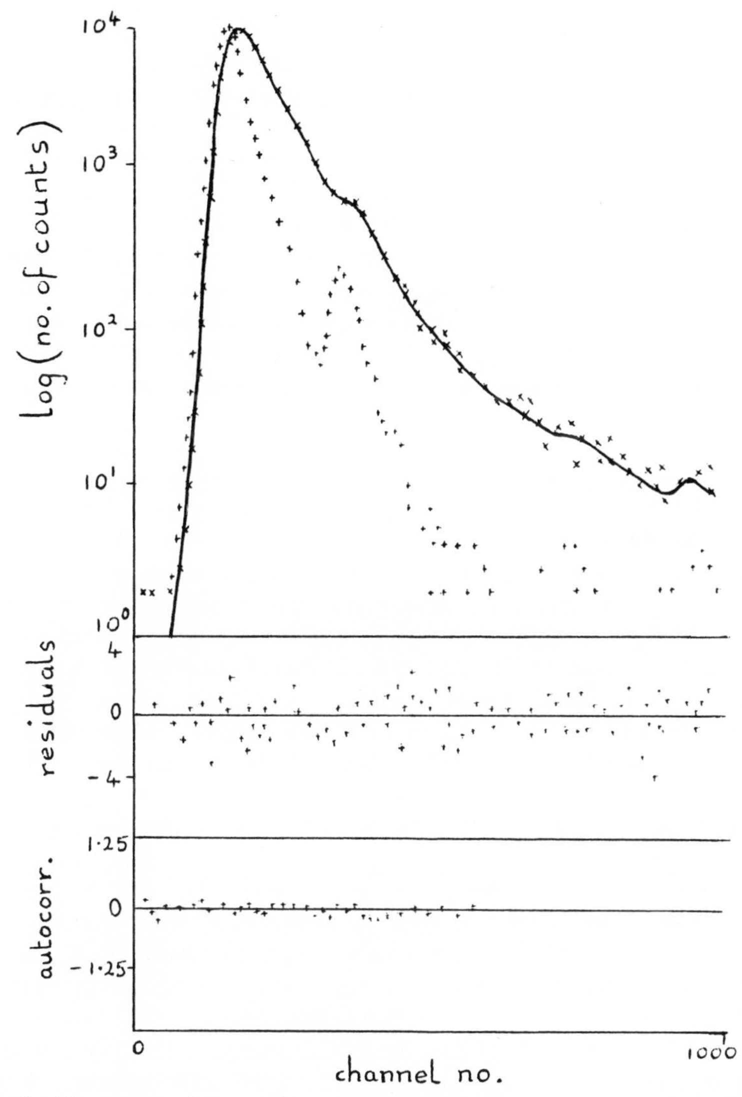

Figure 4.7. Fluorescence decay profiles of lettuce chloroplasts. +, without electron donor; ×, with electron donor.

Fluorescence decay profiles for PS1 from the cyanobacterium *Chlorogloea fritschii* have been measured.[25] The profiles are dominated by a short-lived component of some 20–30 ps, but also exhibit two other components of lifetime 400–650 ps and 3.0–3.6 ns. The short lifetime has been confirmed by picosecond transient absorption measurements on the same sample.[24] The longest lived component is assigned to chlorophyll–protein complexes which have been completely isolated from the rest of the photosynthetic system during the preparation procedure. The other two components represent light-harvesting complexes with different efficiencies of coupling to their reaction center.

A PS2 preparation from lettuce has also been studied and was found to exhibit similar fluorescence characteristics to those of the isolated lettuce chloroplasts. A number of models exist which attempt to explain the primary photochemistry of PS2,[25–27] and we are currently exploring these in the light of our data.

4.4. Membranes

4.4.1. Diphenylhexatriene

The study of phospholipid bilayer membranes by fluorescence methods reveals both dynamic and static properties of these systems.[28] Extrinsic fluorescent probe molecules have been used extensively in time-resolved fluorescence experiments examining the effect on membrane architecture of changes in external conditions such as temperature and pressure and of changes in membrane composition.[29–33] The most common fluorescent probes used in membrane studies are 1,6-diphenyl-1,3,5-hexatriene (DPH) and its analogues.[32, 34] The polarization of DPH fluorescence is sensitive to the microenvironment of the probe, to changes in membrane microfluidity caused by variations in cholesterol content, and to phase changes in the bilayer.[35, 36]

However, doubt has been cast on the predictability of the location of DPH probe molecules incorporated in membranes.[37] In response to this, DPH derivatives which are covalently bound to phospholipid have been synthesized. These derivatives share the overall fluorescence properties of DPH, but the location and orientation of the probes in the bilayer are much better defined.[38, 39] These probes have limited freedom for rotational motion when incorporated into liposomes, yielding site-specific information on the structural dynamics of the membrane.[40] The motion of the probe can be correlated with physical properties of the bilayer, such as acyl chain mobility, gel/liquid-crystalline phase transitions, and disorder due to unsaturation and phase separation.

Fluorescent probes bearing orienting groups which are not close structural analogues of membrane lipids can also be used to reduce uncertainty in probe orientation. The dynamic properties of three homologous carboxyalkyl derivatives of DPH have been studied in host liposomes of dipalmitoyl-phosphatidylcholine (DPPC) using time-resolved anisotropy. Cells are relatively impermeable to these probes, which appear to be localized in the outer leaflet of cell membranes. These molecules share the same chromophore, having identical photophysical properties. Any differences in fluorescence behavior when they are incorporated into lipid dispersions are expected to directly reflect differences of environment and orientational constraints as a consequence of probe location.

Measurements of the fluorescence decay times and anisotropies of the probe/DPPC samples were made above and below the characteristic phase transition temperature of DPPC. A Peltier unit built into the sample compartment was used to control sample temperature. The fluorescence decay profiles of the three homologous carboxyalkyl DPH derivatives incorporated into DPPC liposomes were indistinguishable. This suggests that all three probes assume a similar location in the bilayer. Data could not be adequately fitted by a single-exponential decay law; a double-exponential fit gave reasonable residuals and fitting statistics in each case. Decay times were 3 ns and 6.4 ns at 20 °C (gel phase) and 1.7 ns and 5.9 ns at 50 °C (liquid crystalline). The biexponential form of these decays may be the result of local disordering within the lipid bilayer, induced by the presence of the probe in both liquid-crystalline and gel-phase host lipid environments.

Time-resolved anisotropy measurements also suggest that limited probe mobility is present even in gel-phase lipid environments. In all cases the anisotropy decayed rapidly from a zero-time value with a relaxation time in the nanosecond range. A limiting value is thereafter reached, and this shows a very long relaxation time. Attempts to analyze such data by fitting the long-lifetime component of the anisotropy decay gave poor results. Measurement of long relaxation times cannot be made with any degree of precision when the fluorescent probe has a lifetime of only a few nanoseconds. Another approach is to assume a limiting anisotropy, r_∞, and to fit to the equation

$$r_t = r_\infty + (r_0 - r_\infty)e^{-t/\phi}$$

where ϕ is the rotational correlation time, and r_0 is the zero-time anisotropy value. If the average fluorescence decay time, τ, is known, the initial anisotropy decay time can be obtained from

$$r_s = r_\infty + (r_0 - r_\infty)/[1 + (\tau/\phi)]$$

where r_s is the steady-state anisotropy. This approach gave a relaxation time of approximately 3 ns for the initial anisotropy decay of all three probes incorporated in gel-phase DPPC bilayers. The 3-ns relaxation time probably reflects rapid probe motion within a restricted potential well, while the long component reflects a combination of vesicle tumbling and low-probability motions of the probe within the lipid matrix. Both the fluorescence decay and anisotropy data show that the probes have significant motional freedom even in gel-phase host lipids. Time-resolved anisotropy measurements taken above the gel/liquid-crystalline phase transition temperature show that the major contribution to the motion is a rapid depolarization with a relaxation time of 1.5 ns and a second longer decay in the range 60–90 ns. The time-resolved anisotropy data for the three probes were also very similar.

Another approach to the problem of defining probe location in fluorescence polarization studies of membranes is to prepare aligned phospholipid bilayer films by pressing hydrated lipids between optical flats. The angular distribution of transition dipoles in incorporated DPH molecules can then be measured relative to the bilayer normal, and the average location of the probe molecules can be determined. Time- and angle-resolved depolarization experiments have been carried out on such macroscopically oriented membranes by Levine, van Ginkel, and co-workers at Daresbury.[41–45] This provides a method for testing various theoretical models for reorientational motion in these structurally anisotropic liquid-crystalline materials. Decay curves of the emitted polarized fluorescence were measured using single-photon counting detection for 19 different scattering geometries. A large lighttight sample chamber containing a variable-angle detector mount was constructed for this series of experiments. The first measurements were carried out on DPH and its derivatives embedded in bilayers of various phospholipids. The curves were analyzed using reiterative nonlinear least-squares deconvolution procedures. The decay curves could be fitted for each geometry. Thus, angle-resolved fluorescence depolarization measurements allow separate determination of the average orientation of the probes and, provided the fluorescence decay function is known, of the reorientational dynamics. The results of angle- and time-resolved studies of DPH-labeled, macroscopically oriented phospholipid bilayers, with and without cholesterol, suggest that structural molecular order is independent of reorientational dynamics in lipid bilayers. Cholesterol increases molecular order above the liquid-crystalline transition temperature, but at the same time the rate of reorientational motion increases.

The quantitative analysis of time-resolved anisotropy measurements of DPH incorporated into lipid vesicles was found to yield two distinct but statistically equivalent mathematical solutions, both of which gave values of χ^2 close to unity.[41–46] The two solutions differ with respect to the average orientation (order) of the probe molecules in the bilayer. When cholesterol was incorporated into the vesicles, the mathematically equivalent solutions gave contradictory evidence regarding the effect on membrane fluidity. One predicted a decrease in fluidity due to the presence of cholesterol, consistent with the "wobble-in-cone" model for probe dynamics in a bilayer membrane.[47, 48] The other solution indicated an increase in fluidity above the liquid-crystalline transition temperature in the presence of cholesterol. This contradicts the cholesterol effect postulated in the "wobble-in-cone" model, but is consistent with the results of NMR experiments,[49] and the distribution of the probes is in agreement with that obtained from the angle-resolved anisotropy measurements on oriented bilayers.[42, 43]

The concept of membrane fluidity originally proposed by Chapman and co-workers[50, 51] does not distinguish between structural parameters describing molecular order and motional parameters describing molecular dynamics. Seelig and Seelig[52, 53] were the first to suggest that the membrane fluidity concept should be limited to parameters describing molecular dynamics. It is clear from the analysis of time-resolved fluorescence measurements on DPH-labeled bilayers that the "global" application of the membrane fluidity concept is misleading. Structural molecular order should be considered as distinct from molecular dynamics.

4.4.2. Triazine Dyes

Time-resolved fluorescence anisotropy decay measurements made by Cowley and Pasha have been used to distinguish probe site heterogeneity and to define the rotational modes and rates of triazinylaniline dyes incorporated into phospholipid vesicles.[54, 55] The dyes MTAB and TAGLUC (Figure 4.8), which have a common triazinylaniline chromophore, were found to have very similar photophysical properties in homogeneous solution. MTAB dissolved in *n*-alkanols (C_4, C_6, C_8, and C_{10}) showed a single-exponential fluorescence decay in each case (lifetimes 0.35, 0.7, 0.91, and 1.1 ns, respectively). Anisotropy decays were also good single exponentials, yielding rotational correlation times in the four *n*-alkanols of 0.28, 0.58, 1.04, and 1.44 ns, correlating linearly with solvent viscosity. Thus, this dye probe behaves as a pseudo-isotropic rotor in homogeneous media. Its behavior when it is incorporated into bilayer membranes is expected to be more complex.

The *n*-butyl functionality of MTAB and the hydrophilic glycosyl group of TAGLUC can be expected to impart a differential membrane site preference.

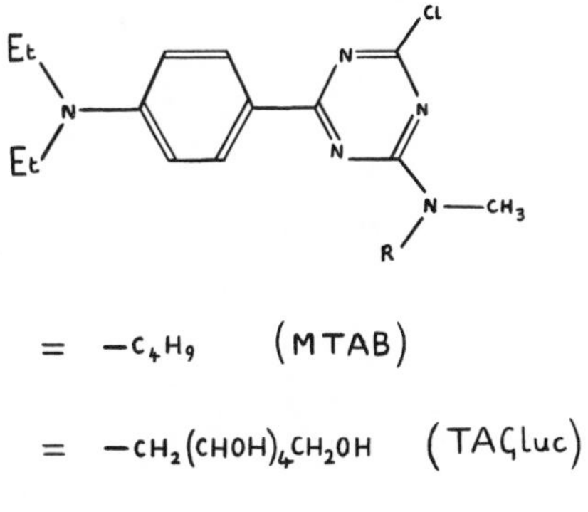

= $-C_4H_9$ (MTAB)

= $-CH_2(CHOH)_4CH_2OH$ (TAGluc)

Figure 4.8. Molecular formulas of the fluorescence probe dyes MTAB and TAGLUC.

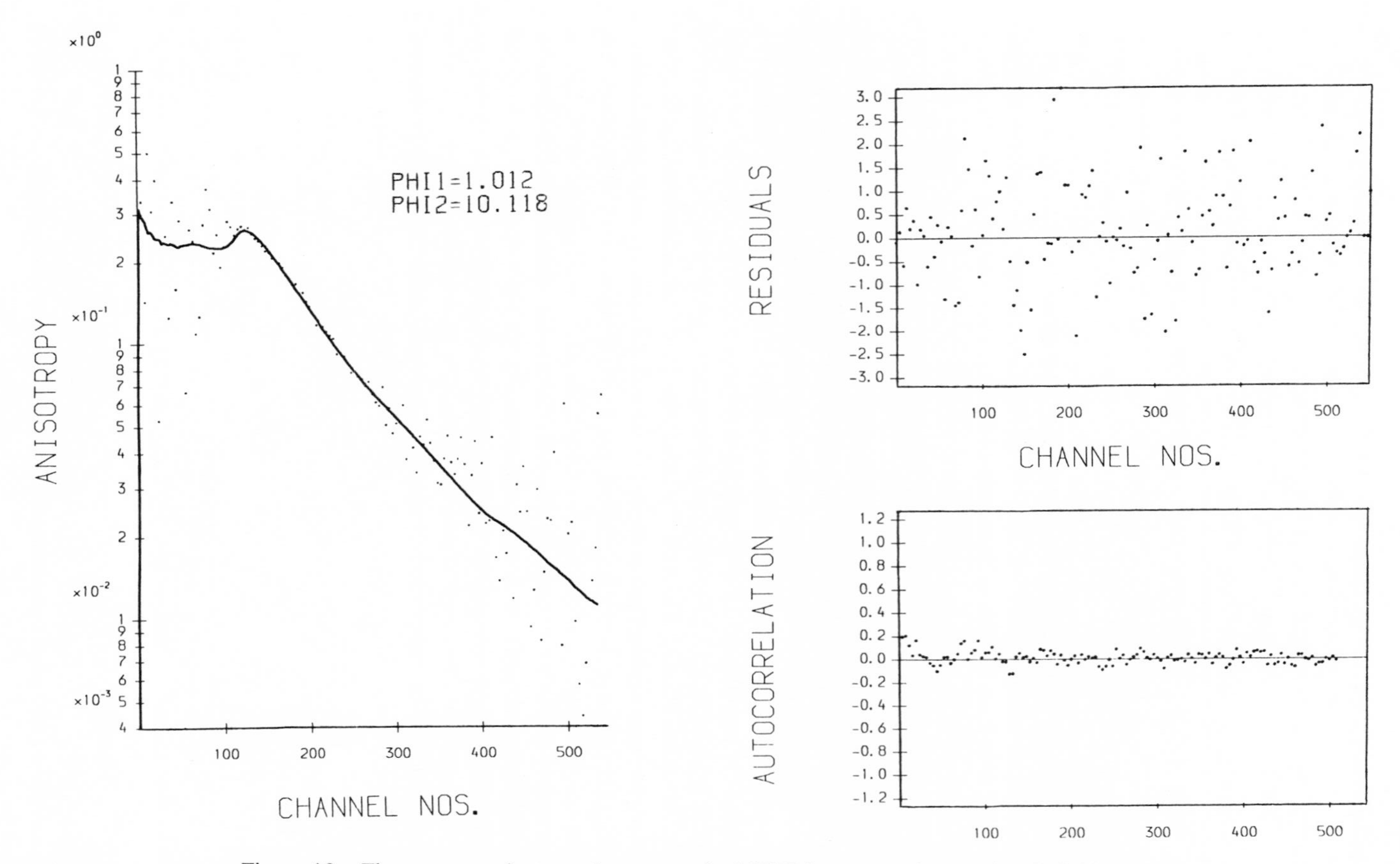

Figure 4.9. Fluorescence anisotropy decay curve for MTAB incorporated onto phospholipid vesicles.

Table 4.3. Temperature Dependence of the Fluorescence Lifetimes of TAGLUC Incorporated into Phospholipid Vesicles

Temperature (°C)	Phase	τ_A (ns)	% A	τ_B (ns)	% B
20.7	Gel	0.65	53	1.8	47
37.5	Transition	0.95	81	2.2	19
40.5	Transition	1.06	95	3.3	5
50.0	Liquid crystalline	0.91	96	3.23	4

MTAB is more lipid soluble, preferring the hydrophobic bilayer interior, whereas TAGLUC prefers the polar, semiaqueous phospholipid head-group environment. The two distinct locations of the probes are indicated by the presence of two fluorescence decay lifetimes. The distribution of probe between the two sites was found to be temperature-dependent, the polar site, A, being highly favored above the liquid-crystalline transition temperature, whereas the probe was almost equally distributed between A and the hydrophobic site, B, in the gel phase (Table 4.3). Here again the experimental results seem to be in conflict with the "wobbling-in-cone" model. Here the inaccessible hydrophobic interior site becomes less accessible as membrane "fluidity" is increased. The excited state lifetimes, τ_A and τ_B, indicate an increase in rotational dynamics above the transition temperature, but the distribution of probe sites indicated is contrary to that predicted by the model. Fluorescence anisotropy decay curves show clearly the existence of two separate rotational correlation times (Figure 4.9). The rapid ($\leqslant 1$ ns) rotation corresponds to that of the probe dye in the polar site, A, while the slow correlation time ($\geqslant 10$ ns) is consistent with the effective microviscosity of the hydrophobic bilayer interior as determined by conventional methods based on isotropic rotation.[56] Introduction of cholesterol into the vesicles also affected the distribution of probe sites (molecular order) without affecting rotational anisotropy (molecular dynamics).

4.5. Excited State Processes

Variations in fluorescence lifetimes yield information about processes that compete with fluorescence such as internal conversion, energy transfer, and quenching.[57] A wide range of chemical, biochemical, and solid-state physics problems can be investigated by using time-resolved fluorescence measurements as a source of information about competing excited state processes. Often photophysics measurements are relatively straightforward, but the results are difficult to interpret. Typically, the intensity of fluorescence is measured as a function of quencher concentration to give a Stern–Volmer plot. The interpretation of such steady-state kinetic measurements is open to abuse. Another method is to use time-resolved measurements to determine the effect of varying the concentration of a quencher, Q, on the excited state fluorescence lifetime.

In theory,

$$\frac{\tau_0}{\tau}=\frac{I_0}{I}=1+K_{sv}[Q]$$

In practice, intensity and lifetime ratios do not always agree. Binding of quencher may affect the steady-state results. Quenching may be affected by diffusion coefficients and by the active diameter of the quenching process. There is a general need to complement steady-state quenching experiments with time-resolved measurements.

4.5.1. Host/Guest Complexes

A range of novel host/guest complexes,[58] involving receptors of the dibenzo-crown-ether type, have been shown to encapsulate the dication of the herbicide diquat, forming 1:1 complexes with high association constants in solution. The complexes are stabilized by hydrogen bonding and electrostatic interactions between the polyether chains and the $CHN^+CH_2CH_2N^+CH$ region of diquat and by charge transfer interactions between π-electron-rich catechol units and the π-electron-deficient bipyridinium ring system. All of these interactions would be expected to compete with diquat fluorescence.

Time-resolved fluorescence quenching experiments have been carried out by Devonshire (personal communication) on diquat encapsulated in crown-ether complexes. The end of the crown-ether molecule is chemically capped so that the complex cannot flop open. This restricts access to the encapsulated diquat by any quencher. Iodide ion and acrylamide quenchers have been used in steady-state and time-resolved quenching studies on these complexes. The results (Figure 4.10) show that quenching is reduced in the complex. Site access is restricted especially in the capped complex. About ten high-quality Stern–Volmer plots could be generated during one 8-h shift of single-bunch beamtime on the SRS, typically measuring about 70 lifetimes, each accumulating 10,000 counts. The Stern–Volmer quenching coefficients were obtained from the gradients of the plots:

$$K_{sv}=\tau_0 k_q$$

where k_q is the second-order rate constant. Iodide ion was found to be the

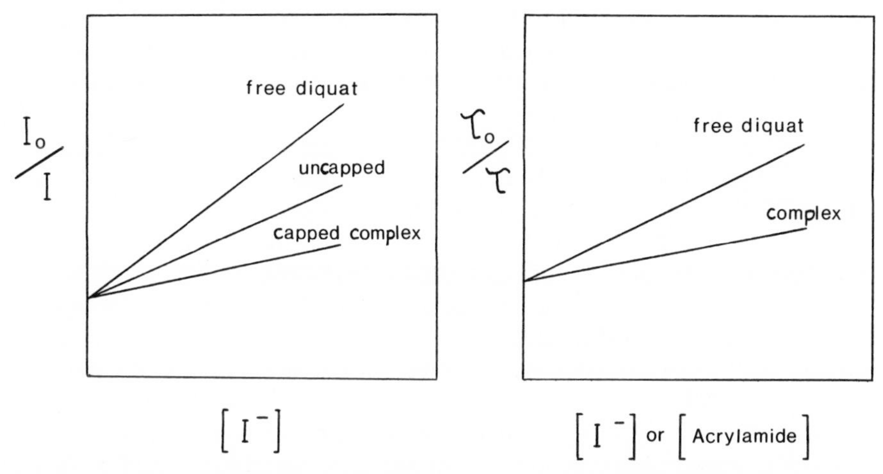

Figure 4.10. Stern–Volmer plots from steady-state and time-resolved quenching experiments for free and encapsulated diquat.

more efficient quencher. Interpretation of the gradients showed very little differences in the second-order rate constants (Table 4.4), implying that the complexed diquat is just as efficiently quenched as the free diquat. Since access to the diquat by iodide or acrylamide is restricted in the complex, then the crown ether itself must be a quencher of excited state diquat. Thus, the experiment was investigating a competitive quenching process.

4.5.2. Stilbene

The temperature dependence of fluorescence decay rates is also a valuable source of information about competing excited state processes in chemical systems. The *cis–trans* photoisomerization of stilbene, which arises from twisting round the central bond, is currently a subject of intense interest, because it is a testing ground for theoretical treatments of molecular motion

Table 4.4. Second-Order Rate Constants (k_q) for Free and Complexed Diquat Determined by Stern–Volmer Quenching Experiments

	k_q (mol^{-1} s^{-1})	
Quencher	Free diquat	Complex
I^-	7.92×10^{10}	8.05×10^{10}
Acrylamide	1.4×10^{10}	1.37×10^{10}

over potential surfaces and internal conversion between them.[59] Stilbene is an optical brightener of commercial importance. It was originally developed for addition to paper, but has since found wide application as a fluorescer in cotton clothing and detergents.

Young *et al.* have used the SRS to study the excited state behavior of 1,3-diphenylallyl anions,[60] the derivatives of which closely resemble those of stilbene. The diphenylallyl carbanion (DPA^-) exists in solution as a mixture of tight and loose ion pairs with Li^+, Na^+, etc., as counterion.[61, 62] The absorption spectra of loose ion pairs and free carbanions are indistinguishable, but tight ion pairs show a marked shift to the blue. Conversion of tight to loose ion pairs is exothermic, so the equilibrium is temperature-dependent. Excited DPA^- is deactivated by three routes: fluorescence, nonradiative decay, and skeletal twisting. This last process is activated and follows an Arrhenius law. Designating the fluorescence lifetime as τ_0 at temperatures low enough to prevent twisting, and as τ at higher temperatures, it can be shown that

$$\tau^{-1} - \tau_0^{-1} = A \exp(-E/RT)$$

The skeletal twisting leads to the existence of two conformers (*trans*, *trans*, 95%, and *cis*, *trans*, 5%).[63] The *cis*, *trans* conformer is less stable because of steric hindrance, but it becomes the more stable form if the allyl 2-position is methyl substituted. Substitution by the bulky *t*-butyl group causes the *cis*, *cis* conformer, which is markedly nonplanar, to become most favored. The conformers can be interconverted by photolysis.[60, 64] The position of the photochemical steady state changes with temperature. Upon cooling, a solution of the tight ion pairs changes color from yellow to red, as conversion to the loose ion pairs takes place. Cooling also increases the intensity of fluorescence. As in the case of stilbene, the main process competing with fluorescence is rotation about one of the central bonds.

Time-resolved fluorescence decay measurements were made in a quantitative study of the kinetics of the processes undergone by the excited states. The fluorescence lifetimes (τ) of these states range from 100 ps or less at room temperature to 5 ns (τ_0) at 77 K. The pulse width from the SRS (200 ps) is ideal for this range. The excellent time resolution available using synchrotron radiation made it possible to detect the rapid decay at the higher temperature.

In the solvents used (ethers such as methyltetrahydrofuran), ion pairing adds an extra dimension to the problem. The loose (solvent-separated) and tight (contact) ion pairs emit at different wavelengths. As Figure 4.11 shows, there is a very striking difference in the excited state behavior of loose and tight ion pairs. Lifetimes range from 4.5 ns at 100 K to 100 ps at ambient temperatures. The Arrhenius plots for the isomerization rate of the *trans*, *trans* diphenylallyl anion into the *cis*, *trans* conformation were obtained by

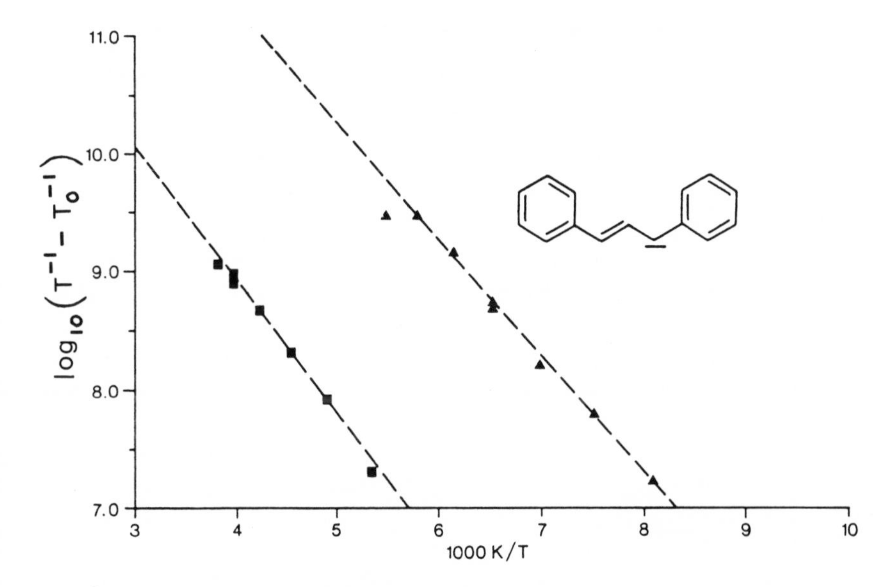

Figure 4.11. Temperature dependence of the twisting process in excited *trans*, *trans* diphenylallyl anions. ■, Loose ion pairs; ▲, tight ion pairs.

subtracting the limiting low-temperature fluorescence rate from the rates at higher temperatures. A striking effect of ion pairing was observed.

In addition to the parent compound and its conformers, related compounds have been studied, showing a great variety of fluorescence behavior. Some show several emissions. Time-resolved excitation spectra are very useful in characterizing the different species.[65] The 2-methyl derivative exists mainly in the *cis*, *trans* form in the ground state, but most of the fluorescence comes from the *trans*, *trans* form in solution. No additional equipment is needed to measure time-resolved excitation spectra, provided that the time-to-pulse height converter (TPHC) has a single-channel analyzer which can select signals falling in a defined time range. Figure 4.12[65] shows the results obtained for the diphenylallyl system by scanning over the excitation wavelengths. The tunability of synchrotron radiation over a wide range is invaluable here. The *cis*, *trans* carbanion has a shorter lifetime than the *trans*, *trans*. Here steric hindrance reduces the barrier to rotation, as for stilbene. This technique also makes it possible to distinguish between very short-lived emission and scattered light transmitted by the cutoff filter. Time-resolved emission spectra can be used similarly to distinguish emitters and also to study relaxation and conversion processes in excited molecules.

There is considerable interest in the general area of *cis*, *trans* photoisomerization. It plays a central role in the visual process (rhodopsin) and is

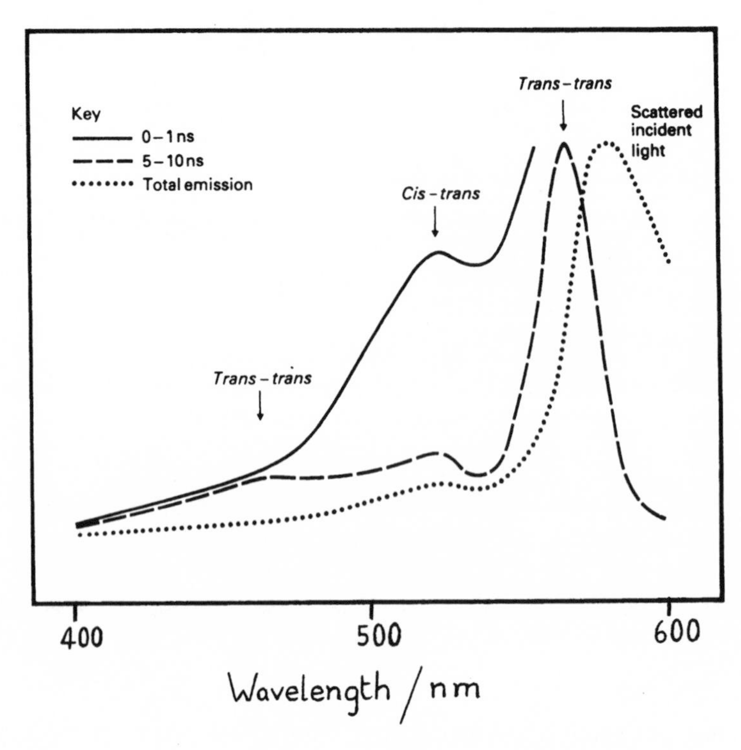

Figure 4.12. Time-resolved excitation spectra of 2-methyldiphenylallyl anions.

of practical importance to the choice of dyes used in mode-locked lasers. Diphenylallyl ion pairs serve as models for the propagating centers in anionic polymerization.

4.5.3. The Vacancy in Diamond

The work of Davies *et al.* on diamonds[66] highlights the variety of work on the high-aperture port of the SRS. The luminescence is associated with a vacant site produced by high-energy radiation. The lifetime shows a strong dependence on temperature (400 ps at 300 K, 2.55 ns at 4 K) which reflects the energy of activation of a process competing with luminescence whose nature is as yet unknown.

Diamond is a wide-gap semiconductor. Almost all the atomic-sized point defects which can be produced in diamond give absorption or fluorescence in the visible or near-UV part of the spectrum. Diamond is the only group IV or III-V semiconductor for which electronic optical transitions have been identified at the vacancy. This vacancy is created by irradiating the diamond

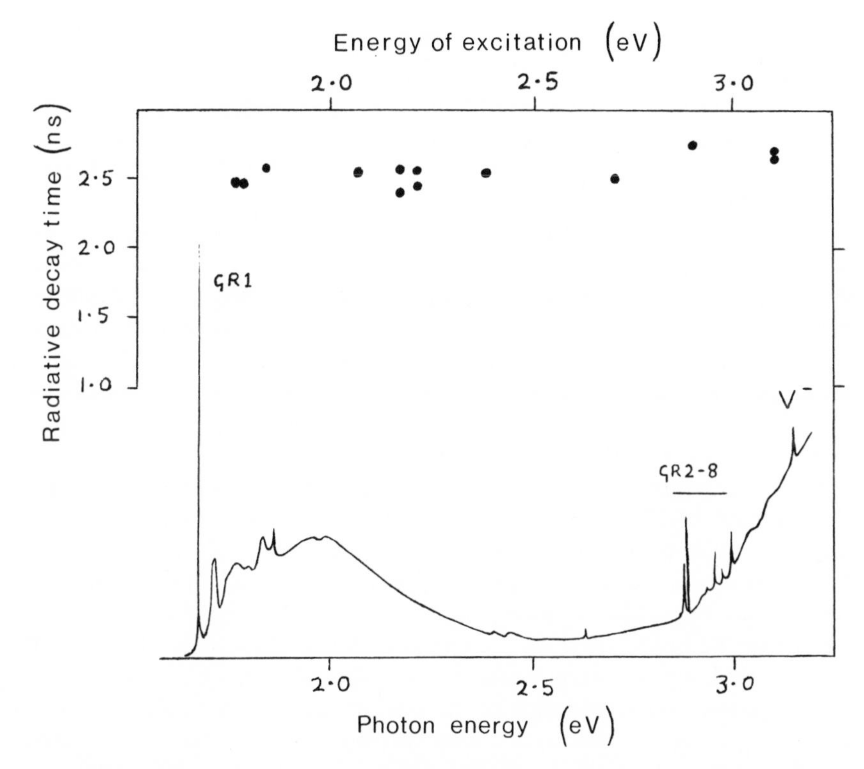

Figure 4.13. Absorption spectrum of an electron-irradiated diamond. The radiative decay time of GRl luminescence (●) as a function of excitation energy is shown at the top of the figure.

with 2-MeV electrons, which knocks out some carbon atoms from their tetrahedral sites. Two charge states of the vacancy are observed, the neutral state, V^0, in which there are four electrons, one donated by each carbon atom adjacent to the vacancy, and the negative state, V^-, containing one more electron. The neutral vacancy, V^0, produces optical absorption in a vibronic band near 2 eV (the GRI band) and in a series of transitions above 2.88 eV (the GR2–8 bands).

The tunability of synchrotron radiation has been used to photoexcite a relatively pure, electron-irradiated, natural diamond into specific excited states. Excitation at 1.673 eV excited the diamond directly into the GR1 absorption band, so that only true GR1 fluorescence was created, not cathodoluminescence. Excitation at photon energies above 2.88 eV generated the many higher states.

The pulsed properties of the light source were used to measure the decay time of this fundamental center.[66] The lowest energy transition, GR1, was found to have a radiative decay time of 1.4 ± 0.2 ns at room temperature. For all the data collected the radiative decay could be described by a single-exponential process.

The wavelength and temperature dependence were found to be consistent with the rate of production of neutral vacancies during electron irradiation of the diamond. Figure 4.13 shows the decay time recorded at 15 K as a function of excitation wavelength.

There is no significant difference in the GR1 decay time measured with either direct excitation in the GR1 absorption band or indirect excitation via the GR2–8 bands. This shows that the deexcitation process from the GR2–8 excited states to the GR1 level occurs more rapidly than the radiative decay, consistent with a fast phonon emission process. Figure 4.14 shows the temperature dependence of the GR1 decay time, which decreases from 2.55 ± 0.1 ns in the limit of low T, through the room temperature value of 1.4 ± 0.2 ns, to 0.4 ns at 200 °C.

The decay time of the GR1 luminescence is short compared with that for other centers in diamond. Typical radiative lifetimes of vacancy-nitrogen aggregates range from 13 ns to 150 ns. The small value of the GR1 decay time and its decrease with increasing temperature suggest that nonradiative processes are important. The nonradiative channel does not involve emission of electrons or holes, since no photoconduction is associated with the GR1 band. Presumably an internal deexcitation takes place from the GR1 excited state to another electronic state of the neutral vacancy. A single-vibrational-coordinate model[67] has been used to describe this assumed internal conversion process. In this model the nonradiative transition rate, $W(T)$, at temperature T, is related to the true radiative lifetime τ_r and the measured decay time τ:

$$\tau(T) = [(1/\tau_r) + W(T)]^{-1}$$

where $\tau_r = 182$ ns.

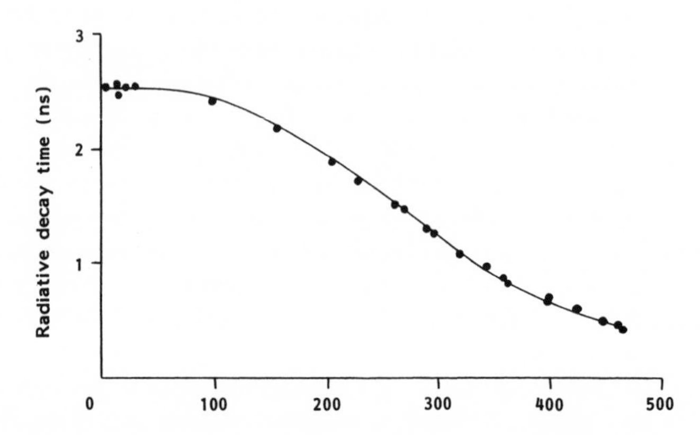

Figure 4.14. Temperature dependence of the decay time of GRl luminescence.

The calculated values of W at different temperatures imply that the GR1 band has a fluorescence quantum efficiency of 1.4% in the limit of low temperature. Detailed balance arguments have been used to convert the measured GR1 absorption into a concentration of vacancies, $[V^0]$. The low quantum efficiency of the GR1 center is in agreement with the strength of vacancy-induced absorption in electron-irradiated diamond. The results show that the displacement energy for an atom in diamond is close to 80 eV.

There are two technical problems associated with these measurements. The diamond sample was very small ($\simeq 1\ mm^3$), and it had to be mounted inside a helium flow cryostat for the low-temperature work. These two factors combined to make alignment of the sample awkward. Lenses were employed to focus the excitation beam onto the sample and to collect and focus as much of the fluorescent light as possible from the small aperture of the exit arm of the cryostat onto the photocathode of the detector. Once aligned using clearly visible green light, the sample and cryostat had to be rigidly mounted so that measurements could be made at UV wavelengths. Some slight adjustment of the sample was then necessary to maximize the count rate. The brightness of the new generation of synchrotron radiation sources allows a greater part of the power output to be channeled into very small samples, without recourse to focusing optics which may have low transmittance. Conversely, the smaller spot size may allow scanning measurements, giving spatial resolution in samples of larger surface area. This improved brightness of recently designed or upgraded synchrotron radiation sources, such as the proposed European Synchrotron Radiation Facility (ESRF), the National Synchrotron Light Source (NSLS) at Brookhaven, or the High Brightness Lattice/SRS at Daresbury, will extend the range of applications of synchrotron radiation.

References

1. P. J. Duke, *Chemistry in Britain 22*, 844–849 (1986).
2. A. J. W. G. Visser and J. Lee, *Biochemistry 19*, 4366–4372 (1980).
3. J. Lee, L. A. Carreira, R. Gast, R. M. Irwin, P. Koka, E. D. Small, and A. J. W. G. Visser, in: *Bioluminescence and Chemiluminescence* (M. DeLuca and W. D. McElroy, eds.), pp. 103–112, Academic Press, New York (1981).
4. A. J. W. G. Visser and J. Lee, *Biochemistry 21*, 2218–2226 (1982).
5. A. J. W. G. Visser, T. Ykema, A. van Hoek, D. J. O'Kane, and J. Lee, *Biochemistry* (Washington) *19*, 4366–4372.
6. A. J. W. G. Visser and A. van Hoek, *J. Biochem. Biophys. Methods 1*, 195–208 (1979).
7. A. J. W. G. Visser and A. van Hoek, *Photochem. Photobiol. 33*, 35–40 (1981).
8. A. van Hoek and A. J. W. G. Visser, *Rev. Sci. Instrum. 52*, 1199–1205 (1981).
9. A. van Hoek, J. Vervoort, and A. J. W. G. Visser, *J. Biochem. Biophys. Methods 7*, 243–254 (1983).
10. J. M. Beechem, M. Ameloot, and L. Brand, *Chem. Phys. Lett. 120*, 466–472 (1985).
11. D. M. Rayner, D. T. Krajcaski, and A. G. Szabo, *Can. J. Chem. 56*, 1238–1245 (1978).
12. W. R. Laws, J. B. A. Ross, H. R. Wyssbrod, J. M. Beechem, L. Brand, and J. C. Sutherland, *Biochemistry 25*, 599–607 (1986).

13. R. B. Smeby and S. Fermandjian, in: *Chemistry and Biochemistry of Amino Acids, Peptides and Proteins*, Vol. 5, (B. Weinstein, ed.), pp. 118–162, Books Demand UMI, Ann Arbor (1978).
14. A. R. Clarke, A. D. B. Waldman, J. J. Holbrook, and I. H. Munro, *Biochim. Biophys. Acta 828*, 375–379 (1985).
15. S. M. Danks, E. H. Evans, and P. A. Whittaker, *Photosynthetic Systems: Stucture and Assembly*, John Wiley & Sons, New York (1983).
16. I. Govindjee, *Photosynthesis: Energy Conversion by Plants and Bacteria*, Vol. 1, Academic Press, New York (1982).
17. A. R. Holzwarth, *Photochem. Photobiol. 43*, 707–725 (1986).
18. F. Pellegrino and R. R. Alfano, *Biological Events Probed by Ultrafast Laser Spectroscopy*, pp. 27–53, Academic Press, New York (1982).
19. D. V. O'Connor and D. Phillips, *Time Correlated Single Photon Counting*, Academic Press, New York (1984).
20. R. Sparrow, R. G. Brown, E. H. Evans, and D. Shaw, *J. Chem. Soc., Faraday Trans. 2*, 82, 2249–2262 (1986).
21. A. R. Holzworth, J. Wendler, and W. Haehnel, *Biochim. Biophys. Acta 807*, 155–167 (1985).
22. M. Hodges and I. Moya, *Biochim. Biophys. Acta 849*, 193–202 (1986).
23. G. H. Schatz, H. Brock, and A. R. Holzworth, *Proc. Natl. Acad. Sci. U.S.A. 84*, 8414–8418 (1987).
24. E. H. Evans, R. Sparrow, R. G. Brown, J. Barr, M. Smith, and W. Toner, Prog. Photosynthesis Research, Proc. 7th International Congress on Photosynthesis, Rhode Island (1986).
25. R. Sparrow, R. G. Brown, E. H. Evans, and D. Shaw, submitted for publication.
26. G. H. Schatz and A. R. Holzworth, *Photosynth. Res. 10*, 309–318 (1896).
27. S. J. Berens, J. Scheele, W. L. Butler, and D. Magde, *Photochem. Photobiol. 42*, 59–68 (1985).
28. M. Shinitzky and Y. Barenholz, *Biochim. Biophys. Acta 515*, 367–394 (1978).
29. J. Yguerabide and L. Stryer, *Proc. Nat. Acad. Sci. U.S.A. 68*, 1217–1221 (1971).
30. G. K. Radda and J. M. Vanderkooi, *Biochim. Biophys. Acta 265*, 509–549 (1972).
31. M. P. Andrich and J. M. Vanderkooi, *Biochemistry 15*, 1257–1261 (1976).
32. S. Kawata, K. Kinosita, Jr., and A. Ikegami, *Biochemistry 16*, 2319–2324 (1977).
33. L. A. Chen, R. E. Dale, S. Roth, and L. Brand, *J. Biol. Chem. 252*, 2163–2169 (1977).
34. R. E. Dale, L. A. Chen, and L. Brand, *J. Biol. Chem. 252*, 7500–7510 (1977).
35. R. P. H. Kooyman, Y. K. Levine, and B. W. van der Meer, *Chem. Phys. 50*, 317–326 (1981).
36. B. W. van der Meer, R. P. H. Kooyman, and Y. K. Levine, *Chem. Phys. 66*, 39–50 (1982).
37. S. M. Johnson and R. Robinson, *Biochim. Biophys. Acta 558*, 282–295 (1979).
38. C. G. Morgan, E. W. Thomas, T. S. Moras, and Y. P. Yianni, *Biochim. Biophys. Acta 692*, 196–201 (1982).
39. C. D. Stubbs, J. Kinosita, F. Munkonge, P. J. Quinn, and A. Ikegami, *Biochim. Biophys. Acta 775*, 374–380 (1984).
40. C. G. Morgan, E. W. Thomas, and Y. P. Yianni, *Biochim. Biophys. Acta 728*, 356–362 (1983).
41. H. van Langen, D. Engelen, G. van Ginkel, and Y. K. Levine, *Chem. Phys. Lett. 138*, 99–104 (1987).
42. F. Mulders, H. van Langen, G. van Ginkel, and Y. K. Levine, *Biochim. Biophys. Acta 859*, 209–218 (1986).
43. G. van Ginkel, L. J. Korstanje, H. van Langen, and Y. K. Levine, *Faraday Discuss. Chem. Soc. 81*, 49–61 (1986).
44. G. Deinum, H. van Langen, G. van Ginkel, and Y. K. Levine, *Biochemistry 27*, 852–860 (1988).
45. H. van Langen, Y. K. Levine, M. Ameloot, and H. Pottel, *Chem. Phys. Lett. 140*, 394–400 (1987).

46. H. van Langen, G. van Ginkel, and Y. K. Levine, *Liq. Cryn. 3*, 1301–1306 (1988).
47. K. Kinosita, Jr., S. Kawato, and A. Ikegami, *Adv. Biophys. 17*, 147–150 (1984).
48. A. Szabo, *J. Chem. Phys. 81*, 150–167 (1984).
49. R. Ghosh and J. Seelig, *Biochim. Biophys. Acta 691*, 151–160 (1982).
50. D. Chapman, P. Byrne, and G. C. Shipley, *Proc. Roy. Soc. London Ser. A 290*, 115–142 (1966).
51. D. Chapman, *Membrane Fluidity in Biology*, Vol. 2, pp. 5–42, Academic Press, New York (1983).
52. A. Seelig and J. Seelig, *Biochemistry 16*, 45–50 (1977).
53. J. Seelig and A. Seelig, *Quart. Rev. Biophys. 13*, 19–61 (1980).
54. D. J. Cowley and I. Pasha, *J. Chem. Soc., Perkin Trans. 2, 1983*, 1139–1145.
55. D. J. Cowley, *J. Chem. Soc., Perkin. Trans. 2, 1984*, 281–285.
56. K. Hildenbrand and C. Nicolau, *Biochim. Biophys. Acta 558*, 365–368 (1979).
57. R. B. Cundall and A. Gilbert, *Photochemistry*, Thomas Nelson & Sons, London (1970).
58. B. L. Allwood, F. H. Kohnke, A. M. Z. Slawin, J. F. Stoddart, and D. J. Williams, *J. Chem. Soc., Chem. Commun. 1985*, 311.
59. M. Lee, A. J. Bain, P. J. McCarthy, C. H. Han, J. N. Haseltine, A. B. Smith, and R. M. Hochstrasser, *J. Chem. Phys. 85*, 4341–4347 (1986).
60. R. N. Young, H. M. Parkes, and B. Brocklehurst, *Makromol. Chem. Rapid Commun. 1*, 65 (1980).
61. J. W. Burley and R. N. Young, *J. Chem. Soc. Perkin Trans. 2, 1972*, 835.
62. G. C. Greenacre and R. N. Young, *J. Chem. Soc. Perkin Trans. 2, 1975*, 1661.
63. J. W. Burley and R. N. Young, *J. Chem. Soc. Perkin Trans. 2, 1972*, 1843.
64. H. M. Parkes and R. N. Young, *J. Chem. Soc. Perkin Trans. 2, 1978*, 259.
65. B. Brocklehurst, *Chemistry in Britain 23*, 853–856 (1987).
66. G. Davies, M. F. Thomaz, M. H. Nazare, M. M. Martin, and D. Shaw, *J. Phys. C: Solid State Phys. 20*, L13–L17 (1987).
67. K. F. Freed and J. Jortner, *J. Chem. Phys. 52*, 6272–6291 (1970).

5

Frequency-Domain Fluorescence Spectroscopy

Joseph R. Lakowicz and Ignacy Gryczynski

5.1. Introduction

It is often of interest to resolve the time-dependent emission of samples. This is because the fluorescence intensity or anisotropy decays are usually characteristic of the sample or the macromolecule. For instance, a protein may contain two tryptophan residues, each in a different local environment. The presence of two tryptophan residues in a protein may result in two decay times, as has been found in the classic reports of Ross *et al.* on liver alcohol dehydrogenase.[1] Alternatively, the dynamic behavior of proteins can result in multiexponential anisotropy decays, where the rapid components reflect segmental motions of the amino acid residues.[2] There are many other molecular origins for the complex emission kinetics displayed by fluorescent samples, including distributions of conformations,[3] energy transfer,[4] and transient effects in collisional quenching[5, 6] to name a few.

Time-resolved fluorescence data are most commonly obtained by measuring the time-dependent emission which follows excitation with a brief pulse of light. Such pulses can be obtained using flashlamps, lasers, or synchrotrons. The emission can be quantified using time-correlated single-photon counting, streak cameras, or transient absorption. In this chapter we describe an alternative method, in which the time-resolved emission is determined from the frequency response of the emission to intensity-modulated light. This method has its origins in the pioneering reports of Gaviola,[7] Birks and Dyson,[8] Bailey and Rollefson,[9] Bonch-Breuvich *et al.*,[10] and Spencer and Weber.[11] While these instruments were technically simple and robust, the phase-modulation or frequency-domain (FD) method did not become generally useful until 1983. Prior to this time, the FD instruments operated

Joseph R. Lakowicz and Ignacy Gryczynski • Department of Biological Chemistry, Center for Fluorescence Spectroscopy, University of Maryland School of Medicine, Baltimore, Maryland 21201.

Topics in Fluorescence Spectroscopy, Volume 1: Techniques, edited by Joseph R. Lakowicz. Plenum Press, New York, 1991.

only at one to three fixed frequencies, and these limited data were not generally adequate to resolve the complex intensity and anisotropy decays which are routinely found for biological macromolecules.

The first useful frequency-domain fluorometers were described in 1983 by Gratton and Limkeman[12] and shortly thereafter by this laboratory.[13] These instruments allowed phase and modulation measurements from 1 to 200 MHz, and this design is the basis of commercially available instruments. These frequency-domain fluorometers are known to provide good resolution of multiexponential intensity[14, 15] and anisotropy decays,[16–19] as well as the ability to resolve nonexponential decays resulting from lifetime distributions,[20–22] conformational distributions,[4, 23, 24] and time-dependent rate constants.[5, 6]

In this chapter we describe the instrumentation for FD measurements and the theory used to interpret the experimental data. Additionally, we provide specific examples which illustrate the many possible applications of the FD method for resolution of complex decay kinetics. Finally, we will describe the present state-of-the art instrumentation, which allows FD data to be obtained to 10 GHz, depending upon the photodetector and associated electronics.[25, 26]

5.2. Comparison of Time- and Frequency-Domain Fluorometry

5.2.1. Intensity Decays

The objective of both the time- and the frequency-domain measurements is to recover the parameters describing the time-dependent decay. Suppose the sample is excited with a δ-function pulse of light. The resulting time-dependent emission is called the impulse response function. It is common to represent decays of varying complexity in terms of the multiexponential model,

$$I(t) = \sum_i \alpha_i e^{-t/\tau_i} \tag{5.1}$$

where the α_i are the preexponential factors, and the τ_i are the decay times.

The fraction of the intensity observed in the usual steady-state measurement due to each component is

$$f_i = \frac{\alpha_i \tau_i}{\sum_j \alpha_j \tau_j} \tag{5.2}$$

Depending upon the actual form of the intensity decay, the values of α_i and

τ_i may have direct or indirect molecular significance. For instance, for a mixture of two fluorophores, each of which displays a single decay time, the τ_i are the two individual decay times, and the f_i are the fractional contributions of each fluorophore to the total emission. If both fluorophores have the same intrinsic rate constant for emission, as might be true for a single fluorophore in two different environments, then the α_i values represent the fraction of the molecules in each environment. However, in many circumstances there is no obvious linkage between the α_i and τ_i values and the molecular heterogeneity of the samples. Nonetheless, the multiexponential model appears to be very powerful and able to account for almost any decay law. Hence, if the decay is nonexponential, or contains more exponential components than can be resolved by the data, then the apparent values of the f_i and τ_i are due to weighted averaging of the components.

If the emission decays with a single decay time (Figure 5.1, solid lines), it is rather easy to measure the decay time with good accuracy. If the single decay time is long relative to the excitation pulse, then log $I(t)$ decays linearly versus time, and the decay time is easily obtained from the slope. The more difficult task is recovery of multiple decay times, which is illustrated for two widely spaced decay times in Figure 5.1 (0.2 and 1.0 ns, dashed line). In this case, log $I(t)$ does not decay linearly with time. Unfortunately, decay times of the emission from macromolecules are often more closely spaced than the fivefold differences shown in Figure 5.1, and resolution of the decay times

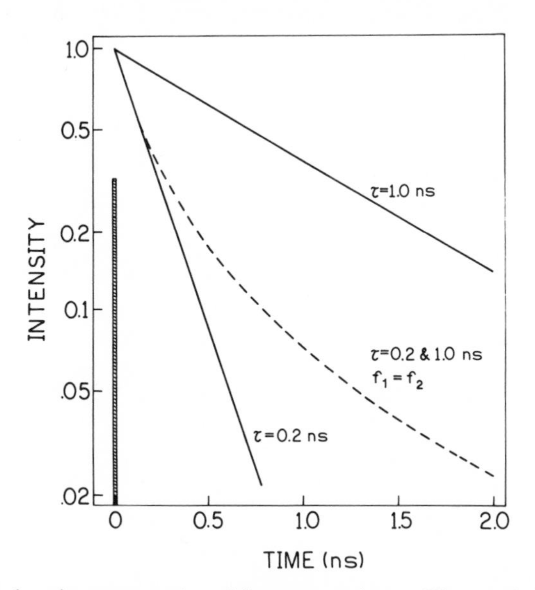

Figure 5.1. Time-domain representation of fluorescence decays. The sample is excited with an instantaneous pulse of light. The solid lines show the intensity decays for single decay times of 0.2 and 1.0 ns. The dashed line shows the decay for $\tau_1 = 0.2$ ns, $\tau_2 = 1.0$ ns, and $f_1 = f_2 = 0.5$.

becomes increasingly difficult as the decay times become more closely spaced. If the decay times are spaced by 20% (e.g., 1 and 1.2 ns or 5 and 6 ns), it is difficult to visually distinguish a single-exponential decay from a double-exponential decay. Furthermore, it is generally difficult to resolve sums of exponentials because the parameters describing the decay are highly correlated. Hence, one requires a high signal-to-noise ratio, or equivalently a large number of photons, to recover the multiple decay times with reasonable confidence. It should be noted that the pulse width of flashlamps is typically near 2 ns, which results in the need for extensive deconvolution of the data from overlap with the excitation pulse. Even if one uses a picosecond laser as the excitation source, the observed instrument response function is often about 0.1 to 0.2 ns in width, due to transit time spread in the photomultiplier tube (PMT) and the timing jitter of the electronics. Hence, it has been rather difficult to obtain adequate resolution of multiexponential decays and hence difficult to resolve the emission from mixtures of fluorophores. In this chapter we show how a complex emission can be resolved from the frequency response of the emission to a modulated light source.

The excitation source and measured values are different for frequency-domain measurements of intensity decays. The pulsed source is replaced with an intensity-modulated light source. Because of the time lag between absorption and emission, the emission is delayed in time relative to the modulated excitation (Figure 5.2). At each modulation frequency this delay is described as the phase shift (ϕ_ω), which increases from 0° to 90° with increasing

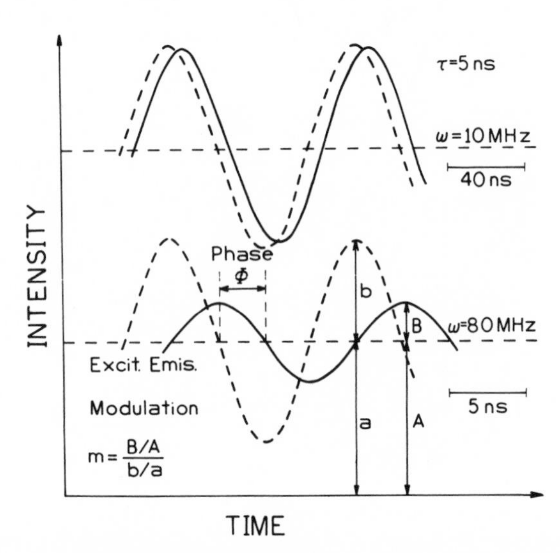

Figure 5.2. Phase and modulation of fluorescence in response to intensity-modulated excitation. The assumed lifetime is 5 ns, and the modulation frequency is 10 (top) and 80 MHz (bottom).

modulation frequency. The finite time response of the sample also results in demodulation of the emission by a factor m_ω, which decreases from 1.0 to 0.0 with increasing modulation frequency. The phase angle and modulation, measured over a wide range of frequencies, constitute the frequency response of the emission.

The characteristic features of the frequency response of a sample are illustrated in Figure 5.3. The shape of the frequency response is determined by the number of decay times displayed by the sample. If the decay is a single exponential (Figure 5.3, top), the frequency response is simple. For a single-exponential decay, the phase and modulation are related to the decay time (τ) by

$$\tan \phi_\omega = \omega\tau \tag{5.3}$$

and

$$m_\omega = (1 + \omega^2\tau^2)^{-1/2} \tag{5.4}$$

At higher modulation frequencies, the phase angles of the emission increase and the modulation decreases. This is shown in Figure 5.3 for decay times of 1 ns and 0.2 ns. At 200 MHz for the 1-ns decay time, the phase shift is 51.5°, and the emission is demodulated by a factor of 0.62 relative to the excitation. At a modulation frequency of 1 GHz the phase angle increases to 81°, and the

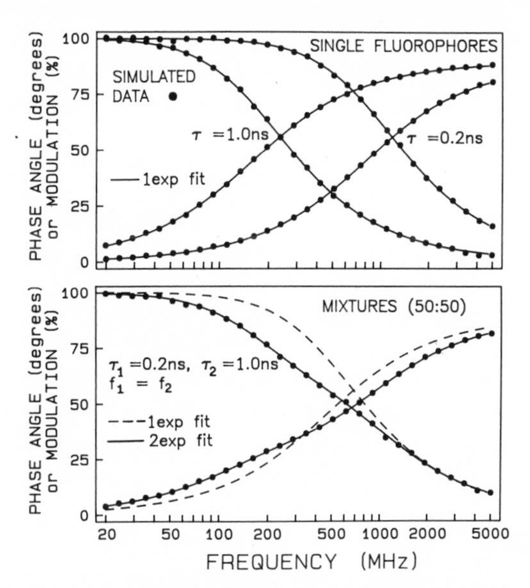

Figure 5.3. Simulated frequency-domain data for single- (top) and double- (bottom) exponential decays. The phase angles increase and the modulation decreases with increasing modulation frequency. The data points indicate the simulated data. *Top*: The solid lines show the best fits to a single decay time. *Bottom*: The dashed and solid lines show the best single- and double-exponential fits, respectively.

modulation decreases to 0.16. Most samples of interest display more than one decay time. In this case the lifetimes calculated from the value of ϕ_ω or m_ω measured at a particular frequency are only apparent values and are the result of a complex weighting of various components in the emission. For such samples it is necessary to measure the phase and modulation values over the widest possible range of modulation frequencies, with the center frequency being comparable to the reciprocal of the mean decay time of the emission.

Simulated data for a multiexponential intensity decay are also shown in Figure 5.3. In this case the assumed decay times are 0.2 or 1.0 ns. The objective is to recover the multiple decay times from the experimentally measured frequency response. This is generally accomplished using nonlinear least-squares procedures.(27) The fitting procedure is illustrated by the solid and dashed lines in Figure 5.3. For the single-exponential decays shown in the top half of the figure, it is possible to obtain a good match between the data and the curves calculated using the single-exponential model. For a double-exponential decay, as shown in the bottom half of the figure, the data cannot be matched using a single-decay time fit, represented by the dashed lines. However, the complex frequency response is accounted for by the double-exponential model, represented by the solid lines, with the expected decay times (0.2 and 1 ns) and fractional intensities ($f_1 = f_2 = 0.5$) being recovered from the least-squares analysis. At present, the frequency-domain method is being widely used in the recovery of multiexponential and nonexponential decay parameters.(14–16, 20–24)

5.2.2. Anisotropy Decays

It is also of interest to recover the fluorescence anisotropy decays for biological molecules. This is because the form of the anisotropy decay is determined by the rotational rate of the molecule about each molecular axis and the orientation of the absorption and emission transitions relative to these axes.(28, 29) Hence, resolution of the anisotropy decay can potentially reveal the size, shape, and segmental mobility of the molecule under investigation. Time-resolved anisotropy decays have been used to detect the multiple rotational rates of asymmetric molecules,(30) the hindered motions of probes in membranes,(31) and the domain motions of immunoglobulins,(32) to name a few examples.

Suppose the sample is excited with a δ-function pulse of vertically polarized light. The decays of the parallel ($\|$) and perpendicular ($\perp$) components of the emission are given by

$$I_{\|}(t) = \tfrac{1}{3} I(t)[1 + 2r(t)] \tag{5.5}$$

$$I_{\perp}(t) = \tfrac{1}{3} I(t)[1 - r(t)] \tag{5.6}$$

where $r(t)$ is the time-resolved anisotropy. Generally, $r(t)$ can be described as a multiexponential decay:

$$r(t) = r_0 \sum_j g_j e^{-t/\theta_j} \tag{5.7}$$

where r_0 is the limiting anisotropy in the absence of rotational diffusion, the θ_j are the individual correlation times, and the g_j are the associated fractional amplitudes ($\sum g_i = 1.0$). The total (rotation-free) intensity decay is given by

$$I(t) = I_{\parallel}(t) + 2I_{\perp}(t) \tag{5.8}$$

In the time domain one measures the time-dependent decays of the polarized components of the emission (Eqs. 5.5 and 5.6). These decays are used to determine the anisotropy decay law which is most consistent with the data. Because of photoselection,[33, 34] and assuming a fundamental anisotropy greater than zero ($r_0 > 0$), the vertically polarized excitation pulse results in an initial population of fluorophores which is enriched in the parallel orientation. This initial difference in populations between the parallel and the perpendicular orientation decays due to both rotational diffusion and decay of the total population of excited molecules. A faster rate of rotational diffusion (shorter correlation time) results in a more rapid decay of the anisotropy $r(t) = (I_{\parallel}(t) - I_{\perp}(t))/I(t)$ (Figure 5.4, dashed line).

The experimental procedures and the form of the data are different for the frequency-domain measurements of the anisotropy decays. The sample is excited with amplitude-modulated light, which is vertically polarized (Figure 5.5). As for the time-domain measurements, the emission is observed through a polarizer, which is rotated between the parallel and the perpendicular orientations. There are two observable quantities which characterize the anisotropy decay. These are the phase shift Δ_ω, at the modulation frequency ω, between the perpendicular ($\phi_{\perp}$) and parallel ($\phi_{\parallel}$) components of the emission,

$$\Delta_\omega = \phi_{\perp} - \phi_{\parallel} \tag{5.9}$$

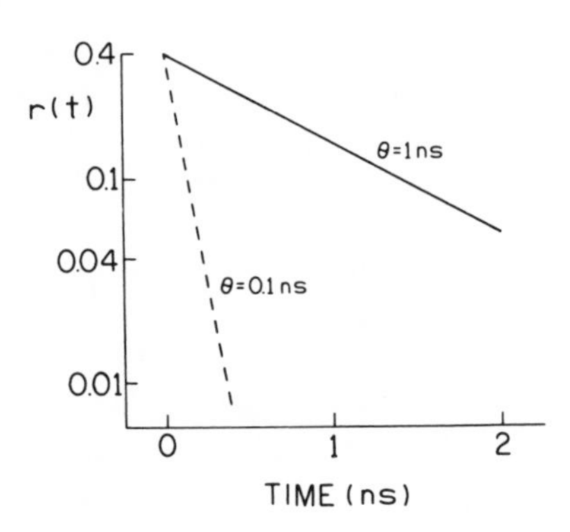

Figure 5.4. Time-domain representation of anisotropy decays with correlation times of 0.1 (– – – –) and 1.0 ns (——).

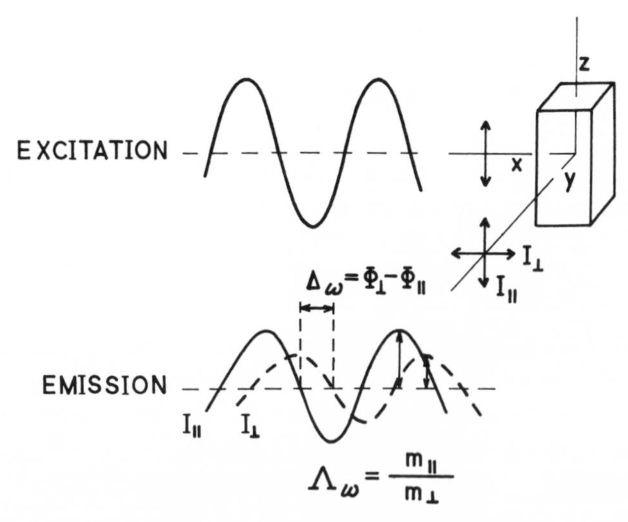

Figure 5.5. Schematic representation of frequency-domain measurements of anisotropy decays.

and the ratio

$$\Lambda_\omega = m_{\parallel}/m_\perp \tag{5.10}$$

of the parallel ($m_{\parallel}$) and the perpendicular ($m_\perp$) components of the modulated emission.[35] To avoid confusion, we stress that Λ_ω is the ratio of the modulated amplitudes, not the ratio of the modulation of each polarized component. The physical meaning of the ratio Λ_ω is made clearer by use of a slightly different form.[36, 37] We use the frequency-dependent anisotropy (r_ω), which is defined by

$$r_\omega = \frac{\Lambda_\omega - 1}{\Lambda_\omega + 2} \tag{5.11}$$

At low modulation frequencies r_ω tends toward the steady-state anisotropy, and at high frequencies r_ω tends toward the limiting anisotropy in the absence of rotational diffusion (r_0).

The origin of the phase shift Δ_ω can be visualized by consideration of the population of molecules which is observed through the polarizer in the parallel or the perpendicular orientation. The parallel orientation selects for those molecules which have emitted prior to rotation, which are those which have emitted sooner during the total decay. The perpendicular orientation selects for those molecules which have rotated further, which are also those which have emitted later within the total decay. Hence, there is a different mean decay time for the parallel and perpendicular components of the emission, which in turn results in the phase difference between the components. However, we caution that the use of a mean decay time to characterize the decay of the polarized components has resulted in confusion in the early literature.[38-40] By similar reasoning, the modulated amplitude of the parallel component is higher than that of the perpendicular component because of the shorter mean decay time of the early-emitting parallel-oriented fluorophores.

The form of the FD anisotropy data is illustrated in Figure 5.6. The differential phase angles appear to be approximately Lorentzian in shape on the log-frequency scale. The modulated anisotropy increases monotonically with frequency. One can notice, that, for $\theta = 1$ ns, the value of r_ω at low frequency is equal to the steady-state anisotropy:

$$r = \frac{r_0}{1 + \tau/\theta} = \frac{r_0}{2} \tag{5.12}$$

At high frequency, r_ω approaches r_0. Longer correlation times shift the Δ_ω and r_ω curves to lower modulation frequencies, and shorter correlation times shift these curves to higher frequencies.

The presence of two or more correlation times has a dramatic effect on the appearance of the FD anisotropy data. This is illustrated in Figure 5.7 for correlation times of 50 ps and 5 ns. Such correlation times are expected for segmental motions of an aromatic ring in a protein which also undergoes overall rotational diffusion with a 5-ns correlation time. The equivalent time-

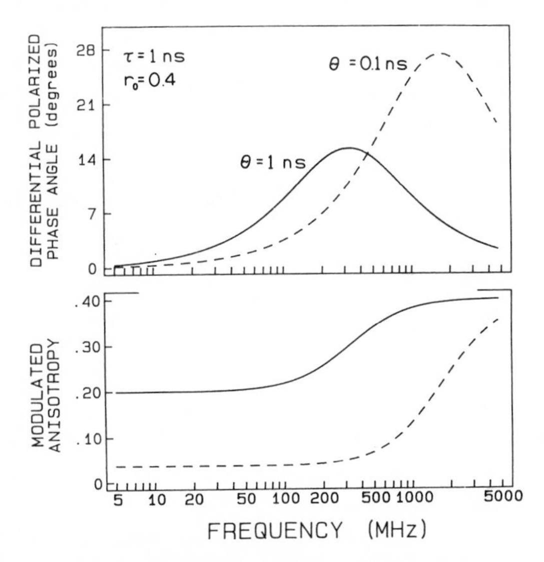

Figure 5.6. Frequency-domain representation of anisotropy decays with correlation times of 0.1 (– – – –) and 1.0 ns (——) and a decay time of 1 ns.

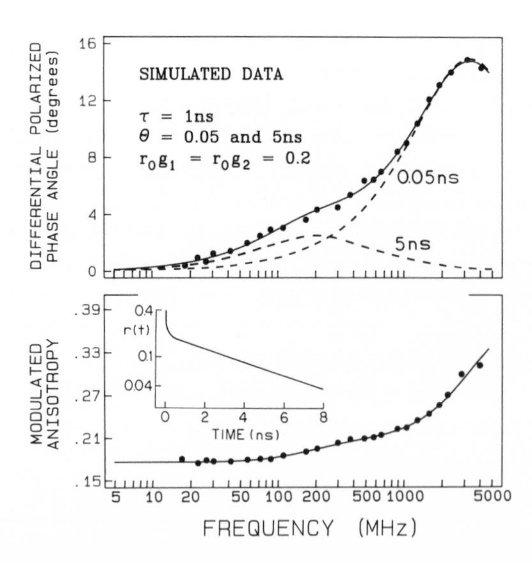

Figure 5.7. Simulated frequency-domain anisotropy data for a double-exponential anisotropy decay, with correlation times of 0.05 and 5 ns and a decay time of 1 ns. The inset shows the equivalent time-dependent anisotropy.

dependent anisotropy is shown as an insert in Figure 5.7. The presence of the rapid motion is evident from the increased phase angles at the higher frequencies. If the amplitude of this rapid motion is increased, there are smaller phase angles at low frequencies and larger phase angles at the higher frequencies, as indicated by the dashed line. If the shorter correlation time is very short, the frequency range of the instrument may not be adequate to detect the presence of this rapid motion. However, one may notice the presence of an unresolved motion by a failure of r_ω to reach the expected limiting value of r_0. Of course, this illustrates the advantage of using both the differential phase (Δ_ω) and modulation (r_ω) data in any attempt to resolved a complex and/or rapid anisotropy decay. As for the intensity decays, the parameters describing the anisotropy decay are recovered by comparison of the data with calculated values obtained using assumed values for the parameters.

5.3. Theory of Frequency-Domain Fluorometry

5.3.1. Decays of Fluorescence Intensity

We now describe the formalism used to recover the intensity decay parameters from the frequency response of the emission. Naturally, if the intensity decay is to be analyzed, it is necessary to use rotation-free measurement conditions. At present, the frequency-domain data are mostly analyzed by the method of nonlinear least squares.[14, 15, 27] The measured data are compared with values predicted from a model, and the parameters of the model are varied to yield the minimum deviations from the data. It is possible to predict the phase and modulation values for any assumed decay law. The frequency-domain data for an intensity decay can be calculated from the sine and cosine transforms of $I(t)$:

$$N_\omega = \frac{\int_0^\infty I(t) \sin \omega t \, dt}{\int_0^\infty I(t) \, dt} \tag{5.13}$$

$$D_\omega = \frac{\int_0^\infty I(t) \cos \omega t \, dt}{\int_0^\infty I(t) \, dt} \tag{5.14}$$

where ω is the circular modulation frequency (2π times the modulation frequency in Hz). Even if the decay is more complex than a sum of exponentials, it is generally adequate to approximate the decay by such a sum. If needed, nonexponential decay laws can be transformed numerically. For a sum of exponentials the transforms are

$$N_\omega = \sum_i \frac{\alpha_i \omega \tau_i^2}{(1 + \omega^2 \tau_i^2)} \Big/ \sum_i \alpha_i \tau_i \tag{5.15}$$

$$D_\omega = \sum_i \frac{\alpha_i \tau_i}{(1 + \omega^2 \tau_i^2)} \Big/ \sum_i \alpha_i \tau_i \tag{5.16}$$

The calculated frequency-dependent values of the phase angle ($\phi_{c\omega}$) and the demodulation ($m_{c\omega}$) are

$$\tan \phi_{c\omega} = N_\omega / D_\omega \tag{5.17}$$

$$m_{c\omega} = (N_\omega^2 + D_\omega^2)^{1/2} \tag{5.18}$$

The parameters (α_i and τ_i) are varied to yield the best fit between the data and the calculated values, as indicated by a minimum value for the goodness-of-fit parameters χ_R^2,

$$\chi_R^2 = \frac{1}{v} \sum_\omega \left[\frac{\phi_\omega - \phi_{c\omega}}{\delta\phi} \right]^2 + \frac{1}{v} \sum_\omega \left[\frac{m_\omega - m_{c\omega}}{\delta m} \right]^2 \tag{5.19}$$

where v is the number of degrees of freedom. The subscript c is used to indicate calculated values for assumed values of α_i and τ_i, and $\delta\phi$ and δm are the uncertainties in the phase and modulation values, respectively. Unlike the errors in photon-counting experiments,[41] these errors cannot be estimated directly from Poisson counting statistics.

The correctness of a model is judged by the values of χ_R^2. For an appropriate model and random noise, χ_R^2 is expected to be near unity. If χ_R^2 is greater than unity, then one should consider whether the χ_R^2 value is adequate to reject the model. Rejection is judged from the probability that random noise could be the origin of the value of χ_R^2.[27, 42] For instance, a typical frequency-domain measurement in this laboratory contains phase and modulation data at 25 frequencies. A double-exponential model contains three floating parameters (two τ_i and one α_i), resulting in 47 degrees of freedom. A value of χ_R^2 equal to 2 is adequate to reject the model with a certainty of 99.9% or higher.[42]

In practice, the values of χ_R^2 change depending upon the values of the uncertainties ($\delta\phi$ and δm) used in its calculation. For consistency, we chose to use constant values of $\delta\phi = 0.2°$ and $\delta m = 0.005$. While the precise values may vary between experiments, the χ_R^2 values calculated in this way indicate to us the degree of error in a particular data set. The use of fixed values of $\delta\phi$ and δm does not introduce any ambiguity in the analysis, as it is the relative values of χ_R^2 which should be used in accepting or rejecting a model. Hence, we typically compare χ_R^2 for the one-, two-, and three-exponential fits. If χ_R^2 decreases twofold or more as the model is incremented, then the data most probably justify inclusion of the additional decay time. However, it should be remembered that the values of $\delta\phi$ and δm might depend upon frequency, either as a gradual increase in random error with frequency or as higher-than-average uncertainties at discrete frequencies due to interference or other instrumental effects. In these cases the values of the recovered parameters might depend upon the values chosen for $\delta\phi$ and δm. We found that the parameter values are generally insensitive to the weighting factors, except at the limits of resolution. When the data are just adequate to determine the parameters, these values can be sensitive to the weighting used in our calculation of χ_R^2.

5.3.2. Decays of Fluorescence Anisotropy

The FD anisotropy analysis is performed in a similar manner, except that there is a somewhat more complex relationship between the data (Δ_ω and Λ_ω) and the transforms. The calculated value ($\Delta_{c\omega}$ and $\Lambda_{c\omega}$) can be obtained from the sine and cosine transforms of the individual polarized decays[35]:

$$N_k = \int_0^\infty I_k(t) \sin \omega t \, dt \tag{5.20}$$

$$D_k = \int_0^\infty I_k(t) \cos \omega t \, dt \tag{5.21}$$

where the subscript k indicates the orientation, parallel ($\|$) or perpendicular ($\perp$). The expected values of Δ_ω ($\Delta_{c\omega}$) and Λ_ω ($\Lambda_{c\omega}$) can be calculated from the sine and cosine transforms of the polarized decays (Eqs. 5.5 and 5.6). The calculated values of Δ_ω and Λ_ω are given by

$$\Delta_{c\omega} = \arctan\left(\frac{D_{\|}N_{\perp} - N_{\|}D_{\perp}}{N_{\|}N_{\perp} + D_{\|}D_{\perp}}\right) \tag{5.22}$$

$$\Lambda_{c\omega} = \left(\frac{N_{\|}^2 + D_{\|}^2}{N_{\perp}^2 + D_{\perp}^2}\right)^{1/2} \tag{5.23}$$

where the N_i and D_i are calculated at each frequency. The parameters describing the anisotropy decay are obtained by minimizing the squared deviations between measured and calculated values, using

$$\chi_R^2 = \frac{1}{\nu}\sum_{\omega}\left(\frac{\Delta_\omega - \Delta_{c\omega}}{\delta\Delta}\right)^2 + \frac{1}{\nu}\sum_{\omega}\left(\frac{\Lambda_\omega - \Lambda_{c\omega}}{\delta\Lambda}\right)^2 \tag{5.24}$$

where $\delta\Delta$ and $\delta\Lambda$ are the uncertainties in the differential phase and modulation ratio, respectively.

5.4. Instrumentation and Applications

5.4.1. FD Instruments with Intensity-Modulated Light Sources

5.4.1.1. Instrumental Design

A schematic of an early (1985) frequency-domain fluorometer constructed in this laboratory[13] is shown in Figure 5.8. This design is similar to that of the commercially available instruments[12] and is rather similar to a standard fluorometer. The main differences are the laser light source, the light modulator, and the associated radio-frequency electronics. Until recently, it was thought to be difficult to obtain wideband light modulation unless the light was highly collimated. Continuously variable frequency modulation of laser sources is possible with electro-optic modulators,[43] and acousto-optic modulators can usually provide modulation at discrete resonances over a limited range of frequencies. However, most electro-optic modulators have long narrow apertures, and electro-optic modulators are not easily used with our lamp sources. Hence, only laser sources seemed practical for use with electro-optic modulators. It is now known that laser sources can be easily modulated to 200 MHz using one of several electro-optic modulators. Surprisingly, it is now possible to modulate arc lamps to 200 MHz, which

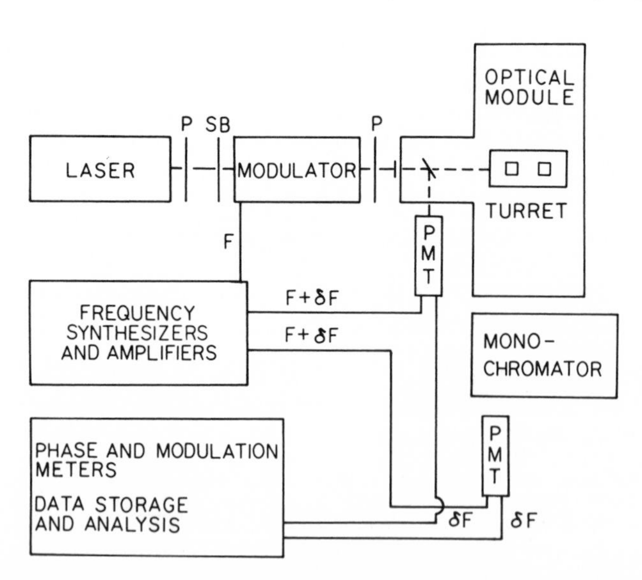

Figure 5.8. Schematic of the variable-frequency phase–modulation fluorometer. P, Polarizer; SB, Soleil–Babinet compensator; F, frequency; δF, cross-correlation frequency; PMT, photomultiplier tube. (From Ref. 53.)

is done in the commercial instruments.[12] The operational principles of the modulators and the electronic parts needed to construct such an FD instrument are discussed in Ref. 13, along with the rationale for selecting the various components.

The light source for an FD instrument can be almost any continuous-wave (cw) light source or a high-repetition-rate pulse laser. The choice of the source is based on the necessary wavelengths, the power level needed, and the experience and/or budget of the investigators. The He–Cd laser is a convenient cw source, at 325 and 442 nm. Unfortunately, these wavelengths are not suitable for excitation of protein fluorescence. A very versatile source is the Ar ion laser, which can now provide deep-UV lines ($\sim$275 nm) for excitation of protein fluorescence and UV lines near 351 nm for excitation of extrinsic fluorescent probes. Additionally, the Ar ion laser is an ideal source for pumping dye lasers and can be mode-locked as part of a picosecond laser system. While the Nd:YAG laser can also be the primary source in a picosecond system, it lacks the variety of wavelengths available from the Ar ion laser. Also, in our hands, the Nd:YAG laser is considerably less stable than an Ar ion laser, both with regard to pulse-to-pulse and long-term stability.

The next apparent difficulty is measurement of the phase angle and modulation at high frequencies. The measurements appear more difficult when one realizes that resolution of multiexponential decays requires accuracy greater than 0.5° in phase and 0.5% (0.005) in modulation. In fact, the measurements are surprisingly easy and free of interference. This is because of cross-correlation detection. The gain of the detector is modulated at a frequency offset $(F+\delta F)$ from that of the modulated excitation (F). The difference frequency $(\delta F = 25\,\text{Hz})$ contains the phase and modulation information. Hence, at all modulation frequencies the phase and modulation can be measured at the low cross-correlation frequency (δF) with a zero-crossing detector and a ratio digital voltmeter. The use of cross-correlation detection results in the rejection of harmonics and other sources of noise. The cross-correlation method is surprisingly robust. The harmonic content (frequency components) of almost any excitation profile can be used if it contains frequency components which are synchronized with the detector. Both pulsed lasers[25] and synchrotron radiation[44, 45] have been used as modulated light sources. Since pulse lasers provide harmonic content to many gigahertz, the bandwidth of the frequency-domain instruments is now limited by the detector and not the modulator.

5.4.1.2. Resolution of Multiexponential Decays

It is instructive to examine data for a mixture of fluorophores. An example is shown in Figure 5.9, which shows the frequency response of a mixture of *N*-acetyl-L-α-tryptophanimide (NATA) (2.9 ns) and indole (4.5 ns). The frequency response is not obviously multiexponential because the decay times are closely spaced. However, the presence of multiple decay times is evident from an attempt to fit the data to a single decay time. The best fit to a single decay time results in an elevated value of $\chi_R^2 = 7.8$ and nonrandom deviations between the measured values and those calculated for the single-decay-time model (lower panels). The need for high precision in the phase and modulation measurements is seen from the close correspondence between the single- and double-exponential curves (dashed and solid lines, respectively, in upper panel). The double-exponential model results in a decreased and acceptable value of $\chi_R^2 = 0.9$ and random deviations (lower panels) between the measured and calculated values. These data and their analysis demonstrate the possibility of resolving even closely spaced decay times from the frequency-domain data.

As a second example, Figure 5.10 shows the frequency response of NADH. Its decay was reported to be doubly exponential, when measured by time-correlated single-photon counting (TCSPC).[46] This is a difficult test sample for the FD instrument because the mean decay time of NADH is near 0.4 ns. The data were not consistent with the single-exponential model (represented by the dashed line; $\chi_R^2 = 116$), whereas the double-exponential model yielded a 50-fold decrease in χ_R^2 to 2.1, good agreement of the calculated curve (represented by the solid line with data, and mostly random

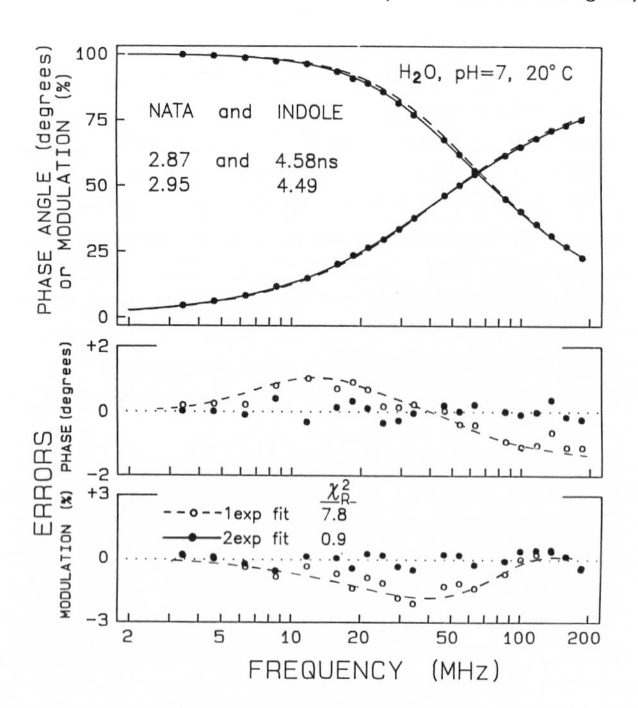

Figure 5.9. Resolution of a two-component mixture using the 200-MHz frequency-domain instrument. The dashed and solid lines show the best single- and double-exponential fits, respectively (see text for details).

deviations (lower panels). The value of χ_R^2 for the accepted fit might be regarded as larger than acceptable by those familiar with analysis of TCSPC data. Alternatively, the value of $\chi_R^2 = 0.9$ for the two-component mixture is smaller than found for TCSPC data. Such changes in the lower limits of χ_R^2 are due to different amounts of random noise and/or systematic errors in the data. This effect causes no ambiguity in the analysis because we also consider the relative value of χ_R^2 as the model is made more complex. To date, we have not found any way around this minor difficulty. We considered and attempted to use the measured standard deviations in the data or the standard deviation of the mean of the measured data. The former seemed to overestimate the actual error (the values of χ_R^2 were too low), and the latter seemed to underestimate the error (the values of χ_R^2 were too large). It should also be noted that it is common practice in FD measurements to perform a rapid series of phase and modulation measurements at each frequency. Under these conditions the successive data points are not independent and uncorrelated, so that the measured standard deviation is not an accurate measure of the amount of error in the measurements.

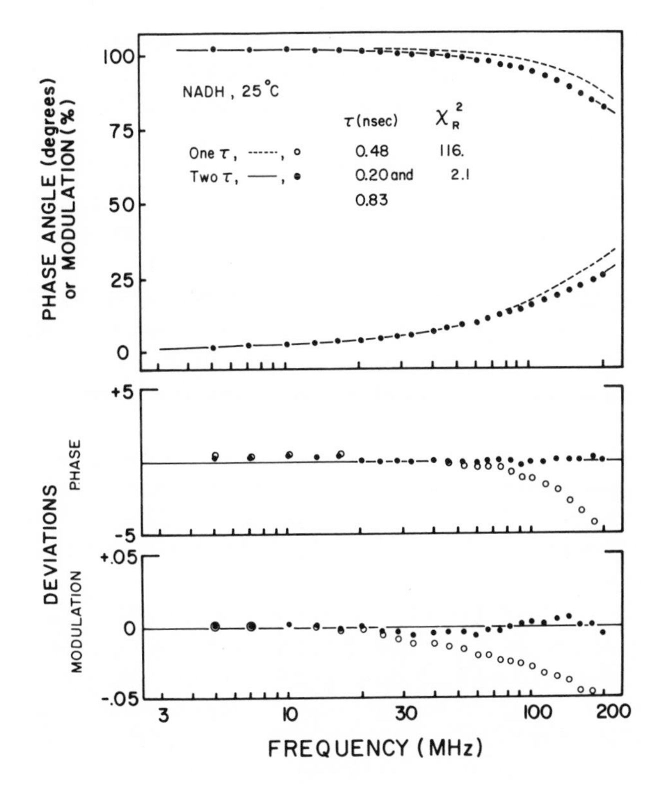

Figure 5.10. Multiexponential analysis of the emission of NADH (see text for details). (From Ref. 13.)

The data for NADH (Figure 5.10) illustrate the need for measurements above 200 MHz. While the double-exponential analysis yielded the two expected decay times, the maximum phase angle is only about 25°, and the extent of modulation is still about 80%. It is obvious that higher certainty in the decay law can only be obtained if the range of frequencies is extended to higher values. For this we developed the gigahertz frequency-domain fluorometers.

5.4.2. A 2-GHz Harmonic Content FD Fluorometer

5.4.2.1. Design of the Instrument

In the frequency domain it is desirable to measure data over the widest possible range of frequencies. Frequencies of 1 GHz or higher are desirable for measurement of subnanosecond lifetimes or correlation times (see Figures 5.3

and 5.6). There were several limitations to the measurements at higher frequencies. The upper frequency limit of most PMTs is near 200 MHz, and it is difficult to obtain useful intensity modulation above 200 MHz. Traveling wave modulators have been described which work to several gigahertz and which have UV transmission and low half-wave voltages.[43, 47, 48] However, we were unable to use a commercially available version of these modulators, due to drifts in transmission through the crossed polarizers.

The need for a light modulator was eliminated by using the intrinsic harmonic content of a laser pulse train. The light source consists of a mode-locked argon ion laser and a synchronously pumped and cavity-dumped dye laser. This source provides 5-ps pulses at a repetition rate of 4 MHz. In the frequency domain this source is intrinsically modulated to many gigahertz, as shown by the schematic Fourier transform in Figure 5.11. The idea of using the harmonic content of a pulse train was originally proposed by Merkelo *et al.*[49] for a pulsed laser source and later by Gratton and Lopez-Delgado[44] and Gratton *et al.*[45] for synchrotron radiation. This source provides light which is intrinsically modulated at each integer multiple of the repetition rate, to about 50 GHz.[50]

A second and significant advantage of this source is that it is rather easy to frequency double the pulse train because of the high peak power. Additionally, the doubled output is used directly to excite the sample. It is no longer necessary to use an electro-optic modulator and nearly crossed polarizers, which results in a significant attenuation of the incident light.[13] During the past three years we performed numerous measurements with the harmonics extending to 10 GHz. There is no detectable increase in noise up

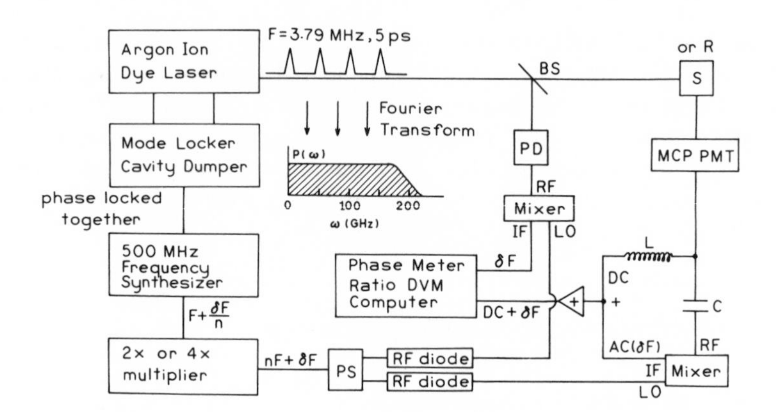

Figure 5.11. Schematic of the 2-GHz frequency-domain fluorometer. PD, Photodiode; PS, power splitter; MCP PMT, microchannel plate photomultiplier tube; BS, beam splitter; DVM, digital voltmeter; F, fundamental frequency of cavity-dumped dye laser output; δF, cross-correlation frequency of 25 Hz; n, number of the harmonic; S, sample; R, reference or scatter.

to 10 GHz, suggesting that there is no multiplication of phase noise at the higher harmonics. We are pleased with this approach and, at present, do not feel the need to identify light modulators for gigahertz frequencies.

The second obstacle to higher frequency measurements was the detector. The PMT in our original instrument (Figure 5.8) was replaced with a microchannel plate (MCP) PMT (Figure 8.11). These devices are 10- to 20-fold faster than a standard PMT,[51, 52] and hence a multi-gigahertz bandwidth was expected. As presently designed, the MCP PMTs do not allow internal cross-correlation, which is essential for an adequate signal-to-noise ratio. This problem was circumvented by designing an external mixing circuit, which preserved both the phase and the modulation data. With these modifications, data have been obtained to 2 GHz[25] and recently to 10 GHz.[26, 72]

The primary pump source is presently a mode-locked argon ion laser. The ion laser pumps a dual-jet dye laser, whose output is cavity dumped at 3.79 MHz. The dye laser with rhodamine 6G provides excitation wavelengths from 570 to 610 nm. For excitation wavelengths in the ultraviolet, its output is frequency doubled to 285–305 nm. The average UV power is near 0.5 mW, which is attenuated 50- to 100-fold prior to the sample. Alternatively, the

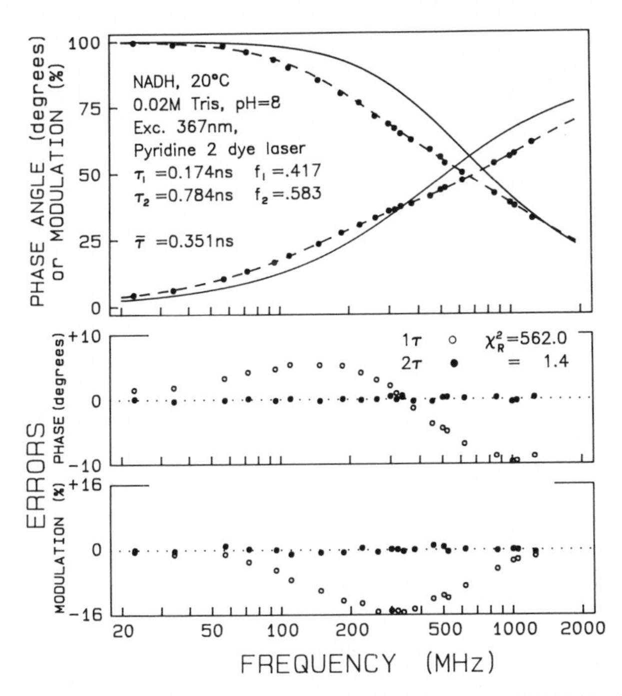

Figure 5.12. Frequency-domain measurements of the intensity decay of NADH with the 2-GHz instrument.

beam diameter is expanded to about 5 mm using a negative lens. For our application the pulse width and shape are not critical, but the pulse width of the visible output of the dye laser was 5 ps or less.

5.4.2.2. *Resolution of Multiexponential Decays in Subnanosecond Range*

We examine NADH once again using the 2-GHz instrument (Figure 5.12). These measurements were performed using the frequency-doubled output of a pyridine 2 dye laser, providing output from 350 to 385 nm. Our attempt to fit the frequency response to a single exponential is shown by the solid line, resulting in an unacceptable χ_R^2 value of 562. It is obvious from this attempted single-exponential fit that the intensity decay of NADH cannot even be approximated by a single decay time. It is perhaps ironic that the intensity decay of NADH was previously regarded as a short-decay-time standard for the fixed-frequency phase fluorometers,[11] and the multiexponential form of the decay may explain some of the difficulties in interpreting the phase and modulation values obtained at the various fixed frequencies. A good fit was found for the double-exponential model ($\chi_R^2 = 1.4$). The decay times and amplitudes we recovered are in relatively good agreement with those

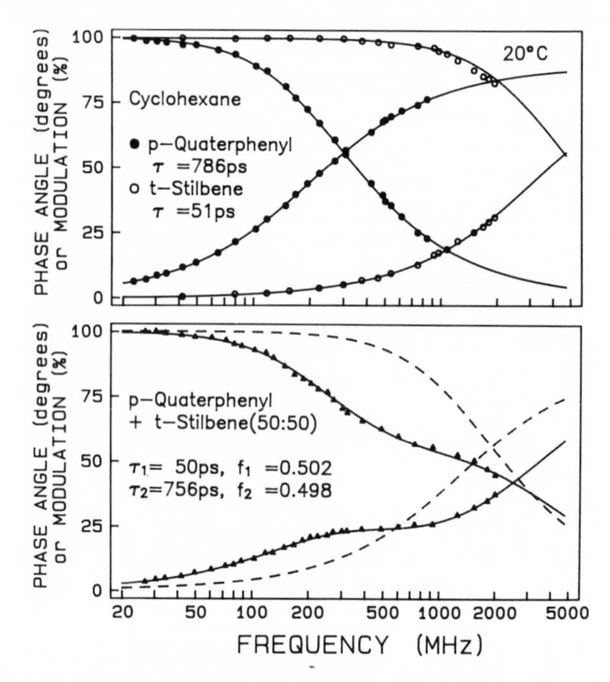

Figure 5.13. Resolution of a mixture of *p*-quaterphenyl and *trans*-stilbene using the 2-GHz fluorometer (see text for details). (From Ref. 53.)

reported in Ref. 46, which were 0.25 ns and 0.82 ns, with nearly equal fractional intensities.

We also examined a mixture of two fluorophores, which individually displayed subnanosecond decay times (*trans*-stilbene, 51 ps, and *p*-quaterphenyl, 786 ps) (Figure 5.13, top). The data measured for each compound individually are well matched by the curves calculated for the single-decay-time model, represented by the solid lines. The lower panel in Figure 5.13 shows the frequency response obtained for the mixture. The presence of multiple decay times is immediately evident from the more complex frequency response and from an attempt to fit the data to a single decay time (shown as the dashed line). The data can be well fitted using a double-exponential model resulting in decay times of 50 and 756 ps (solid line). In other studies we demonstrated that much more closely spaced lifetimes can be resolved, as can the decay times of three-component mixtures.

To compare capabilities in resolution of complex decays of first- (2–200 MHz) and second-generation (2 GHz) frequency-domain fluorometers, we simulated data ($\tau = 0.2$ and 1 ns; $f_1 = f_2$; Figure 5.3, bottom) using 25 frequencies in the range 2–200 MHz, and separately in the range 4–2000 MHz, with random noise at a level comparable to that in our experimental data. These two sets of data were analyzed separately in terms of a double-exponential decay (three floating parameters; two lifetimes and one amplitude), and we performed a rigorous analysis of the uncertainties. The values of χ_R^2 (Eq. 5.19) were examined when one of the parameters was kept constant at different values near the value corresponding to the minimum χ_R^2 while the other parameters were floating. The least-squares analysis was performed again, allowing the floating parameters to vary, yielding the minimum value of χ_R^2 consistent with the fixed parameter value. Since all the other parameters

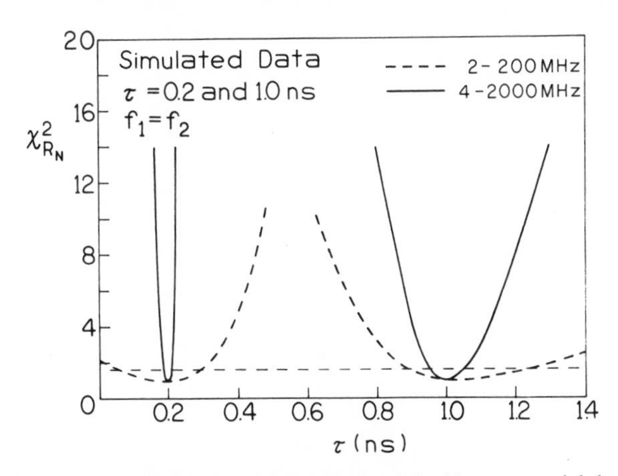

Figure 5.14. χ_R^2 surfaces for the lifetimes in the simulated double-exponential decay shown in the bottom half of Figure 5.3.

were floating during this reanalysis, this procedure should account for all the correlations between the parameters. The result of this procedure for the decay times is shown in Figure 5.14. The horizontal dashed line indicates the value of χ_R^2 expected 95% of the time due to random errors. The χ_R^2 surface for 2-GHz data indicates very good resolution of two simulated lifetimes. In contrast, the χ_R^2 surface for 200-MHz data shows less resolution of the decay times. Because of the lack of more rigorous theory for uncertainties, we propose that the intersection of these χ_R^2 surfaces with the 95% probability line is a measure of the uncertainty of estimated parameters. Hence, the 200-MHz data yields the expected decay times with a range of 200 ± 100 ps and 1000 ± 150 ps, whereas the 2-GHz data result in less uncertainty in the decay times (200 ± 10 ps and 1000 ± 50 ps).

It has been suspected for several years that the intensity decay of tryptophan at neutral pH is at least a double-exponential decay.[54] The frequency-domain data for tryptophan are shown in Figure 5.15. The single-exponential fit is not adequate, yielding systematic deviations (lower panels) and $\chi_R^2 = 59$. The double-exponential model is consistent with the data, and χ_R^2 is decreased to 0.9. The χ_R^2 surface (Figure 5.16) shows excellent resolution of subnanosecond and nanosecond lifetimes from the FD data. It should be

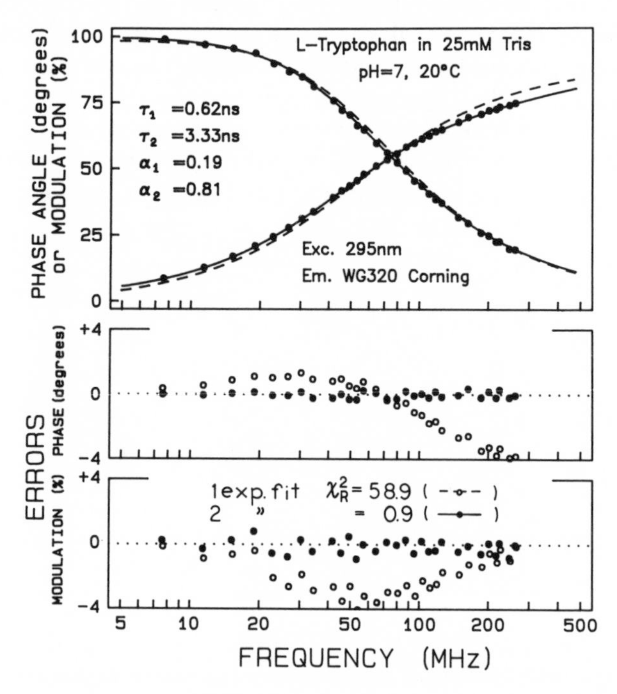

Figure 5.15. Frequency-domain intensity decay of tryptophan in H_2O at 20°C and pH 7.

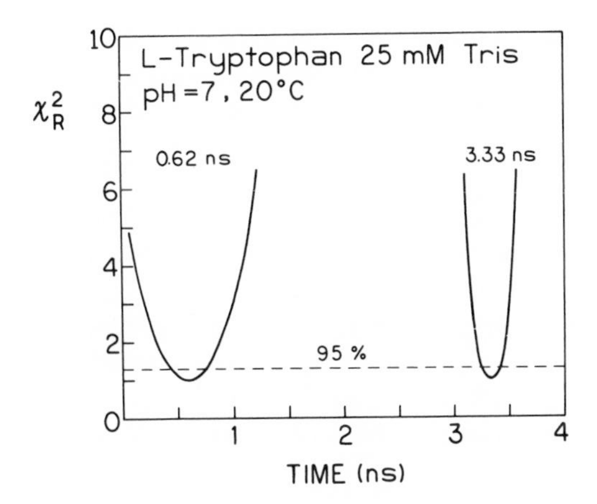

Figure 5.16. χ_R^2 surfaces for lifetimes of tryptophan.

noted that we recently recovered the emission spectra associated with the two decay times of tryptophan at pH 7 by the use of phase angle and modulation spectra.[55]

As a final example of the high sensitivity and resolution of the gigahertz FD measurements, we examined the intrinsic tryptophan fluorescence of human hemoglobin solutions. During the last few years, several laboratories have published studies which indicated that the intrinsic tryptophan emission from hemoglobin (Hb) can be detected.[56–58] Recent time-resolved studies on Hb indicate that the decay times are 100 ps or less, which is consistent with

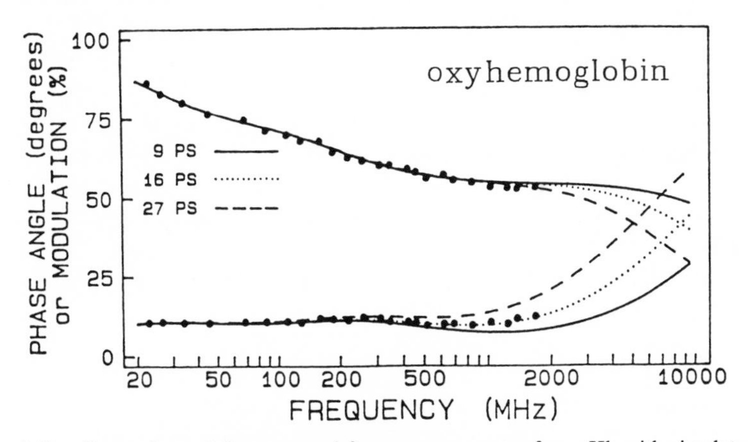

Figure 5.17. Comparison of the measured frequency response of oxy Hb with simulated data. The lines represent simulated data using the parameters observed for oxy Hb (Table 5.1), except for the shortest lifetime, which was varied from 9 to 27 ps. The solid dots represent the experimental data. (From Ref. 62.)

Table 5.1. Decay Times of the Tryptophan Emission from Hemoglobin

Sample[a]	τ_i (ns)	S.D.[b]	α_i	f_i	χ_R^2
Oxy Hb	0.016	0.001	0.992	0.536	
	0.820	0.028	0.007	0.190	
	8.306	0.240	0.001	0.274	2.8 (11.0)[c]
Deoxy Hb	0.009	0.001	0.991	0.349	
	0.901	0.025	0.008	0.226	
	7.479	0.270	0.001	0.385	3.5 (165.8)[c]
CO Hb[d]	0.027	0.001	0.985	0.392	
	1.118	0.037	0.011	0.177	
	7.508	0.160	0.004	0.541	2.4 (90.4)[c]

[a] Excitation at 300 nm and emission at 365 nm by Corning 7-51 and 0-52 filters, 0.05 *M* phosphate buffer at pH 7.0, 4°C.
[b] As calculated from the diagonal elements of the covariance matrix (Ref. 27).
[c] The values in parentheses are the values of χ_R^2 found for analyses of the data using two decay times only.
[d] Excitation at 295 nm; Co Hb, carbonmonoxyhemoglobin.

quenching by energy transfer to the heme.[59–62] Using the 2-GHz instrument, we have found that the tryptophan emission of extensively purified sample of hemoglobin was largely dominated by lifetimes in the picosecond region. These picosecond decay times are sensitive to ligation.[62–64] Longer lived components in the nanosecond range were also present; these accounted for less than 1% of the emitting tryptophans and were not sensitive to ligand binding. It was suggested that only the lifetimes in the picosecond range belong to hemoglobin.[62–64]

Frequency-domain data of oxy Hb are presented in Figure 5.17. The solid, dotted, and dashed lines represent simulated data corresponding to picosecond components of intensity decays (Table 5.1) obtained for deoxy Hb, oxy Hb, and CO Hb, respectively. From this figure it is clear that extension of the frequency range beyond 2 GHz can provide additional resolution of these weak and short-lived components. Additionally, these calculated curves indicate that the data for oxy Hb with its 16-ps decay time are not consistent with the 9-ps or 27-ps decay time found for the other liganded states (Figure 5.17).

5.4.3. Resolution of Anisotropy Decays

The theory, measurements, and simulations of fluorescence anisotropy decays were already described in Sections 5.2.2 and 5.3.2. Now we will present a few particular examples.

5.4.3.1. *Segmental Mobility of Proteins*

As an example of a simple anisotropy decay, we chose ribonuclease T_1 (RNase T_1) at pH 5.5. This protein contains a single tryptophan residue (Trp-59). At pH 5.5 the intensity decay appears to be a single exponential, and the tryptophan residue appears to be rigidly bound within the protein martrix.[3, 65] The tryptophan residue in RNase T_1 is surrounded only by apolar amino acid side chains and appears to be highly shielded from the aqueous phase. The frequency-domain anisotropy data for native RNase T_1 (Figure 5.18) are characteristic of a single correlation time. The single correlation time of 10.7 ns at 5 °C is revealed by a Lorentzian profile of differential phase angles on the log frequency scale. As the temperature or concentration of guanidine hydrochloride (GdnHCl) is increased, the high-frequency (over 200 MHz) differential phase angles increase, which is characteristic of segmental motions of the tryptophan residue.[36] Additionally, the modulated anisotropies decrease with increasing temperature or concentration of GdnHCl. The anisotropy decays become more complex under conditions expected to induce unfolding. This is seen from the elevated values of χ_R^2 for the single-correlation-time fits (Table 5.2). Analysis in terms of two correla-

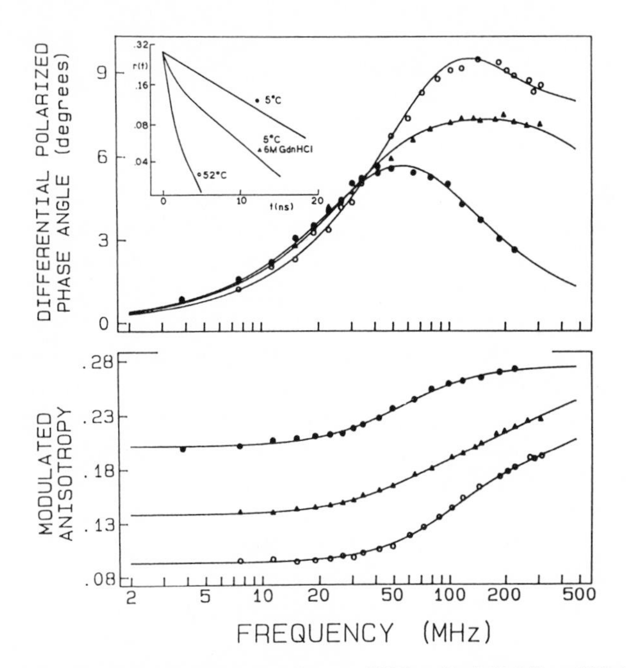

Figure 5.18. Frequency-domain anisotropy decays of RNase T_1 (pH 5.5) at 5°C (●) and 52°C (○) and at 5°C with 6 *M* guanidine hydrochloride (▲). The insert shows equivalent time-dependent anisotropies.

Table 5.2. Anisotropy Decays of RNase T_1

T (°C)	[GdnHCl]	θ_i (ns)	$r_0 g_i$[a]	χ_R^2
5	0	10.68	0.275	1.42
52	0	0.989	0.275	178.57
		0.21, 2.28	0.102, 0.173	1.25
5	6 M	2.41	0.275	356.5
		0.53, 9.27	0.108, 0.167	0.8

[a] For all analyses the $\sum r_0 g_i$ was held fixed at the value (0.275) found for N-acetyl-L-tryptophanamide at −60°C in propylene glycol. The intensity decays were taken as the most appropriate multiexponential fits.

tion times indicates the presence of subnanosecond components at 52 °C and for the sample containing 6 M GdnHCl. To illustrate the nature of the anisotropy decays, the equivalent time-dependent anisotropies are shown as an insert in Figure 5.18. These were calculated using the correlation times and amplitudes resulting from the frequency-domain data. For the native state the plot of log $r(t)$ versus time is linear with a reciprocal slope of 10.7 ns, due to overall rotational diffusion of the protein. The anisotropy decays in the denatured states are much more rapid, which reflects the greater motional freedom of the tryptophan residue in this disordered polypeptide.

5.4.3.2. Picosecond Rotational Diffusion

Molecular dynamics calculations on proteins have indicated that the indole ring may display torsional motions on the time scale of 2 to 20 ps. Such motions could be detected in the anisotropy decays of proteins, if the time resolution of the measurements is adequate. Hence, we evaluated the fastest anisotropy decay which could be detected using our instrumentation. To model the intrinsic fluorescence from proteins, we used indole as the fluorophore. To obtain a rapid rate of rotation, we used methanol as a low-viscosity solvent. The rate of rotation was increased using temperature as high as 80 °C. Since this is above the boiling point of methanol, the sample was placed in an enclosed vessel which could withstand the modest pressure. Finally, we used acrylamide to decrease the mean decay time of indole, which allows measurements to our earlier limit of 2 GHz. Acrylamide acts as a collisional quencher, which selectively eliminates the emission from the longer lived species. This results in an increased proportion of the emission occurring from earlier times, which contain more information on fast rotational processes.[66, 67]

The frequency-domain data for indole in methanol are shown in Figure 5.19. The indole rotations are most easily seen at −55 °C. At this temperature

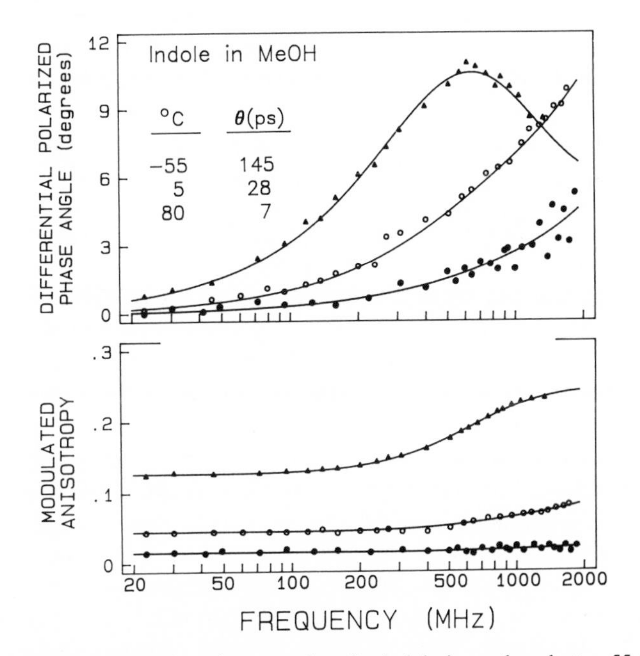

Figure 5.19. Frequency-domain anisotropy data for indole in methanol at −55 (▲), 5 (●), and 80°C (○). At these temperatures the acrylamide concentrations were 1.0, 0.5, and 0.3 *M*, respectively. (From Ref. 67.)

the high-frequency phase angle shows a maximum near 700 MHz. At higher temperatures the maximum is displaced to higher frequencies and cannot be detected with the 2-GHz instrument. At 80 °C the rotational rate is near the measurement limit. The maximum phase angle is near 4°, which is severalfold larger than the noise in the data. The results are satisfactory even at 80 °C, where the correlation time is only 7 ps. To the best of our knowledge, this is the fastest correlation time that has been measured using direct detection of fluorescence, except possibly for upconversion measurements which use optical gating and higher light intensities and which are considerably more complex to execute in practice.

5.4.3.3. Enhanced Resolution of Anisotropic Rotational Diffusion of Rigid Molecules by Global Analysis

It is of interest to recover multiexponential anisotropy decays because such decays reveal the torsional motions of probes within a biological molecule and the distinct rotational rates of anisotropic molecules. The resolution of complex anisotropy decays can be increased by global analysis of data obtained at different excitation wavelengths.[30, 68] By changing the excitation wavelength, one can alter the orientation of the absorption moment with respect to the molecular axis, and thus alter the contribution of rotations about each axis to the anisotropy data.

Another way to enhance resolution of anisotropy decay is to vary the mean lifetime of the fluorophore with the use of collisional quenching.[18, 66] As the lifetime is decreased, the early time portion of the anisotropy decay contributes increasingly to the data (see previous paragraph). To utilize the enhanced information content of these multiple data rates, we performed a doubly global analysis with different excitation wavelengths and collisional quenching.

As an example, we examined anisotropy decays of Y_t-base (4,9-dihydro-4,6-dimethyl-9-oxo-1*H*-imidazo-1,2*a*-purine) in propylene glycol at 10 °C. Asymmetric molecules with absorption and emission moments not directed along the principal axes (as is known to occur for Y_t-base[69] can display three correlation times.[29] To the extent of our knowledge, this example was the first report of the unambiguous detection of three correlation times for a rigid molecule.[18] The absorption and emission spectra of Y_t-base are shown in Figure 5.20, as is the excitation anisotropy spectrum in vitrified solution (−60 °C). We chose the excitation wavelengths 290, 312, and 346 nm. As indicated by the arrows, these wavelengths yield r_0 values of 0.05, 0.19, and 0.32, respectively. The emission was quenched by CCl_4, resulting in about 50 and 75% quenching at 0.5 and 1.0 *M* CCl_4, respectively. The frequency-domain anisotropy data are shown in Figure 5.21. These nine data sets (three wavelengths and three quencher concentrations) were analyzed globally to recover one, two, or three correlation times. The values are seen to depend

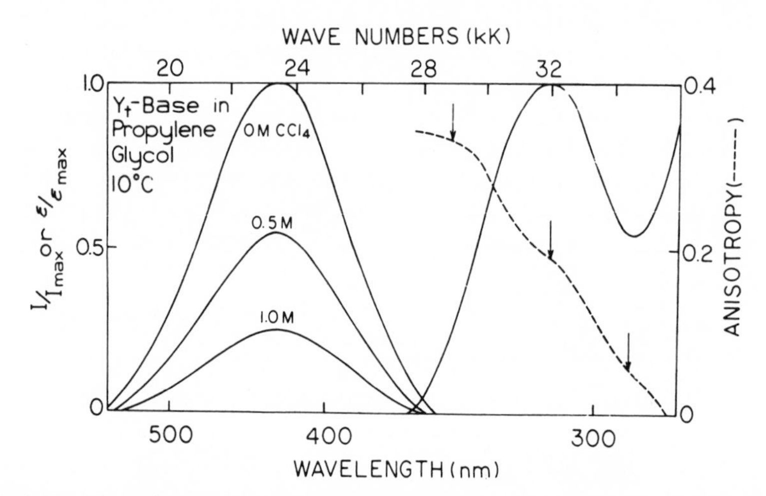

Figure 5.20. Absorption, emission, and anisotropy spectra of Y_t-base. The anisotropy values were measured at −60 °C.

upon both excitation wavelength and quencher concentration. The most dramatic change is caused by variations in the excitation wavelength, with more modest differences between values at different quencher concentrations. At low modulation frequencies the modulated anisotropies (r_ω) increase with [CCl_4], reflecting an increase in the steady-state anisotropy. At high modulation frequencies the r_ω values tend toward the r_0 value for each excitation wavelength. The data could not be fitted using the single-correlation-time model, as is seen from $\chi_R^2 = 34.6$ (Table 5.3). The use of two correlation times results in a 24-fold decrease in χ_R^2. A further 1.45-fold decrease of χ_R^2 was found using three correlation times, and the deviations appear to be somewhat smaller and more random (not shown). With 128 data points and 12 variable parameters (nine r_{0i} and three θ_i values), there are 116 degrees of freedom. For random deviations in the data, and $\nu = 116$, a 1.45-fold increase in χ_R^2 is expected less than 1% of the time. Hence the data almost certainly indicate the need to accept the model with three correlation times. It should be noted that the correlation times are closely spaced and span a range of only 7-fold (i.e., 0.8, 3.0, and 5.6 ns).

We questioned the uncertainties in the correlation times by examining

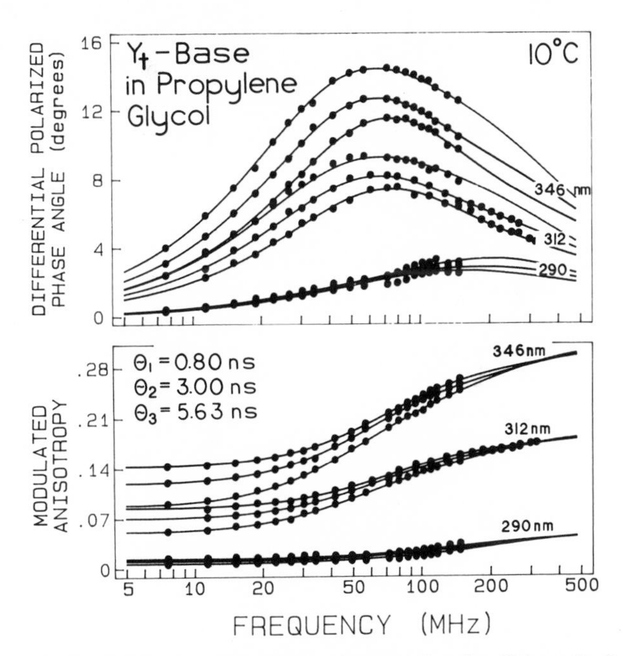

Figure 5.21. Frequency-domain double global anisotropy data for Y_t-base. In the upper panel, at each excitation wavelength, the decreasing phase angles are for 0, 0.5, and 1.0 M CCl_4. In the lower panel, at each excitation wavelength, the increasing modulated anisotropies are for 0, 0.5, and 1.0 M CCl_4. (From Ref. 18.)

Table 5.3. Global Analysis of the Anisotropy Decay of Y_t-Base

Model[a]	θ_i (ns)	r_{0i} 296 nm	312 nm	346 nm	χ_R^2
1θ	3.37	0.037	0.176	0.286	34.55
2θ	0.99	0.040	0.067	0.096	
	4.91	0.007	0.125	0.216	1.45
$\sum r_{0i}^{\lambda}$		0.047	0.192	0.312	
3θ	0.80	0.041	0.048	0.780	
	3.00	0.001	0.068	0.850	
	5.63	0.009	0.076	0.152	0.99
$\sum r_{0i}^{\lambda}$		0.051	0.192	0.315	
Measured r_0		0.05	0.19	0.32	

[a] 1θ is the fit to one correlation time, etc.

the values of χ_R^2 (Figure 5.22). The dashed line in Figure 5.22 indicates the value of χ_R^2 expected 33% of the time due to random errors. We note that the χ_R^2 surfaces do not intersect below the dashed line, which we take to indicate that all three correlation times are determined from the data. The three correlation times for Y_t-base may originate from its asymmetric structure; however, we cannot exclude partial slipping about one or more axes.

5.4.3.4. *Associated Anisotropy Decays*

In some systems, so-called associated ones, the intensity and anisotropy decays can be multiexponential due to the presence of multiple emitting species, each of which displays a discrete intensity decay and a discrete anisotropy decay. For instance, the emission from a mixture of two single-

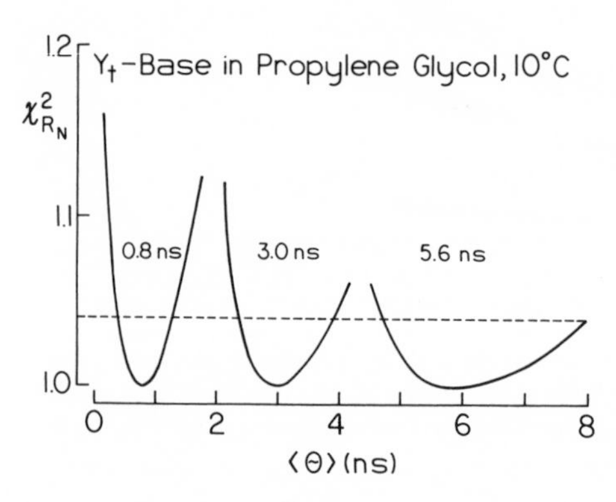

Figure 5.22. χ_R^2 surface for the three correlation times of Y_t-base.

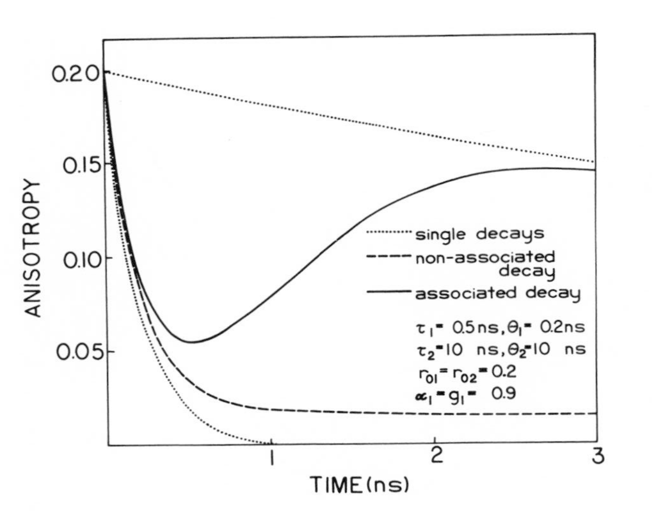

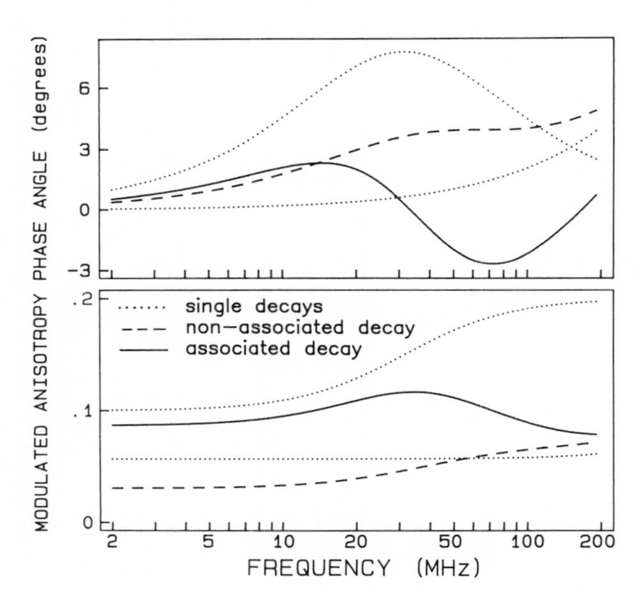

Figure 5.23. Simulated time-domain (left) and frequency-domain (right) decays for associated and nonassociated systems. The assumed decay times are $\tau_1 = 0.5$ and $\tau_2 = 10$ ns, the assumed correlation times being $\theta_1 = 0.2$ and $\theta_2 = 10$ ns. The dotted lines show the anisotropy decays for single correlation times of 0.2 and 10 ns. Also shown are the simulated data for an associated decay with $r_{01} = r_{02} = 0.2$ and $\alpha_1 = 0.90$ (——) and for a nonassociated decay with $g_l = 0.90$ (– – – –) and $\Sigma r_0 g_l = 0.2$ (From Ref. 17.)

tryptophan proteins may be considered to be an associated system, as can two discrete populations of a membrane-bound fluorophore.[70] The meaning of the anisotropy decay parameters is different when there are two emitting species. The observed anisotropy from a mixture is the intensity-weighted average of the anisotropies of the two species. This is true at all times, so

$$r(t) = r_1(t)\, f_1(t) + r_2(t)\, f_2(t) \tag{5.25}$$

The $r_i(t)$ are the individual anisotropy decays, and the $f_i(t)$ are the time-dependent fractional intensities (Eq. 5.2). The associated case can predict time-dependent anisotropies which increase at longer times, which is not possible for the nonassociated case with $r_0 > 0$ (Figure 5.23, left). Depending upon the value of the decay time and the correlation times, the frequency-domain data for an associated anisotropy decay can display unique and unexpected features.[17] If the decay is associated, the differential phase angles can increase with frequency, then decrease to negative values, followed by an increase at still higher frequencies. Additionally, the modulated anisotropies can decrease at the higher frequencies, rather than increasing toward r_0 as occurs for the nonassociated case (Figure 5.23, right).

An associated system is provided by 1-anilino-8-naphthalenesulfonic acid (ANS) in the presence of increasing amounts of apomyoglobin. Emission spectra of ANS with increasing amounts of apomyoglobin are shown in Figure 5.24. In the absence of apomyoglobin, the emission is weak and the maximum is near 513 nm. The addition of low molar ratios of apomyoglobin results in an increase in ANS intensity and a shift of the emission toward

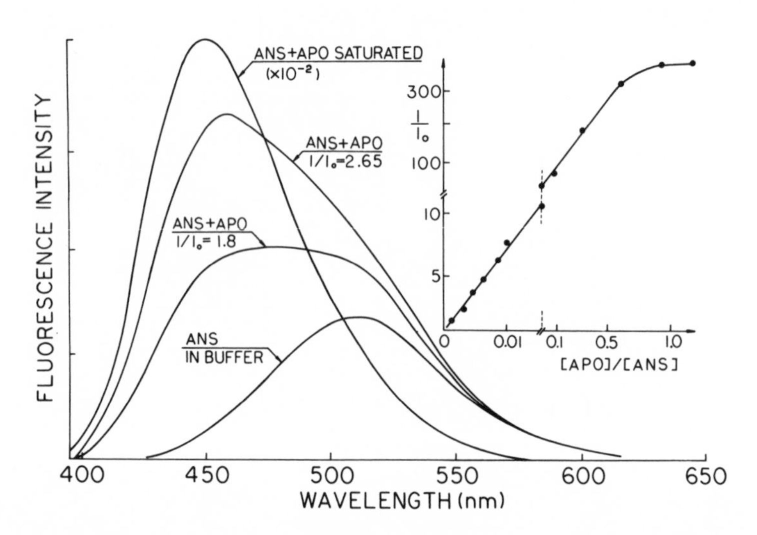

Figure 5.24. Emission spectra of ANS with increasing amounts of apomyoglobin.

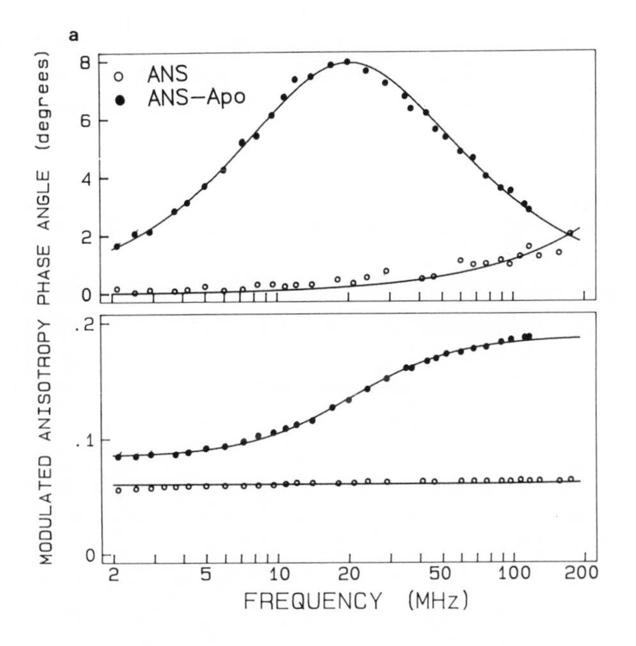

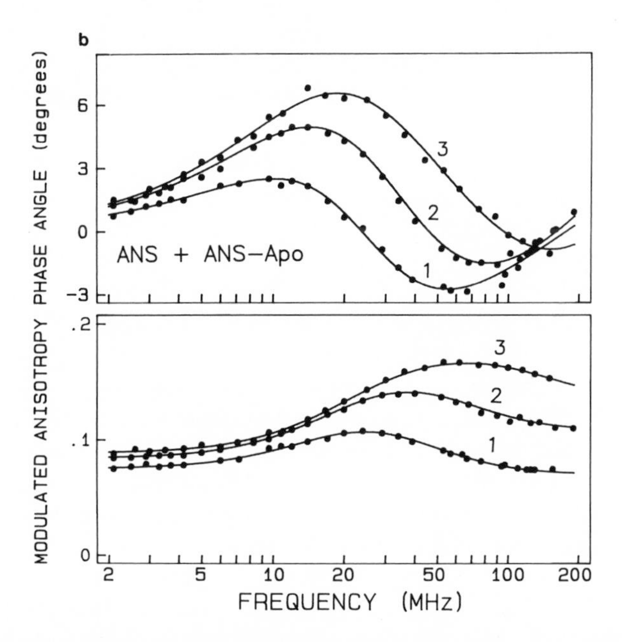

Figure 5.25. (a) Frequency-domain anisotropy data for ANS alone (○) and ANS-apomyoglobin (●); and (b) for ANS-apomyoglobin with molar ratios of (1) 0.0087, (2) 0.0158, and (3) 0.079 (bottom).

shorter wavelengths. This shift is attributed to the binding of ANS to the hydrophobic heme pocket.[71] It should be noted that binding of ANS to apomyoglobin results in a substantial increase in its lifetime. In water the lifetime of ANS is 0.26 ns, which increases to 15.8 ns for ANS bound to apomyoglobin.

Frequency-domain data for ANS in solution and bound to apomyoglobin are shown in the left-hand panel of Figure 5.25. In the bound form the differential phase angles are distributed nearly as a Lorentzian with maximum near 20 MHz. The modulated anisotropy increases monotonically with frequency, and the high-frequency limit of 0.19 is close to the expected value of ANS when excited at 325 nm, which is 0.189. The single-correlation-time analysis provides a good fit and yields a value of 13.2 ns. The data for free ANS are somewhat different due to its shorter correlation time. The small increases in differential phase and modulated anisotropy are consistent with a correlation time near 100 ps. Frequency-domain anisotropy data are remarkably different when both forms of ANS contribute to the emission (Figure 5.25, bottom). Then, the differential phase angles become negative around 20–100 MHz. Additionally, the modulated anisotropies reach a maximum and the decrease at higher frequencies. The data were analyzed to recover the individual correlation times and the apparent values of r_0 for each species. As expected, essentially the same values of the correlation times and r_0 were obtained for a range of apomyoglobin/ANS ratios. The ANS–apomyoglobin system described here is clearly an extreme case which was selected to illustrate differences between associated and nonassociated anisotropy decays.

5.5. Future Developments

The instrumentation for frequency-domain fluorometry was developed only a few years ago. Before this time many individuals thought the technique could not be useful. To date, many instrumental configurations have been tested (a few of them are described in this chapter) by this and other laboratories. Moreover, frequency-domain fluorometers are commercially available and used in many laboratories as standard equipment.

Regarding instrumentation, several developments seem inevitable. First, and probably most important, is the extension of modulation frequency to 10 GHz or higher. For ultrafast processes, such as rapid rotations (see Figure 5.19), it would be desirable to have measurements to frequencies as high as 20 GHz. Recently, using the fastest available MCP PMT 2566 with 6-μm channels (Hamamatsu Corporation), we extended the frequency range up to 6 GHz[72] and then to 10 GHz.[73, 74] Extension beyond this range should depend only on the future development of detectors, because present

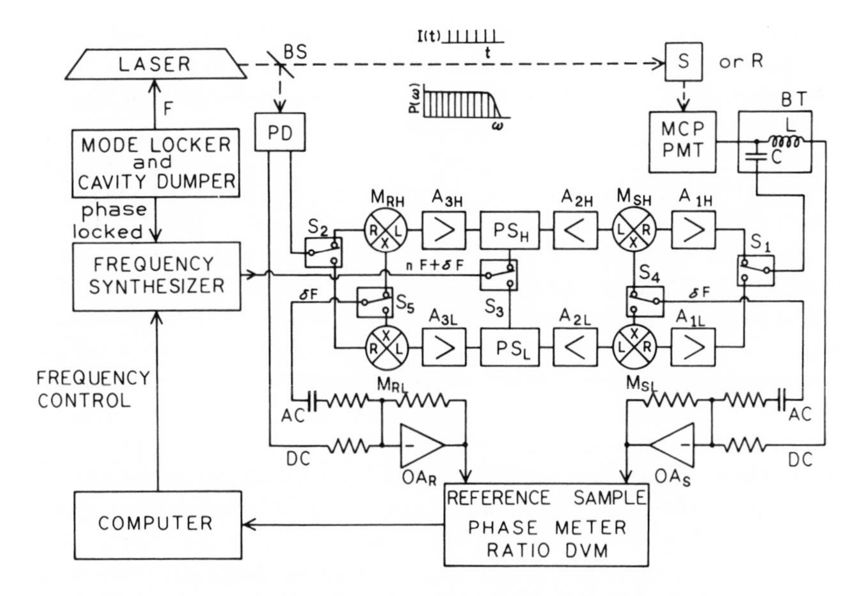

Figure 5.26. A 10-GHz frequency-domain fluorometer. (From Ref. 73.)

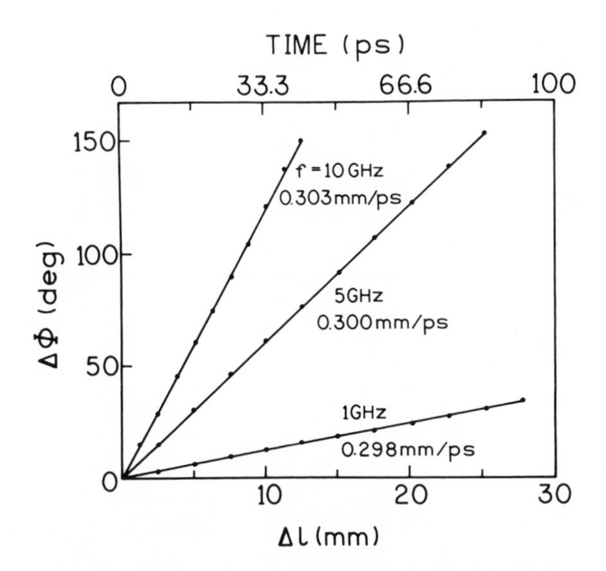

Figure 5.27. Measurement of variable light delays using the 10-GHz instrument.

synchronously pumped and cavity-dumped dye lasers can provide picosecond pulses yielding modulation excitation up to more than 50 GHz.

A schematic of our 10-GHz instrument is shown in Figure 5.26. The light source is once again the cavity-dumped dye laser, because this source is intrinsically modulated to many gigahertz. There are two signal channels, one for the reference photodetector and one for the MCP PMT. Each channel contains electronics for two frequency ranges, 10 MHz–2 GHz and 2–10 GHz. The upper frequency limit of 10 GHz is determined by the detector, and not the electronics, which function to 18 GHz.

The performance of this instrument at 1.5 and 10 GHz is shown in Figure 5.27 for measurement of variable light delays. The measured time delay, which is given by $\Delta t = \omega^{-1} \Delta\varphi$, is linear with distance, and the calculated delays agree with the known speed of light. A still more impressive test of the instruments' performance is shown in Figure 5.28 for a 1.69-ps light delay. In this case the phase angles were measured at frequencies ranging from 0.5 to 10 GHz. The upper panel shows the delay times Δt calculated using the phase angles $\Delta\varphi$ in the lower panel. We observed 1.7 ± 0.2 ps from 2–8 GHz and 1.7 ± 0.4 ps from 0.5–10 GHz. These results are particularly impressive when one recalls that these are low-intensity measurements and do not involve the complex procedure of optical gating or upconversion. We would be justified in claiming that this instrument has femtosecond resolution.

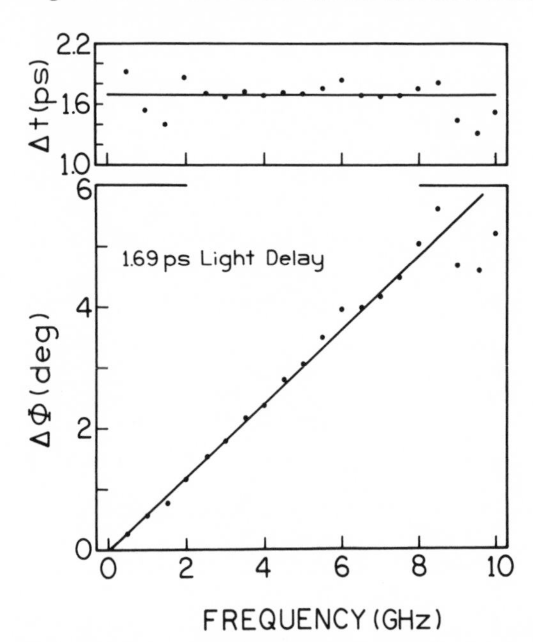

Figure 5.28. Frequency-dependent measurements of an optical delay using the 10-GHz frequency-domain fluorometer.

In our opinion, the most important test of any FD instrument is the measurement of actual fluorescence intensity and anisotropy decays. Frequency-domain measurements on 2,5-Diphenyl-1,3,4-oxadiazole (PPD) to 5 GHz are shown in Figure 5.29. This result is particularly impressive because PPD displays a relatively long decay time of 1.24 ns, which results in a phase angle of 88.5° at 5 GHz, and a relative modulation of only 0.026. In spite of the low modulation, we were able to measure the phase angles with little or no increase in noise to 1 GHz, and with slightly greater noise from 1 to 5 GHz. The random deviations of the measured phase angles indicate that there is little or no systematic error in these measurements. This is particularly important because at high frequencies it is easy to have interferences between the two detection channels, or interferences from external sources. Additional testing of the 10-GHz instrument is shown in Figures 5.30 and 5.31, which show the intensity and anisotropy decays, respectively, of a stilbene derivative. In this case, the decay time is only 61 ps, so that the data could be obtained up to 10 GHz. Similarly, the short decay time permits the FD anisotropy data to be measured up to 10 GHz. This stilbene derivative displays a single-exponential intensity decay and, at the current level of resolution, a single

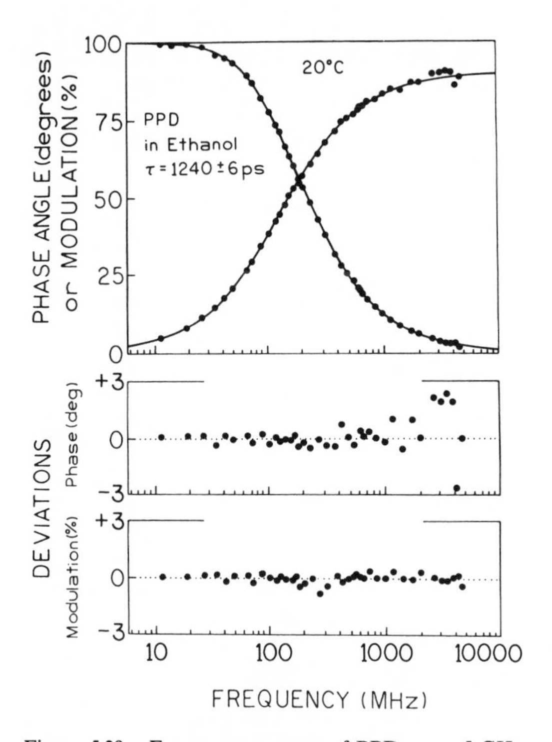

Figure 5.29. Frequency response of PPD up to 5 GHz.

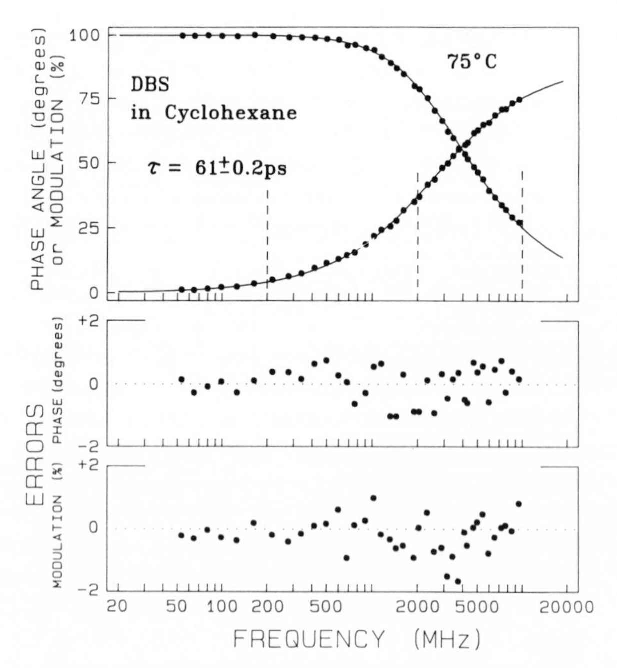

Figure 5.30. Frequency response of 4-dimethylamino-4′-bromostilbene (DBS) up to 10 GHz.

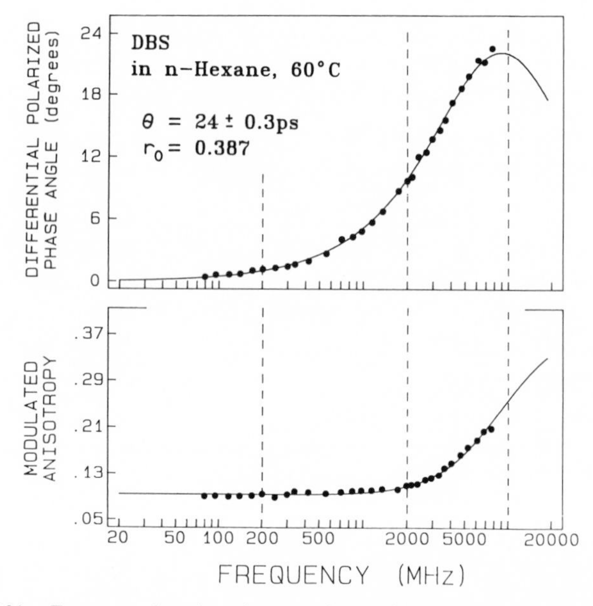

Figure 5.31. Frequency-domain anisotropy decay of DBS in hexane up to 10 GHz.

correlation time of 24 ps. It is apparent that this new instrument will provide increased information on rapid and/or complex time-dependent processes.

Additionally, improvements should be expected in data collection. At present, phase and modulation values are measured one frequency at a time. Simultaneous measurement at many of the frequencies will decrease the time required for data acquisition. Light sources may be modulated simultaneously at many frequencies(43) or intrinsically modulated at many frequencies, as is the case for a picosecond laser. Arc lamps can now be used with electro-optic modulators, and simple acousto-optic modulators can provide simultaneous modulation at several frequencies.(75) Such instruments are now appearing commercially(76) and seem to be the subject of current interest in other laboratories.(77, 78)

One can also expect improvements in resolution as the result of alternative means of data acquisition. In this laboratory we are emphasizing multiwavelength measurements, followed by simultaneous analysis of the multiple data sets. For example, we recently resolved the emission spectra and decay times for three fluorophores with completely overlapping emission spectra.(79) Finally, we expect a significant advance of frequency-domain fluorometry in medical applications, especially in the development of optical sensors and in studies of photon migration in tissues.(80)

5.6. Summary

In this chapter we described instrumentation, basic theory, and applications of first- (1–200 MHz) and second-generation (2 GHz) frequency-domain fluorometers. The frequency-domain data provide excellent resolution of time-dependent spectral parameters. It is possible to resolve closely spaced fluorescence lifetimes, even in the picosecond region, and to determine multiexponential decays of anisotropy. Correlation times as short as 10 ps have been measured. Several novel applications are described, including picosecond fluorescence of hemoglobin and enhanced resolution of anisotropy decays using combination quenching and multiwavelength global measurements and analysis, as well as unusual behavior of phase and modulation data for associated anisotropy decays.

References

1. J. B. A. Ross, K. W. Roussling, and L. Brand, Time-resolved fluorescence of the two tryptophans in horse liver alcohol dehydrogenase, *Biochemistry 20*, 4369–4377 (1981).
2. I. Munro, I. Pecht, and L. Stryer, Subnanosecond motions of tryptophan residues in proteins, *Proc. Natl. Acad. Sci. U.S.A. 65*, 56–60 (1979).

3. I. Gryczynski, M. Eftink, and J. R. Lakowicz, Conformation heterogeneity in proteins as origin of heterogeneous fluorescence decays, illustrated by native and denatured ribonuclease T_1, *Biochim. Biophys. Acta 954*, 244–252 (1988).
4. J. R. Lakowicz, I. Gryczynski, H. C. Cheung, C. Wang, and M. L. Johnson, Distance distributions in native and random-coil troponin I from frequency-domain measurements of fluorescence energy transfer, *Biopolymers 27*, 821–830 (1988).
5. J. R. Lakowicz, M. L. Johnson, I. Gryczynski, N. Joshi, and G. Laczko, Transient effects in fluorescence quenching measured by 2GHz frequency-domain fluorometry, *J. Phys. Chem. 91*, 3277–3285 (1987).
6. J. R. Lakowicz, N. B. Joshi, M. L. Johnson, H. Szmacinski, and I. Gryczynski, Diffusion coefficients of quenchers in proteins from transient effects in the intensity decays, *J. Biol. Chem. 262*, 10907–10910 (1987).
7. Z. Gaviola, Ein Fluorometer, Apparat zur Messung von Fluoreszenzabklingungszeiten, *Z. Phys. 42*, 853–861 (1926).
8. J. B. Birks and D. J. Dyson, Phase and modulation fluorometer, *J. Sci. Instrum. 38*, 282–285 (1961).
9. E. A. Bailey and G. K. Rollefson, The determination of the fluorescence lifetimes of dissolved substances by a phase shift method, *J. Chem. Phys. 21*, 1315–1326 (1953).
10. A. M. Bonch-Breuvich, I. M. Kazarin, V. A. Molchanov, and I. V. Shirokov, An experimental model of a phase fluoremeter, *Instrum. Exp. Tech. (USSR) 2*, 231–236 (1959).
11. R. D. Spencer and G. Weber, Measurement of subnanosecond fluorescence lifetimes with a cross-correlation phase fluorometer, *Ann. N.Y. Acad. Sci. 158*, 361–376 (1969).
12. E. Gratton and M. Limkeman, A continuously variable frequency cross-correlation phase fluorometer with picosecond resolution, *Biophys. J. 44*, 315–324 (1983).
13. J. R. Lakowicz and B. P. Maliwal, Construction and performance of a variable-frequency phase-modulation fluorometer, *Biophys. Chem. 21*, 61–78 (1985).
14. J. R. Lakowicz, G. Laczko, H. Cherek, E. Gratton, and M. Limkeman, Analysis of fluorescence decay kinetics from variable-frequency phase shift and modulation data, *Biophys. J. 46*, 463–477 (1984).
15. E. Gratton, M. Limkeman, J. R. Lakowicz, B. Maliwal, H. Cherek, and G. Laczko, Resolution of mixtures of fluorophores using variable-frequency phase and modulation data, *Biophys. J. 46*, 479–486 (1984).
16. J. R. Lakowicz, G. Laczko, I. Gryczynski, and H. Cherek, Measurement of subnanosecond anisotropy decays of protein fluorescence using frequency-domain fluorometry, *J. Biol. Chem. 261*, 2240–2245 (1986).
17. H. Szmacinski, R. Jayaweera, H. Cherek, and J. R. Lakowicz, Demonstration of an associated anisotropy decay by frequency-domain fluorometry, *Biophys. Chem. 27*, 233–241 (1987).
18. I. Gryczynski, H. Cherek, and J. R. Lakowicz, Detection of three rotational correlation times for a rigid asymmetric molecule using frequency-domain fluorometry, *Biophys. Chem. 30*, 271–277 (1988).
19. J. R. Lakowicz, H. Cherek, B. Maliwal, and E. Gratton, Time-resolved fluorescence anisotropies of fluorophores in solvents and lipid bilayers obtained from frequency-domain phase-modulation fluorometry, *Biochemistry 14*, 376–383 (1985).
20. J. R. Alcala, E. Gratton, and F. G. Prendergast, Resolvability of fluorescence lifetime distributions using phase fluorometry, *Biophys. J. 51*, 587–596 (1987).
21. J. R. Acala, E. Gratton, and F. G. Prendergast, Interpretation of fluorescence decays in proteins using continuous lifetime distributions, *Biophys. J. 51*, 925–936 (1987).
22. J. R. Lakowicz, H. Cherek, I. Gryczynski, N. Joshi, and M. L. Johnson, Analysis of fluorescence decay kinetics measured in the frequency-domain using distribution of decay times, *Biophys. Chem. 28*, 35–50 (1987).
23. I. Gryczynski, W. Wiczk, M. L. Johnson, and J. R. Lakowicz, End-to-end distance

distributions of flexible molecules from steady state fluorescence energy transfer and quenching-induced changes in the Förster distance. *Chem. Phys. Lett. 145*, 439–446 (1988).
24. J. R. Lakowicz, I. Gryczynski, H. C. Cheung, C. Wang, M. L. Johnson, and N. Joshi, Distance distributions in proteins recovered using frequency-domain fluorometry; applications to troponin I and its complex with troponin C, *Biochemistry 27*, 9149–9160 (1988).
25. J. R. Lakowicz, G. Laczko, and I. Gryczynski, A 2 GHz frequency-domain fluorometer, *Rev. Sci. Instrum. 57*, 2499–2506 (1986).
26. G. Laczko, I. Gryczynski, Z. Gryczynski, W. Wiczk, H. Malak, and J. R. Lakowicz, A 10 GHz frequency-domain fluorometer, *Rev. Sci. Instrum. 61*, 2331–2337 (1990).
27. P. R. Bevington, *Data Reduction and Error Analysis for the Physical Sciences*, McGraw-Hill, New York (1969).
28. G. G. Belford, R. L. Belford, and G. Weber, Dynamics of fluorescence polarization in macromolecules, *Proc. Natl. Acad. Sci. 69*, 1392–1393 (1972).
29. T. J. Chuang and K. B. Eisenthal, Theory of fluorescence depolarization by anisotropic rotational diffusion, *J. Chem. Phys. 57*, 5094–5097 (1972).
30. M. D. Barkley, A. A. Kowalczyk, and L. Brand, Fluorescence decay studies of anisotropic rotations of small molecules, *J. Chem. Phys. 75*, 3581–3593 (1978).
31. L. A. Chen, R. E. Dale, S. Roth, and L. Brand, Nanosecond time-dependent fluorescence depolarization of diphenylhexatriene in dimyristoyl-lecithin vesicles and the determination of "microviscosity," *J. Biol. Chem. 252*, 2163–2169 (1977).
32. J. Yguerabide, H. F. Epstein, and L. Stryer, Segmental mobility in an antibody molecule, *J. Mol. Biol. 51*, 573–590 (1970).
33. G. Weber, Polarization of the fluorescence of solutions, in: *Fluorescence and Phosphorescence Analysis* (D. M. Hercules, ed.), pp. 217–240, John Wiley & Sons, New York (1966).
34. G. Weber, Polarization of the fluorescence of macromolecules I: Theory and experimental method, *Biochem. J. 51*, 145–155 (1952).
35. G. Weber, Theory of differential phase fluorometry: Detection of anisotropic molecular rotations, *J. Chem. Phys. 66*, 4081–4091 (1977).
36. B. P. Maliwal and J. R. Lakowicz, Resolution of complex anisotropy decays by variable frequency phase-modulation fluorometry: A simulation study, *Biochim. Biophys. Acta 873*, 161–172 (1986).
37. B. P. Maliwal, A. Hermetter, and J. R. Lakowicz, A study of protein dynamics from anisotropy decays obtained by variable frequency phase-modulation fluorometry: Internal motions of *N*-methylanthraniloyl melittin, *Biochim. Biophys. Acta 873*, 173–181 (1986).
38. R. D. Spencer and G. Weber, Influence of Brownian rotations and energy transfer upon the measurement of fluorescence lifetimes. *J. Chem. Phys. 52*, 1654–1663 (1970).
39. W. Szymanowski, Einfluss der Rotation der Moleküle auf die Messungen der Abklingzert des Fluoreszenzstrahlung, *Z. Phys. 95*, 466–473 (1935).
40. R. K. Bauer, Polarization and decay of fluorescence of solution, *Z. Naturforsch. 18A*, 718–724 (1963).
41. A. Grinvald and I. Z. Steinberg, On the analysis of fluorescence decay kinetics by the method of least-squares, *Anal. Biochem. 59*, 583–598 (1974).
42. J. R. Taylor, *An Introduction to Error Analysis, the Study of Uncertainties in Physical Measurements*, University Science Books, Mill Valley, California (1982).
43. I. P. Kaminov, *An Introduction to Electro-Optic Devices*, Academic Press, New York (1984).
44. E. Gratton and R. Lopez-Delgado, Measuring fluorescence decay times by phase-shift and modulation techniques using the high harmonic content of pulsed light sources. *Nuovo Cimento B56*, 110–124 (1980).
45. E. Gratton, D. M. James, N. Rosato, and G. Weber, Multifrequency cross-correlation phase fluorometer using synchrotron radiation, *Rev. Sci. Instrum. 55*, 486–494 (1984).
46. A. J. W. C. Visser and A. V. Mack, The fluorescence decay of reduced nicotinamides in

aqueous solution after excitation with uv-mode locked Ar ion laser, *Photochem. Photobiol. 33*, 35–40 (1981).
47. C. J. Peters, Gigacycle-bandwidth coherent-light traveling wave amplitude modulator, *Proc. IEEE 53*, 455–460 (May 1965).
48. G. White and G. M. Chin, Traveling wave electro-optic modulators, *Opt. Commun. 5*, 374–379 (1972).
49. H. S. Merkelo, S. R. Hartman, T. Mar, G. S. Singhal, and Govindjee, Mode-locked lasers: Measurements of very fast radiative decay in fluorescent systems, *Science 164*, 301–303 (1969).
50. K. Berndt, H. Duerr, and D. Palme, Picosecond phase fluorometry by mode-locked CW lasers, *Opt. Commun. 42*, 419–422 (1982).
51. S. Kinosita and T. Kushida, Picosecond fluorescence spectroscopy by time-correlated single-photon counting, *Anal. Instrum. 14*, 503–524 (1985).
52. I. Yamazaki, N. Tamai, H. Kume, H. Tsuchiya, and K. Oba, Microchannel-plate photomultiplier: Applicability to the time-correlated photon-counting method, *Rev. Sci. Instrum. 56*, 1187–1194 (1985).
53. J. R. Lakowicz, G. Laczko, H. Szmacinski, I. Gryczynski, and W. Wiczk, Gigahertz, frequency-domain fluorometry: Resolution of complex decays, picosecond processes and future developments, *J. Photochem. Photobiol. B: Biol. 2*, 295–311 (1988).
54. A. G. Szabo and D. M. Rayner, Fluorescence decay of tryptophan conformers in aqueus solution. *J. Am. Chem. Soc. 102*, 554–563 (1980).
55. J. R. Lakowicz, R. Jayaweera, H. Szmacinski, and W. Wiczk, Resolution of two emission spectra for tryptophan using frequency-domain phase-modulation spectra, *Photochem. Photobiol. 47*, 541–546 (1989).
56. B. Alpert, D. M. Jameson, and G. Weber, Tryptophan emission from human hemoglobin and its isolated subunits, *Photochem. Photobiol. 31*, 1–4 (1980).
57. R. E. Hirsch, R. S. Zukin, and R. L. Nagel, Intrinsic fluorescence emission of intact oxy hemoglobins, *Biochem. Biophys. Res. Commun. 93*, 432–439 (1980).
58. R. E. Hirsch and R. L. Nagel, Conformational studies of hemoglobin using intrinsic fluorescence measurements, *J. Biol. Chem. 256*, 1080–1083 (1981).
59. A. G. Szabo, D. Krajcarski, M. Zuker, and B. Alpert, Conformational heterogeneity in hemoglobin as determined by picosecond fluorescence decay measurements, *Chem. Phys. Lett. 108*, 145–149 (1984).
60. J. Albani, B. Alpert, D. Krajcarski, and A. G. Szabo, A fluorescence decay time study of tryptophan in isolated hemoglobin subunits, *FEBS Lett. 182*, 302–304 (1985).
61. R. M. Hochstrasser and D. K. Negus, Picosecond fluorescence decay of tryptophans in myoglobin, *Proc. Natl. Acad. Sci. U.S.A. 81*, 4399–4403 (1989).
62. E. Bucci, H. Malak, C. Fronticelli, I. Gryczynski, and J. R. Lakowicz, Resolution of the lifetimes and correlation times of the intrinsic tryptophan fluorescence of human hemoglobin solutions using 2GHz frequency-domain fluorometry, *J. Biol. Chem. 263*, 6972–6977 (1988).
63. E. Bucci, H. Malak, C. Fronticelli, I. Gryczynski, and J. R. Lakowicz, Resolution at 2GHz of lifetimes and correlation times of highly purified solutions of human hemoglobins, *Proceedings of the International Symposium in Honor of Gregorio Weber's Seventieth Birthday*, held September 9–12, 1986, in Broca di Magra, Italy, Plenum Press, New York (1989).
64. E. Bucci, H. Malak, C. Fronticelli, I. Gryczynski, G. Laczko, and J. R. Lakowicz, Time-resolved emission spectra of hemoglobin on the picosecond time scale, *Biophys. Chem. 32*, 187–198 (1988).
65. D. R. James, D. R. Dremmer, R. P. Steer, and R. E. Verral, Fluorescence lifetime quenching and anisotropy studies of ribonuclease T_1, *Biochemistry 24*, 5517–5526 (1985).
66. J. R. Lakowicz, H. Cherek, I. Gryczynski, N. Joshi, and M. L. Johnson, Enhanced resolution of fluorescence anisotropy decay by simultaneous analysis of progressively quenched samples.

Applications to anisotropic rotations and to protein dynamics, *Biophys. J. 51*, 755–768 (1988).

67. J. R. Lakowicz, H. Szmacinski, and I. Gryczynski, Picosecond resolution of indole anisotropy decays and spectral relaxation by 2GHz frequency-domain fluorometry, *Photochem. Photobiol. 47*, 31–41 (1988).
68. I. Gryczynski, H. Cherek, G. Laczko, and J. R. Lakowicz, Enhanced resolution of anisotropic rotational diffusion by multi-wavelength frequency-domain fluorometry and global analysis, *Chem. Phys. Lett. 135*, 193–199 (1987).
69. I. Gryczynski, Z. Gryczynski, A. Kawski, and S. Paszyc, Directions of the electronic transition moments of synthetic Y_t-base, *Photochem. Photobiol. 39*, 319–322 (1984).
70. L. Davenport, J. R. Knutson, and L. Brand, Anisotropy decay associated fluorescence spectra and analysis of rotational heterogeneity. 1, 6-Diphenylhexatriene in lipid bilayers, *Biochemistry 25*, 1811–1816 (1986).
71. L. S. Stryer, The interactions of a naphthalene dye with apomyoglobin and apohemoglobin: A fluorescent probe of non-polar binding sites, *J. Mol. Biol. 13*, 482–487 (1965).
72. G. Laczko and J. R. Lakowicz, A 6 GHz frequency-domain fluorometer, *Biophys. J. 55*, 190a (Abstr.) (1989).
73. J. R. Lakowicz and G. Laczko, A 10 GHz frequency-domain fluorometer, in: *Time Resolved Laser Spectroscopy II, Proc. SPIE* (J. R. Lakowicz, ed.), *1204*, 13–20 (1990).
74. J. R. Lakowicz and G. Laczko, A 10 GHz frequency-domain fluorometer, *Rev. Sci. Instrum. 61*, 2331–2337 (1990).
75. B. F. Feddersen, D. W. Piston, and E. Gratton, Digital parallel acquisition in frequency domain fluorometry, *Rev. Sci. Instrum. 60*, 2929–2936 (1989).
76. C. Mitchell and K. Swift, The 48000 MHF_{TM}, a dual-domain Fourier transform fluorescence lifetime spectrofluorometer, *Proc. SPIE* (J. R. Lakowicz, ed.), *1204*, 270–274.
77. B. A. Feddersen, D. W. Piston, and E. Gratton, Digital parallel acquisition in frequency domain fluorimetry, *Rev. Sci. Instrum. 60*, 2929–2936 (1989).
78. E. Gratton, B. Feddersen, and M. van de Ven, Parallel acquisition of fluorescence decay using array detectors, in: *Time-Resolved Laser Spectroscopy II, Proc. SPIE* (J. R. Lakowicz, ed.), *1204*, 21–25 (1990).
79. J. R. Lakowicz, R. Jayaweera, H. Szmacinski, and W. Wiczk, Resolution of multicomponent fluorescence emission using frequency-dependent phase angle and modulation spectra, *Analytical Chem. 62*, 2005–2012 (1990).
80. J. R. Lakowicz and K. Berndt, Frequency-domain measurements of photon migration in tissues, *Chem. Phys. Lett. 166*, 246–252 (1990).

6

Fluorescence Correlation Spectroscopy

Nancy L. Thompson

6.1. Introduction

Fluorescence correlation spectroscopy (FCS) is a technique in which temporal fluctuations in the fluorescence measured from a sample of fluorescent molecules are analyzed to obtain information about the processes that give rise to the fluorescence fluctuations. Since the initial development,[1–3] FCS has been used to measure translational diffusion coefficients,[4–18] chemical kinetic rate constants,[1, 3, 8, 19–21] rotational diffusion coefficients,[22–24] flow rates,[25] molecular weights,[17, 26] and molecular aggregation.[16, 27–29, 30] Measurement of these physical parameters has been used for a range of purposes, including an increased understanding of protein, nucleic acid, and lipid diffusion,[5–7, 14, 15, 18, 23] solution and interfacial kinetics,[20, 21, 31] cell-surface receptor[27, 30] and macromolecular[16, 28, 29] clustering, and muscle contraction[8, 22, 32]; the detection of antibodies[10, 12, 31] and viruses[33, 34]; and the characterization of focused laser beams.[11, 35] In this chapter the conceptual basis and theory of FCS, aspects of FCS apparatus and experimental methods, and most of the previous experimental applications of FCS are described. FCS has been the subject of a number of previous review articles and book chapters.[13, 32, 36–51]

6.2. Conceptual Basis and Theoretical Background

The conceptual basis of FCS is illustrated in Figure 6.1. At equilibrium, fluorescent molecules move through a small open region and/or undergo

---

Nancy L. Thompson • Department of Chemistry, University of North Carolina at Chapel Hill, Chapel Hill, North Carolina 27599-3290.
Topics in Fluorescence Spectroscopy, Volume 1: Techniques, edited by Joseph R. Lakowicz. Plenum Press, New York, 1991.

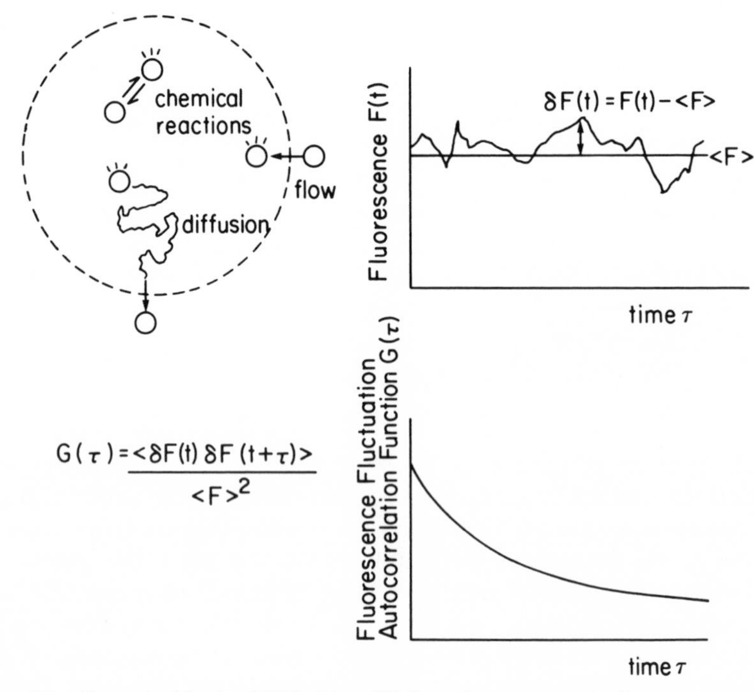

Figure 6.1. Conceptual basis of FCS. At equilibrium, fluorescent molecules are transported by diffusion or flow through an open illuminated and observed region or undergo transitions between states of different fluorescent yields, giving rise to fluctuations in the measured fluorescence. The fluctuations $\delta F(t)$ in the measured fluorescence $F(t)$ from the average fluorescence $\langle F \rangle$ are autocorrelated as $G(\tau)$. The autocorrelation function, which measures the average duration of a fluorescence fluctation, decays with time τ; the rate and shape of decay are related to the mechanisms and rates of the processes that give rise to the fluorescence fluctuations. The magnitude of $G(\tau)$ is related to the number densities and relative fluorescence yields of different chemical species in the sample region.

transitions between different states with different fluorescent yields, resulting in temporal fluctuations in the fluorescence measured from the region. The temporal autocorrelation of the fluorescence fluctuations, which measures the average duration of a fluorescence fluctuation, decays with time. The rate and shape of the decay of the autocorrelation function provide information about the mechanisms and rates of the processes that generate the fluorescence fluctuations. The magnitude of the autocorrelation function provides information about the number densities of fluorescent species in the sample region.

Experimental FCS studies have relied on theoretical predictions of the forms of the autocorrelation functions of fluorescence fluctuations that result from different processes, including chemical reactions between states of different fluorescence yields[2, 20, 52] and translational diffusion[2, 4, 53–55] or flow[25, 56–58] through the sample region. In this section the general features of the fluorescence fluctuation autocorrelation function for translational diffusion, flow, and chemical reaction are outlined. Specific examples of this general case are treated, as needed for the description of experimental results, in later sections. Theories that describe more complex situations, such as rotational motions, interfacial kinetics, and high-order autocorrelation are reserved for discussion in later sections.

In FCS, the fluorescence $F(t)$ as a function of time t is measured. For a stationary ergodic sample, the time average and the thermodynamic ensemble average of $F(t)$ are constant and identical and are denoted by $\langle F \rangle$. Defining the fluctuation in the measured fluorescence $F(t)$ from the average value $\langle F \rangle$ as

$$\delta F(t) = F(t) - \langle F \rangle \tag{6.1}$$

the normalized autocorrelation function $G(\tau)$ of temporal fluctuations in the measured fluorescence $F(t)$ is given by

$$G(\tau) = \frac{\langle \delta F(t+\tau)\, \delta F(t) \rangle}{\langle F(t) \rangle^2} = \frac{\langle \delta F(\tau)\, \delta F(0) \rangle}{\langle F \rangle^2} \tag{6.2}$$

If only one fluorescent chemical species is present in the sample region,

$$F(t) = \kappa Q \int W(\mathbf{r})\, C(\mathbf{r}, t)\, d\Omega \tag{6.3}$$

where $C(\mathbf{r}, t)$ is the number density at position $\mathbf{r}$ and time t, Q is the product of the absorptivity, fluorescence quantum efficiency, and experimental fluorescence collection efficiency of the fluorescent molecules, $d\Omega$ is an area or volume element, the integral is over all two- or three-dimensional space, and κ is a constant. The function $W(\mathbf{r})$ in Eq. (6.3) is the product of a function $I(\mathbf{r})$ that describes the spatial intensity profile of the excitation light, a function $S(\mathbf{r})$ that describes the extent of the sample, and a function $T(\mathbf{r})$ that defines the volume or area of the sample from which fluorescence is measured. Thus,

$$W(\mathbf{r}) = I(\mathbf{r})\, S(\mathbf{r})\, T(\mathbf{r}) \tag{6.4}$$

defines the small sample region. Some FCS theories have assumed that a two-dimensional sample of large extent is illuminated by a focused laser beam of Gaussian intensity profile and that detection is limited to an area slightly larger than the illuminated area by an optical aperture.[28, 57] For this experimental arrangement,

$$\begin{aligned} I(\mathbf{r}) &= I_0 \exp[-2(x^2+y^2)/s^2] \\ T(\mathbf{r}) &= \begin{cases} 1 & \text{for} \quad (x^2+y^2)^{1/2} \lesssim s \\ 0 & \text{otherwise} \end{cases} \\ S(\mathbf{r}) &= 1 \\ W(\mathbf{r}) &\approx I_0 \exp[-2(x^2+y^2)/s^2] \\ d\Omega &= d^2 r \end{aligned} \tag{6.5}$$

where s is the $1/e^2$ radius of the focused laser beam, and I_0 is a constant. Other FCS experimental geometries have assumed that a three-dimensional sample of finite extent in one dimension (a slab) is intersected perpendicularly by a laser beam.(2, 9, 25, 56) For this experimental arrangement,

$$\begin{aligned} I(\mathbf{r}) &= I_0 \exp[-2(x^2+y^2)/s^2] \\ T(\mathbf{r}) &= \begin{cases} 1 & \text{for} \quad (x^2+y^2)^{1/2} \lesssim s,\ 0 \leqslant z \leqslant L \\ 0 & \text{otherwise} \end{cases} \\ S(\mathbf{r}) &= \begin{cases} 1 & \text{for} \quad 0 < z < L \\ 0 & \text{for} \quad z \leqslant 0 \text{ and } z \geqslant L \end{cases} \\ W(\mathbf{r}) &\approx \begin{cases} I_0 \exp[-2(x^2+y^2)/s^2] & \text{for} \quad 0 < z < L \\ 0 & \text{for} \quad z \leqslant 0 \text{ and } z \geqslant L \end{cases} \\ d\Omega &= d^3 r \end{aligned} \tag{6.6}$$

where L is the sample thickness. Other experimental geometries such as a three-dimensional sample of large extent illuminated by a focused laser beam(4, 54) and an a infinite dielectric interface illuminated by a focused, totally internally reflected laser beam(59) have also been treated theoretically.

The fluctuation of the concentration at position $\mathbf{r}$ and time t from the average value $\langle C \rangle$ is

$$\delta C(\mathbf{r}, t) = C(\mathbf{r}, t) - \langle C \rangle \tag{6.7}$$

so that the average fluorescence and the fluorescence fluctuation at time t are

$$\begin{aligned} \langle F \rangle &= \kappa Q \langle C \rangle \int W(\mathbf{r})\, d\Omega \\ \delta F(t) &= \kappa Q \int \delta C(\mathbf{r}, t)\, W(\mathbf{r})\, d\Omega \end{aligned} \tag{6.8}$$

Using Eqs. (6.8) in Eq. (6.2), the normalized fluorescence fluctuation autocorrelation function is found to equal

$$G(\tau)=\frac{\iint W(\mathbf{r})\, W(\mathbf{r}')\, f(\mathbf{r},\mathbf{r}',\tau)\, d\Omega\, d\Omega'}{[\langle C\rangle \int W(\mathbf{r})\, d\Omega]^2} \tag{6.9}$$

where

$$f(\mathbf{r},\mathbf{r}',\tau)=\langle \delta C(\mathbf{r},t+\tau)\, \delta C(\mathbf{r}',t)\rangle=\langle \delta C(\mathbf{r},\tau)\, \delta C(\mathbf{r}',0)\rangle \tag{6.10}$$

is the correlation of a concentration fluctuation at position $\mathbf{r}$ with a concentration fluctuation at position $\mathbf{r}'$ a time τ earlier, and the second equality results from assuming that the sample is stationary.

The value of $G(0)$ depends on the value of the concentration correlation function $f(\mathbf{r},\mathbf{r}',\tau)$ when τ equals 0, which is the correlation of concentration fluctuations at position $\mathbf{r}$ with concentration fluctuations at position $\mathbf{r}'$, *at the same time*. For noninteracting solute molecules, concentration fluctuations are correlated at the same time only at the same position,[2, 28, 52] so that

$$f(\mathbf{r},\mathbf{r}',0)=\langle C\rangle\, \delta(\mathbf{r}-\mathbf{r}') \tag{6.11}$$

where $\delta(\ )$ is the Dirac delta function. Using Eq. (6.11) in Eq. (6.9),

$$G(0)=\gamma/\langle N\rangle \tag{6.12}$$

where the constant γ is given by

$$\gamma=w_2/w_1 \tag{6.13}$$

the constants w_n are defined as

$$w_n=\int [W(\mathbf{r})/W(0)]^n\, d\Omega \tag{6.14}$$

and the average number of molecules in the sample region $\langle N\rangle$, defined to be consistent with the number of molecules contained in a finite but uniformly illuminated region, is†

$$\langle N\rangle=w_1\langle C\rangle \tag{6.15}$$

Thus, $G(0)$ depends on the number of fluorescent molecules in the sample region, $\langle N\rangle$, and weakly depends on the experimental geometry through the

† Some results in this chapter appear different from the form in which they were originally published; the discrepancy arises from different definitions of $\langle N\rangle$ (Eq. 6.15) and $\langle N_i\rangle$ (Eq. 6.24).

constant γ. For the experimental geometries defined in Eqs. (6.5) and (6.6), the value of γ is $\frac{1}{2}$.

For a sample that contains several different chemical species with different relative fluorescence yields, the measured fluorescence as a function of time is

$$F(t)=\kappa \sum_{i=1}^{R} Q_i \int W(\mathbf{r})\, C_i(\mathbf{r}, t)\, d\Omega \tag{6.16}$$

where $C_i(\mathbf{r}, t)$ is the concentration of the ith fluorescent chemical species at position $\mathbf{r}$ and time t, R is the number of fluorescent species, and Q_i is the fluorescence yield of the ith species. The fluctuation in the concentration of the ith species at position $\mathbf{r}$ and time t from the average value $\langle C_i \rangle$ is

$$\delta C_i(\mathbf{r}, t)=C_i(\mathbf{r}, t)-\langle C_i \rangle \tag{6.17}$$

so that the average fluorescence and the fluorescence fluctuation at time t equal

$$\begin{aligned}\langle F \rangle &=\kappa \sum_{i=1}^{R} Q_i \langle C_i \rangle \int W(\mathbf{r})\, d\Omega \\ \delta F(t) &=\kappa \sum_{i=1}^{R} Q_i \int W(\mathbf{r})\, \delta C_i(\mathbf{r}, t)\, d\Omega\end{aligned} \tag{6.18}$$

Using Eqs. (6.18) in Eq. (6.2), the autocorrelation function is found to equal

$$G(\tau)=\frac{\sum_{i=1}^{R} \sum_{j=1}^{R} Q_i Q_j \iint W(\mathbf{r})\, W(\mathbf{r}')\, f_{ij}(\mathbf{r}, \mathbf{r}', \tau)\, d\Omega\, d\Omega'}{[\sum_{i=1}^{R} Q_i \langle C_i \rangle \int W(\mathbf{r})\, d\Omega]^2} \tag{6.19}$$

where

$$\begin{aligned}f_{ij}(\mathbf{r}, \mathbf{r}', \tau) &= \langle \delta C_i(\mathbf{r}, t+\tau)\, \delta C_j(\mathbf{r}', t) \rangle \\ &= \langle \delta C_i(\mathbf{r}, \tau)\, \delta C_j(\mathbf{r}', 0) \rangle\end{aligned} \tag{6.20}$$

In Eqs. (6.19) and (6.20), $f_{ij}(\mathbf{r}, \mathbf{r}', \tau)$ is the correlation function of a concentration fluctuation in the ith chemical species at position $\mathbf{r}$ with a concentration fluctuation in the jth chemical species at position $\mathbf{r}'$ and a time τ earlier.

The value of $G(0)$ depends on the values of the concentration correlation function $f_{ij}(\mathbf{r}, \mathbf{r}', \tau)$ when τ equals 0, which is the correlation of concentration fluctuations in the ith species at position $\mathbf{r}$ with concentration fluctuations in the jth species at position $\mathbf{r}'$, *at the same time*. For ideal solutions containing multiple components, these two fluctuations are correlated at the same time only for the same chemical species and at the same position,[2, 28, 52] so that

$$f_{ij}(\mathbf{r}, \mathbf{r}', 0)=\delta_{ij} \langle C_i \rangle\, \delta(\mathbf{r}-\mathbf{r}') \tag{6.21}$$

where δ_{ij} is the Kronecker delta. Using Eq. (6.21) in Eq. (6.19),

$$G(0) = \frac{\gamma \sum_{i=1}^{R} \alpha_i^2 \langle N_i \rangle}{[\sum_{i=1}^{R} \alpha_i \langle N_i \rangle]^2} \tag{6.22}$$

where γ is defined in Eq. (6.13), the relative fluorescence yield of the ith species α_i is defined as

$$\alpha_i = Q_i/Q_1 \tag{6.23}$$

and the average number of molecules of the ith type in the sample region, $\langle N_i \rangle$, is defined as

$$\langle N_i \rangle = w_1 \langle C_i \rangle \tag{6.24}$$

$G(0)$ depends on the number densities and relative quantum yields of the fluorescent species present in the sample and on the shape of the sample region through the constant γ.

The rate and shape of decay of $G(\tau)$ is determined by the concentration correlation function $f(\mathbf{r}, \mathbf{r}', t)$ (single species) or the concentration correlation functions $f_{ij}(\mathbf{r}, \mathbf{r}', \tau)$ (multiple species), which depend on the process or processes that result in fluctuations in the number of fluorescent molecules in the sample region with time (e.g., diffusion, chemical reaction, flow, or a combination of these processes) and on the shape of the sample region $W(\mathbf{r})$, which is determined by the experimental apparatus. Calculations of $f(\mathbf{r}, \mathbf{r}', t)$ and $f_{ij}(\mathbf{r}, \mathbf{r}', \tau)$ and the corresponding shapes of $G(\tau)$ are given in later sections. At long separation times τ, fluorescence fluctuations are not correlated, so that, referring to Eq. (6.2),

$$\lim_{\tau \to \infty} [G(\tau)] = \lim_{\tau \to \infty} \left[\frac{\langle \delta F(t)\, \delta F(t+\tau) \rangle}{\langle F \rangle^2} \right] = \frac{\langle \delta F(t) \rangle \langle \delta F(t+\tau) \rangle}{\langle F \rangle^2} = 0 \tag{6.25}$$

since the average value of the fluorescence fluctuations is zero.

6.3. Experimental Apparatus and Methods

In a typical FCS apparatus, as shown in Figure 6.2, a light beam illuminates a small region in the sample and excites fluorescent molecules. Emitted fluorescence passes through a filter that blocks scattered excitation light and an aperture that restricts observation to a small region in the sample, and is detected by a photomultiplier. The photomultiplier output is recorded, and fluctuations from the average signal are autocorrelated. As shown, a number of specialized components may also be included in the apparatus. First,

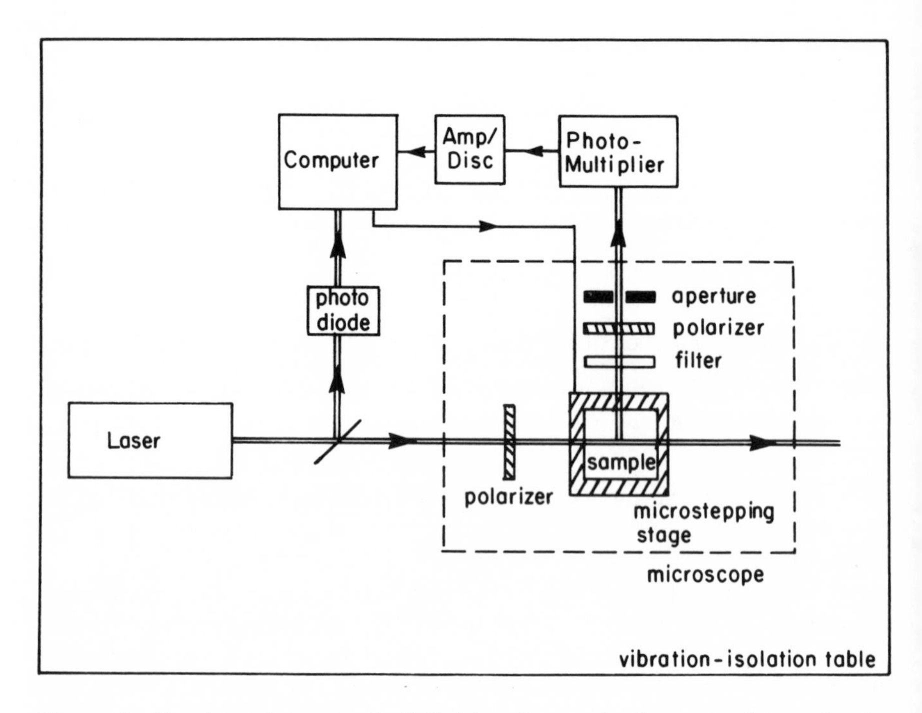

Figure 6.2. Experimental apparatus for FCS. A laser beam excites fluorescence in a small region of the sample. Fluorescence is measured through an aperture which defines the observed volume and reduces background intensity. The measured fluorescence is detected by a photomultiplier interfaced to a computer. Fluctuations in laser intensity are monitored by a photodiode and corrected for electronically (see Figure 6.4). For experiments that employ sample translation, a computer-controlled microstepping stage translates the sample through the illuminated region. For experiments designed to detect orientational molecular motion, polarizers are placed in the excitation and/or emission beams. A number of FCS instruments are designed around optical microscopes (Figure 6.3). The entire apparatus is mounted on a vibration-isolation table.

although a laser source is not necessarily required, most, if not all, successful FCS experiments to date have employed argon ion or krypton ion laser beams as excitation sources. Data may be corrected for fluctuations in laser intensity, which are monitored by a photodiode. Also, in FCS experiments designed to measure orientational motions or rotational diffusion coefficients (Section 6.6.2), polarizers are introduced in the excitation and/or emission paths. In experiments requiring sample translation (Sections 6.4.3 and 6.5.2), a computer-controlled microstepping translator may be included.[58] For fluorescence detection, most FCS instruments have used photomultipliers in the single-photon counting mode, followed by an amplifier, a discriminator, and a pulse-counting computer interface (e.g., Refs. 4, 5, 8, 9, 14, 16, 28, 29, 31, and 60), but photomultipliers in the analog mode[3, 10, 19, 26] have also been used. Finally, the laser, sample, and supporting optics are usually mounted on a vibration-isolated air table.

Many FCS instruments have been designed around optical microscopes (e.g., Refs. 4, 8, 14, 17, 19, 20, 28, 31, 33, and 61), as shown in Figure 6.3. With a microscope, the sample can be inspected visually on a microscopic scale, spatial resolution on a microscopic scale can be obtained, background intensity is usually reduced, and the same instrument can be used for other techniques in fluorescence microscopy such as fluorescence photobleaching recovery[4] and total internal reflection fluorescence.[31] For FCS on a microscope, the laser beam is passed through auxiliary lenses or optical fibers[61] into the epi-illumination port of an upright or inverted optical microscope, reflected by a dichroic mirror, and directed to the sample through

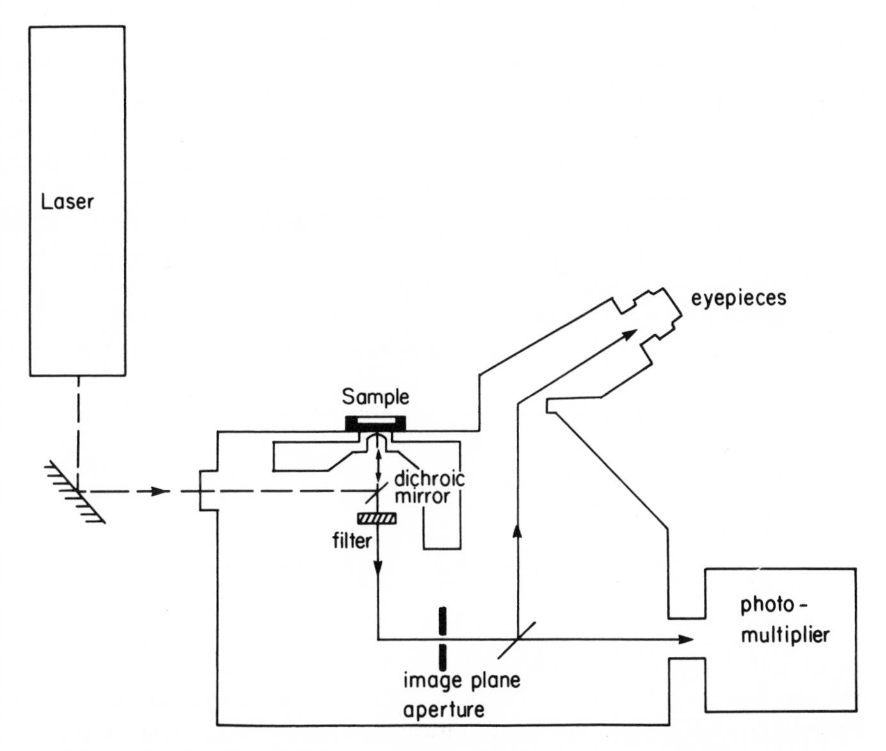

Figure 6.3. Experimental apparatus for FCS on an optical microscope. The laser beam passes through focusing optics and into the epi-fluorescence port of the microscope, is reflected by a dichroic mirror, and is focused by the microscope objective onto the sample. Fluorescence is collected with the microscope objective, passes through the dichroic mirror, an additional filter, an aperture at an intermediate focal plane that reduces background intensity and defines the observed sample region, and is detected by a photomultiplier attached to an output port of the microscope. Shown is an inverted fluorescence microscope; FCS instruments may also be constructed from upright microscopes.

the objective (usually of high numerical aperture to increase the fluorescence collection efficiency). Fluorescence originating from the sample is collected by the objective and passes through a filter block and a small aperture mounted in an intermediate focal plane of the microscope to reduce the detection of background light and fluorescence resulting from scattered excitation light. The photomultiplier is coupled to an output port of the microscope.

Referring to Eqs. (6.12) and (6.22), one sees that the magnitude of $G(\tau)$ is inversely proportional to the number of molecules in the observed volume. Because other sources of noise are always present at some level, the inverse of the number of molecules in the observed volume must be greater than the autocorrelation of the "background" noise, which has typically been on the order of 10^{-4} to 10^{-3}.[31, 58] Thus, a special requirement of FCS is that less than 10^3 molecules reside in the sample region, which means that the sample region must be small. For solution concentrations of 10^{-7} to 10^{-11} M, the sample region must be less than 10^{-14} liter [$\sim(2.2\ \mu\text{m})^3$] or 10^{-10} liter [$\sim(46\ \mu\text{m})^3$], respectively. For two-dimensional samples (e.g., biological membranes or thin films), the requirement on the characteristic *length* of the small sample region is less stringent. Small sample regions in FCS have most frequently been defined by focused laser beams.

Some FCS experiments have employed dedicated correlators for calculating $G(\tau)$.[3, 17, 19, 61] Alternatively, in a number of FCS instruments, the photomultiplier is interfaced to a computer, and the fluorescence fluctuations are autocorrelated in software (e.g., Refs. 5, 14, 16, 28, 29, 31, and 62). In this scheme, for photomultipliers in the single-photon counting mode, the number of photoelectron pulses n_j occurring between consecutive sample times $j\,\Delta T$ and $(j+1)\,\Delta T$ is counted. For photomultipliers in the analog mode, an analog-to-digital converter is employed.[10, 63, 64] The sample time ΔT is specified by the user and typically ranges from 1 μs to several seconds. Using Eq. (6.1) in Eq. (6.2), one finds that

$$G(\tau)=\frac{[\langle F(t)\,F(t+\tau)\rangle-\langle F\rangle^2]}{\langle F\rangle^2}=\frac{[\langle F(t)\,F(t-\tau)\rangle-\langle F\rangle^2]}{\langle F\rangle^2} \qquad (6.26)$$

Using Eq. (6.26), the obtained values of n_j, the user-specified value of ΔT, and the user-specified number of points P in the calculated autocorrelation function (typically 10 to 200), the software estimates $G(i\,\Delta T)$, for $i=0$ to P, according to the following algorithm:

$$G(i\,\Delta T)\approx H(i\,\Delta T)=\frac{[(M-P+1)^{-1}\sum_{j=P}^{M} n_j n_{j-i}]}{(M^{-1}\sum_{j=1}^{M} n_j)^2}-1 \qquad (6.27)$$

In practice, the sum of the products $n_j n_{j-i}$ in Eq. (6.27) becomes large, and this algorithm may require floating-point arithmetic. A faster algorithm that may require only integer arithmetic can be formulated as follows. For random signals,

$$\langle F(t)\rangle^2 = \lim_{\tau \to \infty} \langle F(t)\, F(t+\tau)\rangle \tag{6.28}$$

so that

$$G(\tau) = \frac{[\langle F(t)\, F(t-\tau)\rangle - \langle F(t)\, F(t-\tau_\infty\rangle]}{[\langle F(t)\, F(t-\tau_\infty)\rangle]} \tag{6.29}$$

where τ_∞ is a time much longer than the characteristic time of the fluorescence fluctuations. Thus,

$$G(i\,\Delta T) \approx H(i\,\Delta T) = \frac{\sum_{j=P}^{M} (n_j n_{j-i} - n_j n_{j-k})}{\sum_{j=P}^{M} n_j n_{j-k}} \tag{6.30}$$

where M is the total number of acquired data points n_j, and k is much larger than the ratio of the correlation time to the sample time ΔT. The sum over $n_j(n_{j-i} - n_{j-k})$ becomes large more slowly than the sum $n_j n_{j-i}$ required by the algorithm in Eq. (6.27). Equation (6.30) also has the advantage that slow fluctuations in the mean are accounted for locally. For slow sample times ΔT, $G(i\,\Delta T)$ may be calculated while the data are collected; some FCS instruments display the current value of $G(i\,\Delta T)$ to allow observation of the autocorrelation function as it develops.

For low fluorescent photon count rates, the background intensity may constitute a substantial portion of the measured intensity. If the measured intensity due to number fluctuations is $F(t)$ and to background is $B(t)$, then the autocorrelation function $H(i\,\Delta T)$ calculated by the algorithm of Eq. (6.30) is

$$\begin{aligned} H(i\,\Delta T) &= \frac{\langle [F(t) + B(t)][F(t+\tau) + B(t+\tau)]\rangle - \langle F(t) + B(t)\rangle^2}{\langle F(t) + B(t)\rangle^2} \\ &= \frac{[\langle F(t)\rangle]^2}{[\langle F(t)\rangle + \langle B(t)\rangle]^2}\, G(i\,\Delta T) \end{aligned} \tag{6.31}$$

This expression is based on the assumptions that fluctuations in the background are not correlated with fluorescence fluctuations and are not autocorrelated on the time scale of interest. The corrected autocorrelation function $G(i\,\Delta T)$ is obtained through the multiplicative factor shown in Eq. (6.31).[20, 28, 31]

In some cases, laser intensity fluctuations have durations comparable to the characteristic time of the decay of $G(\tau)$. Koppel *et al.*[4] have described a method for correcting for such fluctuations, illustrated in Figure 6.4, which has been adopted by other experimenters.[8, 9, 31] A small portion of the

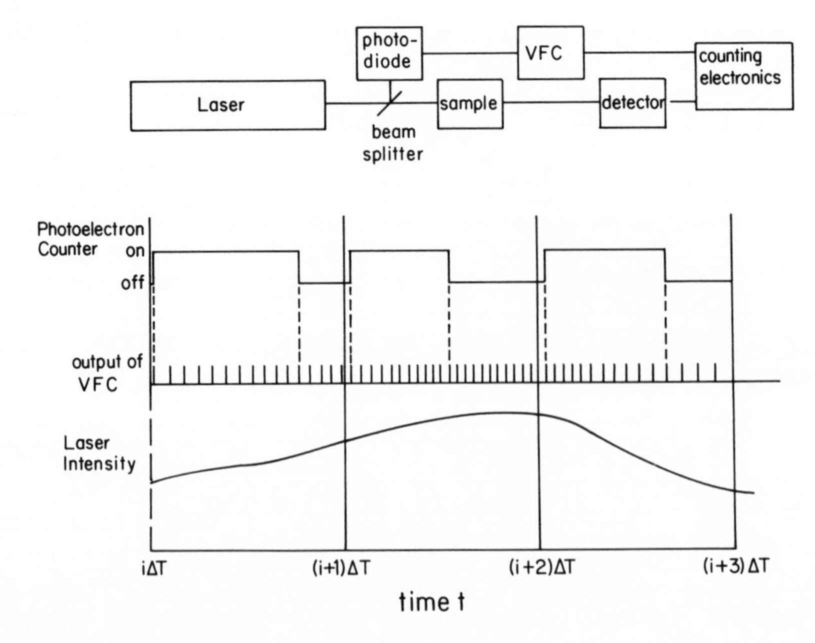

Figure 6.4. Corrections for laser intensity fluctuations using a voltage-to-frequency converter. The laser intensity is monitored by a photodiode whose output is converted to a frequency proportional to the laser intensity by a voltage-to-frequency converter (VFC). The actual count time within a sample time ΔT always begins with the first VFC pulse, but the duration of the count time is determined by the number of VFC pulses, normalizing the actual count time to the integral of the laser intensity. In the example shown here, the number of VFC pulses per count time is 11.

incident laser beam is diverted to a photodiode, the output of the photodiode passes to a voltage-to-frequency converter (VFC), and the output of the VFC passes to the computer interface. Actual count times begin at the first VFC pulse after the beginning of a sample time ΔT and continue for a given number of VFC pulses. The time integral of the laser intensity remains constant, thereby effectively normalizing the actual count time to the excitation light intensity. Methods of correcting for laser intensity fluctuations in FCS when photomultipliers in the analog mode are used have also been developed.[19]

The experimental value of $G(\tau)$ at $\tau = 0$ contains a substantial component due to the noise associated with photon counting (shot noise), which is not included in Eq. (6.12), (6.22), (6.27), or (6.30). In practice, values of $G(0)$ equivalent to Eqs. (6.12) and (6.20) are obtained from experimental functions $G(\tau)$ by extrapolating from times $\tau > 0$.

Of interest is how the statistical accuracy of a fluorescence fluctuation autocorrelation function depends on experimental parameters.[3, 40, 65–68] The signal-to-noise ratio is defined as

$$\sigma = \lim_{i \to 0} \{G_{\text{theor}}(i\,\Delta T)/[\text{variance of } G_{\text{exp}}(i\,\Delta T)]^{1/2}\} \tag{6.32}$$

where $G_{\text{theor}}(i\,\Delta T)$ is given by the numerator of Eq. (6.9), and $G_{\text{exp}}(i\,\Delta T)$ is given by the numerator of Eq. (6.27) or (6.30). Defining τ_c as the characteristic decay time of $G(\tau)$, and assuming that $\Delta T << \tau_c << M\,\Delta T$ and that $\langle N \rangle >> 1$, Koppel[65] has identified the key parameters that determine the value of σ for a single-species FCS autocorrelation function with an exponential dependence on τ and for a case in which the background intensity is negligible:

$$\sigma \approx \begin{cases} M^{1/2}\rho & \rho << 1 \\ (M\,\Delta T/\tau_c)^{1/2} & \rho \geqslant 1 \end{cases} \tag{6.33}$$

Table 6.1. Fluorophores Used in FCS[a]

Fluorophore	$\lambda_{\text{ex}}^{\max}$ (nm)	$\varepsilon_{\text{ex}}^{\max}$ ($10^3\ M^{-1}\ \text{cm}^{-1}$)	Q_f	Characteristics	Reference(s) (applications)
B-Phycoerythrin	545	240	0.98	Light-harvesting protein	16, 35, 70
6-Carboxyfluorescein	490	76	>0.3	Small dye	8, 61
Dioctadecyltetra-methylindocarbocyanine	540	133	0.1–0.3	Lipid membrane probe	4–6, 28
Ethidium bromide	515	3.8[b–d]	0.01[b,d] 0.2[c,d]	Nucleic acid label	1, 3, 18–21, 26, 33, 34
Fluorescein isothiocyanate	490	75	0.4–0.5	Amino reactive	10, 14, 17, 30
Iodoacetamidofluorescein	493	75	0.6–0.9	Sulfhydryl reactive	22
Merocyanine 540	500	5.8		Potential sensitive membrane probe	7
Rhodamine 6G	530[e]	106[e]	≃0.1[d]	Small dye	3, 9, 11, 25, 35, 70, 78
Tetramethylrhodamine iodoacetamide	550	>25	0.1–0.3	Sulfhydryl reactive	22
Tetramethylrhodamine isothiocyanate	550	>55	0.1–0.3	Amino reactive	9, 15, 16, 23, 27, 31
Texas Red	596	85	>0.3	Amino reactive	24

[a] Shown are the wavelengths $\lambda_{\text{ex}}^{\max}$ and the extinction coefficients $\varepsilon_{\text{ex}}^{\max}$ of maximum absorption, the approximate fluorescence quantum efficiencies Q_f, and references to applications of the fluorophores in FCS experiments. Typical photodegradation quantum yields are 2×10^{-5} for rhodamine 6G (Ref. 3), 7×10^{-5} for ethidium bromide (Ref. 3), and 5×10^{-4} for fluorescein-labeled proteins (Ref. 8). Values are sensitive to the environment of probes and are therefore approximate. Numerical values are from Ref. 87 unless otherwise noted.

[b] Free in solution.

[c] Complexed with nucleic acid.

[d] Values from Ref. 3.

[e] Values from Ref. 86.

where the average number of photoelectron pulses detected per sample time per fluorescent molecule ρ is

$$\rho = \langle n \rangle / [\langle N \rangle \Delta T] \tag{6.34}$$

and M is the total number of data points n_i taken, as defined above.

Even though they are derived for a special case, Eqs. (6.33) may be taken as a guide to methods of maximizing σ. Accordingly, the simplest method of

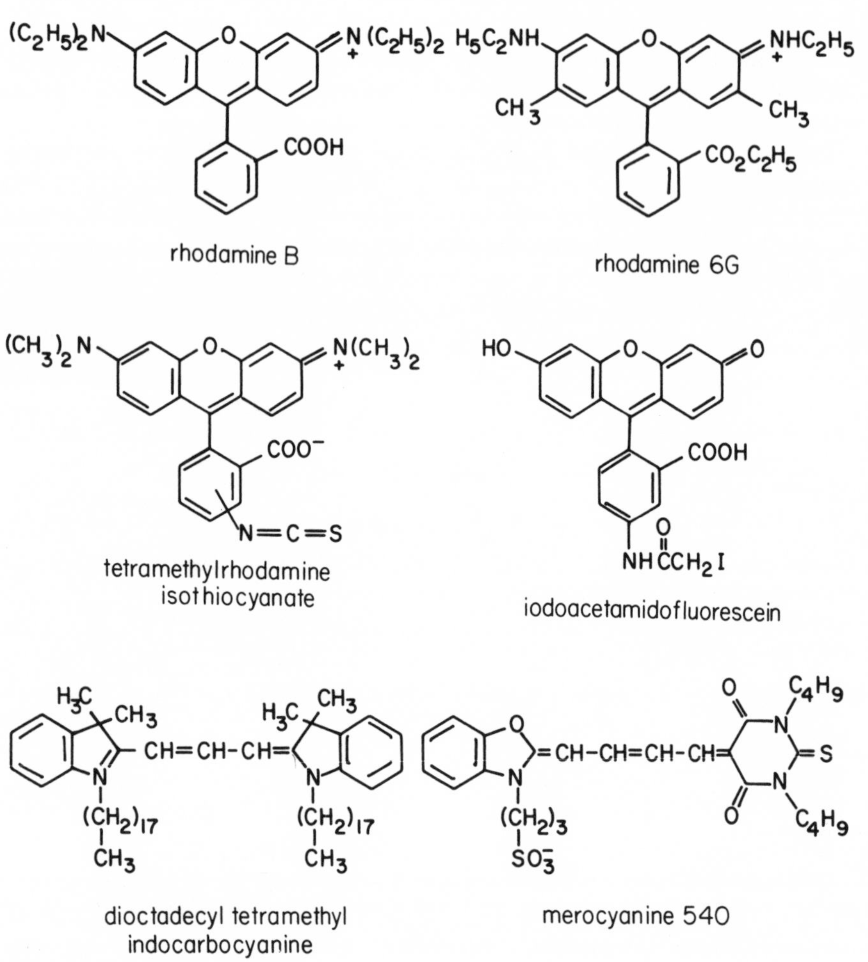

rhodamine B

rhodamine 6G

tetramethylrhodamine isothiocyanate

iodoacetamidofluorescein

dioctadecyl tetramethyl indocarbocyanine

merocyanine 540

Figure 6.5. Fluorophores used in FCS. Shown are examples of fluorescent probes that have been used in FCS. Spectroscopic data for these and other FCS probes are summarized in Table 6.1.

increasing σ, for a given experiment, is to increase the number of data points M by increasing the data acquisition time. Another method is to increase the photoelectron pulse rate per molecule until $\rho \approx 1$; increasing ρ further does not significantly increase σ. The value of ρ may be raised by increasing the laser intensity, unless the maximum count rate of the photomultiplier is exceeded or the fluorophores are photochemically bleached during their residency in the illuminated area. If the latter problem arises, the sample may be translated through the illuminated area to reduce the average residency time. Alternatively, some samples may tolerate the addition of compounds that inhibit bleaching, such as $Na_2S_2O_4$[33, 34] or phenylenediamine[17]; deoxygenation may also decrease the rate of photochemical bleaching.[69] Another method of raising ρ is to increase the fluorescence collection efficiency by placing the sample at the focus of a parabolic mirror.[9]

Because the signal-to-noise ratio of an FCS autocorrelation function, σ, as defined in Eq. (6.33), increases up to a point with the overall fluorescence yield ρ, fluorophores with high absorptivities, high fluorescence quantum yields, and low photodegradation quantum yields have been preferred for use in FCS. Shown in Figure 6.5 are the structures of many of the fluorophores that have thus far been used in FCS. Table 6.1 compares the optical properties of these and other FCS fluorophores.

6.4. Analysis of Autocorrelation Function Magnitudes

FCS autocorrelation functions contain two types of information: the magnitude $G(0)$ and the rate and shape of the temporal decay $G(\tau)/G(0)$. The first property, $G(0)$, contains unique information not readily accessible through other techniques and has been used to measure number densities and molecular weights and to detect and characterize molecular aggregation.

6.4.1. Number Densities

For a single fluorescent species, the extrapolated time-zero value of the fluorescence fluctuation autocorrelation function $G(0)$ depends on the average number of molecules $\langle N \rangle$ in the illuminated and observed region, as shown in Eq. (6.12). If the size and shape of the sample region are known, then, referring to Eqs. (6.12), (6.13), and (6.15), the number density $\langle C \rangle$ is

$$\langle C \rangle = w_2/[w_1^2 G(0)] \tag{6.35}$$

where constants w_n are defined in Eq. (6.14). This relationship has been used to calibrate the concentration of fluorescent molecules in solutions,[9, 11, 16–19, 21, 23, 25, 29] in membranes,[5–7, 17, 18, 30] and at surfaces.[31]†

6.4.2. Molecular Weights

If $G(0)$ is measured in a monodisperse sample for which the weight density $\langle W \rangle$ is known, then, using Eq. (6.35), the molecular weight M_W equals

$$M_W = \langle W \rangle N_0 / \langle C \rangle = [\langle W \rangle N_0 w_1^2 G(0)]/w_2 \tag{6.36}$$

where N_0 is Avogadro's number. For a polydisperse sample, if the relative fluorescence yield α_i of the ith species is proportional to the molecular weight M_i, then using Eqs. (6.13), (6.22), (6.24), and (6.36), the measured quantity M_W equals

$$M_W = \frac{\sum_{i=1}^{R} M_i^2 \langle C_i \rangle}{\sum_{i=1}^{R} M_i \langle C_i \rangle} \tag{6.37}$$

which is the weight-average molecular weight.

In one of the first applications of FCS, this approach was used to determine the molecular weight of DNA molecules stained with ethidium bromide.[26] A focused laser beam excited fluorescence from a small volume in a cylindrical chamber that rotated with constant angular velocity. The period T of the chamber rotation was much shorter than the characteristic time for diffusion of DNA molecules through the illuminated volume. $G(\tau)$ was peaked at times nT, for integral n. The molecular weight was related to the value of $G(nT) = G(0)$ according to Eq. (6.36); accuracy was estimated at 5–10%. A more recent version of this experimental design is one in which the laser beam is focused through a microscope objective and translated so that the beam travels through a circular path in a planar sample.[17] Using this design, the molecular weight of latex beads, several proteins, and fluorescein isothiocyanate were determined with good accuracy.

† In many FCS experiments in which the concentration of fluorescent molecules is independently known, the value of the concentration determined from the extrapolated time-zero value of $G(\tau)$ according to Eq. (6.35) has been lower than the known concentration. (One exception is noted.[3]) This effect has been attributed to uncertainties in the optical geometry, to photochemical bleaching, to the presence of aggregates, or to intermolecular interactions. Uncertainties in the size or shape of the illuminated region can be overcome by calibrating the effective volume with solutions of known concentrations.[17, 35]

6.4.3. Molecular Aggregation and Polydispersity

Macromolecular self-association is of considerable importance in many biological systems. Because of the sensitivity to number densities, FCS is particularly well suited to studying these processes. A review of FCS as a method for examining molecular aggregation has recently been published.[51]

In a sample in which monomeric macromolecules aggregate to form a monodisperse solution of oligomers, the number of monomeric subunits in the sample region, $\langle N_m \rangle$, might be known, but the number of monomers per oligomer, p, might not be known. If the sample is ideal in that interactions between aggregates are negligible, the inverse of $G(0)$ is related to the number of independently mobile oligomers $\langle N_o \rangle$. Referring to Eq. (6.12),

$$p = \langle N_m \rangle / \langle N_o \rangle = \langle N_m \rangle \, G(0)/\gamma \tag{6.38}$$

and the extrapolated time-zero value of $G(\tau)$ yields the value of p.

Even for a polydisperse sample, some information may be obtained about the state of oligomerization of fluorescent or fluorescently labeled molecules in a sample. If the ith species represents an oligomer on i molecules, and if the monomers do not undergo a change in fluorescence yield upon oligomerization, so that α_i is proportional to i, then Eq. (6.22) becomes

$$G(0) = \frac{\gamma \sum_{i=1}^{R} i^2 \langle N_i \rangle}{[\sum_{i=1}^{R} i \langle N_i \rangle]^2} \tag{6.39}$$

Since the total number of monomers in the sample region equals

$$N_m = \sum_{i=1}^{R} i \langle N_i \rangle \tag{6.40}$$

then

$$G(0) = \gamma N_m^{-2} \sum_{i=1}^{R} i^2 \langle N_i \rangle \geqslant \gamma N_m^{-2} \sum_{i=1}^{R} i \langle N_i \rangle = \gamma / N_m \tag{6.41}$$

Thus, the minimum possible value of $G(0)$ is inversely proportional to the average total number of monomers in the observation area, and a measured value of $G(0)$ larger than γ/N_m implies some oligomerization.

The sensitivity of $G(0)$ to molecular aggregation is the basis of a technique called "scanning-FCS," or S-FCS.[27, 30, 46, 58] In S-FCS, a laser beam is focused through a microscope objective to a small spot on a cell membrane that contains or has bound fluorescently labeled molecules in aggregates. The membrane is translated laterally through the beam by a computer-controlled microstepping stage, at a speed such that the characteristic time for diffusion of the labeled molecules through the illuminated area is much larger than the characteristic time for translation through the illuminated area. The measured value of $G(0)$ is analyzed to yield information about the state of oligomerization of the labeled cell-surface molecules.

S-FCS was first used to compare aggregation of Sindbis and vesicular stomatitis virus glycoproteins on cell surfaces[27] and has recently been used to examine fluorescently labeled succinylated concanavalin A (ConA) bound to 3T3 Swiss mouse fibroblasts.[30] The second set of measurements showed that $G(0)$ *increased* with the two-dimensional ConA density on the cell membranes; the explanation for this surprising result was that ConA receptors were present both as small and large aggregates with ConA affinities that were high and low, respectively, so that as the ConA surface density increased, more large aggregates were occupied, giving rise to a larger $G(0)$. A version of S-FCS has also been applied to the aggregation of the matrix protein from *Escherichia coli* outer membranes in planar phospholipid membranes and to Ca^{2+}-dependent aggregation of phospholipid vesicles containing negatively charged phospholipids.[17] These three experimental demonstrations of S-FCS validate this promising new approach toward the difficult problem of detecting and characterizing receptor clusters on the surfaces of viable cells.

6.5. Analysis of Autocorrelation Function Temporal Decays

The rate and shape of the temporal decay of FCS autocorrelation functions contain information about the transport and/or kinetic processes that give rise to the fluorescence fluctuations. Analysis of $G(\tau)/G(0)$ has thus far been most widely applied to the measurement of translational diffusion coefficients, although other temporal processes such as flow, rotational mobility, and chemical kinetics have also been examined.

6.5.1. Translational Diffusion

For a single fluorescent species undergoing simple translational diffusion with coefficient D,[2, 4, 25, 56–58]

$$\partial C(\mathbf{r}, t)/\partial t = D\,\nabla^2 C(\mathbf{r}, t) \tag{6.42}$$

Using Eq. (6.7) in Eq. (6.42) and Fourier transforming $\mathbf{r}$ to $\mathbf{q}$ yields

$$\partial\, \delta C(\mathbf{q}, t)/\partial t = -Dq^2\, \delta C(\mathbf{q}, t) \tag{6.43}$$

or

$$\delta C(\mathbf{q}, t) = \delta C(\mathbf{q}, 0) \exp(-Dq^2 t) \tag{6.44}$$

Multiplying Eq. (6.44) by $\delta C(\mathbf{r}', 0)$, taking an ensemble average, using the Fourier transform of Eq. (6.11), and inverse Fourier transforming $\mathbf{q}$ to $\mathbf{r}$ results in the following expression:

$$f(\mathbf{r}, \mathbf{r}', \tau) = (4\pi D\tau)^{n/2} \exp[-|\mathbf{r} - \mathbf{r}'|^2/(4D\tau)] \tag{6.45}$$

where n is the dimensionality (2 or 3). In Eq. (6.45), $f(\mathbf{r}, \mathbf{r}', \tau)$ is the probability of finding a diffusive molecule at position $\mathbf{r}$ and time t, given that the molecule was at position $\mathbf{r}'$ at time zero. Using Eq. (6.45) in Eq. (6.9) along with the two-dimensional experimental geometry defined in Eq. (6.5) or (6.6) gives the following result for $G(\tau)$:

$$G(\tau) = \frac{G(0)}{1 + (\tau/\tau_D)} \tag{6.46}$$

where the characteristic time τ_D for diffusion through the area illuminated by the focused laser beam is

$$\tau_D = s^2/4D \tag{6.47}$$

and s is the $1/e^2$ radius of the focused laser beam. A plot of $G^{-1}(\tau)$ as a function of τ is linear:

$$G^{-1}(\tau) = G^{-1}(0) + [\tau_D G(0)]^{-1}\,\tau \tag{6.48}$$

and measurement of the slope and intercept yields a measure of τ_D. Independent measurement of the beam's $1/e^2$ radius s yields the diffusion coefficient D as

$$D = s^2/4\tau_D \tag{6.49}$$

For more than one fluorescent species, a similar analysis yields

$$G(\tau) = \frac{\gamma \sum_{i=1}^{R} \alpha_i^2 \langle N_i \rangle [1 + (\tau/\tau_{D_i})]^{-1}}{[\sum_{i=1}^{R} \alpha_i \langle N_i \rangle]^2} \tag{6.50}$$

where

$$\tau_{D_i} = s^2/4D_i \tag{6.51}$$

is the characteristic time for diffusion of the ith species through the illuminated area. A plot of $G^{-1}(\tau)$ as a function of τ is not linear. Also, because terms due to the diffusion of different species are additive, $G(\tau)$ is sensitive to diffusion with a characteristic time in the range of autocorrelation, but diffusional terms with shorter characteristic times appear only in the $G(0)$ channel, and longer diffusion times can be accounted for by the method in which the average of the square of the fluorescence is subtracted (Eq. 6.30). For an experimental geometry in which a three-dimensional sample of large spatial extent is illuminated by a focused Gaussian laser beam, Eqs. (6.46) and (6.50) are not accurate due to contributions from diffusion in out-of-focus planes and to diffusion along the beam propagation direction, so that more complex expressions may be required.[54]

FCS autocorrelation functions for dilute solutions of fluorescent dyes (e.g., rhodamine 6G and ethidium bromide) have yielded reasonable diffusion coefficients ($\sim 3 \times 10^{-6}$ cm^2/s) according to Eqs. (6.46) and (6.47).[3, 9, 11, 35, 70] An FCS autocorrelation function for rhodamine 6G diffusing in ethanol is shown in Figure 6.6. FCS autocorrelation functions for fluorescent or fluorescently labeled proteins and nucleic acids in dilute solutions have also yielded reasonable diffusion coefficients, ranging from 5×10^{-7} cm^2/s to 8×10^{-7} cm^2/s for bovine serum albumin,[13, 17] B-phycoerythrin,[16, 35, 70] a DNA-binding protein,[9] and tRNA.[9] Measurements of the translational diffusion coefficients of latex beads have also agreed well with known values.[17, 54] FCS has also been proposed as a method for measuring the diffusion coefficients of atoms in the gas phase.[71]

The diffusion coefficients of the fluorescent lipid probes dioctadecyltetramethylindocarbocyanine, rhodamine-labeled phosphatidylethanolamine, and merocyanine 540 in planar lipid bilayers have been examined with FCS.[5–7]

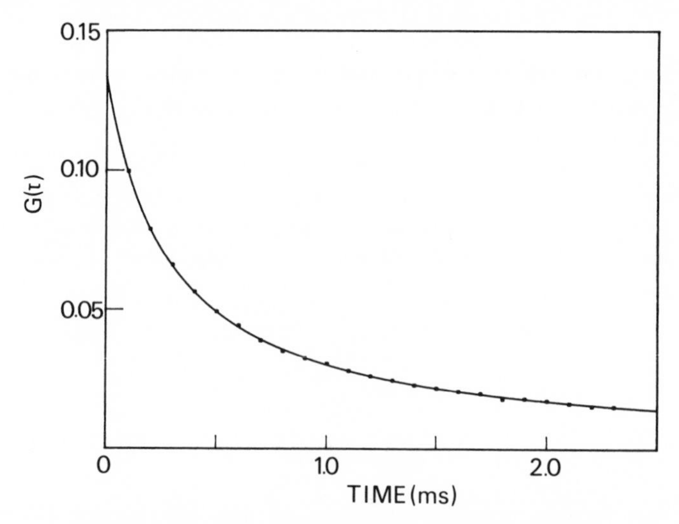

Figure 6.6. FCS autocorrelation function for a fluorescent dye undergoing translational diffusion. Shown in an experimentally obtained $G(\tau)$ for 10^{-9} M rhodamine 6G in ethanol. The solid line is the best fit to Eq. (6.46) plus an additive constant g_0, which yields $G(0) = 0.134$, $\tau_D = 0.288$ msec, and $g_0 = 2.96 \times 10^{-4}$. As estimate of the beam radius as $s \simeq 0.6$ μm gives $D \simeq 3 \times 10^{-6}$ cm^2/s according to Eq. (6.49). The experimental apparatus was as described in Ref. 28.

The measured diffusion coefficients in phospholipid bilayer membranes above their chain melting temperatures were $\sim 10^{-7}$ cm^2/s, which is consistent with measurements made by other techniques.[5-7] A preliminary measurement by FCS of the apparent translational diffusion coefficient of fluorescently labeled monoclonal antibodies bound to fluidlike phospholipid Langmuir–Blodgett films has also been reported and agrees with measurements made by fluorescence pattern photobleaching recovery.[15]

Measurement of translational diffusion coefficients with FCS has been integrated into a method of assaying for immunoglobulin molecules.[10, 12] In the immunoassay, contributions to the FCS autocorrelation function from fluorescently labeled antigen or antibody in solution are distinguished from contributions from fluorescently labeled antigen or antibody bound to specific sites on suspended latex particles, at equilibrium, by the difference in diffusional correlation times (Eq. 6.50). The FCS assay does not require a separation step, is relatively insensitive to the presence of other chemical species (except for those that fluoresce and are of a size similar to that of the beads), and requires only low concentrations of analyte and small volumes (10 μl). The detection of concentrations of gentamicin as low as 10^{-9} M has been reported.[12]

FCS has also been used to characterize the profile of a focused laser beam near the focal minimum.[11] In this application, which employs the experimental geometry in Eq. (6.6), $G(0)$ and τ_D are given by

$$\begin{aligned} G^{-1}(0) &= \pi s^2 L \langle C \rangle [1 + \lambda^2 (\Delta z)^2 / \pi^2 s^4] \\ \tau_D &= [s^2/4D][1 + \lambda^2 (\Delta z)^2 / \pi^2 s^4] \end{aligned} \tag{6.52}$$

where λ is the wavelength of the laser beam, and Δz is the distance from the focal plane. Measuring τ_D and $G(0)$ as a function of Δz and finding the best fit to Eqs. (6.52) gives a simultaneous measure of s, $\langle C \rangle$, and D. This method of measuring $\langle C \rangle$ and D is useful because an independent measure of the beam radius at the focus is not required.

6.5.2. Flow and Sample Translation

For samples in which lateral diffusion is slow, the incident light intensity that would be required to compute a diffusive autocorrelation function with an acceptable accuracy in an experimentally reasonable time (Section 6.3) sometimes photobleaches the fluorescent molecules during data collection. One method of avoiding this difficulty is to translate the sample, the exciting laser beam, or the observation aperture during data collection. In the limit where fluorescent molecules are immobile and are thus not transported through the sample region by diffusion at all, the sample may be translated through the observed region to generate number fluctuations. Sample translation has been employed in the rotating cell[26] and translating beam[17] methods of determining molecular weights, in the FCS immunoassay,[10, 12] and in S-FCS for detecting cell-surface receptor aggregation,[27, 58] as described in Section 6.4.3. As shown in this section, the shape of $G(\tau)$ is different for translational flow than for translational diffusion.[25, 56–58]

Theoretically, the FCS autocorrelation functions for uniform flow and uniform sample translation are equivalent.[25, 27] When fluorescent, non-diffusing molecules flow with constant speed V along the y-direction through the sample region,

$$\partial C_i(\mathbf{r}, t)/\partial t = -V\,\partial C_i(\mathbf{r}, t)/\partial y \tag{6.53}$$

Following an analysis analogous to the one outlined in Eqs. (6.42)–(6.46), one finds that, for the Gaussian-shaped illumination profile in Eq. (6.5) and for a monodisperse sample,

$$\begin{aligned} f(\mathbf{r}, \mathbf{r}', \tau) &= \delta(\mathbf{r} - \mathbf{r}' - \mathbf{j}Vt) \\ G(\tau) &= G(0)\exp[-(\tau/\tau_t)^2] \end{aligned} \tag{6.54}$$

where $\mathbf{j}$ is a unit vector in the y-direction, the characteristic time τ_t for flow through the sample region is

$$\tau_t = s/V \tag{6.55}$$

and $G(0)$ is given by Eq. (6.12). If translational diffusion is also present,

$$\partial C_i(\mathbf{r}, t)/\partial t = D_i \nabla^2 C_i(\mathbf{r}, t) - V\,\partial C_i(\mathbf{r}, t)/\partial y \tag{6.56}$$

and

$$G(\tau) = G(0)[1 + (\tau/\tau_D)]^{-1}\exp\{-(\tau/\tau_t)^2/[1 + (\tau/\tau_D)]\} \tag{6.57}$$

The characteristic decay time of $G(\tau)$ reflects the shorter of τ_D and τ_t; Eq. (6.57) reduces to Eq. (6.46) in the limit of no sample translation ($V \to 0$, $\tau_t \to \infty$) and to Eq. (6.54) in the limit of no diffusion ($D \to 0$, $\tau_D \to \infty$).

The characteristic time of translation or flow is proportional to the beam's $1/e^2$ radius (Eq. 6.55), whereas the characteristic time of diffusion is proportional to the square of the radius (Eq. 6.47). This property has been proposed as a method of distinguishing between transport due to flow and diffusion in samples that contain both.[56] Another method of resolving the two transport processes with FCS, in which the angle between observation and the incident laser beam is varied, has also been considered.[56] The theory outlined above has been experimentally confirmed on solutions of rhodamine 6G.[25]

6.5.3. Kinetic Rate Constants

The original development of FCS included a lengthy theoretical treatment and an experimental demonstration of the measurement of chemical kinetic rate constants.[1–3, 52] In the presence of chemical reaction, uniform translation, and translational diffusion, the concentration fluctuations of molecules in the ith states at positions $\mathbf{r}$ and times t are determined by the following set of differential equations:

$$\partial\, \delta C_i(\mathbf{r}, t)/\partial t = D_i \nabla^2\, \delta C_i(\mathbf{r}, t) - V\, \partial\, \delta C_i(\mathbf{r}, t)/\partial y + \sum_{j=1}^{R} T_{ij}\, \delta C_j(\mathbf{r}, t), \qquad i = 1 \text{ to } R \tag{6.58}$$

where R is the number of fluorescent species, T_{ij} are kinetic coefficients, and the chemical kinetic terms are linear because the concentration fluctuations are small.†[47] In general, $G(\tau)$ derived from Eq. (6.58) has a complex dependence on the eigenvalues and eigenfunctions of a matrix whose elements depend on the reaction rates and transport coefficients.[2] The autocorrelation function for the simplest case of a unimolecular reaction between states of different fluorescence yields has been presented in closed form.[2, 48, 57]

The FCS autocorrelation function for a bimolecular reaction

$$A + B \underset{k_2}{\overset{k_1}{\rightleftharpoons}} C \tag{6.59}$$

has also been treated theoretically.[1–3, 38, 47, 48] In a general form, which is not included here, $G(\tau)$ contains a rich assortment of terms that reflect different combinations of diffusion and reaction mechanisms that can theoretically result in measurable fluorescence fluctuations.‡ With the following set of assumptions, the expression for $G(\tau)$ for a bimolecular reaction simplifies to an experimentally manageable form: A represents ligand binding sites on a large, multivalent, nonfluorescent macromolecule (e.g., DNA); B represents small, weakly fluorescent, unbound ligands (e.g., ethidium bromide); C represents bound ligands where they are strongly fluorescent; the experiments are performed far from saturation, where the macromolecules contain an excess of free binding sites; flow is not present; and the characteristic times for macro-

† Consider the rate law for a bimolecular reaction: $dC/dt = k_1\mathrm{AB} - k_2\mathrm{C}$, where A, B, and C are reactant and product concentrations, and k_1 and k_2 are kinetic rate constants. As in relaxation methods, the rate equation can be written as $d/dt[\langle c \rangle + \delta c] = k_1[\langle \mathrm{A} \rangle + \delta\mathrm{A}][\langle \mathrm{B} \rangle + \delta\mathrm{B}] - k_2[\langle \mathrm{C} \rangle + \delta\mathrm{C}]$, or $d\,\delta\mathrm{C}/\delta t \approx k_1 \langle \mathrm{A} \rangle\, \delta\mathrm{B} + k_1 \langle \mathrm{B} \rangle\, \delta\mathrm{A} - k_2\, \delta\mathrm{C}$, if $\delta\mathrm{A}\, \delta\mathrm{B} << \langle \mathrm{A} \rangle \langle \mathrm{B} \rangle$, and since $k_1 \langle \mathrm{A} \rangle \langle \mathrm{B} \rangle = k_2 \langle \mathrm{C} \rangle$.

‡ Equation (6.58) does not entirely account for reaction–diffusion coupling in a bimolecular reaction. One discussion of the coupling with potential application to FCS is found in Ref. 85.

molecular and ligand diffusion through the sample region are longer than the inverse of the kinetic relaxation rate. Then[38]

$$\frac{G(\tau)}{G(0)}=\left[\frac{K\langle \mathrm{A}\rangle}{1+K\langle \mathrm{A}\rangle}\cdot\frac{1}{1+\tau/\tau_{+}}+\frac{1}{1+K\langle \mathrm{A}\rangle}\cdot\frac{\exp(-R\tau)}{1+\tau/\tau_{-}}\right] \tag{6.60}$$

where

$$\begin{aligned}\tau_{+,-}&=\frac{s^2}{4D}\left[\frac{1+K\langle \mathrm{A}\rangle}{[1,\, K\langle \mathrm{A}\rangle]}\right]\\ R&=k_1\langle \mathrm{A}\rangle+k_2\\ K&=k_1/k_2\end{aligned} \tag{6.61}$$

and D is the ligand diffusion coefficient. An early work on the binding of the dye ethidium bromide to DNA demonstrated not only that FCS autocorrelation functions that reflect chemical kinetic rates could be experimentally obtained but that data analysis according to Eqs. (6.60) and (6.61) could yield reasonable values for k_1 and k_2.[1–3, 38]

The theory above applies only to the case in which the average number of bound dye molecules per DNA molecule is low. Subsequent to the initial treatment under conditions in which the DNA is far from saturated, a more general theory that includes the association of a ligand with a multivalent macromolecule was derived and experimentally demonstrated.[20, 21, 48] FCS has also been used to investigate the association of ethidium bromide with DNA in cell nuclei.[19]

FCS has been applied to the study of the association kinetics of fluorescently labeled heavy meromyosin and its subfragment 1 with actin.[8] With A assigned as the concentration of free actin, B as the concentration of free myosin, and C as the concentration of the complex, and making an approximation consistent with the experimental conditions ($K\langle \mathrm{A}\rangle >> 1$), it was predicted that the first term of Eq. (6.60) would be dominant so that

$$\frac{G(\tau)}{G(0)}\approx\left[1+\frac{\tau}{(s^2/4D)(1+K\langle \mathrm{A}\rangle)}\right]^{-1} \tag{6.62}$$

Equation (6.62) states that $G(\tau)$ primarily reflects the decrease in the diffusion coefficient of myosin due to its association with actin. The slope of a plot of the characteristic time of $G(\tau)$ as a function of the actin concentration gave values for the equilibrium constants equal to $2\times 10^4\, M^{-1}$ for heavy meromyosin and $9\times 10^3\, M^{-1}$ for its subfragment 1. Conceptually, this experiment is similar to the FCS immunoassay described in Section 6.5.1.

Under certain conditions, the transport and reaction terms in $G(\tau)$ take on a simple form.[57] If diffusion is slow or if the diffusion coefficients of all fluorescent species are equal, then $G(\tau)$ separates into a product of the expression in Eq. (6.57) and a factor $X(\tau)$, where

$$X(\tau)=\frac{\sum_{i,j=1}^{R} Q_i Q_j U_{ij}(\tau)}{\sum_{i=1}^{R} Q_i^2 \langle C_i \rangle} \tag{6.63}$$

and the U_{ij} are the solutions to the following set of equations and initial conditions:

$$\begin{aligned} dU_{ik}(\tau)/d\tau &= \sum_{j=1}^{R} T_{ij} U_{jk}(\tau) \\ U_{ik}(0) &= \langle C_i \rangle \, \delta_{ik} \end{aligned} \tag{6.64}$$

6.6. Special Versions of FCS

A number of recent developments promise new applications of FCS to the measurement of dynamic properties such as mutual diffusion, rotational mobility, and surface kinetics that may not be readily accessible by other more conventional techniques.

6.6.1. Nonideal Solutions

The theoretical relationships in Sections 6.2, 6.4, and 6.5 assume that the solutions of interest, although possibly containing multiple fluorescent components, are otherwise ideal in that the motions of the fluorescent molecules or particles are not affected by intermolecular interactions (Eqs. 6.11 and 6.21). A recent theoretical work has outlined a basis for predicting the value of $G(0)$ for a solution that contains a single fluorescent solute in a homogeneous solvent when interactions between solute molecules are not negligible.[72] In this treatment,

$$G(0)=\gamma kT/[\langle N \rangle (\partial\pi/\delta\eta)_T] \tag{6.65}$$

where k is Boltzmann's constant, T is the absolute temperature, γ is defined in Eq. (6.13), π is the osmotic pressure, η is the number density, and $(\partial\pi/\delta\eta)_T$ is the isothermal osmotic compressibility. The quantity of interest is $(\partial\pi/\delta\eta)_T$, which contains information about the nature of the solute–solute interactions.

A special and somewhat related feature of FCS is that it monitors the mutual diffusion coefficient D_m whereas tracer techniques such as fluorescence photobleaching recovery monitor the self-diffusion coefficient D_s. FCS reports the mutual diffusion coefficient when a high fraction of the molecules of interest are fluorescently labeled.[53] At low densities, D_s and D_m are equivalent and equal whereas at higher densities where intermolecular interactions are not negligible D_s and D_m can differ significantly. The values of D_s and D_m can be simply expressed in terms of the isothermal osmotic compressibility $(\partial\pi/\partial\eta)_T$.[55, 72]

In a recent work that employed this theory, the dependence of the translational diffusion coefficient of the bovine eye-lens protein α_L-crystallin on protein concentration and ionic strength was measured using FCS.[14] The concentration dependence of the FCS diffusion coefficients D_m of DNA labeled with ethidium bromide have also been measured and compared with the diffusion coefficients D_s measured by fluorescence photobleaching recovery.[18] The measured values of D_m and D_s displayed significantly different dependencies on the DNA concentration.

6.6.2. Rotational Diffusion

Several groups have theoretically considered the measurement of macromolecular rotational diffusion by FCS (see Refs. 24, 37, 52, and 73–78). If the excitation light is polarized or if the emitted fluorescence is detected through a polarizer, or both, fluorescence fluctuations can arise on the time scale of orientational motions of the absorption or emission dipoles. The theoretical treatments show that, if

$$\tau_D, \tau_t >> \tau_r >> \tau_f \tag{6.66}$$

where τ_D and τ_t are given in Eqs. (6.47) and (6.55), respectively, and τ_f is the fluorescence lifetime, then, in the simplest case, terms like

$$G(\tau) \simeq \exp(-\tau/\tau_r) \qquad \text{for} \quad \tau_f << \tau << \tau_D, \tau_t \tag{6.67}$$

are included in $G(\tau)$, where

$$\tau_r = (6D_r)^{-1} \tag{6.68}$$

General theories that contain terms like Eq. (6.67) and that predict the FCS autocorrelation functions for combined translational and rotational diffusion, flow or sample translation, and chemical kinetics have been developed.[52] Methods for separating contributions to $G(\tau)$ due to rotational motion from those due to other processes have also been presented.[24]

Although, in practice, measurement of rotational diffusion coefficients with FCS may be experimentally more difficult than measurement by other techniques such as fluorescence depolarization or time-resolved fluorescence anisotropy decay, the advantage of the proposed FCS experiments is that information may theoretically be obtained in time ranges greater than the fluorescence lifetime of the fluorophore. Indeed, nanosecond FCS has recently produced rotational correlation times for both tetramethylrhodamine-labeled bovine carbonic anhydrase[23] and Texas Red-labeled porcine pancreatic lipase.[24] In these experiments, a solution of labeled protein flows through a small region illuminated by a focused laser beam. Fluorescence is collected through a polarizer aligned with the polarization of the laser beam. The emitted fluorescence is divided into two beams of equal intensity by a beam splitter and is detected by two single-photon counting photomultipliers whose outputs form the start and stop inputs to a time-to-amplitude converter. The output of the time-to-amplitude converter is passed to a multichannel pulse height analyzer. Because the fluorescence fluctuation autocorrelation function is the probability of detecting a photon at time $t+\tau$, given that a photon was detected at time t, the combination of a time-to-amplitude converter and a pulse height analyzer yields $G(\tau)$. The measured autocorrelation function for bovine carbonic anhydrase is shown in Figure 6.7. Curve fitting to appropriate theoretical forms yielded a value of D_r equal to $(1.1 \pm 0.1) \times 10^7\ \mathrm{s}^{-1}$.

Fluctuations in fluorescence excited by polarized light or detected through a linear polarizer could also arise from directed macromolecular orientational motion that is not true rotational diffusion. Such orientational

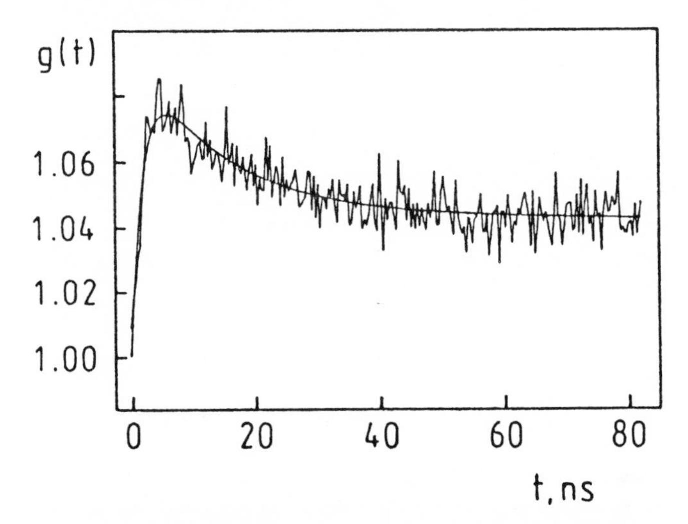

Figure 6.7. Protein rotational diffusion measured by FCS in the nanosecond time range. Shown is the measured autocorrelation function for fluorescently labeled bovine carbonic anhydrase in buffered solution, after 10 h of data collection, for a protein concentration of 4×10^{-10} M and an incident laser intensity of 24 mW. The characteristic length of the sample region was 3.5 μm, and the flow rate was 200 mm/s, so that the conditions in Eq. (6.66) were satisfied. [Reproduced from the *European Biophysical Journal* **14**, 257 (1987), with copyright permission.]

motions might occur in energy-requiring biological processes, especially those that convert ATP into mechanical motion. FCS autocorrelation functions for rotational motion under these more general conditions have been derived using both generalized Fokker–Planck[75] and Langevin[77] approaches. These theories predict that $G(\tau)$ will contain sinusoidal functions of $n\omega\tau$, where ω is the angular speed of the directed rotational motion and n is an integer. In the limit of $D_r \rightarrow 0$, $G(\tau)$ is periodic with angular frequency ω. Directed rotational motion might also be modeled as a progressive succession through distinct chemical states.[57]

The use of FCS to characterize the nonrandom rotational motions of bacterial flagellar filaments[75] and nucleic acids[77] has been proposed. In an earlier but conceptually identical experimental work, FCS autocorrelation functions resulting from the rotational motions of fluorescently labeled myosin cross-bridges in single skeletal muscle fibers under tension were obtained.[22]

6.6.3. Total Internal Reflection Illumination

FCS has been combined with illumination of a macroscopic solid/solution interface by total internal reflection (called TIR/FCS) to measure

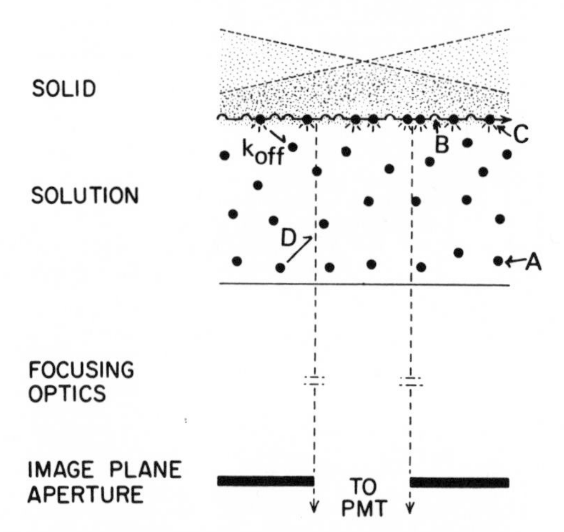

Figure 6.8. Conceptual basis of TIR/FCS. A laser beam is totally internally reflected at the interface of an aqueous solution of fluorescent molecules of concentration $A(\mathbf{r}, z, t)$ and a transparent solid that contains binding sites for the fluorescent molecules. At equilibrium, the concentration of surface-bound fluorescent molecules is $C(\mathbf{r}, t)$ and the concentration of uncomplexed surface sites is $B(\mathbf{r}, t)$. Fluorescence is emitted primarily from surface-bound molecules excited by the evanescent field. The area of the sample region is defined by an aperture placed at an image plane of the microscope. The fluorescence recovery curve is determined in part by the surface dissociation rate k_{off} and the solution diffusion coefficient D.

the surface binding kinetic rates of fluorescent molecules in solution with sites on a surface.[31, 59, 79] In TIR/FCS, as shown in Figure 6.8, fluorescent or fluorescently labeled molecules are in equilibrium between solution and the surface of a transparent solid to which they reversibly bind. A laser beam is incident on the solid/solution interface, from within the solid, at an angle greater than the critical angle,

$$\theta_c = \sin^{-1}(n_1/n_2) \tag{6.69}$$

where n_1 and n_2 are the refractive indices of the solution and the solid, respectively. The beam totally internally reflects at the solid/liquid interface, creating a thin exponentially decaying layer of illumination in the liquid adjacent to the surface, called the "evanescent field." The penetration depth d of the evanescent intensity in the solution is given by

$$d = \lambda_0/[4\pi(n_2^2 \sin^2 \theta - n_1^2)^{1/2}] \tag{6.70}$$

where λ_0 is the vacuum wavelength of the totally internally reflected laser beam ($\simeq 0.5\ \mu$m). In an experiment in which the solid is fused silica ($n_2 \approx 1.5$), the solution is aqueous ($n_1 \approx 1.3$), and the incidence angle θ is $\sim 80°$, the depth d given by Eq. (6.70) is ~ 600 Å. Fluorescent solute molecules that are bound to the surface are selectively excited by the evanescent wave. The measured fluorescence fluctuates as individual molecules bind and dissociate within the observed area or diffuse on the surface and through the observed area. The autocorrelation function of the fluctuations in fluorescence depends on the association and dissociation kinetic rates, on the diffusion coefficients of the fluorescent molecules in the bulk solution and along the surface, and on the concentration of fluorescent molecules in solution.

Derivation of the TIR/FCS autocorrelation function does not follow the method outlined in Section 6.2. The simplest case, as shown in Figure 6.8, is the association of univalent fluorescent molecules with univalent surface sites, represented by

$$A + B \underset{k_2}{\overset{k_1}{\rightleftharpoons}} C \tag{6.71}$$

where $C(\mathbf{r}, t)$ is the concentration of surface-bound fluorescent molecules, $B(\mathbf{r}, t)$ is the concentration of free sites on the surface, $A(\mathbf{r}, z, t)$ is the concentration of molecules in solution, vector $\mathbf{r}$ lies in the x–y plane (the surface), and k_1 and k_2 are the association and dissociation rates. The differential equations that describe the process are

$$\begin{aligned} \partial C(\mathbf{r}, t)/\partial \tau &= D_C \nabla_2^2 C(\mathbf{r}, t) + k_1[A(\mathbf{r}, z, t)]_{z\to 0} B(\mathbf{r}, t) - k_2 C(\mathbf{r}, t) \\ \partial A(\mathbf{r}, z, t)/\partial t &= D_A \nabla_3^2 A(\mathbf{r}, z, t) \end{aligned} \tag{6.72}$$

where ∇_2^2 and ∇_3^2 are two- and three-dimensional Laplacians, and D_C and D_A are the diffusion coefficients on the surface and in solution. The boundary condition that describes the interfacial reaction is

$$D_A[\partial A(\mathbf{r}, z, t)/\partial z]_{z\to 0} = k_1[A(\mathbf{r}, z, t)]_{z\to 0}\, B(\mathbf{r}, t) - k_2 C(\mathbf{r}, t) \quad (6.73)$$

Function $W(\mathbf{r})$ for the optical geometry shown in Figure 6.8 is

$$W(\mathbf{r}) = \begin{cases} I_0 \exp(-z/d) & |x|, |y| \leqslant s/2 \\ 0 & \text{otherwise} \end{cases} \quad (6.74)$$

where s is the length of a square observation area. Equations (6.72) and (6.73) lead to a general expression for $G(\tau)$ that depends on k_1, k_2, D_A and D_C, and s and that has been obtained only in integral form. However, for a shallow evanescent depth and a large observation area, the expression for $G(\tau)$ reduces to

$$G(\tau) = G(0)\{b_1 w[-ib_2(k_2 t/\beta)^{1/2}] - b_2 w[-ib_1(k_2 t/\beta)^{1/2}]\}/(b_1 - b_2) \quad (6.75)$$

where

$$b_{1,2} = -\tfrac{1}{2}[k_2/(\beta k)]^{1/2}\,[1 \pm (1 - 4\beta k/k_2)^{1/2}]$$
$$w(iu) = \exp(u^2)\,\mathrm{erfc}(u) \quad (6.76)$$
$$k = D_A/(\beta\langle C\rangle/\langle A\rangle)^2$$

$\langle C\rangle$ and $\langle A\rangle$ are the equilibrium concentrations of bound and dissolved fluorescent molecules, respectively, and

$$\beta = (1 + k_1\langle A\rangle/k_2)^{-1} \quad (6.77)$$

is the average fraction of surface sites that are not complexed by ligand at equilibrium.

The time-zero value of the autocorrelation function is given by

$$G(0) = \beta/\langle N\rangle = [\beta/(1-\beta)][1/\langle N_T\rangle] \quad (6.78)$$

where $\langle N\rangle$ is the average number of observed molecules, and $\langle N_T\rangle$ is the total number of binding sites in the observed area. Factor γ (Eq. 6.13) equals 1 for the experimental geometry defined in Eq. (6.74), in the limit of shallow evanescent depth. Equation (6.78) does not agree with Eq. (6.12) because the total number of sites in the observed area is not free to fluctuate.[59]

Rate k in Eq. (6.76) is the inverse of a time that is associated with relaxation of a concentration fluctuation in bound molecules by diffusion from the surface. Rate k_2 is the inverse of the average surface residency time. $G(\tau)$ has simpler forms in the limiting cases of $k_2 << \beta k$ (the reaction limit) and $k_2 >> \beta k$ (the bulk diffusion limit):

$$G(\tau) = \begin{cases} G(0)\exp(-k_2\tau/\beta) & k_2 << \beta k \\ G(0)\, w[i(k\tau)^{1/2}] & k_2 >> \beta k \end{cases} \tag{6.79}$$

Also of interest is the TIR/FCS autocorrelation function that would be obtained from a mixture of fluorescent and nonfluorescent molecules that compete for the same surface sites. Interestingly, the autocorrelation function provides information about the binding kinetic rates not only of the fluorescent molecules but also of the nonfluorescent molecules.[79]

Instrumentally, as shown in Figure 6.9, a laser beam is passed through a convex spherical focusing lens into a prism which rests on top of and is optically coupled with glycerin or immersion oil to the substrate of interest. The substrate rests on a spacer that is supported by a glass coverslip; the

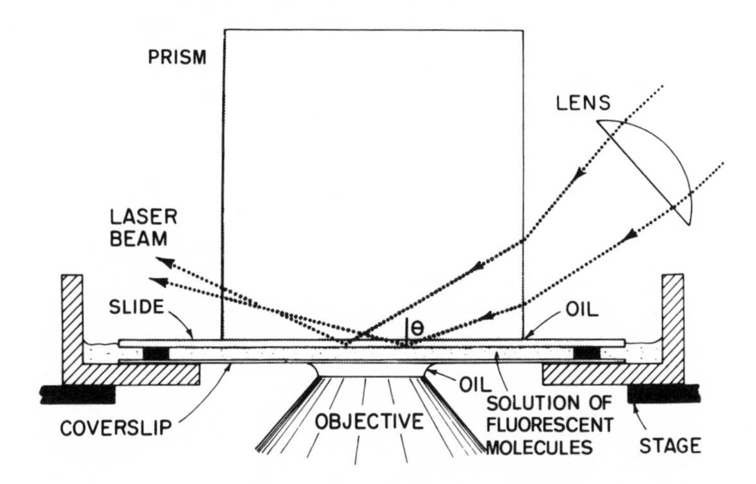

Figure 6.9. TIR/FCS sample chamber. The laser beam, focused by a spherical convex lens, enters a cubic fused-silica prism at an oblique angle. The prism is optically coupled with glycerin or immersion oil to a fused-silica slide. The angle of entrance is such the angle θ at which the beam is incident on the quartz/solution interface is greater than the critical angle θ_c. The beam totally internally reflects, illuminating fluorescent molecules that adsorb from the solution to the underside of the fused-silica slide. The solution is contained by a glass-bottomed plastic tissue culture dish and a thin Teflon spacer between the coverslip and the slide. The sample is on the stage of an inverted microscope, and fluorescence arising from the illuminated surface is collected by a high-numerical-aperture objective.

solution of fluorescent molecules is contained between the substrate and coverslip. Fluorescence is collected from below by a high-numerical-aperture objective on an inverted microscope. The illuminated area is usually larger than a few square microns[80]; thus, a small aperture in an image plane of the microscope blocks the detection of fluorescence arising from all but a small area of the surface (Figure 6.8).

TIR/FCS has been applied to the surface-binding kinetics of tetramethylrhodamine-labeled immunoglobulin G. As shown in Figure 6.10, the experimentally obtained autocorrelation function agreed well with the theoretically predicted shape. Differences in the surface-binding kinetics of IgG and insulin have also been observed with TIR/FCS.[31]

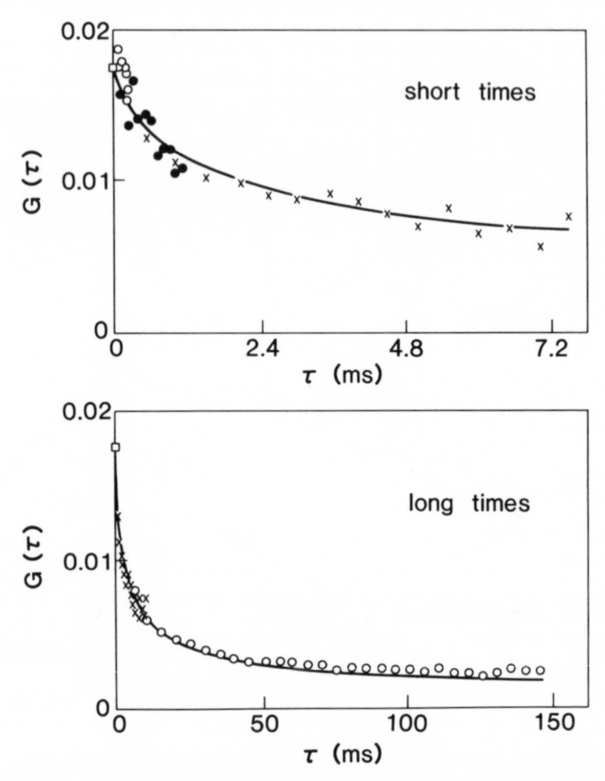

Figure 6.10. FCS autocorrelation function of tetramethylrhodamine-labeled immunoglobulin G adsorption to protein-coated fused silica. Data points are composited from autocorrelation functions accumulated at sample times 0.03 (○), 0.1 (●), and 0.5 ms (×) (upper panel) and 0.5 (×) and 5 ms (○) (lower panel). The line represents a function of the form of Eq. (6.75) with $k = (5\text{ ms})^{-1}$, $k_2 = (0.4\text{ ms})^{-1}$, and $G(0) = 0.0176$. The deviation of the theoretical curve from the data points at very early times arises from IgG moving through the depth of the evanescent field. [Reproduced from the *Biophysical Journal* **43**, 110 (1981), by copyright permission of the Biophysical Society.]

In an earlier work, TIR/FCS was developed as the central part of an instrument called a "virometer."[33, 34] In this instrument, virus particles are stained with ethidium bromide and introduced next to a solid/solution interface at which a laser beam totally internally reflects. The measured fluorescence fluctuates in time as fluorescent virus particles diffuse through the depth of the evanescent field. The characteristic time for diffusion through the evanescent field is

$$\tau_e = d^2/4D \tag{6.80}$$

where d is given in Eq. (6.70), and D is the diffusion coefficient of the virus in solution and depends on the viral size. The value of the viral diffusion time together with a measure of the nucleic acid content is used to identify the virus.

6.6.4. High-Order Autocorrelation

As described in Section 6.4, $G(0)$ depends on the number densities and relative fluorescence yields of difference fluorescent species in the sample region and can provide the number of monomers per aggregate for a monodisperse solution of oligomers. However, if the distribution of oligomers is polydisperse, then more than one measured number, namely, $G(0)$, is required to determine the distribution of aggregates. One method of obtaining more information from a time record of fluorescence fluctuations is to calculate the time-zero values of a series of high-order fluorescence fluctuation autocorrelation functions $G_{m,n}(\tau)$, defined as[†][28]

$$G_{m,n}(\tau) = \frac{\langle \delta F^m(t+\tau)\,\delta F^n(t)\rangle - \langle \delta F^m(t)\rangle\langle \delta F^n(t)\rangle}{\langle F(t)\rangle^{m+n}} \tag{6.81}$$

where m and n are integers. Molecules of the ith species are transported through the sample region, in general, by diffusion in the x–y plane with coefficient D_i and uniform translation along the y-axis with speed V. For systems in equilibrium, the autocorrelation functions are even in time[81]; therefore, $G_{n,m}(t)$ equals $G_{m,n}(t)$, and only autocorrelation functions for which $m \leqslant n$ are considered. Using Eqs. (6.1) and (6.18) in Eq. (6.81), deriving expressions

[†] The mth, nth-order autocorrelation function of fluorescence fluctuations $G_{m,n}(\tau)$ has been defined according to Eq. (6.81) because the choice of offset value ensures that $G_{m,n}(\tau)$ tends toward zero as τ approaches infinity. The choice of normalization in Eq. (6.81) allows simpler algebraic expressions in the denominators of the $G_{m,n}(0)$ (Eqs. 6.82 and 6.83).

for high-order concentration correlation functions, and using the derived expressions evaluated at $\tau \to 0$ in Eq. (6.81) gives†[28]

$$\begin{aligned} G_{1,1}(0) &= h_2 B_2 \\ G_{1,2}(0) &= h_3 B_3 \\ G_{2,2}(0) &= h_4 B_4 + 2h_2^2 B_2^2 \\ G_{1,3}(0) &= h_4 B_4 + 3h_2^2 B_2^2 \\ G_{2,3}(0) &= h_5 B_5 + 9h_2 h_3 B_2 B_3 \\ G_{3,3}(0) &= h_6 B_6 + 15h_2 h_4 B_2 B_4 + 15h_2^3 B_2^3 + 9h_3^2 B_3^2 \end{aligned} \tag{6.82}$$

where

$$B_k = \frac{\sum_{i=1}^{R} \alpha_i^k \langle N_i \rangle}{[\sum_{i=1}^{R} \alpha_i \langle N_i \rangle]^k} \tag{6.83}$$

$$h_n = w_n / w_1 \tag{6.84}$$

and the relative fluorescence yields α_i and constants w_n are defined in Eqs. (6.23) and (6.14).

As shown by Eqs. (6.82) and (6.83), for each value of $m+n$, one new quantity B_{m+n} may be determined from the measured value of $G_{m,n}(0)$; information on the average number of particles of different species in the sample region is obtained from the B_ks. If the relative quantum efficiencies α_i are unknown, then the first species contributes one unknown ($\langle N_1 \rangle$) and each new species after the first contributes two unknowns ($\langle N_i \rangle$ and α_i, for the ith species). Thus, for a single species, only $G_{1,1}(0)$ must be calculated; for two species, the values of $G_{1,1}(0)$, $G_{1,2}(0)$, and $G_{2,2}(0)$ must be calculated; and, for three species, the six $G_{m,n}(0)$ values in Eq. (6.82) must be calculated. Expressions which allow calculation of $\langle N_i \rangle$ and α_i from the B_ks have been derived.[16, 28] If the relative quantum efficiencies α_i are independently known, then each new species requires only one and not two new values of B_k, and the number densities of up to six different species may theoretically be obtained from the values of $G_{m,n}(0)$ in Eq. (6.82).‡

The use of high-order autocorrelation in FCS to determine aggregate distributions does not require separating out fractional decays with different diffusional correlation times; information about the aggregate distribution is contained in the values of $G_{m,n}(0)$. However, the time dependence of the

† See footnote on page 341.

‡ The number densities of the fluorescent species must be small enough so that the fluorescence fluctuations are distributed according to Poisson and not Gaussian statistics. Otherwise, the high-order autocorrelation functions provide no new information over $G_{1,1}(\tau)$.[28]

$G_{m,n}(\tau)$ does indicate appropriate theoretical functional forms for fitting to experimentally obtained autocorrelation functions, which ensures that data have been recorded at separation times τ short enough for accurate extrapolation and that the best extrapolated values of $G_{m,n}(0)$ are obtained. The $G_{m,n}(\tau)$ depend on functions

$$L_{m',n'}(i, \tau) = \frac{\exp\{-2m'n'(\tau/\tau_t)^2/[m'+n'+(2m'n'\tau/\tau_{D_i})]\}}{\langle N_i \rangle^{(m'+n'-1)}[m'+n'+(2m'n'\tau/\tau_{D_i})]} \qquad (6.85)$$

where $m'+n' \leqslant m+n$, and τ_{D_i}, τ_t, and $\langle N_i \rangle$ are defined in Eqs. (6.51), (6.55), and (6.24).[28] Equation (6.85) reduces to Eq. (6.57) when m' and n' equal one. As expected for a system in equilibrium, the time course of the autocorrelation functions given by Eq. (6.85) is symmetric upon an interchange of m' and n'. The time constant for the decay of $L_{m',n'}(i, \tau)$ decreases as $2m'n'/(m'+n')$, which means that higher order correlations have components which decay faster than lower order ones.

That high-order FCS autocorrelation functions with reasonable signal-to-noise ratios may be experimentally obtained has been demonstrated (Figure 6.11). The experimental applicability of high-order FCS to molecular aggregation was first demonstrated by showing that the fluorescent lipid dioctadecyltetramethylindocarbocyanine suspended in various solutions of water and ethyl alcohol yielded different high-order autocorrelation functions corresponding to different aggregate size distributions.[28] Later work showed that high-order analysis could, under some conditions, retrieve the correct concentrations and relative fluorescence yields of different fluorescent species in solutions of known composition.[16, 29] Quantitatively correct interpretation of high-order FCS autocorrelation functions requires measurement of the constants h_n (Eqs. 6.14 and 6.84) that describe the shape and size of the illuminated volume[35] and corrections that account for special high-order effects of the stochastic nature of photon detection.[70, 82]

High-order autocorrelation functions have been proposed as a technique for investigating the time-reversal properties of fluctuating systems[50, 81]; theory predicts that $G_{m,n}(\tau)$ will not equal $G_{n,m}(\tau)$ when $m \neq n$ and when the system is not in equilibrium. Figure 6.12 shows experimental values of $G_{1,3}(t)$ and its time-reverse function $G_{3,1}(t)$ for dioctadecyltetramethylindocarbocyanine in a 4:1 mixture of water and ethyl alcohol. Within statistical accuracy, the two functions are equal, as theoretically predicted.

The terms in Eq. (6.85) decrease as $\langle N_i \rangle^{-(m'+n'-1)}$; thus, for $\langle N_i \rangle >> 1$, the high-order autocorrelation functions have prohibitively small magnitudes as m' and n' increase. Consequently, other generalized autocorrelation functions might be sought which would be less strongly dependent on $\langle N_i \rangle$. For example, the definition in Eq. (6.81) can be extended to include powers a and

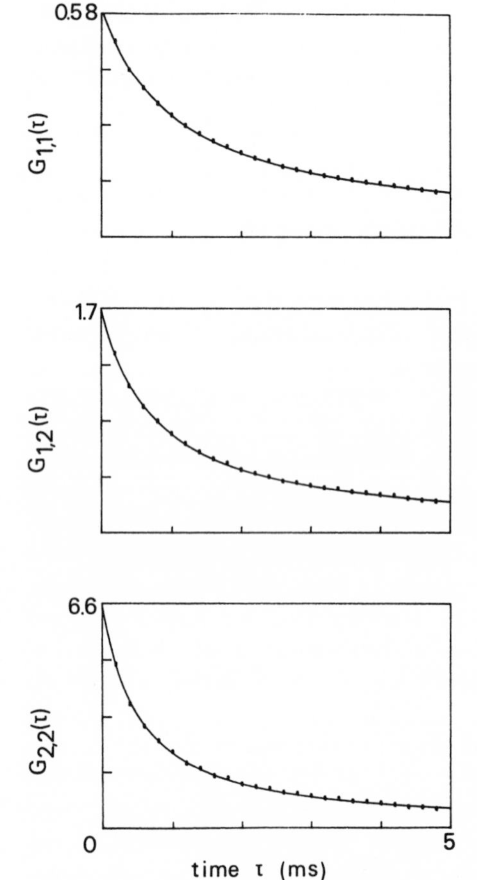

Figure 6.11. High-Order FCS autocorrelation functions for a fluorescent protein in solution. Shown are the experimentally obtained values of $G_{m,n}(\tau)$ for 3×10^{-10} *M* *B*-phycoerythrin in phosphate-buffered saline and the best fits to theoretical functions built from terms given in Eq. (6.85) as described in Ref. 28.

b which are real but not integral (e.g., $a, b \leqslant 1$). Expanding the quantity $\langle F^a(t+\tau)\, F^b(t) \rangle$ in a binomial series gives

$$\begin{aligned} &\langle F^a(t+\tau)\, F^b(t) \rangle \\ &\quad = \langle [\langle F(t+\tau) \rangle + \delta F(t+\tau)]^a\, [\langle F(t) \rangle + \delta F(t)]^b \rangle \\ &\quad = \sum_{j=0}^{\infty} \sum_{k=0}^{\infty} \binom{a}{j} \binom{b}{k} \langle F(t) \rangle^{a+b-j-k} \langle \delta F^j(t+\tau)\, \delta F^k(t) \rangle \qquad (6.86) \end{aligned}$$

where the fluorescence fluctuations are assumed to be small, that is, $\delta F^2 < \langle F \rangle^2$. Since j and k are integers, autocorrelation functions of non-integer order constructed from functions of the form in Eq. (6.81) are linear combinations of the values of $G_{m,n}(t)$ calculated in the above sections.

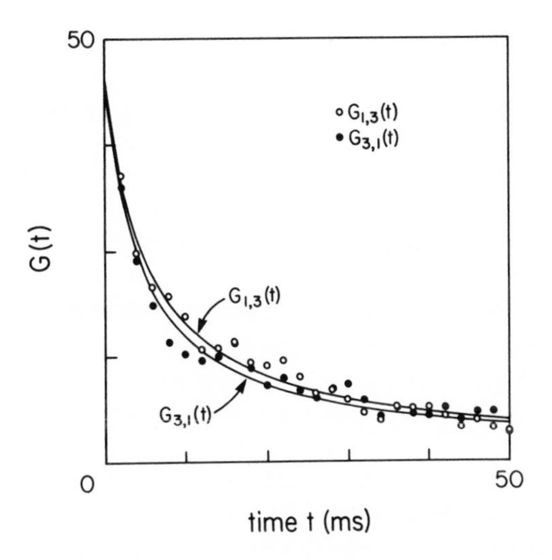

Figure 6.12. Comparison of the experimental values and theoretical best fits for $G_{1,3}(t)$ (○) and $G_{3,1}(t)$ (●) for $10^{-7}\,M$ dioctadecyltetramethylindocarbocyanine in 4:1 water:ethanol. The experimental curves are identical within experimental uncertainty. [Reproduced from the *Biophysical Journal* **52**, 265 (1987), by copyright permission of the Biophysical Society.]

Similarly, autocorrelation functions defined using any functional form which can be expanded in powers will also be linear combinations of the $G_{m,n}(t)$ defined in Eq. (6.81).[81]

6.6.5. Cross-Correlation

A new method for investigating intermolecular interactions in which fluorescence fluctuations arising from samples containing two distinct types of fluorescent molecules are cross-correlated using two detectors with different wavelength sensitivities has been recently proposed.[83] This idea has been experimentally explored by cross-correlating fluorescence fluctuations and scattered light from samples containing latex beads and seems to hold considerable promise for yielding unique information about interacting systems. However, if the emission–absorption overlap integral for the two fluorescent molecules is of a reasonable magnitude, then singlet–singlet energy transfer could significantly complicate data interpretation. A theory that predicts the FCS autocorrelation function due to Förster energy transfer between donor and acceptor fluorophores attached to different sites on a flexible biopolymer has been developed.[84]

6.7. Summary and Future Directions

FCS has a number of advantages and special features and is particularly well suited to the investigation of some physical processes. As is the case for most fluorescence techniques, FCS is highly sensitive to the fluorescent molecules of interest and relatively insensitive to the presence of other nonfluorescent molecules. In addition, FCS does not require an external perturbation such as a temperature jump, a pressure jump, or photochemical bleaching. The time range in which FCS has been the most successful is between 10 μs and 100 ms, and FCS is thus complementary to faster techniques such as time-resolved fluorescence anisotropy and to slower techniques such as fluorescence photobleaching recovery. Although a rather large number of parameters may be measured with FCS, such as chemical kinetic rate constants and transport coefficients, in most cases there are only a few experimental demonstrations of measurement of the parameters. Three areas that seem to be particularly promising for future investigations with FCS are the characterization of kinetic mechanisms in solution and at surfaces, concentration-dependent diffusion, and cell-surface receptor aggregation.

Acknowledgments

I would like to thank Arthur G. Palmer of the University of North Carolina at Chapel Hill for permission to include his unpublished results (Figures 6.6 and 6.11) and for numerous insightful discussions. This work was supported by grant GM37145 from the National Institutes of Health and by Presidential Young Investigator Award DCB8552986 from the National Science Foundation.

References

1. D. Magde, E. L. Elson, and W. W. Webb, Thermodynamic fluctuations in a reacting system—measurement by fluorescence correlation spectroscopy, *Phys. Rev. Lett. 29*, 705–708 (1972).
2. E. L. Elson and D. Magde, Fluorescence correlation spectroscopy. I. Conceptual basis and theory, *Biopolymers 13*, 1–27 (1974).
3. D. Magde, E. L. Elson, and W. W. Webb, Fluorescence correlation spectroscopy. II. An experimental realization, *Biopolymers 13*, 29–61 (1974).
4. D. E. Koppel, D. Axelrod, J. Schlessinger, E. L. Elson, and W. W. Webb, Dynamics of fluorescence marker concentration as a probe of mobility, *Biophys. J. 16*, 1315–1329 (1976).
5. P. F. Fahey, D. E. Koppel, L. S. Barak, D. E. Wolf, E. L. Elson, and W. W. Webb, Lateral diffusion in planar lipid bilayers, *Science 195*, 305–306 (1977).
6. P. F. Fahey and W. W. Webb, Lateral diffusion in phospholipid bilayer membranes and multilamellar liquid crystals, *Biochemistry 17*, 3046–3053 (1978).

7. P. R. Dragsten and W. W. Webb, Mechanism of the membrane potential sensitivity of the fluorescent membrane probe merocyanine 540, *Biochemistry 17*, 5228–5240 (1978).
8. J. Borejdo, Motion of myosin fragments during action-activated ATPase: Fluorescence correlation spectroscopy study, *Biopolymers 18*, 2807–2820 (1979).
9. R. Rigler, P. Grasselli, and M. Ehrenberg, Fluorescence correlation spectroscopy and applications to the study of Brownian motion of biopolymers, *Phys. Scr. 19*, 486–490 (1979).
10. D. F. Nicoli, J. Briggs, and V. B. Elings, Fluorescence immunoassay based on long time correlations of number fluctuations, *Proc. Natl. Acad. Sci. U.S.A. 77*, 4904–4908 (1980).
11. S. M. Sorscher and M. P. Klein, Profile of a focused collimated laser beam near the focal minimum characterized by fluorescence correlation spectroscopy, *Rev. Sci. Instrum. 51*, 98–102 (1980).
12. J. Briggs, V. B. Elings, and D. F. Nicoli, Homogeneous fluorescence immunoassay, *Science 212*, 1266–1267 (1981).
13. D. E. Koppel, Fluorescence techniques for the study of biological motion, in: *Applications of Laser Light Scattering to the Study of Biological Motion*, NATO ASI Series A: Life Sciences, Vol. 59 (J. C. Earnshaw and M. W. Steer, eds.), pp. 245–273, Plenum Press, New York (1983).
14. C. Andries, W. Guedens, J. Clauwaert, and H. Geerts, Photon and fluorescence correlation spectroscopy and light scattering of eye-lens proteins at moderate concentrations, *Biophys. J. 43*, 345–354 (1983).
15. L. L. Wright, A. G. Palmer, and N. L. Thompson, Concentration dependence of the translation diffusion of monoclonal antibodies specifically bound to phospholipid Langmuir–Blodgett films, *Mat. Res. Soc. Symp. Proc. 111*, 419–424 (1989).
16. A. G. Palmer and N. L. Thompson, High-order fluorescence fluctuation analysis of model protein clusters, *Proc. Natl. Acad. Sci. U.S.A. 86*, 6148–6152 (1989).
17. T. Meyer and H. Schindler, Particle counting by fluorescence correlation spectroscopy: Simultaneous measurement of aggregation and diffusion of molecules in solutions and in membranes, *Biophys. J. 54*, 983–993 (1988).
18. B. A. Scalettar, J. E. Hearst, and M. P. Klein, FRAP and FCS studies of self-diffusion and mutual diffusion in entangled DNA solutions, *Macromolecules 22*, 4550–4559 (1989).
19. S. M. Sorscher, J. C. Bartholomew, and M. P. Klein, The use of fluorescence correlation spectroscopy to probe chromatin in the cell nucleus, *Biochim. Biophys. Acta 610*, 28–46 (1980).
20. R. D. Icenogle and E. L. Elson, Fluorescence correlation spectroscopy and photobleaching recovery of multiple binding reactions. I. Theory and FCS measurements, *Biopolymers 22*, 1919–1948 (1983).
21. R. D. Icenogle and E. L. Elson, Fluorescence correlation spectroscpy and photobleaching recovery of multiple binding reactions. II. FPR and FCS measurements at low and high DNA concentrations, *Biopolymers 22*, 1949–1966 (1983).
22. J. Borejdo, S. Putnam, and M. F. Morales, Fluctuations in polarized fluorescence: Evidence that muscle cross bridges rotate repetitively during contraction, *Proc. Natl. Acad. Sci. U.S.A. 76*, 6346–6350 (1979).
23. P. Kask, P. Piksarv, Ü. Mets, M. Pooga, and E. Lippmaa, Fluorescence correlation spectroscopy in the nanosecond time range: Rotational diffusion of bovine carbonic anhydrase B, *Eur. Biophys. J. 14*, 257–261 (1987).
24. P. Kask, P. Piksarv, M. Pooga, Ü. Mets, and E. Lippmaa, Separation of the rotational contribution in fluorescence correlation experiments, *Biophys. J. 55*, 213–220 (1988).
25. D. Magde, W. W. Webb, and E. L. Elson, Fluorescence correlation spectroscopy. III. Uniform translation and laminar flow, *Biopolymers 17*, 361–376 (1978).
26. M. B. Weissman, H. Schindler, and G. Feher, Determination of molecular weights by fluctuation spectroscopy: Application to DNA, *Proc. Natl. Acad. Sci. U.S.A. 73*, 2776–2780 (1976).
27. N. O. Petersen, D. C. Johnson, and M. J. Schlesinger, Scanning fluorescence correlation

spectroscopy. II. Application to virus glycoprotein aggregation, *Biophys. J. 49*, 817–820 (1986).
28. A. G. Palmer and N. L. Thompson, Molecular aggregation characterized by high order autocorrelation in fluorescence correlation spectroscopy, *Biophys. J. 52*, 257–270 (1987).
29. H. Qian and E. L. Elson, Distribution of molecular aggregation by analysis of fluctuation moments, *Proc. Natl. Acad. Sci. U.S.A. 87*, 5479–5483 (1990).
30. P. R. St.-Pierre and N. O. Petersen, Relative ligand binding to small or large aggregates measured by scanning correlation spectroscopy, *Biophys. J. 58*, 503–511 (1990).
31. N. L. Thompson and D. Axelrod, Immunoglobulin surface binding kinetics studied by total internal reflection with fluorescence correlation spectroscopy, *Biophys. J. 43*, 103–114 (1983).
32. J. Borejdo, Application of fluctuation spectroscopy to muscle contractility, *Curr. Topics Bioenerg. 10*, 1–50 (1980).
33. T. Hirschfeld and M. J. Block, Virometer: Real time virus detection and identification in biological fluids, *Opt. Eng. 16*, 406–407 (1977).
34. J. Hirschfeld, M. J. Block, and W. Mueller, Virometer: An optical instrument for visual observation, measurement and classification of free viruses, *J. Histohem. Cytochem. 25*, 719–723 (1977).
35. A. G. Palmer and N. L. Thompson, Optical spatial intensity profiles for high order autocorrelation in fluorescence spectroscopy, *Appl. Opt. 28*, 1214–1220 (1989).
36. E. L. Elson and W. W. Webb, Concentration correlation spectroscopy: A new biophysical probe based on occupation number fluctuations, *Annu. Rev. Biophys. Bioeng. 4*, 311–334 (1975).
37. M. Ehrenberg and R. Rigler, Fluorescence correlation spectroscopy applied to rotational diffusion of macromolecules, *Quart. Rev. Biophys. 9*, 69–81 (1976).
38. D. Magde, Chemical kinetics and fluorescence correlation spectroscopy, *Quart. Rev. Biophys. 9*, 35–47 (1976).
39. W. W. Webb, Applications of fluorescence correlation spectroscopy, *Quart. Rev. Biophys. 9*, 49–68 (1976).
40. D. Magde, Concentration correlation analysis and chemical kinetics, in: *Molecular Biology, Biochemistry and Biophysics 24: Chemical Relaxation in Molecular Biology* (I. Pecht and R. Rigler, ed.), pp. 43–83, Springer-Verlag, Berlin (1977).
41. G. M. Hieftje, J. M. Ramsey, and G. R. Haugen, New laser-based methods for the measurement of transient chemical events, in: *New Applications of Lasers to Chemistry*, ACS Symposium Series No. 85 (G. M. Hieftje, ed.), pp. 118–125, American Chemical Society, Washington, D.C. (1978).
42. G. Horlick and G. M. Hieftje, Correlation methods in chemical data measurement, *Cont. Topics Anal. Clin. Chem. 3*, 153–216 (1979).
43. B. Chu, Dynamics of macromolecular solutions, *Phys. Scri. 19*, 458–470 (1979).
44. J. V. Doherty and J. H. R. Clarke, Noisy solutions: A source of valuable kinetic information, *Sci. Prog. 66*, 385–419 (1980).
45. M. B. Weissman, Fluctuation spectroscopy, *Annu. Rev. Phys. Chem. 32*, 205–232 (1981).
46. N. O. Petersen, Diffusion and aggregation in biological membranes, *Can. J. Biochem. Cell Biol. 62*, 1158–1166 (1984).
47. E. L. Elson, Fluorescence correlation spectroscopy and photobleaching recovery, *Annu. Rev. Phys. Chem. 36*, 379–406 (1985).
48. E. L. Elson, Membrane dynamics studied by fluorescence correlation spectroscopy and photobleaching recovery, in: *Optical Methods in Cell Physiology* (P. De Weer and B. M. Salzburg, ed.), pp. 367–383, Wiley, New York (1986).
49. N. O. Petersen and E. L. Elson, Measurements of diffusion and chemical kinetics by fluorescence photobleaching recovery and fluorescence correlation spectroscopy, *Methods Enzymol. 130*, 454–484 (1986).

50. I. Z. Steinberg, Noise as a source of information about dynamical properties of biological molecules, *Biopolymers 26*, S161–S176 (1987).
51. A. G. Palmer and N. L. Thompson, FCS for detecting submicroscopic clusters of fluorescent molecules in membranes, *Chem. Phys. Lipids 50*, 253–270 (1989).
52. S. R. Aragón and R. Pecora, Fluorescence correlation spectroscopy as a probe of molecular dynamics, *J. Chem. Phys. 64*, 1791–1803 (1976).
53. G. D. J. Phillies, Fluorescence correlation spectroscopy and nonideal solutions, *Biopolymers 14*, 499–508 (1975).
54. H. Qian and E. L. Elson, Three-dimensional fluorescence correlation spectroscopy in bulk solutions, in: *Time-Resolved Laser Spectroscopy in Biochemistry* (J. R. Lakowicz, ed.), *Proc. SPIE 909*, 352–359 (1988).
55. J. R. Abney, B. A. Scalettar, and J. C. Owicki, Mutual diffusion of interacting membrane proteins, *Biophys. J. 56*, 315–326 (1990).
56. H. Asai, Proposal of a simple method of fluorescence correlation spectroscopy for measuring the direction and magnitude of a flow of fluorophores, *Jpn. J. Appl. Phys. 19*, 2279–2282 (1980).
57. A. G. Palmer and N. L. Thompson, Theory of sample translation in fluorescence correlation spectroscopy, *Biophys. J. 51*, 339–343 (1987).
58. N. O. Peterson, Scanning fluorescence correlation spectroscopy. I. Theory and simulation of aggregation measurements, *Biophys. J. 49*, 809–815 (1986).
59. N. L. Thompson, T. P. Burghardt, and D. Axelrod, Measuring surface dynamics by total internal reflection fluorescence with photobleaching recovery or correlation spectroscopy, *Biophys. J. 33*, 435–454 (1981).
60. J. S. Gethner and G. W. Flynn, Evaluation of photomultiplier tubes for use in photon counting correlation spectroscopy, *Rev. Sci. Instrum. 46*, 586–591 (1975).
61. J. Schneider, J. Ricka, and T. Binkert, Improved fluorescence correlation analysis for precise measurements of correlation functions, *Rev. Sci. Instrum. 59*, 588–590 (1988).
62. L. Basano and P. Ottonello, Third-order correlator for point processes, *Rev. Sci. Instrum. 58*, 579–583 (1987).
63. Z. Kam, H. B. Shore, and G. Feher, Simple schemes for measuring autocorrelation functions, *Rev. Sci. Instrum. 46*, 269–277 (1975).
64. S. H. Chen, V. B. Veldkamp, and C. C. Lai, Simple digital clipped correlator for photon correlation spectroscopy, *Rev. Sci. Instrum. 46*, 1356–1367 (1975).
65. D. E. Koppel, Statistical accuracy in fluorescence correlation spectroscopy, *Phys. Rev. 10*, 1938–1945 (1974).
66. H. Geerts, Experimental realization and optimization of a fluorescence correlation spectroscopy apparatus, *J. Biochem. Biophys. Methods 7*, 255–261 (1983).
67. H. Geerts, A note on number fluctuations: Statistics of fluorescence correlation spectroscopy as applied to Brownian motion, *J. Stat. Phys. 28*, 173–176 (1982).
68. S. L. Brenner, R. J. Nossal, and G. H. Weiss, Number fluctuation analysis of random locomotion. Statistics of a Smoluchowski process, *J. Stat. Phys. 18*, 1–18 (1978).
69. D. Axelrod, R. M. Fulbright, and E. H. Hellen, in: *Applications of Fluorescence in the Biomedical Sciences* (D. L. Taylor, A. S. Waggoner, R. R. Birge, R. F. Murphy, and F. Lanni, eds.), pp. 461–476, Alan R. Liss, New York (1986).
70. A. G. Palmer and N. L. Thompson, Intensity dependence of high order autocorrelation functions in fluorescence correlation spectroscopy, *Rev. Sci. Instrum. 60*, 624–633 (1989).
71. M. A. Rebolledo, Fluorescence correlation spectroscopy applied to the measurement of diffusion coefficients of atoms, *Spectrosc. Lett. 13*, 537–541 (1980).
72. J. R. Abney, B. A. Scalettar, and C. R. Hackenbrock, On the measurement of particle number and mobility in nonideal solutions by fluorescence correlation spectroscopy, *Biophys. J. 58*, 261–265 (1990).

73. S. R. Aragón and R. Pecora, Fluorescence correlation spectroscopy and Brownian rotational diffusion, *Biopolymers 14*, 119–138 (1975).
74. M. Ehrenberg and R. Rigler, Rotational Brownian motion and fluorescence intensity fluctuations, *Chem. Phys. 4*, 390–401 (1974).
75. H. Hoskikawa and H. Asai, On the rotational Brownian motion of a bacterial idle motor. II. Theory of fluorescence correlation spectroscopy, *Biophys. Chem. 22*, 167–172 (1985).
76. J. T. Yardley and L. T. Specht, Orientational relaxation by fluorescence correlation spectroscopy, *Chem. Phys. Lett. 37*, 543–546 (1976).
77. B. A. Scalettar, M. P. Klein, and J. E. Hearst, A theoretical study of the effects of driven motion on rotational correlations of biological systems, *Biopolymers 26*, 1287–1299 (1987).
78. P. Kask, P. Piksarv, and Ü. Mets, Fluorescence correlation spectroscopy in the nanosecond time range: Photon antibunching in dye fluorescence, *Eur. Biophys. J. 12*, 163–166 (1985).
79. N. L. Thompson, Surface binding rates of nonfluorescent molecules may be obtained by total internal reflection with fluorescence correlation spectroscopy, *Biophys. J. 38*, 327–329 (1982).
80. T. P. Burghardt and N. L. Thompson, Evanescent intensity of a focussed Gaussian light beam undergoing total internal reflection in a prism, *Opt. Eng. 23*, 62–67 (1984).
81. I. Z. Steinberg, On the time reversal of noise signals, *Biophys. J. 50*, 171–179 (1986).
82. H. Qian and E. L. Elson, On the analysis of high order moments of fluorescence fluctuations, *Biophys. J. 57*, 375–380 (1990).
83. J. Ricka and T. Binkert, Direct measurement of a distinct correlation function by fluorescence cross correlation, *Phys. Rev. A 39*, 2646–2652 (1989).
84. E. Hass and I. Z. Steinberg, Intramolecular dynamics of chain molecules monitored by fluctuations in the efficiency of excitation energy transfer: A theoretical study, *Biophys. J. 46*, 429–437 (1984).
85. N. L. Thompson and T. P. Burghardt, The influence of diffusion on reversible quasi-unimolecular reaction kinetics in one, two or three dimensions, *Biophys. Chem. 21*, 173–183 (1985).
86. Kodak Laboratory and Research Products (Catalog No. 53), Eastman Kodak Co., Rochester, New York (1987).
87. R. P. Haugland, *Handbook of Fluorescent Probes and Research Chemicals*, Molecular Probes, Inc., Junction City, Oregon (1985).

7

Fundamentals of Fluorescence Microscopy

Robert A. White, Karl J. Kutz, and John E. Wampler

7.1. Introduction

The compound microscope, since its introduction around the year 1600, has become one of the most useful tools available for the biological scientist. The basic design features and principles, even though they have been tremendously refined through the use of corrective lenses, filters, artificial light sources, and immersion oils, have remained essentially unchanged since the time of Galileo and Hooke. The modern compound microscope, as depicted in Figure 7.1, consists of a regulated light source, a condenser lens, an objective lens, and an ocular (eyepiece) lens. The ocular and objective lenses are usually mounted at a fixed position with reference to each other. Focusing is then achieved by moving the barrel or tube containing these elements or by moving the specimen holder. The condenser and light source rigidly control the illumination of the specimen field.

Microscopy of biological material, especially live cells, has always been challenging because of the lack of contrast and definition between cellular organelles. One of the earliest refinements in modern biological microscopy was the use of stains to enhance visibility of preserved subcellular structures. Many of the biological stains commonly used in microscopy are also fluorescent (e.g., eosin, Acridine Orange, acriflavine, etc.), a property which enhances the visibility of stained material even under transmitted illumination. In addition, many of the components of living cells are themselves fluorescent.

With appropriate filters on the illumination and viewing sides of microscopic samples, fluorescence from stains and intrinsic components can be

Robert A. White • Department of Biochemistry, University of Georgia, Athens, Georgia 30602. *Karl J. Kutz* • Georgia Instruments, Inc., Atlanta, Georgia 30366. *John. E. Wampler* • Department of Biochemistry, University of Georgia, Athens, Georgia 30602.
Topics in Fluorescence Spectroscopy, Volume 1: Techniques, edited by Joseph R. Lakowicz. Plenum Press, New York, 1991.

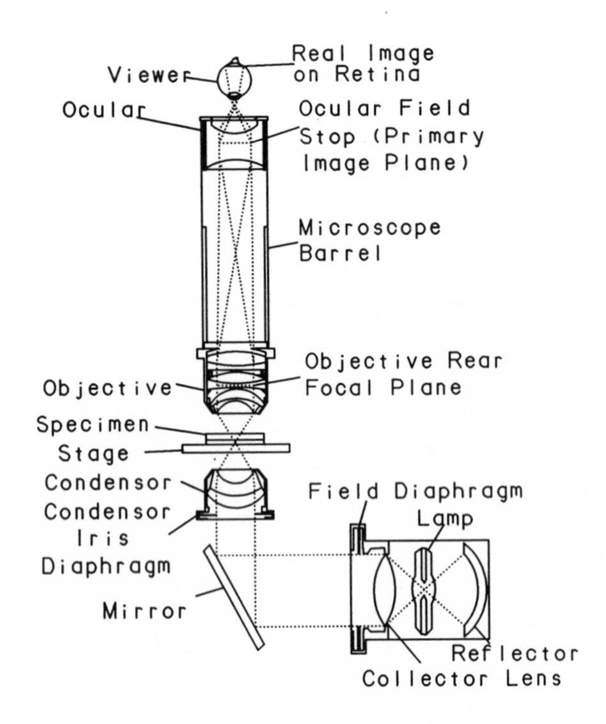

Figure 7.1. Cross-sectional view of a modern compound microscope.

isolated and viewed separately. Fluorescence microscope geometries (pictured in Figure 7.2) all make provisions for separating the stray exciting light from the emitted light in the imaging light path through the use of optical filters. The simplest configuration places these filters in the light path of the standard compound microscope geometry (Figure 7.2A). However, typically with such an arrangement, the fluorescence is viewed against a significant background of stray exciting light. In dark-field illumination (Figure 7.2B), the fluorescence stands out against a darker background, and the early applications of fluorescence to microscopy involved adaptation of dark-field illumination with filtration to enhance the contrast of the fluorescence emission.[1] However, for several reasons discussed below, the epi-illumination geometry[2] of Figure 7.2C is the one used in most modern microscopes and is the most easily adapted for quantitative spectral and spatial measurements. For a detailed analysis of the fundamentals and history of fluorescence microscopy, the reader is referred to the recent Ploem and Tanke monograph[3] and the earlier work by Ellinger,[1] Chapters 5 and 6 of Culling's book,[4] and Goldman's monograph.[5]

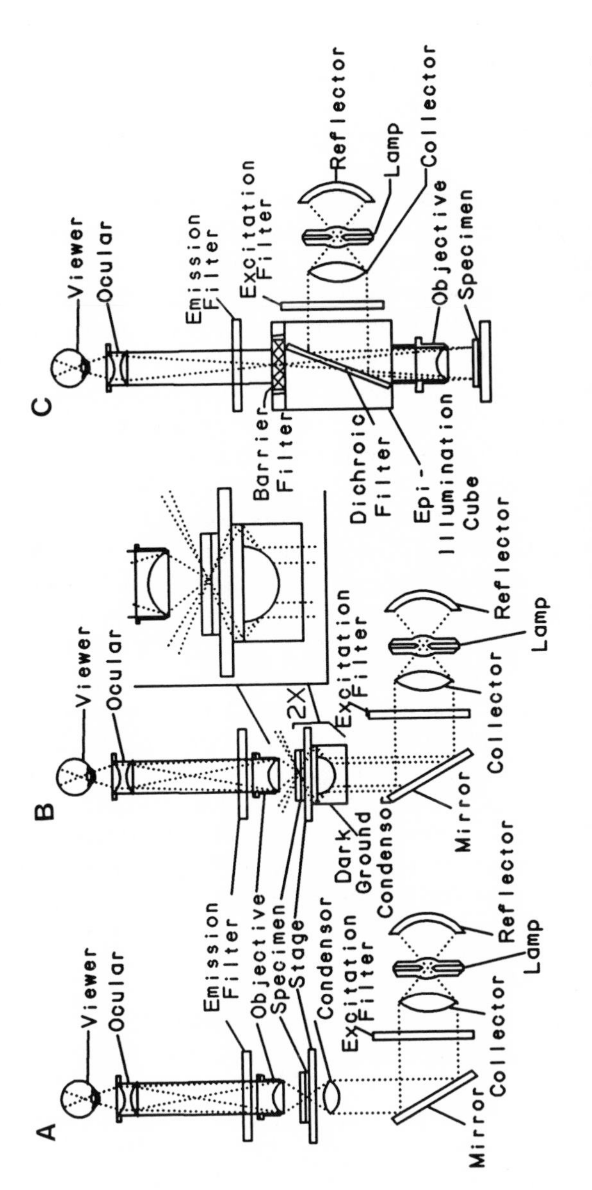

Figure 7.2. The three most common fluorescence microscope geometries: (A) conventional fluorescence; (B) dark ground fluorescence; (C) epi-fluorescence.

7.1.1. Sensitivity and Its Limitations

As with all fluorescence measurements, the limit to sensitivity in fluorescence microscopy is the stray light from the excitation. However, because of the very small size of the irradiated area and the very short optical path length in a microscopic sample, the common approaches which are used macroscopically to increase signal often fail, particularly in biological microscopy. For example, increasing the excitation irradiation level in order to increase the emission signal usually has deleterious effects on the sample because of photoirradiation damage to both the biological material and the fluorophore. Narrow band filtering to enhance contrast (i.e., separation of emission and excitation wavelengths and/or tightly delimiting emission wavelengths) often reduces the net signal to a quantum-limited level (see Wampler[6] for discussion), thus causing losses in resolution or detectability that more than offset gains in contrast. Therefore, from a practical point of view, when the eye is the detector, fluorescence microscopy is limited to fairly large signals and high fluorophore concentrations as compared to macroscopic fluorescence analysis.

Changing to electronic or film detectors does not necessarily result in improved performance, since a typical microscopic view must be critically focused within very narrow tolerances. With a quantum-limited image the long exposure integration needed to build the image on film or an integrating electronic detector will not be successful if there is no way to dynamically focus that image. However, the recent application of sensitive electronic imaging devices and real-time image processing to fluorescence microscopy has opened up many new avenues of research allowing, for the first time, dynamic visualization of weak, even quantum-limited, fluorescence signals. The pioneer in this area is Dr. George Reynolds of Princeton University, whose critical, detailed early work[7–10] has introduced researchers in the field of fluorescence microscopy to many new technologies. The broad range of applications of video technologies to microscopy has recently been covered in depth by Inoué.[11]

7.1.2. Visualization versus Quantitation

The largest single influence on biological fluorescence microscopy was the development of fluorescently labeled antibodies and the techniques for using them to localize specific sites within microscopic samples. This important tool was developed mainly through the initial efforts of Coons and coworkers[12, 13] and allows visualization of precise loci for a wide variety of biochemicals. Immunofluorescence technology, instrumentation, and techniques, reviewed in detail by Goldman,[5] are continuing to expand the sensitivity and specificity of this approach.

Obviously, visualization of fluorescent antibodies is less demanding on instrumentation, wavelength selection, and stray light control than is quantitation. If the fluorescent probe exhibits a good Stokes' shift, the contrast is enhanced by the color difference seen optically (green fluorescence against a deep-blue excitation background, for example). With image detectors and photometric measurements that have no color discrimination, however, the requirement to exclude exciting light from the image is more rigorous.

While the early use of fluorescence biological microscopy was primarily to visualize fluorescently stained components of tissues and cells, a natural extension of this instrumentation and technology was to add photometric detection (see Thaer[14] for review). While the literature on photometric quantitation of fluorescence signals is large and, like the macroscopic fluorescence literature, replete with descriptions of how to standardize and correct the signals obtained,[15–17] to a large extent photometric measurements of fluorescence from biological samples have been difficult to analyze and, therefore, of limited practical value. Until recently, advances in this area were primarily technical, marked by increases in the precision of the measurement and spatial resolution (for a review, see Piller[18]).

The emerging techniques in quantitative fluorescence microscopy use temporal or spectral resolution to extract quantitative information from signals on a relative scale rather than depending on absolute signal levels. Such measurements are less corrupted by optical and physical artifacts. Typical analytical measurements now involve analysis of the difference or ratio of signals at multiple wavelengths or, in the more sophisticated approaches, complete spectral analysis of excitation or emission spectra, or both.[6, 19–21] Computer control and coordination of the entire fluorescence microscope means that modern experimenters can now manipulate the optical environment of the sample using complicated, multistep, repetitive measurements. In many cases, visualization and measurement can be performed within an extremely narrow time frame, thus allowing complex kinetic studies.

Modern technical advances—developments of laser light sources, computer automation and control, and digital imaging—have allowed development of systems which combine quantitative analysis using temporal or spectral selectivity with spatial resolution. Such instrument systems have the potential for measurement of a vast array of information on a microscopic sample, including, but not limited to, mapping a fluorescent property to obtain a spatially resolved analysis. With photometric systems combining laser excitation beam scanning or computer-controlled stage positioning, the spatially resolved signal can have impressive measurement precision with somewhat limited temporal resolution. On the other hand, digital, video-based microspectrofluorometers are able to analyze entire images for spectral properties with less precision in the measurement, but with more impressive temporal response.

7.2. The Illumination Light Path of the Fluorescence Microscope

As discussed in the introduction and pictured diagrammatically in Figure 7.2C, the typical geometry for modern fluorescence microscopy is epi-illumination geometry, where the exciting light is delivered to the sample through the objective lens. The other geometries pictured in Figure 7.2 are now primarily of historical interest. Transillumination and dark-field geometries both suffer from similar limitations in that excitation light is delivered to areas and sections of the sample which are not in the focal plane or the image field being viewed. In addition, separate excitation and emission optics require separate efforts in focusing and adjustment. Transillumination and, to a lesser extent, dark-field illumination geometries give more stray light in general than obtained with epi-illumination.

The cross section diagram of the illuminating light path for epi-illumination from Figure 7.2C is enlarged in Figure 7.3A. The component parts are: (1) the lamp, including source, mirrors, housing, and lenses; (2) the auxiliary components; (3) the intervening optics; (4) the epi-illumination cube with wavelength-selecting components and beam combiner; and (5) the objective lens used as a condenser. While the illumination optics diagramed in Figure 7.3A are typical, they are not all necessary to achieve good epi-illumination. Special-purpose geometries separate wavelength-selecting components from the beam-combining function of the epi-illumination cube. One possible geometry is shown in Figure 7.3B.

In the geometry of Figure 7.3A, the epi-illumination cube usually has a removable core, slider, or turret assembly so that components for different fluorophores can be easily substituted. The typical epi-fluorescence microscope uses a dichroic mirror as a reflector/beam splitter in which the glass surface coatings are reflective over one wavelength range and transmissive over another. In the optical literature, this type of device is often referred to as a cold mirror or edge interference filter. It is essential that the coated surface of such a beam splitter be on the side toward the source and the objective, not the eyepiece side. If the mirror is reversed, then refractive effects when illumination passes through the glass can result in wavelength-dependent image position shifts.

The reflection/transmission transition line on the dichroic mirror must be extremely sharp for quantitative fluorescence microscopy. Furthermore, the wavelength position of this edge should match the fluorescence properties of the probe; if this edge is not sharp, then there will be significant mixing of stray light with the fluorescence, degrading sensitivity and the quality

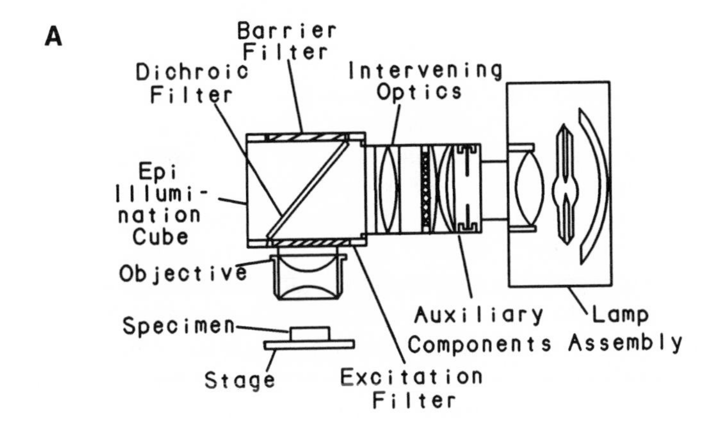

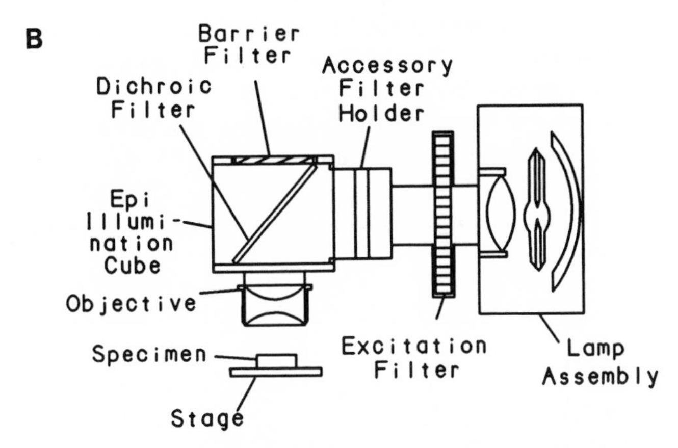

Figure 7.3. Two epi-illumination fluorescence filter arrangements: (A) excitation filter included in the epi-illumination cube; (B) excitation filter placed between the light source and the cube.

of the signal. In addition to the beam splitter/dichroic filter, a typical epi-illumination cube contains other filters. Ideally, the excitation filter permits full fluorophore excitation but cuts off on both the blue and red sides to limit fluorescence from other biological materials. The barrier filter is included as an emission filter to separate the emitted fluorescence from unabsorbed and residual excitation light as well as fluorescence at wavelengths other than those of the fluorophore of interest.

Regardless of the details, the purpose of the illumination system is to uniformly irradiate the sample, through the objective, with a cone of intensity-regulated, wavelength-selected light which fills the field of view (this is called Koehler illumination). Spatial inhomogeneity in either the irradiance or wavelength can be a major problem in both quantitation and visualization.

7.2.1. Lamps

The typical lamp for epi-fluorescence is selected primarily for brightness at selected excitation wavelengths rather than for broad spectral output or uniformity. This, high-pressure mercury arc lamps are often used. High-pressure arcs have slightly broader spectral lines than low-pressure lamps, and a larger portion of the light is in the continuum. Other types of arc lamps (xenon and xenon–mercury) can also be used. Typical output spectra for several different types of microscope illuminators are shown in Figure 7.4.

The common choice of the mercury arc lamp is partially due to the match between the arc's output and the spectral properties of the most common fluorescent probes, fluorescein and rhodamine. Both dyes have absorption spectra which give good absorption of major spectral lines from a mercury arc (the 435.8-nm line for fluorescein; the 546-nm line for rhodamine). In addition, portions of their emission bands (half-bandwidth from 520 to 560 nm for fluorescein; 575 to 620 nm for rhodamine) are in regions of the mercury arc spectrum where there is less output and, thus, stray light contributions can be reduced with appropriate filtration. The xenon arc lamp is typically weaker but has a more continuous spectrum than a mercury

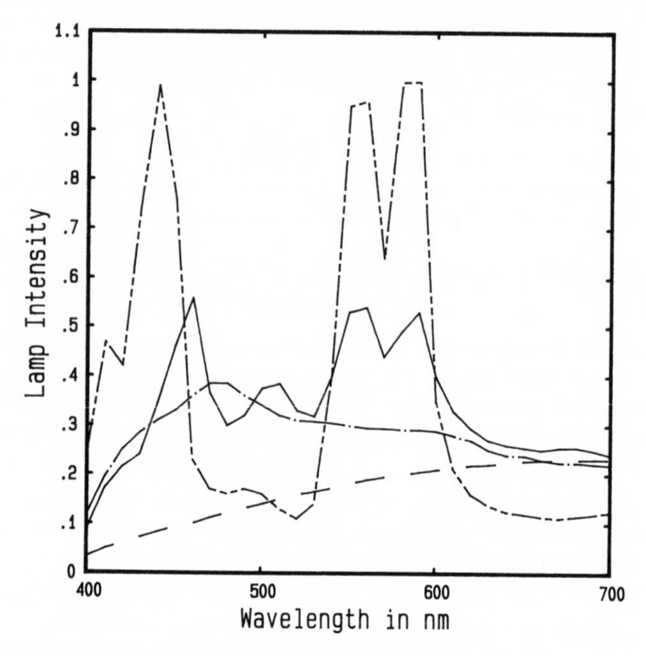

Figure 7.4. Emission spectra for different microscope lamps: 12 V, 100-W tungsten lamp (– – –), tin halide lamp (——), xenon arc (75 W) (– · – · –), and high-pressure mercury arc lamp (100 W) (— — — —).

arc. Thus, in applications requiring spectral scanning or excitation wavelength flexibility, a xenon arc is frequently preferred. The advent of a variety of fluorescence-based measurement protocols[20, 21] which use the ratio of fluorescence excited at two or more wavelengths as the analytical parameter favors the xenon arc over the mercury arc. The continuum of the xenon arc emission spectrum allows for measurements at different wavelengths without adjustment for differences in lamp intensity whereas wavelengths on the sides of mercury lamp emission bands are often very different in intensity. Of course, the increased flexibility of the continuous source is offset somewhat by increased stray light.

The other continuous sources such as those included in Figure 7.4 are not generally bright enough for broad application in fluorescence microscopy. Goldman[5] and Ploem[22] have discussed the selection of light sources in these applications.

One thing not shown by the data of Figure 7.4 is the dependence of spectral output on environmental and electrical factors. For example, tungsten filament-based lamps have a spectral output dependent on the current flowing in the filament. This effect is responsible for the week-to-week variation in measurement calibration curves discussed by Bright *et al.*[20] and can be a source of considerable variation, if such lamps are used without rigorous procedures to set their input current before each use.

Arc lamps, on the other hand, have more stable spectral output but suffer from arc wander, which contributes noise to the fluorescence signal they excite. Arc lamps can also be the source of electronic problems. Radio-frequency noise from the arc can be picked up by electrophysiological equipment. In addition, the circuits that are used to start arc lamps often send high-voltage transients through ground loops, and these starting transients can cause failure of digital electronic components that are improperly grounded or not electrically isolated. A standard procedure of this laboratory is to turn off all digital electronic components prior to starting an arc lamp.

In spite of its limitations, it is our opinion that the best general-purpose source for fluorescence measurements is a xenon arc lamp supplied by a particularly stable power supply with very precise current regulation. Using power supplies of the design of DeSa,[23] we have experienced extremely stable performance and very long bulb lifetimes. Most commercial microscope power supplies, however, are not very stable or well regulated. Other alternatives for stabilizing arc lamps have been discussed in the literature.[24–28]

7.2.2. Lamp Housing

As discussed by Goldman,[5] the main requirements of the lamp housing can be easily categorized:

1. It must be mechanically sound, providing for user safety, including provisions for containment of potentially explosive high-pressure arc lamps, protection from stray UV light, and ventilation of the photochemically formed ozone. In addition, lamp housing ventilation must be properly designed so that the bulb temperature is maintained in the optimum operating zone and not allowed to become excessive. High housing temperatures can rapidly shorten bulb life. However, forced air cooling is not recommended.
2. It must provide means for easy alignment of the bulb, lenses, and mirrors, preferably from outside the housing.
3. It should collect a large portion of the total flux from the lamp over a large solid angle by means of reflectors and lenses and direct that light to an adjustable focusing lens element.
4. It should provide for adjustment of its position relative to the other mechanical components of the microscope for gross alignment and centering.
5. Finally, it should allow replacement of the bulb with minimal re-alignment.

The lamp housing in Figure 7.5 illustrates the location of the major optical components of a typical fluorescence illuminator. It should be noted that very small design differences such as bulb clamp placement and the materials used for insulation, holders, and contacts can be major factors in attaining stable, worry-free operation. Therefore, even though the basic lamp housing features among the different commercial models available are very

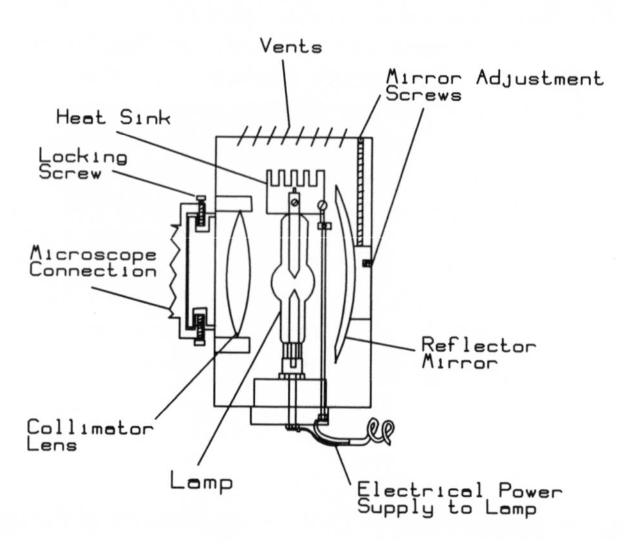

Figure 7.5. Diagrammatic representation of a typical fluorescence lamp housing.

similar to those depicted in Figure 7.5, the experimenter should be aware of characteristics and limitations inherent in the different individual products.

7.2.3. Auxiliary and Alternative Optics

The output of the very bright lamps employed as fluorescence microscope illuminators generally requires some filtration. At minimum, a heat filter is needed to protect the subsequent optics and lens coatings from irradiation damage due to excessive absorption of infrared wavelengths. As discussed in detail below, this area of the optical path typically includes two additional elements: (1) some form of aperture to reduce light scatter and glare, and (2) a wavelength selection device or filter. These are pictured diagrammatically in Figure 7.6.

7.2.3.1. Illumination Apertures and Stray Light Control

As with all fluorescence systems, the fluorescence microscope must rigorously control stray light. Apertures and diaphragms are placed in the excitation light path to reduce off-angle light, spurious reflections, etc., while controlling the primary signal. Surfaces are anodized or painted black. Mechanical joints in the light path equipment can be externally sealed with matte black photographic tape (available from the Eastman Kodak Com-

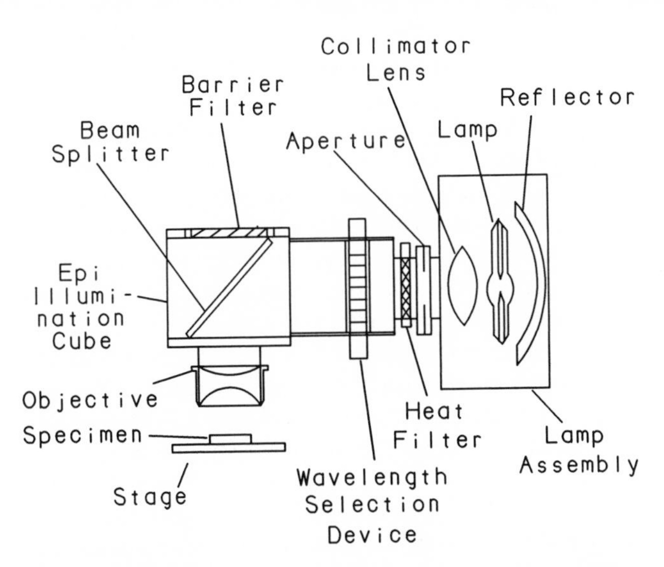

Figure 7.6. Epi-illumination diagram with aperture, heat filter, and wavelength selection device placed next to lamp housing.

pany, Rochester, New York, or 3M, St. Paul, Minnesota, for example) for additional lighttight integrity.

Depending upon the light path location of an aperture, it can reduce internal reflections, control glare, or bias data. Apertures are used in both the illumination and the imaging light paths, each having different ramifications in fluorescence image quantification. The most frequently used locations in the illumination light path are at the luminous field diaphragm, at the rear focal plane of the objective, and at other nonimage planes. Similarly, apertures are most often used in the imaging light path in the rear focal plane of the objective, at the primary image, and before the detector. It should be noted, generally speaking, that an aperture at the plane of the luminous field diaphragm or the rear focal plane of the objective serves a dual function in epi-illumination geometries since it is part of both the illumination and imaging light paths.

In the illumination light path, the first location for an aperture is that of the luminous field diaphragm. This position is coincident with the image of the specimen and defines the amount of the field to be illuminated. For full-field illumination the aperture is limited to just outside the field of view. This reduces stray light or flare in the body of the microscope and allows for full-field, low-magnification viewing with the maximum illumination. More frequently, however, only the area of the specimen to be quantified is illuminated; this eliminates undesirable photobleaching of adjacent areas and reduces extraneous or scattered light contributions from areas of the specimen outside the region of immediate interest. The illuminated field stop or luminous field diaphragm aperture can be a pinhole, slit, or any other shape which enhances the measurement signal by limiting extraneous light contributions.

The other main location for an illumination light path aperture is the rear focal plane of the condenser, which is also the rear focal plane of the objective in epi-fluorescence. An aperture placed here limits the size of the illuminating cone angle of light passing through the condenser/objective, thereby reducing the angle of light available at the specimen plane and, therefore, the area imaged by the objective. Narrowing the illumination cone of light results in a decrease in the effective numerical aperture (N.A.) of the condenser/objective lens, something which may be an advantage to spatial resolution and reduced corruption of the spatially resolved fluorescence signal by out-of-focus light as discussed later in Section 7.3.2. Modulators and phase rings are often placed in this plane, limiting the amount of light transmittance. For fluorescence, it is usually not advantageous to limit light transmittance with apertures in this plane. Modulators are especially restrictive since they limit the objective aperture and introduce another air-to-glass interface which can contribute to reflective and absorptive light losses.

As was mentioned previously, apertures in the field diaphragm plane serve the same function of area selection in both the imaging and illumination light paths. Similarly, rear focal plane modulators and apertures will affect light transmission both to and from the specimen. An aperture at the primary image plane (also referred to as the ocular field diaphragm), however, is only in the imaging light path; other than at the specimen plane, it is here that the light rays are most closely focused. An aperture in this image plane is used to restrict the field size presented to the ocular and, in photometry, to limit the measured spot in the image. In a typical situation involving green light and a 1.3-N.A. objective, the diffraction limit for unambiguous measurements requires an aperture diameter equivalent to about 0.5 μm at the specimen.[28, 18] An aperture in the primary image plane also serves to eliminate glare and stray light. In addition, it is used to further restrict the size of the image presented to the detector and can limit parasitic light from striking the detector and biasing the measurements. The ocular field diaphragm aperture, like the illumination field diaphragm aperture, can be a pinhole, a slit, or other shapes which conform to the specimen area of interest. Furthermore, the effects of spherical and chromatic aberrations on measurements taken through this aperture can be minimized by making the measurements directly on the optical axis.

7.2.3.2. *Wavelength Selection, Beam Combining, and Light Collimation*

In most modern commercial epi-fluorescence microscopes, a bandpass filter and dichroic mirror in an illuminator cube within the microscope body select the excitation wavelength band or spectral line from the source. This type of wavelength selection limits the system to one illumination wavelength at a time, eliminating the possibility of rapid spectral scanning or dynamic wavelength comparisons. Ideally, the lenses in the system match the illumination cone to the acceptance angle of the objective lens, allowing full-field irradiation. The role of these lenses is to deliver the light to the limiting aperture and control its uniformity. The limiting aperture (luminous field diaphragm) must be uniformly filled with light so that the sample is uniformly illuminated. Efficiency is gained by using a large collection angle of the source and a cone-matching focusing lens at the other end. However, if an image of the source is precisely focused on the field at the local plane of the sample, illumination will not be uniform. Thus, it is generally necessary to either defocus, "homogenize," or diffuse the illumination image, trading optimum irradiation levels for more uniform excitation across the field of view. As discussed by Inoué,[11] the most efficient way to achieve uniform illumination and optimum irradiation levels with visible light is with a bent optical fiber which can be used to "scramble" the source image with little loss in the total radiation level. A commercial scrambling device of this type is marketed by Leitz (E. Leitz, Inc., Rockleigh, New Jersey).

In order to perform meaningful quantitative fluorescence measurements, it is necessary to incorporate into the microscope a mechanism which would allow multiple-wavelength selection. One approach to this problem is to automate the interchange of dichroic filters. Multiple-position filter sliders are made by several microscope manufacturers. Leitz, for example, uses a rotating carousel in some of its microscopes. When the measurement is based on very different wavelengths of two fluorophores such as the pH measurement technique described by Geisow,[29] this approach is certainly feasible. More typically, however, we are interested in defining the shape of the spectrum of a single fluorophore or of closely related fluorescent species. In such cases, fluorescence measurements at multiple, closely spaced wavelengths become necessary, and the excitation wavelength selection components become sufficiently complex that they must be removed from the beam splitter compartment. The additional mechanical and electrical complexity incurred with these changes is nicely balanced by a considerable increase in system flexibility. The dichroic filter is replaced with a beam splitter, and the light is wavelength regulated at the source, resulting in a simpler illuminator cube and a wavelength selector which is faster to operate and much more flexible in its applications.

The simplest source geometry for a multiwavelength illuminator system involves a lamp, a collimator, and a wheel containing multiple interference filters (Figure 7.7). Another solution is to use multiple lamps, each with its

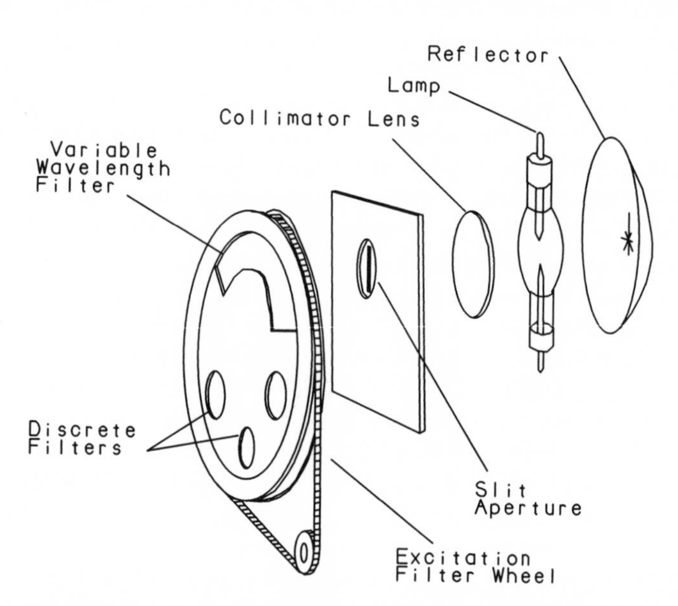

Figure 7.7. Exploded isometric drawing of a typical fluorescence light source which includes a filter wheel and slit.

own filtration, combine the beams from these lamps optically, and select between them with fast electronic shutters. This approach has recently been employed for fluorescence ratio measurements by Chu and Kutz (unpublished results). More sophisticated measurements requiring continuously variable wavelength selection employ various types of monochromators.

Several laboratories currently use discrete filters mounted in a wheel for fluorescence ratio measurements (see Bright *et al.*[20, 21] for references). Typically, a collimator lens ensures that the light beam strikes the filters normal to their surface, ensuring accuracy of wavelength position selection and minimum bandwidth. This allows for optimum use of interference filters since, if such filters are placed in diverging or converging light rays, their performance is degraded, with a broadening of the bandpass.

Replacing the discrete interference filters with a graded or variable spectrum interference filter such as those manufactured by Ealing Corporation (South Natick, Massachusetts) and Optical Coatings Laboratories (Santa Rosa, California) increases the usefulness of the wavelength selector. These devices allow continuously variable selection of wavelengths with relatively broad bandwidths over most of the visible wavelength range and can be mounted as in Figure 7.7 in the same type of wheel as discrete filters. With the variable spectrum interference filter, one can perform rapid scanning operations or follow fluorescence flux at specific wavelengths.[6] Ratiometric measurements and corrections for naturally occurring fluorophores become routine as a result of the flexibility of such a system.

7.2.3.2a. Fixed Wavelength Selection (Filters). There are four types of wavelength-selecting filters that can be used in a microspectrofluorometer: (1) gelatin filters, (2) colored glass filters, (3) interference filters, and (4) cold and hot edge filters. There are three components which contribute to the transmission characteristics of all of these filters. These are: (1) reflection loss at the front surface, (2) absorption or interference loss within the filter, and (3) reflection loss at the back surface. These filters are usually not coated, and, therefore, reflection losses in transmitted light can be as much as 5–10%, even at normal incidence angle.

Gelatin filters consist of a gelatin film containing dissolved organic dyes, coated with a protective lacquer. The Kodak Wratten filters are typical of this type of wavelength-selecting material. Since the components of gelatin filters can be easily varied in composition, these filters are available with a wide range of transmission properties, broad bandpass, and long- and short-wavelength transmission (cut-on and cutoff, respectively). The gelatin foundation has a low melting temperature, is hygroscopic, and must be protected from harsh environments. These filters will not tolerate temperatures above 50 °C or conditions of high moisture. Since gelatin filters operate on the principle of wavelength absorption, intense light, especially in their absorption bands, can cause local heating and melting. Prolonged exposure to ultraviolet light, of course, can bleach the dyes and change the filter's characteristics. These filters are, as optical filters go, rather delicate and must be treated accordingly. This limits their application to conditions of moderate light intensity, low heat, and average moisture.

As with the gelatin filters, colored glass filters also offer a wide range of transmission characteristics. Corning, Hoya, and Schott colored glass filters are the best known of such filters. Both the Hoya and Schott filters are made of optically ground and polished optical/quality glass. These filters can withstand more severe conditions than can the gelatin filters and are more uniform in their optical features. However, since they function by absorbing light, they can overheat and even crack under extreme cases of high light intensity and/or exposure duration.

Interference filters typically have a thin film, composed of partially reflecting layers, applied to optical glass. These layers have a Fabry–Perot interference effect on the light passing through them, selectively enhancing the transmission of some wavelengths and greatly diminishing transmittance of others through destructive interference. These filters are much more precise in their optical characteristics than are the gelatin or colored glass filters. They are generally available in series of narrow-bandpass filters with fixed half-bandwidths. Since one side of these filters is usually highly reflective to the off-band wavelengths, it should face the source to reduce internal heating of the filter and optimize filtration.

There are two aspects of interference filters which are of some concern. These are related to the fact that the center band wavelength of interference filters is dependent on the angle of incidence. This is seldom a concern near the normal ($<10°$), but at severe angles it can cause a significant shift in the position of the filtered wavelength (typically 10% lower wavelength at 45°). The second problem, which occurs for the same reason, is that interference filters placed in uncollimated light will have a broader bandwidth than described in their specifications since this light contains rays that contact the filter at a variety of angles.

In contrast to other filters, edge interference filters are designed with very well-defined characteristics, which include a sharp transition from reflection to transmission. Therefore, they make ideal cut-on and cutoff filters for fluorescence studies, much more so than corresponding colored glass cut-on filters. They have been used with success as wavelength-selective beam splitters in fluorescence microscopes, dramatically simplifying the required optics.

7.2.3.2b. Variable Wavelength Selection (Monochromators). In fluorescence microscopy, monochromators are primarily chosen based on their efficiency. Stray light characteristics of monochromators are of importance, but high spectral resolution is not a strict requirement. Prisms were used as some of the earliest fluorescence microscopy monochromators,[30–34] but grating monochromators are now much more likely to be the system of choice. Their high efficiency and simple triangular transmission curves (which tend to be independent in shape regardless of the selected wavelength) make them much more useful to the fluorescence microscopist. Unfortunately, with grating monochromators, the light is dispersed in several overlapping orders, resulting in contamination of the selected light with light from other orders. For example, for an instrument operating in the first grating order set so that the detector "sees" a specific wavelength, the second-order stray light is from the spectral region at one-half the selected wavelength. Stray light of this nature is much more of a problem if the monochromator is on the emission side of the spectrofluorometer than if it is on the excitation side.

Unfortunately, grating monochromators are not uniformly efficient; they have a peak efficiency wavelength, called the blaze wavelength, and their efficiency drops off fairly quickly on either side of this peak. The blaze wavelength is determined by the angle and shape of the grating's grooves. Efficiency can fall off as much as 50% at two-thirds of this wavelength and at twice the blaze wavelength in traditional ruled diffraction gratings. Holographically recorded diffraction gratings,[36] on the other hand, have an efficiency curve that is less wavelength-dependent, but their peak efficiency is typically less than that of a similar ruled grating.

Imaging detector systems, especially the low-light video types, require a wavelength selector which maintains strict image quality. Therefore, applications of this nature impose constraints on the type of wavelength selection mechanism which can be employed in the imaging light path. A simple, variable-spectrum interference filter is one solution which has been effectively used with a low-light video detector[6, 35] to obtain excellent quantitative measurements of biological and organic fluorophore spectra. Since this type of monochromator has good image quality with a relatively broad bandwidth, it is relatively efficient and effective when examining *in vivo* fluorophore samples. A circular, graded interference filter can easily be inserted into a microscope (Figure 7.7), driven by stepper motors, and, via computer control, implement rapid, repetitive wavelength selection or spectral scanning.[6, 36]

Rapid wavelength scanning in a photometric microscope system can also be accomplished using a prism monochromator with a motor-driven spiral slit.[32] The prism monochromator disperses the image of the photometric aperture, and the spiral slit is then used at the plane of the dispersed image to select the wavelengths seen by a photomultiplier tube. The slit is incorporated into a wheel which can be rapidly turned in order to continuously scan spectral data.

A slightly more exotic approach to microscopic spectral analysis has been the use of optical multichannel analyzers (OMA) with a spectrograph.[37] Rapid and sensitive spectral analysis can be performed with the newer, more sensitive OMAs such as those manufactured by Princeton Applied Research

(Princeton, New Jersey). The microscopic area to be measured with this system should be sharply delimited by an aperture in the specimen image plane to remove stray light and enhance the image quality.

A recent innovation in wavelength selection is the acousto-optic tuned filter,[38] where high-frequency modulation of a birefringent tellurium oxide crystal mounted on a piezoelectric transducer is used to set up a standing wave pattern. This pattern, in turn, diffracts a narrow wavelength band of light. The acousto-optic tuned filter has been reported to provide 1-nm spectral resolution with rapid wavelength selection or scanning. Since it involves no moving parts and the output wavelength can be easily and rapidly selected in automated control, this system may represent the next generation of developments in optical monochromators for microscopy.

7.2.3.2c. Light Collimation. Optimally, selection of excitation wavelengths should occur with the light rays parallel to optimize narrow bandpass and avoid shifts in the spatial distribution of excitation at the specimen at different wavelengths. Such shifts and bandpass fluctuations are dependent on the angle at which the light strikes dichroic reflectors or wavelength-selecting components. Frequently, commercial systems use auxiliary lenses (sometimes referred to as telan or tube lenses) to overcome this problem in their fluorescence illuminators; these lenses, employed in pairs, act first to collimate the exciting light, keeping the light rays parallel in the vicinity of the beam splitter/reflector, and then recombine the wavelength-selected light into a focused beam. Unfortunately, each additional air/glass interface of such a lens system is a source of lowered transmission efficiency.

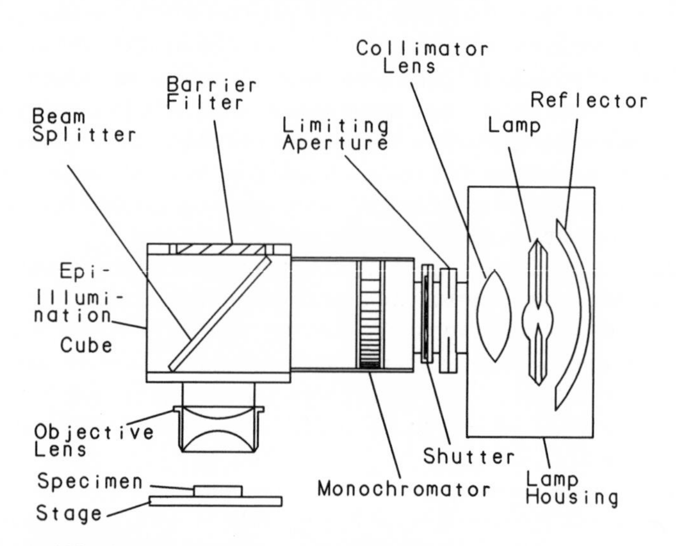

Figure 7.8. Epi-illumination diagram with limiting aperture, shutter, and monochromator placed next to lamp housing.

Overall, more flexibility is obtained by using a lamp assembly which produces collimated light (Figure 7.5). Such an assembly allows for insertion of a variety of optional/additional wavelength-selecting components into the light path. Interference filters inserted into a collimated light path exhibit optimum bandpass and exclusion. Light ray collimation can be maintained through monochromators placed in the light path (Figure 7.8), with their collector lens systems matching the light beam to the proper collection angle and then recollimating the selected light. Wavelength-selected collimated light is then focused, forming a light source image plane at which the rays are bundled closely together. Both a shutter and a limiting aperture (for stray light control) can be installed here for maximum effect. The illumination beam is then collected by a second auxiliary lens, matching it to the collection angle of the objective and focusing an image of the source at the back focal plane of the objective. A simple beam splitter (half-silvered mirror, pellicle, etc.) can be used as a beam combiner in these systems, but a dichroic mirror will give higher efficiency, albeit with limited excitation wavelength range.

7.2.4. The Objective Lens as Condenser

The wavelength-selected light travels through the objective lens (Figure 7.8) and is condensed by the objective onto the specimen. The objective should be selected on the basis of glass compound, number of glass elements, degree of correction, numerical aperture, and maximum transmittance. Generally, fluorite lens are preferred over crown or flint type glass lens. However, new technologies in lens fabrication now make it possible to use more highly corrected lens (planapochromatic) in fluorescence applications over a broad wavelength range. The old rule that planapochromatic lenses should not be used at excitation ranges below 400 nm does not apply. Lenses having both high numerical apertures and high transmittance are now being offered by a variety of manufacturers.

In the epi-fluorescence microscope with the objective functioning also as a condenser, there is a gain in brightness of the fourth power of the numerical aperture of the objective.[11] This is one of several reasons that high-N.A. objectives are often used in fluorescence microscopy. However, as discussed below (Section 7.3.2), there are also several negative aspects in the use of high-N.A. objectives.

7.2.5. Specimen and Mount

The specimen should be prepared with attention to details such as the light transmission of mounting medium, cleanliness, thickness of cover slips, and sample thickness. Light scattering, reflection, and refraction in biological samples have even greater effects on fluorescence emission than on visible light. This means that something which has a small, insignificant effect on the visible image may have a measurably distorting effect on the fluorescent signal from the same sample. Moreover, biological structures (cell walls and membranes, especially) which appear as slight distorters of visible light can act in fluorescence detection to concentrate light and alter wavelengths so that nonfluorescent objects appear to be fluorescing.

Naturally occurring fluorophores widely occur in biological samples. It is very important that these contributions to the fluorescence signal be accounted for. Therefore, it is essential that control specimens be routinely measured under equally rigorous conditions of sample preparation and handling. This is especially important when the sample feature of interest changes its optical and/or fluorescence characteristics following alterations in the surrounding environment (such as ionic or pH changes).

7.3. The Imaging Light Path in Fluorescence Microscopy

7.3.1. The Specimen as an Optical Component

In microscopy, the slide, specimen, mounting medium, coverslip, and immersion oil (if used) are integral components of the optical system, as discussed above. Refractive and reflective effects within the mounted specimen and the components of the mount alter the signal being measured even before it reaches the objective lens. The image seen depends, of course, on the plane of focus as illustrated by the drawing in Figure 7.9; two planes of focus give two completely different fluorescent images. In addition, the image obtained at different focal planes may not exactly represent the locations and intensities of fluorescence from these planes since refraction and reflection can give both false signals where there is no fluorescence and loss of signal where there is fluorescence. Moreover, it is essential that focusing be performed not merely on the transmitted light image of a biological specimen but on the fluorescence image as well, since the plane of focus of fluorescence emission may not be the same.

Both the overall fluorescence flux and the accompanying spectra of subcellular particles can change as a result of the influence of cellular optical effects; this has been demonstrated decisively on a theoretical basis by Kerker *et al.*[39] using model systems. The fluorescence signal can also be significantly contaminated by light from areas both above and below the plane of focus.[40] High-N.A. lenses may contribute to this out-of-focal plane light since they can collect a large percentage of their total fluorescence signal from emitters out-

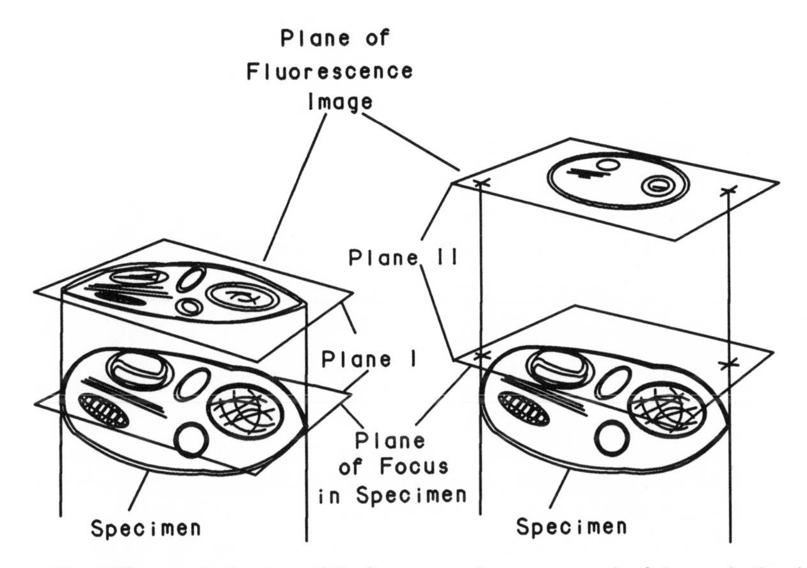

Figure 7.9. Differences in the plane of the fluorescence image as a result of changes in the plane of specimen fluorescence focus.

side the plane of focus. Immersion oils and mounting media may contribute fluorescence signals of their own. Refractive effects between a low-refractive-index mounting medium (such as water, $n_D = 1.33$) and the glass of the slide and coverslip (typically, $n_D \approx 1.5$) can change the collected light signals.

These effects combine to make correlation of a measured fluorescence signal with a specific chemical or environmental factor within a cell difficult; in extreme cases, quantitative flux measurements may become almost useless. This makes fluorometric measurements without wavelength sensitivity of little, if any, value and has brought about the movement in research efforts toward microspectrofluorometric analysis. Additionally, the fact that high-N.A. lenses collect large amounts of out-of-focus emission reinforces the arguments for increased detector sensitivity (see Section 7.3.5) and supports alternative approaches to spatial resolution in fluorescence such as laser excitation scanning and confocal microscopy.

7.3.2. The Objective

The ideal objective lens must satisfy the following four criteria in order to perform optimally in quantitative fluorescence microscopy:

1. It must have high transmission efficiency.
2. It must contribute no fluorescence from its internal parts.

3. For best results it should have a high light collection efficiency.
4. In order to maintain accuracy and precision of measurement, it should have good resolution.

Based on conventional wisdom, it could be concluded that these criteria are best satisfied with objective lenses having high numerical aperture (N.A.). The numerical aperture of a lens is defined as the effective collection angle of that lens. A lens viewing a sample through air can only capture that portion of the light limited by the refraction angle (Snell's law). Using the central point in the specimen as a reference point, this angle is defined by the product of the refractive index of the intervening media (air, $n = 1$) and the half-angle subtended by the lens itself; that is,

$$\text{N.A.} = n \sin \theta \tag{7.1}$$

With air as the medium, the maximum N.A. is 1.0, since $\sin \theta$ in air has its largest value (1.0) at 90°. It is obvious then that larger effective collection angles and, therefore, N.A. values larger than 1.0 can only be obtained by filling the space between the lens and the specimen with a medium of larger refractive index such as water ($n = 1.33$), glycerol ($n = 1.47$), or microscope immersion oil ($n = 1.5$). For example, a 1.4-N.A. objective viewing a sample through immersion oil intercepts light through a plane half collection angle of 69° and a solid angle of 4 steradians. If light emission from the specimen is isotropic, then only about one-third of the emitted light is within the collection cone of the lens (since there are 4π steradians in the closed sphere). The actual amount of this light that is finally transmitted back through the lens is even less, however, since some of it is partially reflected by the surface of the lens, some is lost in the imperfect transmission through lens components, and some is scattered by dust and surface imperfections on the lenses.

Another way of describing the effect of immersion oil is shown pictorially in Figure 7.10. Assuming that the physical geometry of the lens shown limits the collection angle to 120° when it is focused on a specimen, without immersion oil (left-hand side of Figure 7.10) the maximum plane collection angle of light from the specimen is limited to 78°. Light leaving the coverslip at 78° from the sample is refracted at the air interface to the 120° limiting angle of the objective lens. With immersion oil, however, the light coming from the sample is not appreciably refracted, and the full view angle of 120° is possible (right-hand side of Figure 7.10). Thus, more light collection and a greater field of view is possible because the refractive index of the immersion oil (1.5) is close to that of the glass surfaces of the coverslip and objective.

In conventional, bright-field microscopy, high N.A. is translated as good resolution since microscope resolution is limited in direct proportion to the reciprocal of the N.A. One negative consequence of the use of high-N.A. lenses is immediately apparent to the microscopist—close working distance (as the

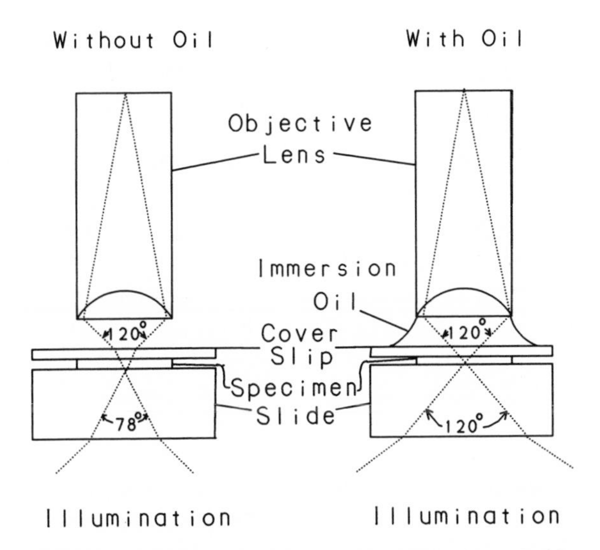

Figure 7.10. The effective acceptance angle at the objective lens with and without immersion oil.

result of the very short focal distance of these lenses). Since it is necessary that the plane angle be as large as possible, the focal point of a lens must be very close to the sample. This is the direct consequence of the fact that as any lens of a given physical size moves further from its focal point, the plane collection angle decreases and so, therefore, would the N.A. by Eq. (7.1).

A second limiting characteristic of high-N.A. objectives is a shallow depth of field, roughly proportional to the inverse square of the N.A. (in units of microns). For example, a 1.4-N.A. lens would have a depth of focus of approximately 0.5 μm.

In fluorescence microscopy both of these limitations of high-N.A. lenses have significant consequences. For instance, the shallow depth of field does not exclude collection of large amounts of out-of-focus fluorescence when emitters are present within the volume illuminated by the illumination cone.[40] This extraneous light contributes to a loss of contrast and can result in considerable degradation in fluorescence microscopy resolution. Furthermore, if the luminous field diaphragm is to be varied in size in order to isolate different size particles for photometric measurement, then high-N.A. lenses are also not recommended.[5] It may be most effective in some fluorescence microscopy situations, therefore, to use lower N.A. lenses in order to obtain increased photometric flexibility or resolution even though some loss in detector sensitivity is concomitantly incurred.

It should be pointed out that high N.A. and good light transmission and collection efficiency do not necessarily directly correlate. Absorption, refraction, and reflection light losses within lenses vary widely so that two lenses with similar N.A. values may differ greatly in light gathering and transmitting efficiency. For example, a Zeiss (Carl Zeiss, Inc., Thornwood, New York) epi-Pol 40× oil immersion lens with an N.A. of 0.85 is 25% more efficient than the 1.0-N.A. planapochromatic lens of the same manufacturer (Wampler and Rich, unpublished). In this regard, manufacturers' literature has been of little aid, and the best recommendation is to quantitatively test the "brightness" of lenses using the samples and optical systems specific to each application.

Unlike the case in normal bright-field microscopy, highly corrected lenses may not typically be advantageous in fluorescence microscopy. This is especially so where the imaging light path involves highly filtered light of only a relatively narrow wavelength range. Thus, chromatic aberrations in the focused image will be minimal. Plan or flat field optics are desirable from the aesthetic point of view and certainly necessary if spatial measurements are to be made without correction for lens distortion. If images from multiple wavelengths are sampled, then even a highly corrected objective will require geometric correction before direct spatial comparisons can be made with high resolution.[41, 42] Mathematical correction has the advantage that it does not degrade the signal, whereas optical correction involves light losses. For example, a planapochromatic objective may contain a dozen or so lens elements, all of which have some light reflection. These are usually coated with the best antireflection materials (see discussion in Section 7.3.3 on coatings), but, even so, the transmission loss due to reflections may be as much as 10–15% with additional losses from the absorption of the glues and coatings.

7.3.3. Optics between Objective and Eyepiece

In a conventional epi-fluorescence microscope, the light emitted by the specimen must return through the beam splitter as well as any intermediate optics installed by the manufacturer to control the internal light path. As mentioned earlier, some light loss occurs at each glass/air interface, regardless of the efficiency of the antireflection coatings and the size of the optical elements. Typical broad-band multilayer antireflection coatings reduce on-axis (normal angle of incidence) reflection to around 0.5% while also reducing off-axis losses, with a typical value of, say, 1% at 45° incidence. Therefore, ten such lens surfaces in a light path would reduce the transmission by approximately 5% to 10%.

As pointed out in Section 7.1, typically the tube length of a microscope is fixed in order to simplify the design and complementarity of eyepieces and objectives. However, the fixed tube length often limits design of complex optical paths. Thus, it is typical for manufacturers to obtain more linear working distance between the objective and eyepiece by using pairs of auxiliary lenses (sometimes referred to as tube or telan lenses) to first collimate and then recombine the imaging light rays. This approach maintains the corrections for optical aberrations designed into the objective and ocular lenses and allows insertion of additional optics such as wavelength-selecting components without image shifts or distortions that otherwise might be seen as image spatial displacements or parasitic images.

In fluorescence, the critical components added to the imaging light path are for emission wavelength selection. In the typical epi-illuminator cube, two filters are used to make this selection—the dichroic mirror and a barrier filter. In custom-designed systems, a single filter or a complex monochromator may be used. In either case the discussion of Section 7.2.3 is pertinent.

7.3.4. Role and Position of Ocular

The ocular is composed of a series of lenses designed to collect the light from the primary image and convert it to a virtual image for viewing or a primary image for detection. In order to perform this function, the elements of the ocular magnify the image, control the field of view, and correct or compensate for aberrations in the objectives. In many systems, therefore, the primary image is not fully corrected, and an ocular is necessary to provide the detector with a fully corrected image for quantification. Increased magnification of the primary image by the ocular also allows for higher spatial resolution in both photometric and image-based systems as well as lower N.A. objective lenses since the overall magnification in the system is the product of the magnifying power of the objective and the ocular lenses.

All oculars delimit and define the field of view seen by the detector. Each ocular has an internal ocular field diaphragm, located at the primary image position within the ocular, which limits this field. Pinholes and apertures may also be placed at this location in order to reduce the area of the field presented to a detector. As mentioned previously, oculars also are required to provide compensation for chromatic aberrations found in the objectives of most microscopes. Recently, a new device—a color-compensation tube lens—has been reported as a chromatic aberration-compensating replacement for the eyepiece ocular. Purportedly, the tube lens yields better corrected images and increased microscope flexibility.

7.3.5. Detector Placement for Electronic Imaging and Photometry

For fluorescence image quantification, detector position is critical. Depending upon whether the information desired is mainly temporal or spatial, detector placement for correct data acquisition can vary tremendously. The typical detector arrangement places the detector at the real image plane formed by the ocular or projection eyepiece. There are instances, however, in which it is preferable to place the detector directly at the primary image plane. In conditions of very low light, for example, it may be necessary to remove all nonessential air-to-glass interfaces in order to increase optical efficiency. The image at the primary image plane theoretically should yield a much more intense light concentration since the light rays are most closely compacted at that location and the magnification is smaller. Placement of the detector, therefore, is determined by whether the investigator is measuring broadly emitted light or quantifying the signal from spatially complex, distinct, subcellular fine structures.

There is both an advantage and a disadvantage to placing the detector at the primary image plane of a projection eyepiece. The benefit is an increase in precision resulting from the magnified image; a single picture element (pixel) represents a smaller area of the specimen. A typical vidicon-based image digitizing system uses either a (512×512) or a (1024×1024) matrix to delimit the number of information points for quantification. Obviously, increasing the magnification to such a detector results in more measurement points per actual unit area of the specimen.

The disadvantage of this geometry is the loss of light resulting from the introduction of more lens elements (and, therefore, more glass-to-air interfaces) and the magnification losses. This loss of signal may be significant since, as discussed previously, uncoated glass surfaces yield a loss of 5% per air-to-glass interface due to reflection, and coated glass, even with modern coatings, can lose as much as 0.5–1.0%.[18] Therefore, it is usually a good idea to limit the number of air-to-glass interfaces to only those necessary to accomplish the desired degree of spatial separation for quantification.

7.3.5.1. Detector Types

Two classes of detectors are used for quantitative fluorescence microscopy measurements: (a) photomultiplier tubes, and (b) a variety of types of imaging devices. If measurement precision is the primary goal, then photomultiplier tubes are the detector of choice, provided that no spatial information is desired or that another system component can be used to accomplish this function. Laser excitation beam scanning or mechanical stage scanning are methods which can be employed to secure spatial information in this type of system arrangement. If rapid acquisition of spatially detailed information is required, then image detectors become the detector of preference.

7.3.5.1a. Photomultiplier Tubes. Photomultiplier tubes (PMTs) have one distinct advantage over other types of detectors; they are excellent amplifiers as well as detectors. While there are several different types of PMTs (for recent reviews of this subject, see Boutot *et al.*[43] and Candy[44]), commercial microscope photometers usually use side-window tubes since they are compact and relatively inexpensive. Side-window tubes are not highly regarded in the physics and engineering literature, but their specifications do compare favorably with those of end-window tubes. Since end-window tubes offer the most flexibility as well as extended performance, they are often the choice for custom-designed instrument systems. With the increased demand for higher instrument sensitivity and photon-counting electronics, the use of end-window tubes in microscope photometers is certain to continue to grow. As photon counting is applied to microscopy, there will, no doubt, be increased use of advanced photomultiplier tubes such as microchannel plate devices (see Boutot *et al.*[(43)]).

With respect to their use in general microfluorometry, PMTs have three performance characteristics which are of importance: (1) their quantum efficiency, (2) the overall gain of the device, and (3) the dark current. Using a high-gain PMT allows the use of low-cost, easy-to-build amplifiers. In a typical PMT, quantum efficiency is very wavelength-dependent, peaking at less than 30% in the blue region of the spectrum. A PMT's spectral sensitivity curve is determined by two factors: (1) the wavelength-dependent quantum efficiency of the photocathode material, and (2) how that material is deposited on the light-collecting surface. Unfortunately, there are few, if any, PMTs with quantum efficiency above 10% either in the spectral region above 600 nm or below 300 nm. Even the best quantum efficiency spectra gleaned from those published by PMT manufacturers show this glaring limitation. Special and more expensive window materials such as quartz and fused silica can improve the efficiency below 300 nm, but there is little that can be done to better the efficiency at wavelengths above 600 nm.

The most flexible PMT would be one in which the quantum efficiency spectrum extends from the ultraviolet to the red part of the visible spectrum. Such a photocathode would allow spectral studies on a tremendous range of fluorophores. In practice, however, this flexibility is gained at a cost. First, extended response tubes (both ends of the spectrum) are more expensive. Second, this broad response includes a broad sensitivity to stray light. Since stray light is usually limiting in the typical fluorescence analysis, this is a decidedly negative feature. Finally, red-sensitive PMTs give more noise in the signal (even in the absence of stray light) since their work function is lower. Thermal events in these tubes eject electrons spontaneously, and, therefore, for optimum performance, red-sensitive tubes are usually cooled, with the cooling system adding to both the cost and the system's complexity. Thus, for broad-range, high-gain response one might choose a nine-dynode, end-window tube with an extended S-20 photocathode and a UV window. If, however, only limited spectral information is required or extreme measurement refinement is not necessary, then a PMT with a more limited response or a compact, inexpensive side-window tube might be more desirable.

7.3.5.1b. Imaging Detectors. In contrast to PMTs, the available imaging detectors are not high-gain devices. This places a greater burden on the signal amplifier, but, unfortunately, the amplifiers used in most commercial video equipment are engineered more for speed than for low noise or high gain. The resulting combinations of image detector and amplifier are therefore usually much noisier than the conventional PMT/amplifier combination in reading out the same irradiance level from equivalent detector areas.

The response time requirement for readout of the spatial information at video repetition rates (typically 30 frames/s) is also a significant factor in the noise of a video signal. At these frame rates, one line of a 500-line image must be read in about 67 μs, and one pixel of a 500-pixel resolution line in 0.13 μs. Thus, the amplifier must have a submicrosecond time constant and cannot be used to filter much of the detector noise.

The two main types of imaging detectors are vidicons and solid-state array detectors. The current state of the art in image detectors is the result of the synthesis of a wide variety of technologies to solving specific optical and electronic problems. Progress in this area has been monitored by the contributions to *Photo-Electronic Image Devices* published by Academic Press as part of their *Advances in Electronics and Electron Physics* series. Recent volumes in this series[45, 46] covered advances in image intensifiers, solid-state detectors, and a variety of different types of camera tubes.

The sensitivity of conventional solid-state detectors and target vidicon tubes has been insufficient for most low-light applications. With the advent of intensified and cooled detectors, however, single-photon sensitivity has now been realized in advanced detector systems.[10, 47–49] Most commercial image cameras are very limited in this respect as a result of high dark-current noise, which limits their usable integration time for weak signals. Cooling has been used to reduce this dark signal to allow long-term integrations in the case of silicon intensified target (SIT) and ISIT vidicons, charge-injection device (CID) and charge-coupled device (CCD) cameras. Readout noise can then be minimized if image information is slowly scanned using a highly filtered amplifier. The overall result is considerable gain in sensitivity; however, it comes at the expense of temporal resolution. The optimum system would have photomultiplier-like gain and dynamic range in the detector system itself, that is, single-photon sensitivity in real time.

The earliest and most extensive development efforts to deal with this detector impasse were in astronomy (see Meaburn[50]). Detector systems were formed by coupling video cameras and solid-state detectors to high-gain image intensifiers. Such systems have been applied to several fields since that time. The first applications for biology and microscopy were developed by Reynolds and co-workers.[51–55] The system designed by Reynolds for fluorescence and bioluminescence applications and microscopy uses a very high gain ($\sim 4 \times 10^6$), low-noise (from 6 to 20 dark events per square centimeter per second) four-stage image intensifier, coupled to either an SIT or secondary electron conductor (SEC) vidicon. Reynolds' system is limited primarily as the result of the large physical size of the intensifier itself and the size and voltage requirements (35 kV) of its supporting electronics. A much more compact system, the IDG low-light video system,[35] was designed to be similar in performance to the Reynolds' system. Basically, it is composed of a small two-stage microchannel plate (MCP) intensifier requiring less sophisticated support electronics and a fraction of the voltage of the Reynolds' apparatus. It has considerably higher noise ($\sim$21,000 dark events per square centimeter per second), but does achieve single-photon sensitivity. A similar system, developed for astronomical applications, has been described by Airey *et al.*[56] It employs a high-gain double MCP intensifier optically coupled to a CCD solid-state camera. This combination gives good noise performance ($<$200 dark events per square centimeter per second) and moderate to low resolution (244×190 pixels). Following the development of these systems, commercial manufacturers have introduced low-light video cameras with single-photon sensitivity (see Wampler[49] for a review of these cameras).

Both the commercial and custom-designed intensifier/camera systems have a very broad dynamic operating range. In this way they are indeed similar to PMTs. However, video frame rates still limit amplifier performance and make the measurement resolution limited at any one combination of camera and intensifier gain. Most intensifier/camera systems are able to form an image with illumination levels varying over at least one-millionfold. Using a variable iris diaphragm to limit exposure of the detector, it is possible to attain even wider operating ranges. The gain of these systems can be translated in fluorescence microscopy into lower excitation irradiation levels and narrower wavelength selection without loss of signal fidelity or measurement precision. They allow use of lower output, broad-band light sources and monochromators for convenient and accurate wavelength isolation.

Acknowledgments

This work was supported by grants from the National Institutes of Health (GM-37255), the United States Department of Agriculture (86-CRCR-1-1954), and the University of Georgia Research Foundation. We would also like to acknowledge Mr. Ned Rich and Mr. John Gilbert of the Instrument Design Group, University of Georgia, for their work in the design and fabrication of the IDG low-light video microscope system.

References

1. P. Ellinger, Fluorescence microscopy in biology, *Biol. Rev. 15*, 323–350 (1940).
2. E. M. Brumberg, Fluorescence microscopy of biological objects using light from above, *Biophysics 4*, 97–104 (1959).

3. J. S. Ploem and H. J. Tanke, *Introduction to Fluorescence Microscopy*, Oxford University Press, Oxford (1987).
4. C. F. A. Culling, *Modern Microscopy. Elementary Theory and Practice*, Butterworths, London (1974).
5. M. Goldman, *Fluorescence Antibody Methods*, Academic Press, New York (1968).
6. J. E. Wampler, Microspectrofluorometry with an intensified vidicon detector and whole image spectral scanning, in: *Applications of Fluorescence in the Biomedical Sciences* (D. L. Taylor, A. S. Waggoner, R. F. Murphy, F. Lanni, and R. R. Birge, eds.), pp. 301–319, Alan R. Liss, New York (1986).
7. G. T. Reynolds, Evaluation of an image intensifier system for microscopic observations, *IEEE Trans. Nucl. Sci. NS-11*, 147–151 (1964).
8. G. T. Reynolds, Image intensification applied to microscope systems, *Adv. Opt. Elect. Microsc. 2*, 1–40 (1968).
9. G. T. Reynolds, Image intensification applied to biological problems, *Quart. Rev. Biophys. 5*, 295–347 (1972).
10. G. T. Reynolds and D. L. Taylor, Image intensification applied to light microscopy, *Bioscience 9*, 586–592 (1980).
11. S. Inoué, *Video Microscopy*, Plenum Press, New York (1986).
12. A. H. Coons, H. J. Creech, and R. N. Jones, Immunological properties of an antibody containing a fluorescent group, *Proc. Soc. Exp. Biol. Med. 47*, 200–202 (1941).
13. A. H. Coons, The beginnings of immunofluorescence, *J. Immunol. 87*, 499–503 (1961).
14. A. A. Thaer, Instrumentation for microfluorometry, in: *Introduction to Quantitative Cytochemistry* (G. L. Wied, ed.), pp. 409–426, Academic Press, New York (1966).
15. M. Sernetz and A. Thaer, A capillary fluorescence standard for microfluorometry, *J. Microsc. 91*, 43–52 (1970).
16. J. S. Ploem, Quantitative Immunofluorescence, in: *Standardization in Immunofluorescence* (H. J. Holborow, ed.), pp. 63–73, Blackwell Scientific Publications, Oxford, England (1970).
17. J. S. Ploem, Quantitative fluorescence microscopy, in: *Analytical and Quantitative Methods in Microscopy* (G. A. Meek and H. Y. Elder, eds.), pp. 55–89, Cambridge University Press, Cambridge (1977).
18. H. Piller, *Microscope Photometry*, Springer-Verlag, New York (1977).
19. E. Kohen, B. Thorell, J. G. Hirschberg, A. W. Wouters, C. Kohen, P. Bartick, J.-M. Salmon, P. Viallet, D. O. Schachtschabel, A. Rabinovitch, D. Mintz, P. Meda, H. Westerhoff, J. Nestor, and J. S. Ploem, Microspectrofluorometric procedures and their applications in biological systems, in: *Modern Fluorescence Spectroscopy* (E. L. Wehry, ed.), pp. 295–339, Plenum, New York (1981).
20. G. R. Bright, G. W. Fisher, J. Rogowska, and D. L. Taylor, Fluorescence ratio imaging microscopy: Temporal and spatial measurements of cytoplasmic pH, *J. Cell Biol. 104*, 1019–1033 (1987).
21. G. R. Bright, J. Rogowska, G. W. Fisher, and D. L. Taylor, Fluorescence ratio imaging microscopy: Temporal and spatial measurements in single living cells, *Bio/Techniques 5*, 556–562 (1987).
22. J. S. Ploem, A study of filters and light sources in immunofluorescence microscopy, *Ann. N.Y. Acad. Sci. 177*, 414–428 (1971).
23. R. J. DeSa, A simple precision current-regulated power supply for laboratory lamps, *Anal. Biochem. 35*, 293–303 (1970).
24. M. Green, R. H. Breeze, and B. Ke, Simple power supply system for stable xenon arc lamp operation, *Rev. Sci. Instrum. 39*, 411–412 (1968).
25. R. H. Breeze and B. Ke, Some comments on xenon arc lamp stability, *Rev. Sci. Instrum. 43*, 821–823 (1972).
26. A. Katzir and M. Rosmann, Xenon arc lamp intensity stabilizer, *Rev. Sci. Instrum. 45*, 453 (1974).

27. P. B. Oldham, G. Patonay, and I. M. Warner, Evaluation of AC stabilization of a DC arc lamp for spectroscopic applications, *Anal. Instrum. 16*, 263–274 (1987).
28. H. G. Zimmer, Microphotometry, in: *Micromethods in Molecular Biology* (V. Neuhoff, ed.), pp. 297–328, Springer-Verlag, New York (1983).
29. M. J. Geisow, Fluorescein conjugates as indicators of subcellular pH, *Exp. Cell Res. 150*, 29–35 (1984).
30. M. Rousseau, Spectrophotométrie de fluorescence en microscopie, *Bull. Microsc. Appl. 7*, 92–94 (1957).
31. R. A. Olson, Rapid scanning microspectrofluorometer, *Rev. Sci. Instrum. 31*, 844–849 (1960).
32. A. G. E. Pearse and F. W. D. Rost, A microspectrofluorometer with epi-illumination and photon counting, *J. Microsc. 89*, 321–328 (1968).
33. F. W. D. Rost, A microspectrofluorometer for measuring spectra of excitation, emission and absorption in cells and tissues, in: *Fluorescence Techniques in Cell Biology* (A. A. Thaer and M. Sernetz, eds.), pp. 57–63, Springer-Verlag, New York (1973).
34. J. S. Ploem, J. A. de Sterke, J. Bonnet, and H. Wasmund, A microspectrofluorometer with epi-illumination under computer control, *J. Histochem. Cytochem. 22*, 668–677 (1974).
35. E. S. Rich and J. E. Wampler, The spatial distribution of light emission from liquid phase bio- and chemiluminescence. *Photochem. Photobiol. 33*, 727–736 (1981).
36. G. S. Hayat and G. Pieuchard (eds.), *Handbook of Diffraction Gratings, Ruled and Holographic*, Jobin Yvon Optical Systems, Metuchen, New Jersey (undated).
37. M. M. Jotz, J. E. Gill, and D. T. Davis, A new optical multichannel microspectrofluorometer, *J. Histochem. Cytochem. 24*, 91–99 (1976).
38. I. Kurtz and P. Katzka, Rapid scanning excitation microfluorometry using an acousto-optic tunable filter, *Biophys. J. 53*, 197a (1988).
39. M. Kerker, H. Chew, P. J. McNulty, J. P. Kratohvil, D. D. Cooke, M. Sculley, and M.-P. Lee, Light scattering and fluorescence by small particles having internal structure, *J. Histochem. Cytochem. 27*, 250–263 (1979).
40. D. M. Benson and J. A. Knopp, Effect of tissue absorption and microscope optical parameters on the depth of penetration for fluorescence and reflectance measurements of tissue samples, *Photochem. Photobiol. 39*, 495–502 (1984).
41. Z. Jericevic, D. M. Benson, J. Bryan, and L. C. Smith, Geometric correction of digital images using orthonormal decomposition, *J. Microsc. 149*, 233–245 (1988).
42. Z. Jericevic, B. Wiese, L. Rice, J. Bryan, and L. C. Smith, Statistical criteria for multiwavelength comparisons of digital fluorescence images, in: Time-Resolved Laser Spectroscopy in Biochemistry (J. R. Lakowicz, ed.), *Proc. SPIE*, Vol. 909 (1988).
43. J. P. Boutot, J. Nussli, and D. Vallat, Recent trends in photomultipliers for nuclear physics, in: *Advances in Electronics and Electron Physics*, No. 60, pp. 223–305, Academic Press, New York (1983).
44. H. Candy, Photomultiplier characteristics and practice relevant to photon counting, *Rev. Sci. Instrum. 56*, 183–193 (1985).
45. B. L. Morgan (ed.), *Photo-Electronic Image Devices, Advances in Electronics and Electron Physics*, No. 64A, Academic Press, New York, pp. 1–298 (1985).
46. B. L. Morgan (ed.), *Photo-Electronic Image Devices, Advances in Electronics and Electron Physics*, No. 64B, Academic Press, New York, pp. 299–687 (1985).
47. J. L. Lowrance, A review of solid state image sensors, in: *Advances in Electronics and Electron Physics*, No. 52, pp. 421–429, Academic Press, New York (1979).
48. J. E. Wampler, Instrumentation. Seeing the light and measuring it, in: *Chemiluminescence and Bioluminescence Today* (J. Burr, ed.), pp. 1–44, Marcel Dekker, New York (1985).
49. J. E. Wampler, Low-light video systems, in: *Bioluminescence and Chemiluminescence: Instruments and Applications*, Vol. II (K. Van Dyke, ed.), pp. 123–145, CRC Press, Boca Raton, Florida (1985).
50. J. Meaburn, *Detection and Spectrometry of Faint Light*, D. Reidel, Boston (1976).

51. G. T. Reynolds and P. Botos, Image intensification and magnetic tape recording system for microscopic observations of bioluminescence and fluorescence, *Biol. Bull. 139*, 432–433 (1970).
52. G. T. Reynolds, J. Milch, and S. Gruner, High sensitivity image intensifier–TV detector for X-ray diffraction studies, *Rev. Sci. Instrum. 49*, 1241–1249 (1978).
53. J. Milch, Slow scan SIT detector for X-ray diffraction studies using synchrotron radiation, *IEEE Trans. Nucl. Sci. NS-26*, 338–345 (1979).
54. S. M. Gruner, J. R. Milch, and G. T. Reynolds, Evaluation of area photon detectors by a method based on detective quantum efficiency, *IEEE Trans. Nucl. Sci. NS-25*, 562–565 (1978).
55. G. T. Reynolds, Applications of image intensification to low level fluorescence studies of living cells, *Microsc. Acta 83*, 55–62 (1980).
56. R. W. Airey, D. J. Lees, B. L. Morgan, and M. J. Traynar, The Imperial College system for photon event counting, in: *Advances in Electronics and Electron Physics*, No. 64A, pp. 49–59, Academic Press, New York (1985).

8

Flow Cytometry and Cell Sorting

László Mátyus and Michael Edidin

8.1. Introduction/History

Characterization of cell populations has always been a primary interest of cell biologists. Traditionally, some kind of microscopy—combined with biochemical or immunological techniques—was used to discriminate different cell types or otherwise characterize a new cell population. Microscopy gives very accurate morphological information about the cells, and it may also be used for quantitative measurement of cellular properties. However, these measurements are time-consuming, and so the accuracy of the determined parameter is usually limited by the low number of the cells observed and the consequently high statistical errors. Rare cell types and variants remain undetected in such analyses.

The drawbacks of quantitative microscopy can be overcome by automated sampling of cell populations. The first big step in this direction was made by Caspersson *et al.*, who solved the problem of measuring DNA content of cells on slides automatically.[1] High-resolution image analysis microscopy has developed from his approach.

A second approach to automated cytology is the analysis of cells in suspension flowing past a detector, known as flow cytometry.[2–16] Flow cytometry is now a widely applied tool in cell biology and immunology and as well in clinical laboratories, though its potential for characterization of cells is far from fully realized. This chapter deals with the flow cytometric approach to analysis of cell populations. In it we summarize past advances in instrumentation and methodology and also discuss some recent applications of the method to cell biology and biophysics.

László Mátyus and Michael Edidin • Department of Biology, The Johns Hopkins University, Baltimore, Maryland 21218. László Mátyus's permanent address is Department of Biophysics, University Medical School of Debrecen, H-4012 Debrecen, Hungary. László Mátyus was partially supported by the Hungarian Academy of Sciences (OTKA 112 and OKKFT 1.1.4.2.). The authors thank the Epics Division of Coulter Electronics (Hialeah, Florida) for permission to reproduce diagrams from their technical manuals.
Topics in Fluorescence Spectroscopy, Volume 1: Techniques, edited by Joseph R. Lakowicz. Plenum Press, New York, 1991.

Flow cytometry utilizes so many physical principles and technical solutions that it would be very hard to list all the important discoveries which contributed to the development of the field. Instead we begin with Coulter's cell counter and volume analyzer, the forerunner of flow cytometry.[17] In this instrument, cells flow through an aperture of a few microns thickness, and the electrical conductance is measured across the two sides of the aperture. A cell in passing displaces buffer in the aperture, and the conductance decreases. The number of pulses of conductance change is proportional to the cell number, while the height of the peak or the area under it is proportional to the cell volume. Cell orientation relative to the electrodes has little influence on the area of the peaks; the so-called "Coulter volume" is a good estimate of the real cell volume.[18–21] The application of the principle of laminar flow for cell suspensions greatly improved the performance of Coulter counters, and the use of multichannel pulse height analyzers considerably decreased the time of the analysis and increased the accuracy of volume measurements.[22, 23] Both improvements are also important to optical flow cytometry.

The second great advance in the development of modern flow cytometers was made by Kamentsky *et al.* in 1965.[24] They applied optical methods to quantitate constituents of cells in suspension and were simultaneously able to measure multiple parameters of single cells. They introduced two-dimensional data display and the use of computers for storage and analysis of multiparameter data. Flow cytometers evolved further when Van Dilla and his group designed an instrument with orthogonally placed flow, detectors, and illumination.[25] This arrangement became standard in most commercial flow cytometers. Probably the most important contribution to the field was Fulwyler's invention. In 1965 he described a sorting device, which was attached to a Coulter analyzer and was able to sort cells based on their volume.[26] The principle is very similar to that applied by Sweet in his ink jet printer.[27] This technique, following several modifications, became standard in almost all commercial flow cytometers.

These basic designs have been developed and elaborated. Multiple laser excitation and the use of several detectors have opened the field of multiparameter, multistation flow cytometry.[28–30] This approach makes possible the application of numerous fluorescent dyes at the same time, thus providing information about several cellular parameters at once. Computers with fairly large memories are a prerequisite for successful application of such systems.

Another important modification of flow cytometers aims to overcome the disadvantage of "zero" resolution systems; that is, it tries to get morphological information from each cell. The so-called slit-scan technique allows one to scan cells during the time of illumination. The excitation beam is focused to the diffraction limit; therefore, only a small part of a cell is illuminated at a time. The shape of the detected signal carries the morphological information, which can be evaluated using sophisticated computer programs. Though there are limitations in this technique, this branch of flow cytometry may solve several problems of automated cytology in the near future.[31]

8.2. Operation of Flow Cytometers

The basic principle of operation is similar in every flow cytometer (Figures 8.1 and 8.2). The sample, a single-cell suspension, is delivered to a nozzle or flow cell through sample tubing under positive pressure. In a typical system the sample suspension is led through the center of the orifice of the nozzle. Since direct contact of the cells with the orifice would destroy them, usually another fluid system, the so-called sheath fluid, is used; this surrounds the sample stream and keeps it focused. The dimensions of both systems and

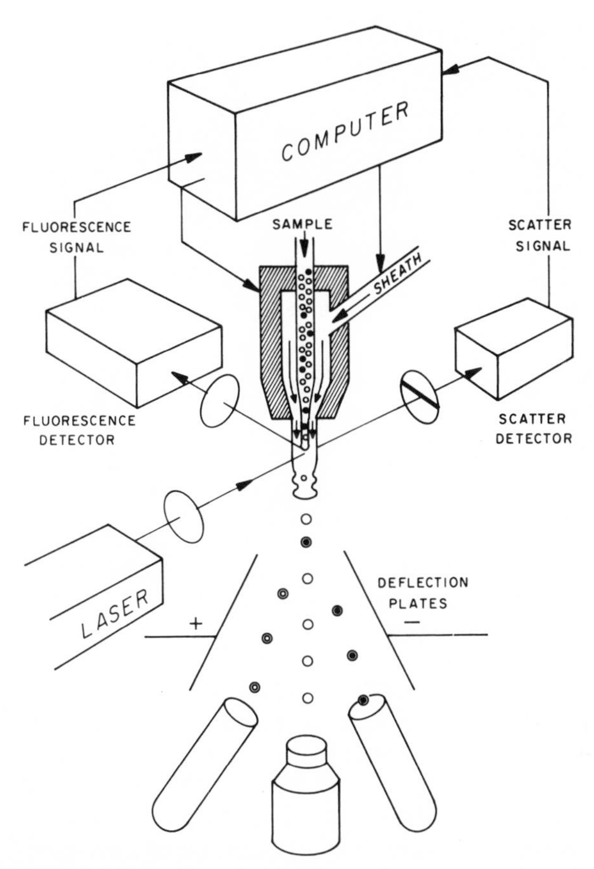

Figure 8.1. Generalized diagram of a fluorescence-activated cell sorter. The functions of the components are described in the text.

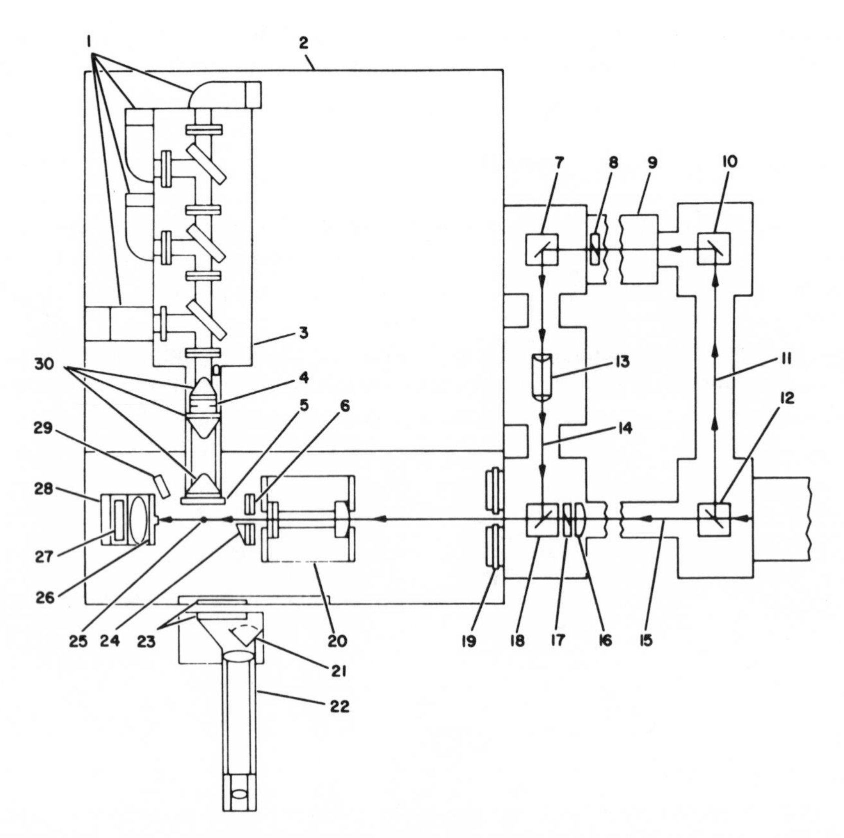

Figure 8.2. Upper view of the optical system of a Coulter Epics 752 flow cytometer. 1, Photomultiplier tubes; 2, detector compartment; 3, filter holder; 4, pinhole; 5, beam stop; 6, pinhole; 7, mirror; 8, dye laser power sensor; 9, dye laser; 10, mirror; 11, argon ion laser beam (514-nm green beam); 12, 500-nm dichroic mirror; 13, beam expander; 14, dye laser beam; 15, argon ion laser beam (blue beam); 16, 488-nm bandpass filter; 17, argon ion laser power sensor (488-nm beam); 18, half mirror; 19, mechanical shutter interlock; 20, beam-shaping lens assembly; 21, movable mirror; 22, microscope; 23, 550-nm bandpass filter; 24, strobe mirror; 25, laser fluid stream intersection point; 26, beam stop; 27, filter slot; 28, forward-angle light scatter detector; 29, stroboscope; 30, aspheric lenses.

the pressures in them—and therefore the velocities of both fluid streams—are carefully chosen, so that no turbulence will occur in either of them. Since the cells are in suspension, they leave the orifice in single file. A few millimeters away from the nozzle the cells are illuminated with focused light from a laser or from a noncoherent light source. For each cell, optical signals—light scatter, absorption, or fluorescence—are measured by respective detector systems: photodiodes or photomultiplier tubes. The light-scatter detector can be placed either parallel to the excitation beam, in which case we speak about forward-angle light scatter (FALS), or orthogonal to the beam, to measure 90° light scatter or side scatter. The absorption detector, naturally, is always placed in the path of the excitation beam while the fluorescence detector is orthogonal or, in some very rare situations, parallel to the excitation beam. The analog signals coming from the detectors are amplified and are translated to digital signals using analog-to-digital converters. Either the height of the pulses (peak signal) or the area under them (integrated signal) can be used for the conversion (Figure 8.3). The data, in digital form, may be output as relative frequency distributions, where the x axis is proportional to the intensity of the signal, and the y axis indicates the number of cells having a given signal. The data are usually stored in computer memories and processed further if necessary.

Since the duration of illumination typically is a few microseconds, considerably high speeds can be achieved, with data collected on a few thousand cells per second. It is therefore possible to collect data on seven different parameters of each of 100,000 cells in a few minutes. Indeed, the present limits to data collection lie in the flow cytometer computers rather than in sample and detector systems.

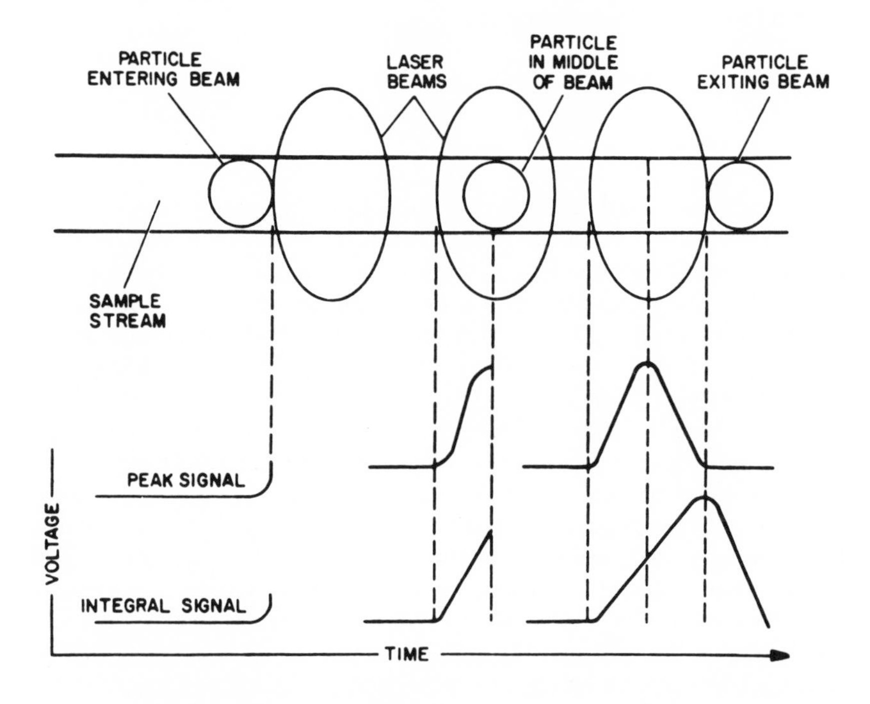

Figure 8.3. Mechanism of generation of peak and integrated signals.

8.2.1. Sample Handling and Delivery Systems[32]

The cell concentration of the sample should be adjusted to between 10^5 and 1×10^6–2×10^6 cells/ml. This range is essential for the optimal performance of the flow cytometers, because both extremes of flow velocities must be avoided. When a pressurized sample delivery system is used, usually 0.5–2-kPa overpressure is applied relative to the pressure of the sheath fluid, which is generally 90 kPa. The stability of the flow is better with smaller sample volume, because the diameter of the sample stream decreases in the nozzle. Since the intensity distribution of the laser beam is not uniform through the cross section of the fluid jet, smaller sample core diameter results in a higher accuracy in the measurement. This would seem to suggest that a high cell concentration combined with lower sample-sheath differential pressure is best, but unfortunately the currently available pressure systems are not stable enough below a certain limit. To overcome this disadvantage, most laboratories requiring high accuracy of measurement, such as for chromosome analysis and DNA content measurement, change the gas-pressurized sample delivery system to a syringe, driven by a motor. This solution has two advantages: the sample flow can be well regulated down to low sample speeds, and in some cases the length of the sample tubing can be substantially reduced, which is very useful for kinetic measurements.

8.2.2. The Nozzle and the Sheath Fluid[32–34]

The nozzle is a conically shaped chamber into which both the sample and the sheath fluids are introduced. The sample insertion tube is placed centrally, while the sheath fluid surrounds it (Figures 8.1 and 8.4). The dimensions of the nozzle and tubing and the pressures in them are designed to result in a continuously accelerating flow, which remains entirely laminar. The sample stream, surrounded by the sheath stream, leaves the nozzle through the orifice. The orifice is round, with a diameter of 50–200 μm. The cells do not make contact with the orifice, which would destroy them. The velocity of the sample stream, when it leaves the nozzle, is around 10 m/s. The sheath fluid is usually distilled water. Ionic buffers are used when cells are to be sorted. The sheath fluid is transported from a container to the nozzle by a pressurized system, similar to that used for delivery of the sample. In some cases a dual-sheath system is used, which gives a better stability to the flow. Late models of flow cytometers incorporate another tube system connected to the nozzle or the flow chamber, which is kept under negative pressure relative to the sheath fluid. This system serves to remove plugs of cell aggregates or debris from the nozzle orifice.

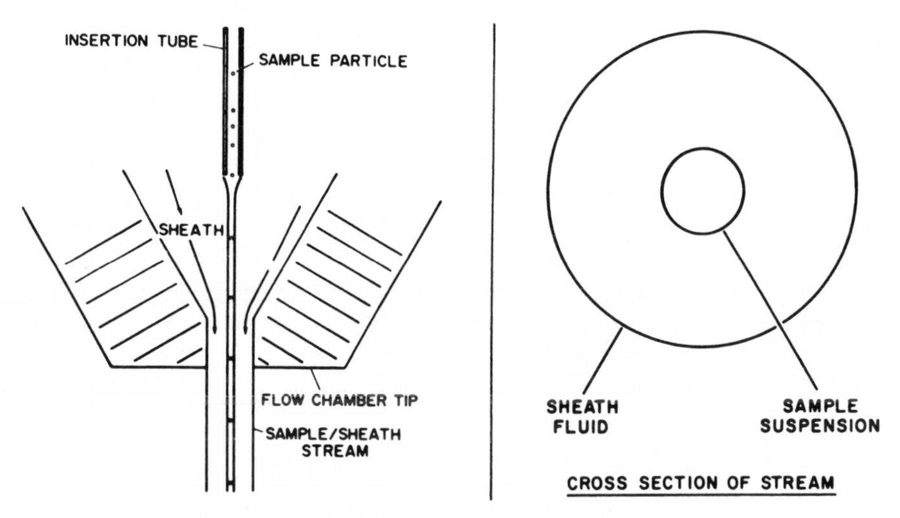

Figure 8.4. Vertical and horizontal sections of a flow chamber tip.

8.2.3. Light Sources[35]

8.2.3.1. Lamps

Recent flow cytometers use arc lamps or lasers as light sources. Both have advantages and disadvantages, and the choice depends primarily on the application. Arc lamps have the real advantage that they are much cheaper than lasers and do not require water cooling or special electric circuits to operate. The mercury arc lamp or the xenon lamp is used in flow cytometry. The mercury arc lamp has several strong peaks in its emission spectrum, which can be extremely useful if one coincides with the absorption spectrum of the fluorophore. Since the mercury arc lamp has a strong peak at 365 nm, it is suitable for most of the DNA-specific dyes. The xenon lamp has a broad emission range with a more uniform light intensity distribution than that of the mercury arc lamp. This is more suitable for exciting fluorescein, which is one of the most commonly used dyes. Use of arc lamps requires special optical arrangements, where optical elements with high numerical aperture are used for the illumination and for the detector systems.

8.2.3.2. Lasers[35]

Most of the commercially available flow cytometers use argon ion lasers as primary light sources. Lasers produce highly monochromatic and coherent light with low divergence. They offer light-output stabilized mode of operation, in which case a negative feedback circuit is applied to keep the output light intensity as constant as possible. In this case their light intensity does not vary more than 1 %, which is sufficient for most applications. The laser beam can be focused to a diffraction-limited spot. All these properties are very useful in flow cytometry. The argon ion laser can be tuned, providing a wide range of wavelengths from 457 to 528 nm. The 488-nm and the 514-nm laser lines are most often used. Argon lasers can be operated in an all-line mode, if the investigator is more concerned about light intensity than about selective excitation of the fluorophores, but intensity of excitation is rarely a problem in flow cytometers. Some argon lasers with higher power offer light output in the UV range of the spectrum, around 353–362 nm. Power in the UV usually is not more than 100–200 mW, and the laser tube must be run at its highest current, which decreases the lifetime of the plasma tube. However, UV is essential for certain applications, for example, detecting supravital DNA staining by Hoechst 33258/33342 dyes[36, 37] or measuring intracellular Ca^{2+} concentration with some Ca^{2+}-sensitive fluorophores.[38]

The intensity profile of the laser beam has a Gaussian distribution in TEM_{00} mode. Sometimes a minor misalignment of the reflecting mirrors, especially if the laser aperture is kept wide open, can lead to modes other than TEM_{00}. This causes serious artifacts and must be avoided.

In multistation flow cytometers, two or more separate laser beams are used to illuminate the cells. The first laser beam usually is the 488-nm line of an argon ion laser, and the other beam can be the 514-nm line or that of another laser emitting at longer wavelength. This other laser can be dye laser pumped by the 514-nm line of an argon ion laser or a red-emitting krypton or He–Ne laser. The dye lasers are tunable, though in a smaller range than argon ion lasers, but by changing the dyes and some optics, the investigator can range over the entire visible and near-IR spectrum. We should note, though, that changing the dye is a time-consuming, expensive, and potentially hazardous procedure. It is also a disadvantage of the dye lasers that they cannot be operated in light-stabilized mode. The krypton laser also can be tuned, nicely covering the spectrum from 350 nm up to almost 800 nm. Both these and argon lasers are costly, and they need special electrical power sources and water cooling. Some of the newest flow cytometers use low-power (25 mW), air-cooled argon and He–Ne lasers, which operate from regular electric circuits. Though they do not provide as much power as their bigger brothers, this disadvantage is compensated for by more sensitive detector configurations. He–Cd lasers also can be used for UV and blue excitations. Their application in flow cytometry is expected in the future.[3]

8.2.4. The "Intersection Point"

Whatever the light source in a flow cytometer, it must be focused to the so-called "intersection point," where the cells cross the illumination beam. In flow cytometers with a nozzle, this point is a few millimeters from the orifice of the nozzle. Here the cells travel in the center of a cylindrical fluid stream. Though its size depends on the differential pressure, the diameter of the sample core is around 15–20 μm. The diameter of the sheath fluid stream is similar to that of the orifice, typically 50–100 μm. The excitation beam is focused to the sample stream. The shape of the focused beam depends upon the purpose of the investigator. If maximum sensitivity is important, as much laser power as possible should be concentrated onto the cells. In this case the laser beam is focused with spherical lenses to a round spot, whose size is comparable to the size of the sample stream. The fluorescence signal is maximal, but because the beam intensity distribution is Gaussian, the smallest variations in the position of the sample stream will lead to an altered fluorescence signal, and therefore will reduce the accuracy. This strategy is good for detection of small numbers of fluorescent molecules, for example, receptors present at low density on the cell surface, but it is not recommended when accuracy is essential. Greater spatial homogeneity of the excitation beam is necessary for quantitative analysis. This can be achieved by a combination of a spherical and a cylindrical lens. This lens system results in an elliptical beam with a Gaussian intensity distribution along both axes. The intensity is nearly uniform at the middle of the long axis of the ellipse, and therefore differences in the position of the cells will not substantially change the detected fluorescence intensities. This configuration allows very accurate measurements, though sensitivity is sacrificed to some extent. It is important in applications such as measurement of cellular DNA content, where the

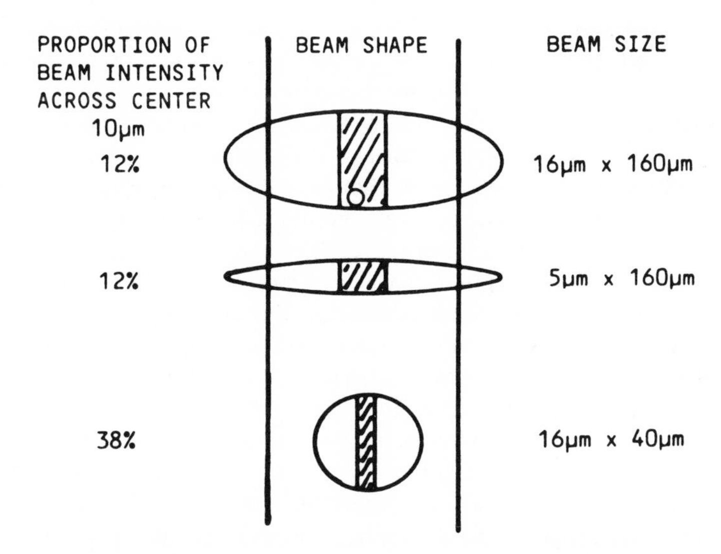

Figure 8.5. Laser beam profiles for Coulter epics for cytometer with different focusing lenses.

signal is usually strong and a high degree of accuracy is most important (Figure 8.5).

The "intersection point" is not always at a fluid jet, outside of a nozzle. There are other configurations in which the cells leave the nozzle through a flow chamber. The flow chamber is optically transparent and usually rectangular, with a rectangular channel inside. The sample stream surrounded by the sheath fluid travels through this channel. The point of intersection is somewhere at the middle of the chamber. This arrangement is not as sensitive to external mechanical disturbances as the fluid jet system. The optical paths of the illuminating and emitted light are more simple, since the chamber surfaces are parallel to each other, and therefore there are no unwanted reflections present in the system. The flow chamber can be effectively combined with mirrors and lenses to collect and relay light emitted 180° away from the detectors, resulting in a much more effective signal collection. This modification incorporated into some of the latest flow cytometer models, mainly used in clinical laboratories, allows use of lower-powered and therefore cheaper lasers. Unfortunately, cell sorting with cytometers equipped with flow chambers is more complicated than sorting on conventional flow cytometers.

8.2.5. Detectors[39, 40]

8.2.5.1. Light Scatter[41]

Even unlabeled cells or particles illuminated by the light beam scatter light. The light is scattered in all directions; therefore, there are several possibilities for its detection. The most common way is to place the detector in the path of the illuminating beam measuring forward-angle light scatter (FALS).[42] The light detected parallel to the illumination consists of several components, and only a part of it, though the most important part, is the scattered light. The most intense part, the direct light coming from the light source, is blocked by obscuration bars. The theory of the light scattering is very complex.[43–45] Absorption, refractive indices of cellular components, cell size, shape, and orientation relative to the detectors, and intracellular structure affect scatter differently,[46] and therefore it is almost impossible to predict the light scattering behavior of a given cell with a given detector. The interpretation of these signals is mainly empirical.

The scattered light is collected by a focusing lens. Since the light intensities are relatively high, neutral-density optical filters are usually applied in order to attenuate the signal. An adjustable aperture makes it possible to choose the angle range of the detection. The detector itself is a photodiode, which tolerates much greater light intensities than photomultipliers. More complex FALS detector systems have also been developed. Salzman and coworkers designed a detector which consisted of several photodiodes placed concentrically.[47,48] This made it possible to measure light scattering as function of angle. Though they found significant differences between different cell types, this approach did not become popular, mainly because more specific cellular markers have been discovered. A scatter detector may be also placed orthogonal to the exciting beam, measuring 90° light scatter or side scatter. The basic features of this detector are similar to those of the fluorescence detector, which we will discuss later. The side scatter signal contains information about intracellular structure and helps to differentiate between certain cell types.[49]

8.2.5.2. Fluorescence

8.2.5.2a. Optics. The fluorescence detector system is different in different flow cytometers. The design and the position of the detectors depend strongly on the source of the illuminating beam. Microscope-based systems equipped with arc lamp epi-illumination have fluorescence detectors placed parallel to the excitation beam. When a laser is used as a light source, the direction of the detection is usually orthogonal to both the excitation beam and the fluid jet. Since there are no basic differences in the structure of the detectors, but only in their positioning, we treat them together.

In some instruments, the emitted light is collected by a modified microscope objective with high numerical aperture and long working distance. The focal point of the objective is adjusted to the center of the fluid stream. An alternative collecting method uses confocal optics, which do not produce an image. The alignment of the system becomes easier; however, the signal-to-noise ratio decreases. Following the lens, the light beam is led through an aperture, in order to block unwanted light. If the system incorporates a side scatter detector, a 488-nm dichroic mirror is placed in the path of the light. It is positioned at 45° to the optical axis of the system; it reflectsthe scattered light while transmitting the fluorescent light. The side scatter detector is placed in the path of the reflected part of the beam. It is important to eliminate the rest of the scattered laser light, and so a laser-blocking long-pass filter is placed after the beam splitter. Figure 8.6 shows the optical configuration of a system detecting three-color fluorescence and forward-angle and side light scatters.

8.2.5.2b. Filters. Most flow cytometers are designed to detect more than one region of the fluorescence spectrum.[35] Therefore, the emitted light has to be further divided. Dichroic mirrors with different characteristic wavelengths separate the whole visible spectrum into several bands, allowing simultaneous measurement of fluorescence of a single cell at several different wavelengths.

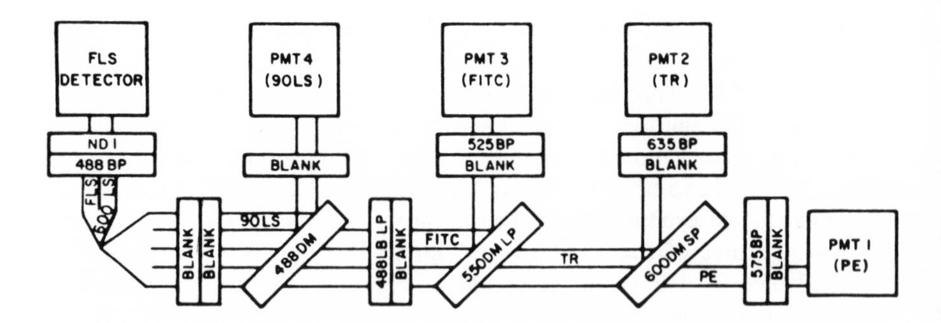

Figure 8.6. Filter configuration for the detection of three-color immunofluorescence and forward-angle and side light scatters.

Since specific cell fluorescence is weak and must be differentiated from scattered light, further filtering is usually necessary. There are two major groups of filters: the dichroic filters and colored glass filters. The dichroic filters are manufactured with different central wavelength and full width at half-maximum (fwhm). They are characterized by the slope of the bandpass, the region where they change their transmittance. The other type of filters, glass filters, can be short-wavelength pass or long-wavelength pass filters. They are characterized by their cutoff wavelength, that is, the wavelength at which their transmission is 50%. Before selecting filters, it is important to examine carefully their spectral characteristics, taking into account the spectra of the fluorophores to be used. Sometimes a combination of dichroic and glass filters gives the best result. The dichroic filter should be placed closer to the light source than the glass filters. This placement reduces any fluorescence of the glass filters, since it decreased the intensity of the light reaching them.[50]

8.2.5.2c. Photomultipliers. Photomultipliers (PMTs) are best for detecting the very weak light signals coming from cells.[39, 40] Their sensitivity allows us to use them even in single-photon counting mode. PMTs convert the light into electric current, which, within a range, is proportional to the incident light intensity. Unfortunately, their detection efficiency is a function of the photon energy, that is, of the wavelength of the light. The lower wavelength limit of detection depends on the window material, while the composition of the photocathode determines the detection efficiency at long wavelengths. Recently, red-sensitive PMTs have been developed. They have special importance in flow cytometry, because the use of several fluorophores simultaneously requires the frequent use of red-emitting dyes. The light-sensitive surface can be placed either at the end or the side of the tube. Different PMT designs are available according to the purpose of application.

8.2.6. Electronics

A very broad range of light intensity is measured in flow cytometry. This makes great demands on amplifiers.[39, 40] They should be linear over the entire range with low noise levels.[51] Logarithmic amplifiers are particularly useful since they can handle a wide range of fluorescence intensity on a single cell. Amplifiers capable of handling 3.5–4 decades are usually satisfactory for flow cytometry. Figure 8.7 shows the signal processing system of Coulter flow cytometers.

The amplified signals from the PMTs and the unamplified signals from the photodiodes are converted to digital signals by analog-to-digital converters (ADCs) or by multichannel pulse height analyzers (PHAs).[39, 40] Both of these translate analog information to numbers. The electrical pulse collected from the PMTs can be characterized by the height of the pulse or by the area under it, and in theory either can be used as a signal to be digitized. Assuming symmetric cells with even fluorescence distribution, the two signals will be proportional to each other. If cells are asymmetric, the area describes the total fluorescence intensity, and the height shows the maximal density of the fluorophore. There are several interesting possible applications in the

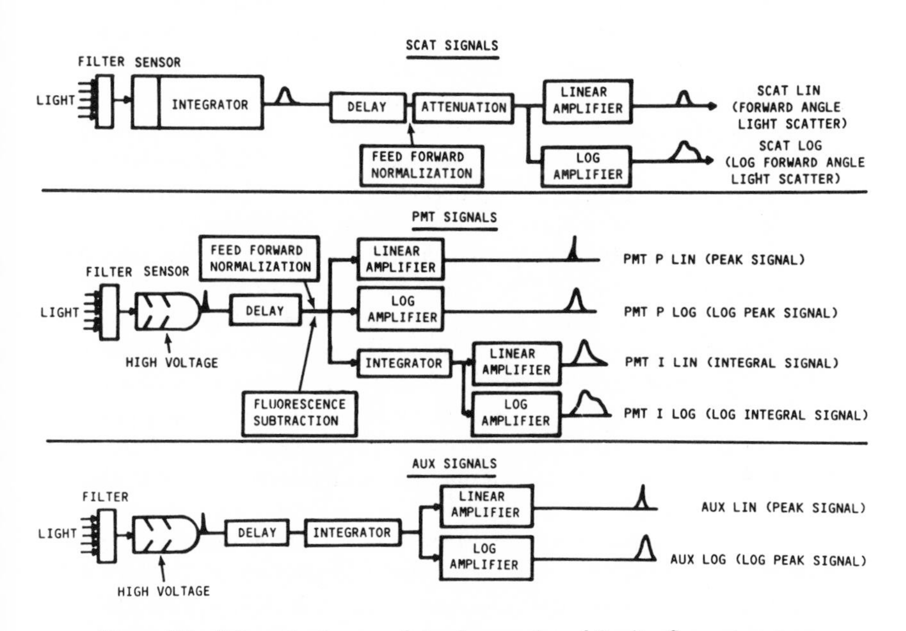

Figure 8.7. Schematic diagram of signal processing of Coulter flow cytometers.

simultaneous analysis of both signals. The first flow cytometers generated a signal which was a mixture of the two parameters, but newer instruments are able to handle both signals separately. Early flow cytometers contained 128-channel PHAs, sufficient for most applications. Recent models have 256-channel resolution for general purposes and one or two 1024-channel resolution ADCs as an option. Higher resolution than that is sometimes required. For example, for chromosome analysis 2K or 4K resolution might be useful.

8.2.7. Data Analysis and Storage

There are several methods of data storage and display.[39, 40, 52, 53] The simplest way is to treat each parameter separately and to store and display each as a single-parameter frequency distribution histogram. Here the abscissa is proportional to the size of the signal, and the ordinate shows the relative frequency of the event. Such histograms can be characterized by secondary parameters for easier comparison. The simplest parameter is the arithmetic average. In certain cases the modal values of peaks are satisfactory. Symmetry of peaks can be described by more complicated parameters, such as skewness and curtosis. Sharpness of peaks is usually described by their coefficient of variation (CV), which is the ratio of the standard deviation to the arithmetic mean and is usually expressed in percent. The CV is used to determine the alignment of the instrument, by running standards with known CVs. In computerized systems all these parameters are calculated automatically.

Correlated data can be handled as two-parameter distributions. These data are stored as a matrix. Each line of the file corresponds to one cell at the matrix, a unique intersection of one row and one column. Since the storage of matrices requires extensive computer memory, usually only 64-channel resolution is used. The two-parameter distributions can be displayed in several forms. The traditional, still widely used, form is the so-called dot plot. Each cell is represented by a spot on the screen, and the coordinate values along the two axes are proportional to the corresponding parameter. The same data can be displayed as contour plots, where matching values in the matrix are connected by a continuous line, resulting in a plot similar to topographical maps. In systems with better graphics capability, three-dimensional plots can also be created. Figure 8.8 shows two-parameter histograms of the same sample presented in several forms. In multiparameter data collection there are many combinations of the parameters. To store all the corresponding parameters as two-parameter distributions would require an extremely large computer memory. It is much more economical to store data in the so-called list mode. In this case all the data measured from each cell are stored as a list of numbers. This storage mode makes it possible to do further calculations on the corresponding parameters. For example, fluorescence profiles can be analyzed with different gatings (see below) according to other parameters, or when measuring fluorescence polarization, anisotropy or degree of polarization can be calculated. These calculations also can be made in real time, using fast computers, offering the possibility of creating histograms or even sorting cells based on secondary, computed parameters. Recently, such computers have been adapted to flow cytometers.

"Gating" is a technique applied in multiparameter measurements. Here the data collection is restricted to those cells which fall into a region on a one-parameter histogram or into an area on a two-parameter histogram. In recent systems one can gate on many parameters simultaneously. Gating makes it possible to analyze a cell population selectively defined by a combination of parameters. The gating technique is utilized also in cell sorting.

For long-term data storage, hard or floppy magnetic disks or magnetic tapes can be used, or drawings can be made with plotters or X–Y recorders. Data storage on magnetic media is preferable when further data analysis is required.

8.2.8. Slit-Scan Flow Cytometry[31, 35, 54, 55]

Conventional flow cytometry gives highly accurate quantitative information about cellular constituents, but little morphological information. Slit-scan

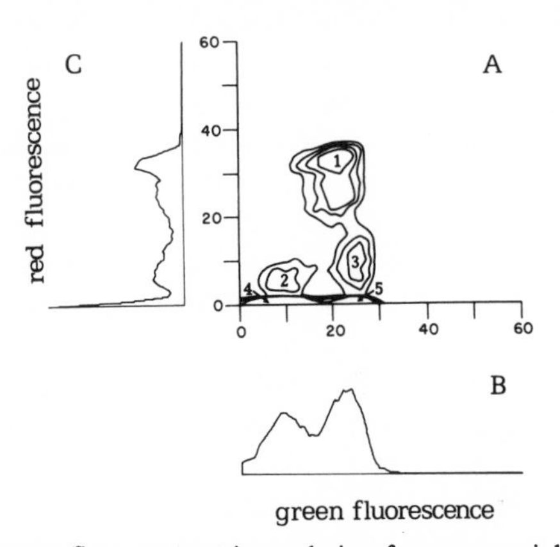

Figure 8.8. Two-parameter flow cytometric analysis of mouse peripheral lymph node cells stained with phycoerythrin (PE)-conjugated anti-L3T4 antibody and with fluorescein-conjugated anti-Qa2 antibody. The x axis shows the logarithm of the green fluorescence (fluorescein), and the y axis shows the logarithm of the red fluorescence (PE). Five subpopulations can be distinguished: (1) bright green–bright red; (2) dim green–dim red; (3) bright green–dim red; (4) dim green–no red; (5) bright green–no red. A, Dot plot; B, contour plot; C, three-dimensional plot of the same data; D, green fluorescence intensity distribution; E, red fluorescence intensity distribution.

flow cytometry has been developed to gain morphological information about cells or particles.[56, 57] In conventional flow cytometers the illumination and the detection aperture are bigger than the investigated object, resulting in near-zero spatial resolution. In slit-scan techniques either the excitation or the detection is limited to a slit, smaller than the object being analyzed. Since the optical signals measured come only from a small part of the object, and the object moves with constant speed, the spatial distribution of the material along the scanning axis is translated to a time-dependent intensity distribution. The shape of this curve carries information about morphology.

Two different optical configurations are used for scanning: object plane scanning and image plane scanning.[31] In the former design the exciting beam is focused to a narrow line, thus illuminating only a portion of the object at a time. In the latter, the objects pass a relatively wide illuminating beam, and the light-collecting optics produces an image of the object on a narrow aperture. In both cases the pulse profile is a measure of the distribution of fluorescence along the scanned axis, but it is distorted, because of the finite size of the scanning aperture. The pulse is a convolution of the scanning profile and the spatial distribution of the fluorophore along the scanned axis. To reconstitute the "ideal" pulse profile from the measured profile, a deconvolution algorithm should be used. The spatial resolution can be improved using the Fourier transform method for deconvolution.[31]

Though these techniques have resolution limited to around 1 m, they are useful for studying many biological problems. In chromosome analysis the centromere position can be measured in addition to the DNA content. The evaluation of nuclear diameter or the detection of multiple nuclei can be applied to detect transformed cells.

Multidimensional x–y–z slit-scan flow cytometers have also been developed and are used successfully for automated detection of different type of cancer cells.[58–62]

A possible way to improve the resolution is to apply more tightly focused illuminating beams, but diffraction limits beam diameters to around 1.5 μm, depending on the wavelength of the laser light. A recently introduced approach uses interference between two laser beams, which results in a series of parallel intense beams, called a fringe field. The diameter of the lines may approach 0.3 μm, greatly improving the resolution. The reconvolution procedure becomes more complicated, and the system is very sensitive to vibrations.[31] Another direction for improvement is to increase the speed of the pulse profile analysis, which seems easy with the introduction of faster computers.

The slit-scan technique is not widely applied yet, mainly because it requires a special optical system, sophisticated computers, and highly skilled operators, but the appearance of "user-friendly" slit-scan flow cytometers is expected in the future.[31]

8.2.9. Cell Sorting[32, 63, 64]

Cell sorting is an attractive feature offered by most flow cytometers.[26] Cell sorting makes it possible to investigate further certain subpopulations, for example, to correlate morphology to different parts of the histograms or do biochemical analysis on selected cell populations. Recent sorting devices even allow us to clone cells.

The principle of operation is the following. The nozzle or the flow chamber is vibrated with a piezoelectric crystal, at a frequency of about 30–40 kHz, resulting in droplet formation from the continuous fluid stream. Having laminar flow and constant velocity and vibration puts the droplet breakoff point at a constant fixed distance from the intersection point. The time required for the cells to travel that distance is called the delay time. If a cell passing the detector has the desired properties—the gating technique is utilized for the decision—and must be sorted, then at the end of the delay time the whole fluid stream is charged. The forming droplets keep the charge on their surface. These charged droplets pass between charged plates and are deflected and can be collected in a collecting vessel or on a slide. By applying positive and negative charge, two subpopulations can be sorted at the same time. There are systems which are able to put different amounts of positive or negative charge onto the fluid stream, allowing one to sort up to four subpopulations.

Present commercial flow cytometers are able to sort cells with a speed of 3–5000 cells/s. This speed is enough for most applications, but in certain cases, such as chromosome separation, much higher speed is necessary. The speed of the separation is determined by instrumental factors: the speed and the pressure of the sheath fluid, the diameter of the orifice, the frequency of the piezoelectric crystal, and the speed of the analyzing electronics. For a given configuration, the coincidence between cells limits the speed of the separation. If two cells follow too close to each other, neither of them can be safely separated. There is a built-in circuit which aborts the sorting signal in such cases. If the cell population to be sorted is a small fraction of the sample and the recovery rate is more important than the purity, it is better to override the anticoincidence circuit and repeat the sorting in order to get a pure sample. The speed of sorting can also be increased by increasing the velocity of the fluid stream and the frequency of droplet formation. Since the fluid velocity is proportional to the square root of the applied pressure, extremely high pressure is necessary to get a significant increase in the velocity. At the Lawrence Livermore National Laboratory a high-speed sorter, built for chromosome separation, operates at about 15 atm sheath pressure and 220-kHz nozzle drive frequency. The machine operates about 20-fold faster than commercial sorters.[65]

8.3. Parameters of Flow Cytometry

8.3.1. Light Scatter[66]

The light scattered by cells is measured in every flow cytometer. The FALS signal is used as a trigger signal to initiate the collection of fluorescence. It helps to discriminate between different cell types and to eliminate data collection from debris and unwanted cells. Triggering based on light scatter substantially improves the signal-to-noise ratio in the detection of fluorescence. The intensity of the scattered light in the forward direction, detected at low angles, is related to cell size, assuming spherical cells. The internal refractive index of the cells affects their scatter signal. The refractive index of dead cells is reduced; therefore, they can be excluded from the analysis. Asymmetric cells can pass the detector at several orientations, resulting in different FALS signals. For example, a population of uniformly sized but asymmetrically shaped erythrocytes gives a bimodal light-scatter distribution.[67]

The light scatter at forward angles is relatively insensitive to the internal structure of cells, but that detected at 90° increases with the cells' granularity. Side scatter plotted against FALS makes it possible to resolve different leukocyte populations.[49] The light-scatter signal is also a function of the wavelength of the illuminating beam. Two-color light-scatter analysis resolves different types of blood cells,[68] and even T from B lymphocytes,[69] but the approach has become unimportant in the era of specific cell surface markers for these cells. Light-scatter standards, such as plastic microspheres or fixed chicken red blood cells, are used to control for the proper alignment of the system. However, the light scattering behavior of the microspheres is different from that of the cells; therefore, different alignments or systems can produce different ratios of cell to microsphere scatter.

8.3.2. Fluorescence

Measuring fluorescence intensities in flow cytometers is based on the proposition that the measured signal is proportional to the concentration or amount of fluorophore of interest. This is called the "central dogma of flow cytometry."[70] It is generally true, but the investigator should always consider the possibility of an artifact. There are so many possible applications of fluorescence measurements in flow cytometry that we do not even attempt to review all of them. We have selected a few of them for illustrative purposes; some are included because they are often used; some others, though not widely used, because they may have importance in the future.

8.3.2.1. Cellular DNA Content[71]

Cellular DNA content is one of the most often measured parameters. Since the DNA content of cells changes during the cell cycle, by measuring the DNA content one can determine the fractions of the cells of a population in different phases of the cell cycle. The cells should be stained with a DNA-specific dye, such as chromomycin A_3, propidium iodide (PI),[72] mithramycin,[73–75] or Hoechst 33258 or 33342.[36] All these dyes to some extent give a fluorescence signal proportional to the DNA content, though there are slight differences in their binding specificity. Hoechst 33258 binds preferentially to A–T-rich regions of DNA, while chromomycin A_3 prefers G–C-rich regions. PI does not show any base pair preference. Usually the resolution is better with the Hoechst dyes than with chromomycin A_3 or PI. PI has the advantage that it can be excited by the 488- or 514-nm line of an argon ion laser, while the Hoechst dyes and chromomycin A_3 need UV and 457-nm excitation, respectively.[72]

Since the DNA content in the G_0 and the G_2M compartments is very well determined, in an ideal case these subpopulations would correspond to a single channel on a DNA distribution. This would simplify the cell cycle analysis, but in real life this ideal distribution is distorted by staining inhomogeneities and instrumental factors, resulting in a distribution with 1–2% CV. These factors cause overlap between the S and G_1 or S and G_2M cell cycle compartments, and therefore analytical methods are necessary to evaluate the true percentages of cells in each part of the cycle. Several graphical and mathematical models have been described.[53, 76, 77] The simpler graphical methods give relatively accurate results if the CV and the S phase fraction are low and the cell population is not synchronized. For histograms with higher CVs or S phase fractions, more sophisticated, computer-based methods should be used. Since none of the methods is unequivocally superior to the others, the selection depends on the accuracy required and the speed of the available computer. Dean *et al.*,[72] however, showed that variations in sample preparation and staining usually cause bigger errors than the mathematical analysis. These factors limit the accuracy of flow cytometric determinations of stage in cell cycle of cells stained with a single dye; however, the relatively low cost and the high speed compensate for all these disadvantages. The accuracy can be improved by repeated experiments.

Another promising method has recently been developed by Gratzner and co-workers.[78–80] The cells are incubated in a medium containing bromodeoxyuridine (BrdUrd), which is incorporated into the cellular DNA during the S phase of the cell cycle. The BrdUrd is visualized by directly conjugated antibody raised against BrdUrd, typically using a fluorescein-conjugated second antibody, thus giving a fluorescence signal proportional to the rate of the incorporation of BrdUrd into the DNA. S phase cells can be distinguished from G_1/G_2M cells on the basis of the green fluorescence. Propidium iodide counterstaining differentiates between the G_1 and G_2M phase cells. This method gives an extremely precise determination of the fraction of a cell population in each part of the cell cycle, and furthermore the rate of the DNA synthesis can be determined as a function of DNA content. Unfortunately, bivariate data analysis should be applied, and the sample preparation is more complicated and must be adapted individually to every cell type.[81–83] The cost of the experiment is considerably higher than that of an experiment using only one dye, but this method replaces the more expensive and dangerous tritiated thymidine incorporation assay.

Acridine Orange (AO) is a cationic dye and has widespread application in cytochemical determination of nucleic acids.[84] AO complexes with double-stranded nucleic acids to exhibit green fluorescence, while its interaction with single-stranded nucleic acids results in metachromatic red fluorescence. These properties are utilized in flow cytometric staining techniques as well.[85–90]

Simultaneous DNA/RNA determination can be performed after selective denaturation of double-stranded RNA. The resistance of DNA to heat or acids can also be monitored by AO following RNase treatment of the cells.[91] In this technique the ratio of green to red fluorescence is a measure of the denaturation of DNA. AO staining combined with preincubation with 5-bromodeoxyuridine distinguishes cycling from noncycling cells in terms of their lowered green fluorescence.[92] Through these techniques, several cellular parameters can be measured simultaneously: total DNA content, total RNA content, and sensitivity of DNA to denaturation.[93]

Quantitative cellular DNA determinations can be combined with other parameters. A cytochemical method has recently been developed to stain cellular DNA, RNA, and protein simultaneously. DNA is labeled with Hoechst 33342, RNA with pyronin Y, and protein with fluorescein isothiocyanate.[94, 95] This method has great prospects for the classification of transformed cells or for following metabolic changes during the cell cycle and the mechanism of action of different drugs. However, the analysis requires modified flow cytometers equipped with three excitation laser beams and with three photomultipliers. This could limit its widespread application.

8.3.2.2. Chromosome Analysis and Sorting[96, 97]

Chromosomes are organizational units of DNA in eucaryotic cells. Several hundred genes have been localized on chromosomes—about 50 of them are known to be associated with diseases—but this represents only a small fraction of the entire human genome. Isolation of individual chromosomes would greatly facilitate the construction of recombinant gene libraries and the identification of new genes with known chromosomal localization.

Chromosome analysis became routine as a result of improved staining and analytical procedures. The conventional approach is to isolate cells in mitosis, stain the chromosomes with different DNA-specific dyes, and examine them under a microscope. This is time-consuming, so only small numbers of chromosomes can be examined. Also, individual chromosomes cannot be isolated. The high speed and resolution offered by flow cytometry is useful for chromosome analysis and isolation, though it is not yet sufficient for karyotyping. In this approach the isolated chromosomes[98] are stained with DNA-specific fluorescent dyes and classified and sorted individually according to DNA content[99, 100] and base pair composition.[101]

There are many fluorescent dyes which bind to DNA, but only a few of them are suitable for chromosome analysis. Phenanthridium dyes such as propidium iodide and ethidium bromide give satisfactory results for univariate chromosome analysis. Both dye intercalate into DNA and do not show base pair preferences; therefore, they are good indicators of the total nucleic acid content. They bind to RNA as well, but RNase treatment eliminates this artifact.

Bisbenzimidazol dyes such as Hoechst 33258 show brighter fluorescence than phenanthridium dyes and do not bind to RNA, but they are excited in the UV, which is hard on laser plasma tubes. These dyes bind preferentially to A–T-rich regions of the DNA, so the histograms are somewhat different from those obtained with PI or EB staining.

Univariate chromosome analysis does not resolve all the chromosomes into separate peaks,[96] because there are chromosome groups with the same DNA content and similar base pair composition. Much better resolution is achieved by simultaneous staining with Hoechst 33258 and chromomycin A_3, utilizing the different base pair preferences of the dyes (chromomycin A_3 preferentially binds to G–C-rich regions of the DNA).[101] Though these dyes bind independently, fluorescence resonance energy transfer may occur between them, so the fluorescence intensities from each dye cannot be measured separately. However, with dual-laser excitation (UV for Hoechst, 457 nm for chromomycin) this problem can be eliminated. The method resolves all the human chromosomes, except the 9–12 group and chromosomes 14 and 15 of some individuals.

This method is used in the National Laboratory Gene Library Project, which aims to construct human chromosome-specific DNA libraries.[102, 103] In order to separate those chromosomes which cannot be resolved by the method, human × hamster hybrid cell lines which contain only one of the unresolvable human chromosomes are used. The hamster and human chromosomes are distinguished by DNA content. A high-speed cell sorter, separating about 10 times faster than conventional machines, is used so that significant amounts of chromosomes can be produced in a reasonably short time.[65]

8.3.2.3. Immunofluorescence Measurements[104–106]

Cells express many antigens on the outer surface of their cytoplasmic membrane which are characteristic of the cell type or functional state of the cell. Determination of the presence or the density of specific cell surface molecules has great importance in cell biology, immunology, and clinical laboratory practice. Cell surface antigens are defined by specific antibodies.[107] A fluorophore can be conjugated directly to the antibody (direct immunofluorescence) or to a second antibody which reacts with the first (indirect immunofluorescence). Useful fluorophores have strong absorbance at the wavelength of the available light source, a high quantum efficiency, and a large Stokes shift. Of course, the dye should have a functional group which reacts with immunoglobulins.

Fluorescein isothiocyanate (FITC) is the most commonly used dye for immunofluorescence studies. It can be excited by the 488-nm line of an argon ion laser and with a mercury arc lamp (at 415 nm) as well. It can easily be conjugated to antibodies without a substantial loss of their binding activity, and with little change in the dye's fluorescence. FITC-conjugated antibodies can be applied in combination with PI staining. Both dyes excite well at the 488-nm line of argon lasers. The relatively small spectral overlap allows one to detect these dyes independently.[108–110]

Tetramethylrhodamine isothiocyanate (TRITC) can be excited very well with mercury arc lamps (565-nm line), and while the 514-nm line of an argon ion laser is not optimal, it gives satisfactory results. For multiparameter studies, when more than one surface antigen is labeled at a time, other dyes are necessary.[111–115] Other rhodamine derivatives, such as Texas Red or X-RITC, are suitable dyes, because both their excitation and emission spectra are shifted toward the red. X-RITC may decrease the water solubility of immunoglobulin G; Texas Red is a better choice from this point of view.

A new family of red-emitting dyes was introduced a few years ago. Phycobiliproteins are products of cyanobacteria and red algae. The major members of this family are B-phycoerythrin, R-phycoerythrin, C-phycocyanin, and allophycocyanin.[116–119] Their molecular masses are 100–200 kDa, and their molar extinction coefficients are typically over 1 million. Their quantum efficiency is almost 1. They have several other properties which make them valuable for immunofluorescence studies. Their excitation spectra are broad; they have a large Stokes shift; they are water-soluble; and their fluorescence intensity is not sensitive to pH. Thus, they are extremely useful in visualizing antigens expressed at very low density on the cell surface. Phycobiliproteins can be coupled to avidin as well.[120] Avidin binds biotin with extremely high affinity, and the latter can be conjugated to antibodies; thus, avidin can be used instead of a second antibody. Phycobiliproteins, because of their large size, are less useful than fluorescein or rhodamines in fluorescence energy transfer measurements.

Sample preparation for flow cytometric immunofluorescence measurements is similar to that for microscopy. The only difference comes from the different detection configuration of flow cytometers. Compared to microscopy or fluorescence measurements in a cuvette, a much smaller extracellular volume is excited in flow cytometry, so the elimination of the unbound free ligand is not as critical.[121] This makes possible the measurement of ligand binding without seriously disturbing equilibrium conditions.

8.3.2.3a. Autofluorescence. Autofluorescence is a general property of cells. It causes a background fluorescence signal, which becomes disturbing when specific low-intensity signals are to be detected, for example, weak surface immunofluorescence. Though flow cytometers have a sensitivity limit of about 2000 fluorescein molecules per particle (plastic beads), in practice this limit is around 10,000 fluorescein molecules per cell, because of the autofluorescence. Cellular autofluorescence has a broad excitation spectrum; the emission spectrum is also quite broad.[122] Autofluorescence is mainly caused by nicotinamide adenine dinucleotide (NADH), riboflavin, and flavin coenzymes.[123] Unfortunately, it cannot be removed by any nontoxic procedure, though its intensity varies with different cell growth conditions.

There are various ways of making a correction for autofluorescence. The simplest method is to subtract the average background fluorescence intensity from the mean, resulting in errors in the correction. Two methods have recently been developed which make the correction on a cell-by-cell basis. The first method uses dual-laser excitation and detects the fluorescence twice at a single emission wavelength peak. Since the ratios of the molar extinction coefficients of the fluorophore and that of the autofluorescence are different, the contribution of autofluorescence to the specific fluorescence signal can be calculated and subtracted individually.[124] The other method uses a similar principle; one laser is used for excitation, and the fluorescence is measured at two wavelengths.[125, 126] Since the emission spectra of the fluorophore and the autofluorescence are different, the nonspecific contribution is calculable. The latter method does not require any instrumental modifications and will probably be widely used.

A recent method for background fluorescence correction is based on mathematical modeling of autofluorescence. The deconvolution of the histograms obtained from labeled and unlabeled samples is calculated to generate corrected histograms.[127]

8.3.2.3b. Correction for Spectral Overlap.[105] When two fluorophores are excited by a single laser beam, the detectors and the filters are selected to detect light emitted by only one dye. However, if the emission spectra of the dyes overlap or the optical filters are not perfect, both detectors will detect some light emitted by the inappropriate dye. The extent of the spillover can be so high that it disturbs the analysis. The detection efficiency in the inappropriate channels can be measured by running samples labeled with only one dye. Each measured intensity can be treated as a linear combination of the theoretical intensities and the proper correction factors, and therefore the real intensities can easily be calculated. Most commercial instruments have built-in fluorescence compensation circuits. Since the detection efficiency is a function of the photomultiplier voltage, the voltage should not be changed once the correction factor is determined. Fluorescence compensation can be also done by computers on data collected in list mode. It is hard to decide which method is better. The real-time compensation is better for sorting, while we prefer the latter method for analytical purposes, because the raw intensities are recorded, and any error in the compensation can be corrected.

8.3.2.4. Fluorescence Resonance Energy Transfer (FRET) Measurements in Flow Cytometry

FRET is useful technique for investigating intra- and intermolecular distance relationships in the range 1–10 nm.[128–132] Chapter 3 in Vol. 2 of this series deal with FRET in more detail. We give a brief description of the theory for easier understanding of its application in flow cytometry. According to the theory of Förster, a donor molecule is an excited state can transfer energy to an acceptor molecule by nonradiative energy transfer. The donor and acceptor molecules should meet several criteria: (1) the donor should have sufficiently high quantum efficiency; (2) the emission spectrum of the donor should overlap the excitation spectrum of the acceptor; and (3) the distance between the donor and the acceptor molecules should be in a given range depending on the lifetime and the spectral overlap integral. The energy transfer efficiency E can be expressed as follows:

$$E = R_0^6/(R_0^6 + R^6) \tag{8.1}$$

where R is the actual distance between the donor and acceptor molecules, and R_0 is a characteristic distance for a given donor–acceptor pair, with 50% transfer efficiency. R_0 is a function of the spectral overlap integral between the donor emission and the acceptor absorption, the refractive index of the medium, the quantum yield of the donor, and an orientation factor between the molecules.

Energy transfer efficiency is a very sensitive parameter for small distance changes in a range of $0.5R_0$ to $2R_0$. E can be determined by measuring the quenching of the donor fluorescence, the enhancement of the acceptor fluorescence, the decrease of emission anisotropy in the acceptor fluorescence, or the decrease of the lifetime of the donor fluorescence. Of these possibilities, the quenching of the donor and/or the enhancement of the acceptor fluorescence can be used in flow cytometric measurements. It is difficult to directly relate transfer efficiencies measured on cell surfaces to distance relationships, because, among other factors, the localization of the fluorophores is restricted to two dimensions and the relative orientations are not known.[133, 134] Despite these limitations, the FRET method is useful for differentiating random from nonrandom distributions of cell surface molecules or for showing interaction between different cell surface components.

The FRET method has been applied successfully to several biological systems using steady-state fluorimeters or microscopes[135] (reviewed by Szöllősi *et al.*[136] and Matkó *et al.*[137]). Recently, the FRET method has been adapted to flow cytometers as well.[138] Without describing the experimental conditions, we summarize the major advantages of flow cytometric FRET determinations over other methods. Energy transfer efficiency can be calculated based on averages computed from flow cytometric histograms. This procedure is very similar to spectrofluorimetric methods; however, it allows us to eliminate several error sources. The presence of free dye in the medium has a much smaller effect on flow cytometric measurement than on cuvette measurement, and the fluorescence collection can be specifically gated on cell populations of interest, eliminating the effect of cell debris or dead cells and revealing inhomogeneities in the sample. Recently, another method, called flow cytometric energy transfer (FCET), has been elaborated for the determination of FRET efficiency values on a cell-by-cell basis using dual-laser flow cytometers.[139–142] Data are collected in list mode, and FRET efficiency values are calculated for each cell. A simplified version of the FCET method permits—in the case of homologous ligands or of ligands binding to the same molecule but different epitopes—the determination of FRET efficiency values on a cell-by-cell basis employing flow cytometers equipped with single-laser excitations.[143] If the measurement is done on a flow cytometer with a high-speed computer, real-time FRET efficiency histograms can be collected, and, moreover, cell populations with different FRET efficiencies can be sorted.

The flow cytometric energy transfer method has certain limitations. Autofluorescence is often a problem; if it is more than 5–10% of the specific intensity, it causes unacceptable error in the FRET efficiency. For presently used donor–acceptor pairs, such as fluorescein/rhodamine or fluorescein/Texas Red, 1×10^5–5×10^5 binding sites are necessary to get detectable transfer efficiency. Unfortunately, most of the cell surface receptors of interest are expressed at much lower density. In order to study them by FRET, we should select variants with higher receptor number or develop dye pairs with better sensitivity. This technique is useful for studying interactions between cell surface proteins[144–146] or cell surface topography[147] or following receptor cluster formation or cell fusion.

8.3.2.5. Membrane Potential Measurements by Flow Cytometry

There is an electric potential difference between the cell membrane interior and exterior, the membrane potential. The origin of the membrane potential is related to the impermeability of a cell membrane to protein molecules, to the action of the [Na, K]ATPase, and to the differential premeability of the membrane to different ions. Many cellular events are accompanied by membrane potential changes, and therefore monitoring membrane potential has great importance.

There are three main methods of membrane potential measurement: direct electrophysiological measurements (with electrodes), determination of the distribution of radioisotope-labeled charged molecules, and fluorescence measurements. The membrane potential-sensitive fluorescent dyes can be divided into three groups: merocyanines, carbocyanines, and oxonols.[148–150] Merocyanines change their spectroscopic properties with changing electric field; however, these changes are too small to be detected in spectrofluorimeters or in flow cytometers. Their response time is extremely fast, so they are suitable dyes for following even action potentials, but the measurement requires a specially designed detection system. Carbocyanines and oxonols are so-called translocating dyes; that is, their concentration in the membrane depends upon and therefore follows the transmembrane electrochemical gradient.

Carbocyanines[151] are widely applied for measuring potential differences in cells and in different organelles. Carbocyanines are positively charged symmetric molecules. They are lipophilic compounds, and their membrane permeability varies with the length of the side chain. The equilibrium distribution of the dye across the cell membrane depends upon the lipid/water partition coefficient of the dye (resulting in membrane potential-independent background fluorescence) and the membrane potential of the cell (potential-sensitive part of the fluorescence). The calibration of the fluorescence signal to the membrane potential is done by changing the membrane potential by using buffers with different potassium concentrations and by using different ionophores.[152–154] When the membrane potential is increased, that is, when the cells are hyperpolarized, they take up more dye, while depolarized cells release the dye. The direction of the change in the fluorescence intensity depends on the dye concentration; when relatively high dye concentrations are applied, the dye uptake leads to dye aggregation in the membrane, resulting in self-quenching and a decrease in the fluorescence intensity[136]; but at lower concentrations (nanomolar), the aggregation can be neglected, and the higher quantum efficiency in the apolar environment results in a higher fluorescence signal. The dye concentration should be individually titrated for every cell type, and for every experimental configuration, because carbocyanine dyes stick to surfaces, and the dye distribution is affected by mitochondria as well as by surface membrane potential.[155] Mitochondrial potential is usually higher than the plasma membrane potential, and it also may respond to drugs changing the plasma membrane potential. Since the optical properties of the dye can be altered by chemical modification of the structure, carbocyanines for flow cytometric measurements are also available.[156–164] Carbocyanines were successfully applied to follow early membrane potential changes caused by various drugs, chemotactic factors, ionophores, and other ligands. Though there are doubts regarding the interpretation of the data,[165–167] carbocyanines seem to be suitable dyes for membrane potential determination, mainly for cells with no or few mitochondria.

Oxonols, like carbocyanines, have symmetrical structures, but their delocalized charge is negative, so their uptake differs from that of carbocyanines. Oxonols are excluded from hyperpolarized membranes, whereas depolarized cells accumulate them. This property results in a great advantage for oxonols compared to carbocyanines, because the highly polarized mitochondria do not affect the dye distribution. The application of oxonols in cuvette was somewhat restricted, because of the smaller intensity changes. Their flow cytometric application has recently started, after oxonol derivatives with spectral properties suitable for argon ion laser excitation were synthesized.[168] Though the application of oxonols in flow systems has just started, their widespread use is expected.[169, 170]

Flow cytometric measurements are superior to spectrofluorimetric measurements here, as in general, because the free extracellular dye has less influence on the measured signal, and inhomogeneities in the population can also be revealed. Because of the higher sensitivity, much lower dye concentrations are necessary for flow cytometry, thus decreasing the toxic effect of the dye. Gating makes it possible to eliminate the effect of dead cells and cell debris, and with the use of a cell surface marker, membrane potential changes can be correlated with defined subpopulations.[136]

8.3.2.6. Measurement of Intracellular Ca^{2+} Concentration[171]

A wide variety of cellular processes (secretion, proliferation, metabolism, phototransduction) are controlled by a second messenger, inositol triphosphate, in many cell types. As a part of the signal transduction mechanism, inositol triphosphate is released into the cytoplasm, which mobilizes intracellular Ca^{2+}.[172, 173] Therefore, measuring intracellular Ca^{2+} levels in individual cells and correlating them to other cellular parameters is useful.

Intracellular Ca^{2+} concentration can be measured by Ca^{2+}-sensitive microelectrodes and by the use of dyes that their absorption or fluorescence characteristics upon Ca^{2+} binding. The electrode measurements cannot be carried out on large numbers of cells.

The first Ca^{2+}-sensitive fluorescent dye, quin-2, was widely applied in spectrofluorimetric measurements, but it was not suitable for flow cytometric measurements, because of its weak fluorescence signal and the lack of the appropriate exciting laser line.(174, 175)

Indo-1 and fura-2 belong to a newer family of Ca^{2+}-sensitive fluorescent dyes.(176) These dyes exhibit much stronger fluorescence and can be used for single-cell measurements. The cells are incubated with an ester form of the dye, which does not bind calcium and easily penetrates the cell membrane. The ester bond is hydrolyzed by intracellular esterases, resulting in a Ca^{2+}-sensitive form of the dye, which, being less lipid-soluble, is trapped intracellularly. One of these dyes, indo-1, is suitable for flow cytometry, because it can be excited by the UV line of the argon ion laser. The Ca^{2+} free form emits at 485 nm, and after Ca^{2+} binding the emission maximum shifts to 405 nm. The ratio of the two intensities can be related to the intracellular Ca^{2+} concentration. For more accurate calibration the indo-1-loaded cells are permealized in the presence of different extracellular Ca^{2+} concentrations.(38)

In multiple-laser flow cytometers, indo-1 can be combined with other dyes emitting in the red or yellow, allowing one to correlate the intracellular Ca^{2+} concentration with cell surface antigens or other parameters. The major disadvantage of indo-1 is that it requires UV excitation. Fluo-3,(177, 178) a recently synthesized fluorescein derivative, might be the calcium-sensitive dye of the future, since it has an excitation maximum of 488 nm. Its fluorescence signal depends not only on the cytosolic calcium concentration, but also on the amount of intracellular dye. This drawback can be overcome by simultaneously loading the cells with another fluorescent label, SNARF-1. Since the uptake and intracellular conversion is similar for both dyes, and the SNARF-1 is insensitive to calcium concentration changes, the ratio of the fluo-3 and SNARF-1 fluorescence intensity is independent of the cell volume.(179)

8.3.2.7. Membrane Permeability Measurements

Controlling cell viability has great importance in many biological experiments. Increased cell membrane permeability is one of the first indicators of cell death. There are several flow cytometric methods for monitoring the cell membrane permeability. All these methods are based on the measurement of intracellularly accumulated fluorescent dyes.

The first approach takes advantage of the so-called fluorogenic substrates. Conversion of nonfluorescent fluorescein diacetate (FDA) into fluorescent fluorescein was described by Rotman and Papermaster.(180) They called the phenomenon fluorochromasia. FDA is a nonpolar molecule, so it easily penetrates the cell membrane. Intracellular esterases convert it into fluorescein, which is a highly polar compound, and therefore less membrane permeable. The result of these processes is the intracellular accumulation of fluorescein. The intracellular amount of fluorescein is determined by relative speeds of uptake, hydrolysis, and release of the product; thus, the measured parameter is complex, partly reflecting an intracellular enzyme activity and the integrity of the cell membrane.[181, 182] Damaged cells do not accumulate fluorescein, because the intracellular esterases lose their activity and the cell membrane is leaky. Thus, a mixture of live and dead cells stained with FDA shows a bimodal fluorescence intensity distribution, where the left peak corresponds to dead and the right peak to live cells. For most cells the separation between the two peaks is large enough to unequivocally evaluate viability.

From measurements of the intracellular fluorescein content as a function of time, the efflux half-time can be calculated. This parameter was found to be different in different cell types and with altered metabolic conditions.[182] The method is suitable for following the effect of drugs or treatments affecting membrane permeability. These measurements are very simple with flow cytometers equipped with so-called time cards, which correlate fluorescence intensities to the elapsed time, generating time-dependent fluorescence histograms.[183]

Another method uses dyes which do not penetrate the intact cell membrane, such as propidium iodide, a widely used DNA stain. PI brightly stains the nuclei of damaged cells, while the intact cells remain nonfluorescent. This method has the advantage that it can be combined with immunofluorescence staining (with fluorescein-conjugated antibodies), allowing one to determine subset-specific cytotoxicity.[184, 185]

The combination of FDA and PI staining is extremely effective. Both dyes can be excited with the 488-nm line of an argon ion laser, and the spectral separation is large enough to detect them separately.[186, 187] The live cells fluoresce green but not red, while dead cells fluoresce only red. If FDA or PI is used separately, contaminating particles showing similar light scatter to that of cells are considered as dead or live cells, respectively. When both dyes are used, every cell fluoresces, so the nonfluorescent contaminants are disregarded. (A small fraction of the cells may show both red and green fluorescence; they are considered to be in a borderline state.)

8.4. Conclusion

In the previous sections we have discussed the origin and design of flow cytometers and some applications of these machines. Though we expect that other methods of analysis and single-cell sorting will evolve, growing out of developments in digital video microscopy, we also expect to see further evolution of the design and use of flow cytometers.

It is our impression, based on the use of a number of these machines, that the applications of flow cytometers are largely computer-limited. This limit is being extended as commercial computers become available which take control of the cytometer, allowing greatly increased sorting rates and analytical calculations. We expect that, if development continues in this direction, within a few years flow cytometer performance will be constrained by the design of the flow system and that this will in turn be improved to match the new standards of computer control.

Detectors are also susceptible to improvement. In particular, it should be possible to implement the suggestion that spectrometers can be coupled to flow cytometers,[188, 189] allowing detailed analysis of the spectra of cell-bound fluorophores. Though cells travel through the flow chamber at high velocity, it may even be possible to make crude measurements of lifetimes of fluorophores excited through optics similar to those used for slit scanning.

The changes in hardware described will enlarge the applications of flow cytometry and in particular ought to increase the use of these instruments in the study of cell biophysics. Besides spectral analyses and real-time determination of resonance energy transfer, we expect to see measurements of fluorescence polarization in cell populations, though this may require synthesis of new fluorophores for satisfactory results. Other applications are already being realized, in the use of flow cytometry to measure ligand binding at equilibrium as well as the kinetics of ligand binding. Indeed, if a sufficient number of fluorescent ligands, for example, peptide hormone analogues, can be developed, these ought to replace radioligands as the method of choice for a wide variety of assays.

Though this brief summary stresses new analytical applications of the machines, we conclude with the reminder that once a detection scheme is devised for a particular cell property, we may isolate the cells of interest and move back from the biophysics of single cells in suspension to the biology of cell populations.

References

1. T. Caspersson, G. Lomakka, and O. Caspersson, Quantitative cytochemical methods for the study of tumour cell populations, *Biochem. Pharmacol. 4*, 113–127 (1960).
2. M. R. Melamed, P. F. Mullaney, and M. L. Mendelsohn (eds.), *Flow Cytometry and Sorting*, John Wiley & Sons, New York (1979).
3. H. M. Shapiro, *Practical Flow Cytometry*, Alan R. Liss, New York (1985).
4. M. A. Van Dilla, P. N. Dean, O. D. Laerum, and M. R. Melamed (eds.), *Flow Cytometry: Instrumentation and Data Analysis*, Academic Press, London (1985).
5. M. R. Melamed, T. Lindmo, and M. L. Mendelsohn (eds.), *Flow Cytometry and Sorting*, 2nd ed., Wiley & Liss, New York (1990).
6. L. A. Herzenberg, R. G. Sweet, and L. A. Herzenberg, Fluorescence-activated cell sorting, *Sci. Am. 234*(3), 108–117 (1976).

7. D. J. Arndt-Jovin and T. M. Jovin, Automated cell sorting with flow systems, *Annu. Rev. Biophys. Bioeng. 7*, 527–558 (1978).
8. M. J. Fulwyler, Flow cytometry and cell sorting, *Blood Cells 6*, 173–184 (1980).
9. O. D. Laerum and T. Farsund, Clinical application of flow cytometry: A review, *Cytometry 2*, 1–13 (1981).
10. J. A. Steinkamp, Flow cytometry, *Rev. Sci. Instrum. 55*, 1375–1400 (1984).
11. K. A. Ault, Clinical applications of fluorescence-activated cell sorting techniques, *Diagn. Immunol. 1*, 2–10 (1983).
12. E. J. Lovett III, B. Schnitzer, D. F. Keren, A. Flint, J. L. Hudson, and K. D. McClatchey, Application of flow cytometry to diagnostic pathology, *Lab. Invest. 50*, 115–140 (1984).
13. F. Traganos, Flow cytometry: Principles and applications. I., *Cancer Invest. 2*(2), 149–163 (1984).
14. F. Traganos, Flow cytometry: Principles and applications. II., *Cancer Invest. 2*(3), 239–258 (1984).
15. K. A. Muirhead, P. K. Horan, and G. Poste, Flow cytometry: Present and future, *Bio/Technology 3*, 337–356 (1985).
16. D. R. Parks, L. L. Lanier, and L. A. Herzenberg, Flow cytometry and fluorescence activated cell sorting (FACS), in: *Handbook of Experimental Immunology* (D. M. Weir, C. C. Blackwell, L. A. Herzenberg, and L. A. Herzenberg, eds.), Vol. 1, pp. 29.1–29.21, Blackwell Scientific Publications, Edinburgh (1986).
17. W. H. Coulter, Means for counting particles suspended in a fluid, U. S. Patent 2,656,508 (1953).
18. M. R. Melamed and P. F. Mullaney, An historical review of the development of flow cytometers and sorters, in: Ref. 2, pp. 3–9.
19. M. R. Melamed, P. F. Mullaney, and H. M. Shapiro, an historical review of the development of flow cytometers and sorters, in: Ref. 5, pp. 1–10.
20. V. Kachel, Electrical resistance pulse sizing (Coulter sizing), in: Ref. 2, pp. 61–104.
21. V. Kachel, Electrical resistance pulse sizing: Coulter sizing, in: Ref. 5, pp. 45–80.
22. L. Spielman and S. L. Goren, Improving resolution in Coulter counting by hydrodynamic focusing, *J. Colloid Interface Sci. 26*, 175–182 (1968).
23. J. T. Merrill, N. Veizades, H. R. Hulett, P. L. Wolf, and L. A. Herzenberg, An improved cell volume analyzer, *Rev. Sci. Instrum. 42*, 1157–1163 (1971).
24. L. A. Kamentsky, M. R. Melamed, and H. Derman, Spectrophotometer: New instrument for ultrarapid cell analysis, *Science 150*, 630–631 (1965).
25. M. A. Van Dilla, T. T. Trujillo, P. F. Mullaney, and J. R. Coulter, Cell microfluorometry: A method for rapid fluorescence measurement, *Science 163*, 1213–1214 (1969).
26. M. J. Fulwyler, Electronic separation of biological cells by volume, *Science 150*, 910–911 (1965).
27. R. G. Sweet, High frequency recording with electrostatically deflected ink jets, *Rev. Sci. Instrum. 36*, 131–136 (1965).
28. H. M. Shapiro, E. R. Schildkraut, R. Curbelo, R. B. Turner, R. H. Webb, D. C. Brown, and M. J. Block, Cytomat-R: A computer-controlled multiple laser source multiparameter flow cytophotometer system, *J. Histochem. Cytochem. 25*, 836–844 (1977).
29. H. M. Shapiro, Multistation multiparameter flow cytometry: A critical review and rationale, *Cytometry 3*, 227–243 (1983).
30. H. M. Shapiro, D. M. Feinstein, A. S. Kirsch, and L. Christenson, Multistation multiparameter flow cytometry: Some influences of instrumental factors on system performance, *Cytometry 4*, 11–19 (1983).
31. L. S. Cram, M. F. Bartholdi, L. L. Wheeless, Jr., and J. W. Gray, Morphological analysis by scanning flow cytometry, in: Ref. 4, pp. 163–193.
32. D. Pinkel and R. Stovel, Flow chambers and sample handling, in: Ref. 4, pp. 77–128.

33. V. Kachel and E. Menke, Hydrodynamic properties of flow cytometric instruments, in: Ref. 2, pp. 41–59.
34. V. Kachel, H. Fellner-Feldegg, and E. Menke, Hydrodynamic properties of flow cytometry instruments, in: Ref. 5, pp. 27–44.
35. L. L. Wheeless, Jr., and D. B. Kay, Optics, light sources, filters, and optical systems, in: Ref. 4, pp. 21–76.
36. D. J. Arndt-Jovin and T. M. Jovin, Analysis and sorting of living cells according to deoxyribonucleic acid content, *J. Histochem. Cytochem. 25*, 585–589 (1977).
37. G. Szabó, Jr., A. Kiss, and S. Damjanovich, Flow cytometric analysis of the uptake of Hoechst 33342 dye by human lymphocytes, *Cytometry 2*, 20–23 (1981).
38. T. M. Chused, H. A. Wilson, D. Greenblatt, Y. Ishida, L. J. Edison, R. Y. Tsien, and F. D. Finkelman, Flow cytometric analysis of murine splenic B lymphocyte cytosolic free calcium response to anti-IgM and anti-IgD, *Cytometry 8*, 396–404 (1987).
39. R. D. Hiebert and R. G. Sweet, Electronics for flow cytometers and sorters, in: Ref. 4, pp. 129–162.
40. R. D. Hiebert, Electronics and signal processing, in: Ref. 5, pp. 127–144.
41. G. C. Salzman, S. B. Singham, R. G. Johnston, and C. F. Bohren, Light scattering and cytometry, in: Ref. 5, pp. 81–108.
42. P. F. Mullaney, M. A. Van Dilla, J. R. Coulter, and P. N. Dean, Cell sizing: A light scattering photometer for rapid volume determination, *Rev. Sci. Instrum. 40*, 1029–1032 (1969).
43. J. R. Hodkinson and I. Greenleaves, Computations of light-scattering and extinction by spheres according to diffraction and geometrical optics, and some comparisons with the Mie theory, *J. Opt. Soc. Am. 53*, 577–588 (1963).
44. P. F. Mullaney, Application of the Hodkinson scattering model to particles of low relative refractive index, *J. Opt. Soc. Am. 60*, 573–574 (1970).
45. T. M. Jovin, S. J. Morris, G. Striker, H. A. Schultens, M. Digweed, and D. J. Arndt-Jovin, Automatic sizing and separation of particles by ratios of light scattering intensities, *J. Histochem. Cytochem. 24*, 269–283 (1976).
46. T. K. Sharpless, M. Bartholdi, and M. R. Melamed, Size and refractive index dependence of simple forward angle scattering measurements in a flow system using sharply-focused illumination, *J. Histochem. Cytochem. 25*, 845–856 (1977).
47. G. C. Salzman, J. M. Crowell, C. A. Goad, K. M. Hansen, R. D. Hiebert, P. M. LaBauve, J. C. Martin, M. L. Ingram, and P. F. Mullaney, A flow-system multiangle light-scattering instrument for cell characterization, *Clin. Chem. 21*, 1297–1304 (1975).
48. P. F. Mullaney, J. M. Crowell, G. C. Salzman, J. C. Martin, R. D. Hiebert, and C. A. Goad, Pulse-height light-scatter distributions using flow-systems instrumentation, *J. Histochem. Cytochem. 24*, 298–304 (1976).
49. S. M. Watt, A. W. Burgess, D. Metcalf, and F. L. Battye, Isolation of mouse bone marrow neutrophils by light scatter and autofluorescence, *J. Histochem. Cytochem. 28*, 934–946 (1980).
50. K. A. Kelley and J. L. McDowell, Practical considerations for the selection and use of optical filters in flow cytometry, *Cytometry 9*, 277–280 (1988).
51. W. Gandler and H. Shapiro, Logarithmic amplifiers, *Cytometry 11*, 447–450 (1990).
52. T. K. Sharpless, Cytometric data processing, in: Ref. 2, pp. 359–379.
53. P. N. Dean, Methods of data analysis in flow cytometry, in: Ref. 4, pp. 195–221.
54. L. L. Wheeless, Jr., Slit-scanning and pulse width analysis, in: Ref. 2, pp. 125–135.
55. L. L. Wheeless, Jr., Slit-scanning, in: Ref. 5, pp. 109–126.
56. L. L. Wheeless, Jr., and S. F. Patten, Slit-scan cytofluorometry, *Acta Cytol. 17*, 333–339 (1973).
57. D. B. Kay, J. L. Cambier, and L. L. Wheeless, Jr., Imaging in flow, *J. Histochem. Cytochem. 27*, 329–334 (1979).

58. J. L. Cambier, D. B. Kay, and L. L. Wheeless, Jr., A multidimensional slit-scan flow system, *J. Histochem. Cytochem. 27*, 321–324 (1979).
59. L. L. Wheeless, S. F. Patten, T. K. Berkan, C. L. Brooks, K. M. Gorman, S. R. Lesh, P. A. Lopez, and J. C. S. Wood, Multidimensional slit-scan prescreening system: Preliminary results of a single blind clinical study, *Cytometry 5*, 1–8 (1984).
60. L. L. Wheeless, T. K. Berkan, S. F. Patten, J. E. Reeder, R. D. Robinson, M. M. Eldidi, W. C. Hulbert, and I. N. Frank, Multidimensional slit-scan detection of bladder cancer: Preliminary clinical results, *Cytometry 7*, 212–216 (1986).
61. R. D. Robinson, D. M. Wheeless, S. J. Hespelt, and L. L. Wheeless, System for acquisition and real-time processing of multidimensional slit-scan flow cytometric data, *Cytometry 11*, 379–385 (1990).
62. R. D. Robinson, J. E. Reeder, and L. L. Wheeless, Technique for cellular fluorescence distribution analysis, *Cytometry 10*, 402–409 (1989).
63. R. G. Sweet, Flow sorters for biologic cells, in: Ref. 2, pp. 177–189.
64. T. Lindmo, D. C. Peters, and R. G. Sweet, Flow sorters for biological cells, in: Ref. 5, pp. 145–170.
65. D. Peters, E. Branscomb, P. Dean, T. Merrill, D. Pinkel, M. A. Van Dilla, and J. W. Gray, The Livermore high speed sorter: Design features, operational characteristics, and biological utility, *Cytometry 6*, 290–301 (1985).
66. G. C. Salzman, P. F. Mullaney, and B. J. Price, Light-scattering approaches to cell characterization, in: Ref. 2, pp. 105–124.
67. M. R. Loken, D. R. Parks, and L. A. Herzenberg, Identification of cell asymmetry and orientation by light scattering, *J. Histochem. Cytochem. 25*, 790–795 (1977).
68. M. R. Loken and D. W. Houck, Light scattered at two wavelengths can discriminate viable lymphoid cell populations on a fluorescence-activated cell sorter, *J. Histochem. Cytochem. 29*, 609–615 (1981).
69. G. R. Otten and M. R. Loken, Two color light scattering identifies physical differences between lymphocyte subpopulations, *Cytometry 3*, 182–187 (1982).
70. M. Kerker, M. A. Van Dilla, A. Brunsting, J. P. Kratohvil, P. Hsu, D. S. Wang, J. W. Gray, and R. G. Langlois, Is the central dogma of flow cytometry true: That fluorescence intensity is proportional to cellular dye content?, *Cytometry 3*, 71–78 (1982).
71. H. A. Crissman, A. P. Stevenson, R. J. Kissane, and R. A. Tobey, Techniques for quantitative staining of cellular DNA for flow cytometric analysis, in: Ref. 2, pp. 243–261.
72. P. N. Dean, J. W. Gray, and F. A. Dolbeare, The analysis and interpretation of DNA distributions measured by flow cytometry, *Cytometry 3*, 188–195 (1982).
73. H. A. Crissman and R. A. Tobey, Cell-cycle analysis in 20 minutes, *Science 184*, 1297–1298 (1974).
74. V. T. Hamilton, M. C. Habbersett, and C. J. Herman, Flow microfluorometric analysis of cellular DNA: Critical comparison of mithramycin and propidium iodide, *J. Histochem. Cytochem. 28*, 1125–1128 (1980).
75. I. W. Taylor and B. K. Milthorpe, An evaluation of DNA fluorochromes, staining techniques, and analysis for flow cytometry. I. Unperturbed cell populations, *J. Histochem. Cytochem. 28*, 1224–1232 (1980).
76. P. N. Dean and J. H. Jett, Mathematical analysis of DNA distributions derived from flow microfluorometry, *J. Cell Biol. 60*, 523–527 (1974).
77. J. Fried, Analysis of deoxyribonucleic acid histograms from flow cytofluorometry: Estimation of the distribution of cells within S phase, *J. Histochem. Cytochem. 25*, 942–951 (1977).
78. H. G. Gratzner, Monoclonal antibody to 5-bromo- and 5-iododeoxyuridine: A new reagent for detection of DNA replication, *Science 218*, 474–475 (1982).
79. F. Dolbeare, H. Gratzner, M. G. Pallavicini, and J. W. Gray, Flow cytometric measurement

of total DNA content and incorporated bromodeoxyuridine, *Proc. Natl. Acad. Sci. U.S.A. 80*, 5573–5577 (1983).
80. P. N. Dean, F. Dolbeare, H. Gratzner G. C. Rice, and J. W. Gray, Cell-cycle analysis using a monoclonal antibody to BrdUrd, *Cell Tissue Kinet. 17*, 427–436 (1984).
81. W. Beisker, F. Dolbeare, and J. W. Gray, An improved immunocytochemical procedure for high-sensitivity detection of incorporated bromodeoxyuridine, *Cytometry 8*, 235–239 (1987).
82. B. Schutte, M. M. J. Reynders, C. L. M. V. J. van Assche, P. S. G. J. Hupperets, F. T. Bosman, and G. H. Blijham, An improved method for the immunocytochemical detection of bromodeoxyuridine labeled nuclei using flow cytometry, *Cytometry 8*, 372–376 (1987).
83. R. Moran, Z. Darzynkiewicz, L. Staiano-Coico, and M. R. Melamed, Detection of 5-bromodeoxyuridine (BrdUrd) incorporation by monoclonal antibodies: Role of the DNA denaturation step, *J. Histochem. Cytochem. 33*, 821–827 (1985).
84. Z. Darzynkiewicz, Acridine Orange as a molecular probe in studies of nucleic acids *in situ*, in: Ref. 2, pp. 285–316.
85. Z. Darzynkiewicz, F. Traganos, T. Sharpless, and M. R. Melamed, Lymphocyte stimulation: A rapid multiparameter analysis, *Proc. Natl. Acad. Sci. U.S.A. 73*, 2881–2884 (1976).
86. Z. Darzynkiewicz, F. Traganos, and M. R. Melamed, New cell cycle compartments identified by multiparameter flow cytometry, *Cytometry 1*, 98–108 (1980).
87. Z. Darzynkiewicz, T. Sharpless, L. Staiano-Coico, and M. R. Melamed, Subcompartments of the G_1 phase of cell cycle detected by flow cytometry, *Proc. Natl. Acad. Sci. U.S.A. 77*, 6696–6699 (1980).
88. Z. Darzynkiewicz, Metabolic and kinetic compartments of the cell cycle distinguished by multiparameter flow cytometry, in: *Growth, Cancer, and the Cell Cycle* (P. Skehan and S. J. Friedman, eds.), pp. 249–278, The Humana Press, Clifton, New Jersey (1984).
89. Z. Darzynkiewicz, Cytochemical probes of cycling and quiescent cells applicable to flow cytometry, in: *Techniques in Cell Cycle Analysis* (J. W. Gray and Z. Darzynkiewicz, eds.), pp. 255–290, The Humana Press, Clifton, New Jersey (1986).
90. Z. Darzynkiewicz and J. Kapuscinski, Acridine Orange: A versatile probe of nucleic acids and other cell constituents, in: Ref. 5, pp. 291–314.
91. Z. Darzynkiewicz, F. Traganos, T. Sharpless, and M. R. Melamed, Thermal denaturation of DNA *in situ* as studied by Acridine Orange staining and automated cytofluorometry, *Exp. Cell Res. 90*, 411–428 (1975).
92. Z. Darzynkiewicz, F. Traganos, and M. R. Melamed, Distinction between 5-bromodeoxyuridine labeled and unlabeled mitotic cells by flow cytometry, *Cytometry 3*, 345–348 (1983).
93. M. Piwnicka, Z. Darzynkiewicz, and M. R. Melamed, RNA and DNA content of isolated cell nuclei measured by multiparameter flow cytometry, *Cytometry 3*, 269–275 (1983).
94. H. A. Crissman, Z. Darzynkiewicz, R. A. Tobey, and J. A. Steinkamp, Correlated measurements of DNA, RNA, and protein in individual cells by flow cytometry, *Science 228*, 1321–1324 (1985).
95. H. A. Crissman, Z. Darzynkiewicz, R. A. Tobey, and J. A. Steinkamp, Normal and perturbed Chinese hamster ovary cells: Correlation of DNA, RNA, and protein content by flow cytometry, *J. Cell Biol. 101*, 141–147 (1985).
96. J. W. Gray and R. G. Langlois, Chromosome classification and purification using flow cytometry and sorting, *Annu. Rev. Biophys. Biophys. Chem. 15*, 195–235 (1986).
97. J. W. Gray and L. S. Cram, Flow karyotyping and chromosome sorting, in: Ref. 5, pp. 503–530.
98. R. Sillar and B. D. Young, A new method for the preparation of metaphase chromosomes for flow analysis, *J. Histochem. Cytochem. 29*, 74–78 (1981).
99. J. W. Gray, A. V. Carrano, L. L. Steinmetz, M. A. Van Dilla, D. H. Moore II, B. H. Mayall, and M. L. Mendelsohn, Chromosome measurement and sorting by flow systems, *Proc. Natl. Acad. Sci. U.S.A. 72*, 1231–1234 (1975).

100. K. E. Davies, B. D. Young, R. G. Elles, M. E. Hill, and R. Williamson, Cloning of a representative genomic library of the human X chromosome after sorting by flow cytometry, *Nature 293*, 374–376 (1981).
101. R. G. Langlois, L.-C. Yu, J. W. Cray, and A. V. Carrano, Quantitative karyotyping of human chromosomes by dual beam flow cytometry, *Proc. Natl. Acad. Sci. U.S.A. 79*, 7876–7880 (1982).
102. M. A. Van Dilla, L. L. Deaven, K. L. Albright, N. A. Allen, M. R. Aubuchon, M. F. Bartholdi, N. C. Brown, E. W. Campbell, A. V. Carrano, L. M. Clark, L. S. Cram, B. D. Crawford, J. C. Fuscoe, J. W. Gray, C. E. Hildebrand, P. J. Jackson, J. H. Jett, J. L. Longmire, C. R. Lozes, M. L. Luedemann, J. C. Martin, J. S. McNinch, L. J. Meincke, M. L. Mendelsohn, J. Meyne, R. K. Moyzis, A. C. Munk, J. Perlman, D. C. Peters, A. J. Silva, and B. J. Trask, Human chromosome-specific DNA libraries: Construction and availability, *Bio/Technology 4*, 537–552 (1986).
103. J. W. Gray, P. N. Dean, J. C. Fuscoe, D. C. Peters, B. J. Trask, G. J. van den Engh, and M. A. Van Dilla, High-speed chromosome sorting, *Science 238*, 323–329 (1987).
104. D. R. Parks and L. A. Herzenberg, Fluorescence-activated cell sorting: Theory, experimental optimization, and applications in lymphoid cell biology, *Methods Enzymol. 108*, 197–241 (1984).
105. M. R. Loken, R. D. Stout, and L. A. Herzenberg, Lymphoid cell analysis and sorting, in: Ref. 2, pp. 505–528.
106. M. R. Loken, Immunofluorescence techniques, in: Ref. 5, pp. 340–354.
107. M. R. Loken, Cell surface antigen and morphological characterization of leucocyte populations by flow cytometry, in: *Monoclonal Antibodies* (P. C. L. Beverley, ed.), pp. 132–144, Churchill Livingstone, Edinburgh (1986).
108. S. Lakhanpal, N. J. Gonchoroff, J. A. Katzmann, and B. S. Handwerger, A flow cytofluorometric double staining technique for simultaneous determination of human mononuclear cell surface phenotype and cell cycle phase, *J. Immunol. Methods 96*, 35–40 (1987).
109. G. Szabó, Jr., and L. Mátyus, A follow-up study of the development of Rauscher erythroleukemia, *Cytometry 5*, 92–95 (1984).
110. P. Surányi, L. Mátyus, I. Sonkoly, and Gy. Szegedi, Cellular DNA content of T helper, T suppressor and B lymphocytes in SLE, *Clin. Exp. Immunol. 58*, 37–41 (1984).
111. L. L. Lanier, E. G. Engleman, P. Gatenby, G. F. Babcock, N. L. Warner, and L. A. Herzenberg, Correlation of functional properties of human lymphoid cell subsets and surface marker phenotypes using multiparameter analysis and flow cytometry, *Immunol. Rev. 74*, 143–160 (1983).
112. D. R. Parks, R. R. Hardy, and L. A. Herzenberg, Dual immunofluorescence—new frontiers in cell analysis and sorting, *Immunol. Today 4*, 145–150 (1983).
113. D. R. Parks, R. R. Hardy, and L. A. Herzenberg, Three-color immunofluorescence analysis of mouse B-lymphocyte subpopulations, *Cytometry 5*, 159–168 (1984).
114. M. R. Loken and L. L. Lanier, Three-color immunofluorescence analysis of Leu antigens on human peripheral blood using two lasers on a fluorescence-activated cell sorter, *Cytometry 5*, 151–158 (1984).
115. R. Festin, B. Bjorklund, and T. H. Totterman, Detection of triple antibody-binding lymphocytes in standard single laser flow cytometry using colloidal gold, fluorescein and phycoerythrin as labels, *J. Immunol. Methods 101*, 23–28 (1987).
116. V. T. Oi, A. N. Glazer, and L. Stryer, Fluorescent phycobiliprotein conjugates for analyses of cells and molecules, *J. Cell Biol. 93*, 981–986 (1982).
117. A. N. Glazer and L. Stryer, Fluorescent tandem phycobiliprotein conjugates: Emission wavelength shifting by energy transfer, *Biophys. J. 43*, 383–386 (1983).
118. A. N. Glazer and L. Stryer, Phycofluor probes, *Trends Biochem. Sci. 9*, 423–427 (1984).

119. S. W. Yeh, L. J. Ong, J. H. Clark, and A. N. Glazer, Fluorescence properties of allophycocyanin trimer, *Cytometry 8*, 91–95 (1987).
120. M. R. Loken, J. F. Keij, and K. A. Kelley, Comparison of helium–neon and dye lasers for the excitation of allophycocyanin, *Cytometry 8*, 96–100 (1987).
121. L. A. Sklar, Real-time spectroscopic analysis of ligand-receptor dynamics, *Annu. Rev. Biophys. Chem. 16*, 479–506 (1987).
122. J. E. Aubin, Autofluorescence of viable cultured mammalian cells, *J. Histochem. Cytochem. 27*, 36–43 (1979).
123. R. C. Benson, R. A. Meyer, M. E. Zaruba, and G. M. McKhann, Cellular autofluorescence —is it due to flavins? *J. Histochem. Cytochem. 27*, 44–48 (1979).
124. J. A. Steinkamp and C. C. Stewart, Dual-laser, differential fluorescence correction method for reducing cellular background autofluorescence, *Cytometry 7*, 566–574 (1986).
125. M. Roederer and R. F. Murphy, Cell-by-cell autofluorescence correction for low signal-to-noise systems: Application to epidermal growth factor endocytosis by 3T3 fibroblasts, *Cytometry 7*, 558–565 (1986).
126. S. Alberti, D. R. Parks, and L. A. Herzenberg, A single laser method for subtraction of cell autofluorescence in flow cytometry, *Cytometry 8*, 114–119 (1987).
127. J. P. Corsetti, S. V. Sotirchos, C. Cox, J. W. Cowles, J. F. Leary, and N. Blumburg, Correction of cellular autofluorescence in flow cytometry by mathematical modeling of cellular fluorescence, *Cytometry 9*, 539–547 (1988).
128. T. M. Jovin, Fluorescence polarization and energy transfer: Theory and application, in: Ref. 2, pp. 137–165.
129. L. Stryer, Fluorescence energy transfer as a spectroscopic ruler, *Annu. Rev. Biochem. 47*, 819–846 (1978).
130. R. E. Dale, J. Novros, S. Roth, M. Edidin, and L. Brand, Application of Förster long-range excitation energy transfer to the determination of distributions of fluorescently-labelled concanavalin A–receptor complexes at the surfaces of yeast and of normal and malignant fibroblasts, in: *Fluorescent Probes* (G. S. Beddard and M. A. West, eds.), pp. 159–189, Academic Press, London (1981).
131. C. R. Cantor and P. R. Schimmel, *Biophysical Chemistry*, pp. 448–454, W. H. Freeman and Company, San Francisco (1980).
132. J. R. Lakowicz, *Principles of Fluorescence Spectroscopy*, pp. 303–339, Plenum Press, New York (1983).
133. P. K. Wolber and B. S. Hudson, An analytic solution to the Förster energy transfer problem in two dimensions, *Biophys. J. 28*, 197–210 (1979).
134. T. G. Dewey and G. G. Hammes, Calculation of fluorescence resonance energy transfer on surfaces, *Biophys. J. 32*, 1023–1036 (1980).
135. S. M. Fernandez and R. D. Berlin, Cell surface distribution of lectin receptors determined by resonance energy transfer, *Nature 264*, 411–415 (1976).
136. J. Szöllősi, S. Damjanovich, S. A. Mulhern, and L. Trón, Fluorescence energy transfer and membrane potential measurements monitor dynamic properties of cell membranes: A critical review, *Prog. Biophys. Mol. Biol. 49*, 65–87 (1987).
137. J. Matkó, J. Szöllősi, L. Trón, and S. Damjanovich, Luminescence spectroscopic approaches in studying cell surface dynamics, *Quart. Rev. Biophys. 21*, 479–544 (1988).
138. S. S. Chan, D. J. Arndt-Jovin, and T. M. Jovin, Proximity of lectin receptors on the cell surface measured by fluorescence energy transfer in a flow system, *J. Histochem. Cytochem. 27*, 56–64 (1979).
139. L. Trón, J. Szöllősi, S. Damjanovich, S. H. Helliwell, D. J. Arndt-Jovin, and T. M. Jovin, Flow cytometric measurement of fluorescence resonance energy transfer on cell surfaces: Quantitative evaluation of the transfer efficiency on a cell-by-cell basis, *Biophys. J. 45*, 939–946 (1984).

140. L. Trón, J. Szöllősi, G. Szabó, Jr., L. Mátyus, and S. Damjanovich, Cell surface dynamics and distance relationship of integral membrane proteins, in: *Membrane Dynamics and Transport of Normal and Tumor Cells* (L. Trón, S. Damjanovich, A. Fonyó, and J. Somogyi, eds., *Symp. Biol. Hung. 26*, 307–328 (1984).
141. L. Trón, J. Szöllősi, G. Szabó, Jr., L. Mátyus, and S. Damjanovich, Cell surface dynamics and information transfer, in: *Tissue Culture and RES* (P. Röchlich and E. Bácsy, eds.), pp. 75–83, Akadémiai Kiadó, Budapest (1984).
142. J. Szöllősi, L. Trón, S. Damjanovich, S. Helliwell, D. J. Arndt-Jovin, and T. M. Jovin, Fluorescence energy transfer measurements on cell surfaces: A critical comparison of steady-state fluorimetric and flow cytometric methods, *Cytometry 5*, 210–216 (1984).
143. J. Szöllősi, L. Mátyus, L. Trón, M. Balázs, I. Ember, M. J. Fulwyler, and S. Damjanovich, Flow cytometric measurements of fluorescence energy transfer using single laser excitation, *Cytometry 8*, 120–128 (1987).
144. J. H. Hochman and M. Edidin, Association between class I MHC antigens and viral glycoproteins detected by fluorescence resonance energy transfer, in: *Major Histocompatibility Genes and Their Role in Immune Function* (D. S. Chella, ed.), Plenum Press, New York (1989).
145. J. Szöllősi, S. Damjanovich, C. K. Goldman, M. J. Fulwyler, A. A. Aszalos, G. Goldstein, P. Rao, M. A. Talle, and T. A. Waldmann, Flow cytometric resonance energy transfer measurements support the association of a 95-kDa peptide termed T27 with the 55-kDa Tac peptide, *Proc. Natl. Acad. Sci. U.S.A. 84*, 7246–7250 (1987).
146. J. Szöllősi, F. M. Brodsky, M. Balázs, P. Nagy, L. Trón, M. J. Fulwyler, and S. Damjanovich, Physical association between MHC class I and class II molecules in cytoplasmic membranes. A flow cytometric energy transfer study, *J. Immunol. 143*, 208–213 (1989).
147. D. M. Jenis, A. L. Stepanowski, O. C. Blair, D. E. Burger, and A. C. Sartorelli, Lectin receptor proximity on HL-60 leukemia cells determined by fluorescence energy transfer using flow cytometry, *J. Cell. Physiol. 121*, 501–507 (1984).
148. A. Waggoner, Optical probes of membrane potential, *J. Membr. Biol. 27*, 317–334 (1976).
149. A. S. Waggoner, Dye indicators of membrane potential, *Annu. Rev. Biophys. Bioeng. 8*, 47–68 (1979).
150. A. S. Waggoner, Fluorescent probes for analysis of cell structure, function, and health by flow and imaging cytometry, in: *Application of Fluorescence in the Biomedical Sciences* (D. L. Taylor, A. S. Waggoner, R. F. Murphy, F. Lanni, and R. R. Birge, eds.), pp. 3–38, Alan R. Liss, New York (1986).
151. P. J. Sims, A. S. Waggoner, C.-H. Wang, and J. F. Hoffman, Studies on the mechanism by which cyanine dyes measure membrane potential in red blood cells and phosphatidylcholine vesicles, *Biochemistry 13*, 3315–3330 (1974).
152. S. B. Hladky and T. J. Rink, Potential difference and the distribution of ions across the human red blood cell membrane: A study of the mechanism by which the fluorescent cation, diS-C_3-(5), reports membrane potential, *J. Physiol. 263*, 287–319 (1976).
153. J. F. Hoffman and P. C. Laris, Determination of membrane potentials in human and *amphiuma* red blood cells by means of a fluorescent probe, *J. Physiol. 239*, 519–552 (1974).
154. J. C. Freedman and J. F. Hoffman, The relation between dicarbocyanine dye fluorescence and the membrane potential of human red blood cells set at varying Donnan equilibria, *J. Gen. Physiol. 74*, 187–212 (1979).
155. H. A. Wilson, B. E. Seligmann, and T. M. Chused, Voltage-sensitive cyanine dye fluorescence signals in lymphocytes: Plasma membrane and mitochondrial components, *J. Cell. Physiol. 125*, 61–71 (1985).
156. H. M. Shapiro, P. J. Natale, and L. A. Kamentsky, Estimation of membrane potentials of individual lymphocytes by flow cytometry, *Proc. Natl. Acad. Sci. U.S.A. 76*, 5728–5730 (1979).

157. J. G. Monroe and J. C. Cambier, B cell activation. I. Anti-immunoglobulin-induced receptor cross-linking results in a decrease in the plasma membrane potential of murine B lymphocytes, *J. Exp. Med. 157*, 2073–2086 (1983).
158. J. G. Monroe and J. C. Cambier, B cell activation. II. Receptor cross-linking by thymus-independent and thymus-dependent antigens induces a rapid decrease in the plasma membrane potential of antigen-binding B lymphocytes, *J. Immunol. 131*, 2641–2644 (1983).
159. J. G. Monroe and J. C. Cambier, B cell activation. III. B cell plasma membrane depolarization and hyper-Ia antigen expression induced by receptor immunoglobulin cross-linking are coupled, *J. Exp. Med. 158*, 1589–1599 (1983).
160. L. Mátyus, M. Balázs, A. Aszalos, S. Mulhern, and S. Damjanovich, Cyclosporin A depolarizes cytoplasmic membrane potential and interacts with Ca^{2+} ionophores, *Biochim. Biophys. Acta 886*, 353–360 (1986).
161. S. Damjanovich, A. Aszalos, S. Mulhern, G. Marti, M. Balázs, and L. Mátyus, Cyclosporin A influences membrane potential of human and mouse lymphocytes. A critical comparison of steady state fluorimetric and flow cytometric measurements, *Biophys. J. 47*, 271a (1985).
162. S. Damjanovich, A. Aszalos, S. Mulhern, M. Balázs, and L. Mátyus, Cytoplasmic membrane potential of mouse lymphocytes is decreased by cyclosporins, *Mol. Immunol. 23*, 175–180 (1986).
163. S. Damjanovich, A. Aszalos, S. A. Mulhern, J. Szöllősi, M. Balázs, L. Trón, and M. J. Fulwyler, Cyclosporin depolarizes human lymphocytes: Earliest observed effect on cell metabolism, *Eur. J. Immunol. 17*, 763–768 (1987).
164. B. E. Seligmann and J. I. Gallin, Comparison of indirect probes of membrane potential utilized in studies of human neutrophils, *J. Cell. Physiol. 115*, 105–115 (1983).
165. R. M. Johnstone, P. C. Laris, and A. A. Eddy, The use of fluorescent dyes to measure membrane potentials: A critique, *J. Cell. Physiol. 112*, 298–301 (1982).
166. T. C. Smith, The use of fluorescent dyes to measure membrane potentials: A response, *J. Cell. Physiol. 112*, 302–305 (1982).
167. T. M. Chused, H. A. Wilson, B. E. Seligmann, and R. Tsien, Probes for use in the study of leukocyte physiology by flow cytometry, in: Ref. 150, pp. 531–544.
168. H. A. Wilson and T. M. Chused, Lymphocyte membrane potential and Ca^{2+}-sensitive potassium channels described by oxonol dye fluorescence measurements, *J. Cell. Physiol. 125*, 72–81 (1985).
169. L. Mátyus, C. Pieri, R. Recchioni, F. Moroni, L. Bene, L. Trón, and S. Damjanovich, Voltage gating of Ca^{2+}-activated potassium channels in human lymphocytes, *Biochem. Biophys. Res. Commun. 171*, 325–329 (1990).
170. G. Vereb, Jr., G. Panyi, M. Balázs, L. Mátyus, J. Matkó, and S. Damjanovich, Effect of cyclosporin A on the membrane potential and Ca^{2+} level of human lymphoid cell lines and mouse thymocytes, *Biochim. Biophys. Acta 1019*, 159–165 (1990).
171. P. S. Rabinovitch and C. H. June, Measurement of intracellular ionized calcium and membrane potential, in: Ref. 5, pp. 651–668.
172. M. J. Berridge and R. F. Irvine, Inositol triphosphate, a novel second messenger in cellular signal transduction, *Nature 312*, 315–321 (1984).
173. M. C. Sekar and L. E. Hokin, The role of phosphoinositides in signal transduction, *J. Membr. Biol. 89*, 193–210 (1986).
174. R. Y. Tsien, T. Pozzan, and T. J. Rink, T-cell mitogens cause early changes in cytoplasmic free Ca^{2+} and membrane potential in lymphocytes, *Nature 295*, 68–71 (1982).
175. R. Y. Tsien, T. Pozzan, and T. J. Rink, Measuring and manipulating cytosolic Ca^{2+} with trapped indicators, *Trends Biochem. Sci. 9*, 263–266 (1984).
176. G. Grynkiewicz, M. Poenie, and R. Y. Tsien, A new generation of Ca^{2+} indicators with greatly improved fluorescence properties, *J. Biol. Chem. 260*, 3440–3450 (1985).
177. J. P. Y. Kao, A. T. Harootunian, and R. Y. Tsien, Photochemically generated cytosolic calcium pulses and their detection by fluo-3, *J. Biol. Chem. 264*, 8179–8184 (1989).

178. A. Minta, J. P. Kao, and R. Y. Tsien, Fluorescent indicators for cytosolic calcium based on rhodamine and fluorescein chromophores, *J. Biol. Chem. 264*, 8171–8178 (1989).
179. G. T. Rijkers, L. B. Justement, A. W. Griffioen, and J. C. Cambier, Improved method for measuring intracellular Ca^{2+} with fluo-3, *Cytometry 11*, 923–927 (1990).
180. D. Rotman and B. W. Papermaster, Membrane properties of living mammalian cells as studied by enzymatic hydrolysis of fluorogenic esters, *Proc. Natl. Acad. Sci. U.S.A. 55*, 134–141 (1966).
181. J. V. Watson, S. H. Chambers, P. Workman, and T. S. Horsnell, A flow cytofluorimetric method for measuring enzyme reaction kinetics in intact cells, *FEBS Lett. 81*, 179–182 (1977).
182. E. Prosperi, A. C. Croce, G. Bottiroli, and R. Supino, Flow cytometric analysis of membrane permeability properties influencing intracellular accumulation and efflux of fluorescein, *Cytometry 7*, 70–75 (1986).
183. J. C. Martin and D. E. Swartzendruber, Time: A new parameter for kinetic measurements in flow cytometry, *Science 207*, 199–201 (1980).
184. P. Surányi, L. Mátyus, I. Sonkoly, and Gy. Szegedi, Subset specificity of lupus antilymphocyte antibodies studied by two-colour microfluorimetry, *Immunol. Lett. 10*, 91–93 (1985).
185. D. T. Sasaki, S. E. Dumas, and E. G. Engleman, Discrimination of viable and non-viable cells using propidium iodide in two color immunofluorescence, *Cytometry 8*, 413–420 (1987).
186. L. Mátyus, G. Szabó, Jr., I. Resli, R. Gáspár, Jr., and S. Damjanovich, Flow cytometric analysis of viability of bull sperm cells, *Acta Biochim. Biophys. Acad. Sci. Hung. 19*, 209–214 (1984).
187. T. Takács, J. Szöllősi, M. Balázs, R. Gáspár, L. Mátyus, G. Szabó, L. Trón, I. Resli, and S. Damjanovich, Flow cytometric determination of the sperm cell number in diluted bull semen samples by DNA staining method, *Acta Biochim. Biophys. Acad. Sci. Hung. 22*, 45–57 (1987).
188. C. G. Wade, R. H. Rhyne, Jr., W. H. Woodruff, D. P. Bloch, and J. C. Bartholomew, Spectra of cells in flow cytometry using a vidicon detector, *J. Histochem. Cytochem. 27*, 1049–1052 (1979).
189. H. B. Steen and T. Stokke, Fluorescence spectra of cells stained with a DNA-specific dye, measured by flow cytometry, *Cytometry 7*, 104–106 (1986).

Index

Acousto-optic device, 117
  Bragg reflection, 117, 122
  Debye–Sears effect, 117, 119
  mode locker, 137
  Raman–Nath, 120
Angiotensin, 266
Anisotropy, 241: *see also* Anisotropy decay; Associated anisotropy decays
  hindered rotation, 68, 277
  limiting anisotropy, 277
  modulated, 300
  order parameter, 69
  T-format, 81
  time-resolved, 26, 250
Anisotropy decay, 26, 177, 293, 304
  associated, 322
  cone angle, 68
  frequency domain, 298, 304
  modulated anisotropy, 300
  tryptophan, 265
  tyrosine, 256
Apomyoglobin, 324
Argon ion laser, 123
Associated anisotropy decays, 322
Autofluorescence, 433
Avalanche photodiodes, 47; *see also* Detectors

Brewster's angle, 112
Bragg reflection, 117, 122

Calcium, 437
Carbonic anhydrase, 363
CCD: *see* Charge coupled device; Detectors
Cell sorting, 427; *see also* Flow cytometry
Charged-coupled device (CCD), 406
Chromosome, 430
Collisional quenching, 282
Cone angle, 68, 277
Constant fraction discriminator, 48, 170
Convolution, 14
Correlation spectroscopy, *337*
  instrumentation, 343
  total internal reflection, 364

Data analysis, 16, 168
  anisotropy, 27
  flow cytometry, 424
Debye–Sears effect, 117, 119
Deconvolution, 14
Detectors, *41*
  charge-coupled device, 406
  flow cytometry, 420
  imaging detectors, 406
  microchannel plate PMT, *146*
  MCP photomultiplier tube, *146*
  photocathode, 149
  photomultiplier tubes, 404, 422
  SIT, 406
Differential phase fluorometry, 298
Diffusion
  rotational, 362
  translational, 338, 354
Dye laser, 128
  cavity-dumped, 140
  synchronous-pumped, 138

Energy windowing, 159
Energy transfer
  flow cytometry, 434
  micelles, 64
  random distribution, 60
Excimer formation, 54
Exponential depression, 169

Flashlamp, 33
Flow cytometry, *411*

Italic numbers indicate a detailed description of the topic.

Flow cytometry (*Cont.*)
calcium, 437
cellular DNA, 429
chromosomes, 430
detectors, 420
energy transfer, 434
fluorescence, 421, 428
instrumentation, 413
light scatter, 420, 428
membrane permeability, 438
mebrane potential, 436
multicolor, 421
Fluorescence resonance energy transfer (FRET): *see* Energy transfer
Fluorescence correlation spectroscopy: *see* Correlation spectroscopy
Fluorescence microscopy, *379*
epi-illumination, 385
detectors, 404
filters, 393
monochromators, 394
sensitivity, 382
Förster transfer: *see* Energy Transfer
Frequency domain fluorescence, 4, *293*
anisotropy decay, 298, 304
differential phase, 299
harmonic content, 309
instrumentation, *305*
intensity decays, 302
microchannel plate PMT, 309
modulated anisotropy, 300
FRET: *see* Energy transfer

Harmonic content fluorometer: *see* Frequency-domain fluorescence
Hemoglobin, 315
Host–guest complexes, 282

Image intensifier, 185, 205
Imaging detector, 406
Immunofluorescence, 432
Instrumentation
correlation spectroscopy, 343
flow cytometry, 413
time-correlated single photon counting, 10, *28*
Intensity decay, 175

Lactate dehydrogenase, 270
Lasers, *97*
Argon ion, 123
Brewster's angle, 112
Bragg reflection, 117, 122
Lasers (*Cont.*)
cavity dumping, 140
Debye–Sears effect, 117, 119
diagnostics, 245
dye, 128
flow cytometry, 417
frequency-doubling, 113
mode-locking, 133
modes, 109
Nd:YAG, 126
optical resonator, 107
regenerative feedback, 106
Raman–Nath, 120
synchronous pumping, 138
three- and four-level, 103
Least-squares analysis, 19, 168, 303
anisotropy, 27
Lifetime, 2, 226
definition, *2*
measurement, *1*
Light scatter, 420
Light sources, 28, 336
flashlamp, 32
laser, 29, *97*
synchrotron, 29, *261*
Limiting anisotropy, 277
Lumazine protein, 263

Membrane potential, 436
Membranes, 276
Micelles, quenching, 64
Microchannel plate PMT, 46, *146*, 153
gain saturation, 166
proximity-focused, 147
pulse height distribution, 159
wavelength response, 162
Microscopy: *see* Fluorescence microscopy
Mode-locking, 133
acousto-optic, 137
Monochromators, 36
Multi-exponential decay, 24

Nd:YAG laser, 126
Nucleic acids, 251

Order parameter, 69

Phase-modulation fluorescence: *see* Frequency domain fluorescence
Photocathode, 195
Photomultiplier tubes, 41
microchannel plate, 46, 146

Photosynthesis, 272
Phycobiliproteins, 249, 432
Polydispersity, 353
Protein dynamics, 250, 318
Proteins
  apomyoglobin, 324
  carbonic anhydrase, 363
  lactate dehydrogenase, 270
  hemoglobin, 315
  phycobiliproteins, 249, 432
  ribonuclease T_1, 317
  tryptophan, 314
Pulse pileup, 11
Pulse height distribution, 159
Pulse lifetime measurements, *1*

Ribonuclease T_1, 317
Rise time, 231
Rotational diffusion, anisotropic, 319

Solvent relaxation, 247
Stilbene, 283
Streak cameras, *183*
  anisotropy, 241
  averaging, 222
  corrected and uncorrected emission spectra, 238
  dynamic range, 217
  image intensifier, 205
  multi-scan, 192
  phosphor, 204
  polarization, 221
Streak cameras (*Cont.*)
  signal-to-noise, 217, 244
  single-scan, 190
  synchroscan, 192
  time resolution, 214, 226
  trigger, 197
  wavelength dependence, 222, 233
  x-ray, 251

Time-to-amplitude converter (TAC), 173
Time-correlated single photon counting, *1*; *see also* Time domain fluorescence
Time-dependent emission, 237
Time domain fluorescence, *1*, *97*, 294
  anisotropy decay, 177
  array fluorometer, 83
  constant fraction discriminator, 173
  differential pulse fluorometry, 75
  intensity decay, 175
  multi-photon events, 174
  multiplexing, 73
  streak cameras, *183*
  T-format, 81
  time-to-amplitude converter, 173
Time-resolved anisotropy, 241: *see also* Anisotropy decay
Time-resolved emission spectra, 74
Total internal reflection, 364
Tryptophan, 314

Y_t-base, 320